Silicon-on-Insulator Technology and Devices 13

Editor:

G. K. Celler
SOITEC
Summit, New Jersey, USA

Assistant Editors:

S. Cristoloveanu
IMEP-INPG
Grenoble, France

F. Gámiz
Universidad de Granada
Granada, Spain

Y. Omura
Kansai University
Osaka, Japan

S. W. Bedell
IBM T. J. Watson Research Center
Yorktown Heights, New York, USA

B.-Y. Nguyen
Freescale Semiconductor
Austin, Texas, USA

Sponsoring Division:

 **Electronics and Photonics**

Published by
The Electrochemical Society
65 South Main Street, Building D
Pennington, NJ 08534-2839, USA

tel 609 737 1902
fax 609 737 2743

www.electrochem.org

ecstransactions™
Vol. 6 No. 4

Published by:

The Electrochemical Society, Inc.
65 South Main Street
Pennington, New Jersey 08534-2839, USA

Telephone 609.737.1902
Fax 609.737.2743
e-mail: ecs@electrochem.org
Web: www.electrochem.org

ISBN 978-1-56677-553-3

Printed in the United States of America

PREFACE

The Thirteenth International Symposium on Silicon-On-Insulator Technology and Devices was part of the 211th Meeting of the Electrochemical Society, held on May 6 – May 11, 2007, in Chicago, Illinois. The Electronics Division of the Electrochemical Society sponsored the Symposium.

The Symposium was first established when the field of SOI was immature and focused primarily on discovering and developing methods of forming SOI structures and their characterization. As SOI technology matured and moved into the commercial arena, technical presentations focused more and more on device and circuit architecture in SOI, on electrical characterization of novel devices that are expected to become important in future technology nodes, on reliability and radiation hardness, and on extensions of SOI to new engineered materials such as strained SOI (sSOI) and GeOI.

The Symposium consisted of 12 invited reviews and 41 contributed papers. In addition, professor Yue Kuo, recipient of the 2007 Electronics Division Award, delivered his Award presentation in this Symposium. The Symposium has preserved its international nature, with 19 papers from North America, 17 from Europe, 10 from Asia, 6 from South America and one from Australia. Many of the papers include collaborations between researchers from institutions on different continents.

This issue of *ECS Transactions* contains 53 papers (all but one) presented at the conference. I would like to thank the authors for submitting excellent manuscripts and doing it on time in spite of a very tight schedule. Papers in these proceedings have been arranged in exactly the same sequence as the symposium presentations in order to make it convenient to follow the text in the conference room.

It is my great pleasure to acknowledge the symposium co-chairs – Sorin Cristoloveanu, Stephen Bedell, Francisco Gámiz, Bich-Yen Nguyen, and Yasuhisa Omura – for their active involvement and efforts. Thanks to the ECS staff: John Lewis, Stephanie Plassa, and Mary Yess for their support and advice. We greatly appreciate the participation of invited speakers who set the tone for the meeting and we thank all symposium speakers and co-authors for coming to Chicago to share their recent data and their technical expertise. Their contributions made the conference a success. We thank the Electronics Division for their support. Finally, we would like to express gratitude to SOITEC for being a cosponsor and for providing a generous contribution in support of the Symposium.

George Celler
Summit, New Jersey
May 2007

iii

ECS Transactions, **Volume 6, Issue 4**
Silicon-on-Insulator Technology and Devices 13

Table of Contents

Session 4
Poster Papers

Session 5
Material Characterization

Session 6
Technology and Materials

** Plenary paper*
*** Invited paper*
**** Electronics and Photonics Division Award Address*

Facts about ECS

The Electrochemical Society (ECS) is an international, nonprofit, scientific, educational organization founded for the advancement of the theory and practice of electrochemistry, electrothermics, electronics, and allied subjects. The Society was founded in Philadelphia in 1902 and incorporated in 1930. There are currently over 7,000 scientists and engineers from more than 70 countries who hold individual membership; the Society is also supported by more than 100 corporations through Corporate Memberships.

The technical activities of the Society are carried on by Divisions. Sections of the Society have been organized in a number of cities and regions. Major international meetings of the Society are held in the spring and fall of each year. At these meetings, the Divisions and Groups hold general sessions and sponsor symposia on specialized subjects.

The Society has an active publications program that includes the following.

Journal of The Electrochemical Society — JES is the peer-reviewed leader in the field of electrochemical and solid-state science and technology. Articles are posted online as soon as they become available for publication. This archival journal is also available in a paper edition, published monthly following electronic publication.

Electrochemical and Solid-State Letters — ESL is the first and only rapid-publication electronic journal covering the same technical areas as JES. Articles are posted online as soon as they become available for publication. This peer-reviewed, archival journal is also available in a paper edition, published monthly following electronic publication. It is a joint publication of ECS and the IEEE Electron Devices Society.

Interface — *Interface* is ECS's quarterly news magazine. It provides a forum for the lively exchange of ideas and news among members of ECS and the international scientific community at large. Published online (with free access to all) and in paper, issues highlight special features on the state of electrochemical and solid-state science and technology. The paper edition is automatically sent to all ECS members.

Meeting Abstracts (formerly Extended Abstracts) — Abstracts of the technical papers presented at the spring and fall meetings of the Society are published on CD-ROM.

ECS Transactions — This online database provides access to full-text articles presented at ECS and ECS-sponsored meetings. Content is available through individual articles, or as collections of articles representing entire symposia.

Monograph Volumes — The Society sponsors the publication of hardbound monograph volumes, which provide authoritative accounts of specific topics in electrochemistry, solid-state science, and related disciplines.

For more information on these and other Society activities, visit the ECS website:

www.electrochem.org

SESSION 1

TECHNOLOGY, PROCESS INTEGRATION, RADIATION HARDNESS

ECS Transactions, 6 (4) 3-13 (2007)
10.1149/1.2728835, ©The Electrochemical Society

Embedding Device Solutions in Engineered Substrates

Carlos Mazure

SOITEC, Parc Technologique des Fontaines, 38190 Bernin, France

A review of the advances in Smart Cut engineered substrates and the impact on device architecture and IC design is given. Primarily focusing on CMOS applications, mobility enhancing substrates, fully depleted device compatibility to SOI product capability and multi-gate FinFETs will be reviewed.

Introduction

A simple structure like silicon on insulator (SOI) has allowed the IC industry to optimize the MOSFET architecture by reducing parasitic capacitances to the substrate, cutting device leakage, improving device speed while reducing power consumption, decreasing considerably soft error rate, and so forth. The benefits and potential of SOI have been known since the 60's [1]. The real issue for an industrial SOI development was SOI substrate availability and wafer quality. This roadblock vanished with the invention of Smart Cut[TM] [2, 3], a single crystal thin layer transfer technology. The simplicity of this technique underlines simultaneously the power and potential of Smart Cut technology and explains why it has become today the industrial standard for thin layer transfer [4].

Smart Cut technology consists in defining a splitting region within the donor substrate by ion implantation (e.g. H), which allows a thin film to transfer to a handle wafer after bonding and splitting. The thickness removed from the donor wafer is negligible compared to the total wafer thickness. Consequently, the donor wafer can be reused many times. A range of 10 nm up to a few 10^3 nm in top Si and buried oxide (BOX) thickness is easily covered by this technology.

Beyond SOI, Smart Cut technology opens the door to substrate engineering, i.e. the capability to design new substrates tailored to specific applications [5]. It is possible to amplify certain device properties while weakening others by combining different materials or crystal orientations, adding strained layers, or simply creating a new composite substrate.

High performance ICs continue to be the driver for a specialized family of advanced substrates. Ultra-thin (UT) SOI, mobility enhancing substrates like strained SOI (sSOI) in addition to local strain techniques [6], as well as improved thermal dissipation to reduce the impact of hot spots on MOSFET performance are among the most obvious engineered substrate solutions [5]. Direct silicon Bonding (DSB) adds a high-end mobility enhancing substrate to the bulk Si IC architectures [7]. Non-planar device architectures like FinFETs take advantage of the buried oxide or nitride as an etch stop for fin height definition, thus reducing fin variability [5].

A review of the advances in Smart Cut engineered substrates and the impact on device performance and IC design is given below.

High Performance Substrates

Beyond the 90nm technology node, the electron and hole mobility enhancement have become essential to assure device performance increase while scaling. Two approaches are clearly identified: strained silicon, and dual crystal orientation surfaces. The first one is evolutionary and takes advantage of strain inducing CMOS processes which can be further amplified by combining them with a strained silicon substrate (sSOI) [8,9]. The alternative and probably complementary approach is the dual crystal orientation surface with (100) and (110) Si surfaces for the n and p-channels, respectively [7, 10]. Beyond 32nm, a hybrid material surface with tensile Si and compressive Ge or SiGe for the n and p-channels, respectively, may be the path to further mobility increase [11, 12].

<u>Strained Silicon on Insulator</u>

Mobility enhancement has become a standard scaling parameter to assure a device performance increase for sub-90nm device geometries. Different types of stressors are used to induce the appropriate strain in the n and p channel regions. Probably the most effective stressors are the embedded SiGe (eSiGe) in the p+ source and drain regions which induce a strong compressive stress [13], and the contact etch stop layers (ESL) encapsulating the gate which can induce a tensile or compressive stress depending on the nature of the film [14]. Stressors are most effective for the smallest geometries but, paradoxically, their process window is limited by gate and active area pitch. As schematically shown in figure 1, decreasing the gate-to-gate spacing limits the effective ESL thickness.

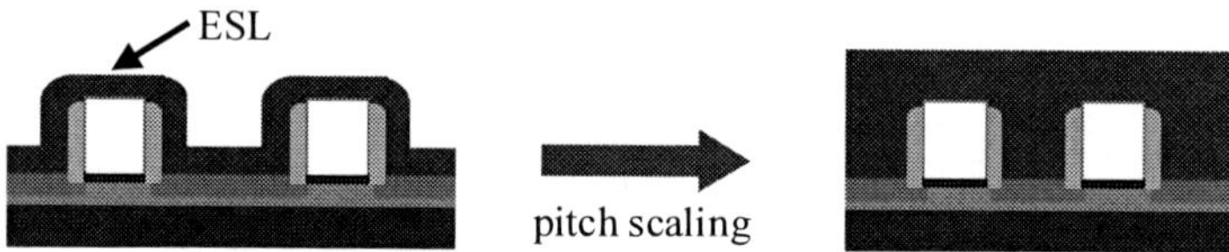

Figure 1. Schematic representation of the gate pitch imposed limitation on ESL thickness. The gap filling limits the ESL efficiency in transferring the stress into the channel.

Transistor layout and final printed features and the gate to S/D misalignment also affect strongly the resulting channel stress for deep sub-µm geometries. Figure 2 shows the impact of gate pitch on the efficiency of ESL stressors in the range of –650 MPa to –450 MPa for compressive ESL, and in the range of 350MPa to 400MPa for tensile ESL. The way to overcome these limitations is to build the devices on a uniformly strained Si surface that is transferred onto a Si substrate with a buried oxide in between to stabilize the strained Si film and isolate it from the substrate. The resulting substrate is strained silicon on insulator (sSOI). The wafer level strain is generated by epitaxial techniques. The principle for sSOI fabrication is to generate a relaxed $Si_{1-x}Ge_x$ epitaxial template on a Si substrate followed by a Si epitaxial step [6]. The epitaxial Si will retain the in-plane lattice constant of the template, resulting in a biaxial strained Si film if the thickness is below a certain critical value. Typical built-in stress is around 1.3GPa. sSOI has been developed as the solution that offers higher carrier mobility, combining the advantages of SOI with those of strained silicon [6]. The fundamental difference between the uniaxial stressors and the wafer level (biaxial) strained Si is that the former is introduced during CMOS processing while the latter is built into the substrate. As a consequence, sSOI wafers offer tensile strain for large geometry devices like in ESD and I/Os circuits, giving IC design a further degree of freedom.

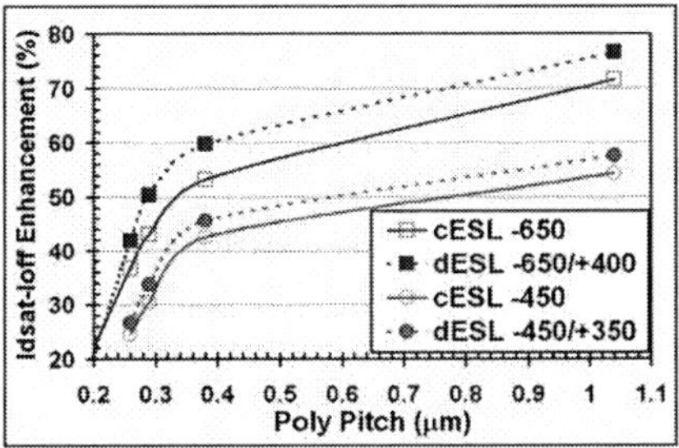

Figure 2. Impact of gate poly pitch on gate encapsulating contact etch stop layers (ESL) stressor efficiency in terms of current drive IDsat enhancement. cESL stands for compressive strain inducing ESL and dESL for dual compressive/tensile ESL. Courtesy of B.Y. Nguyen, Freescale.

On-going work by IC makers and research institutes [9, 12, 15] show that the combination of uniaxial stressors and sSOI amplifies the mobility enhancement of both n- and p-channel devices. Published data [15] for Lg=35nm show that sSOI combined to tensile ESL results in an on-current (IDsat) enhancement of 18% over tensile-ESL optimized n-channel transistors, which is equivalent to an overall IDsat boost of 27% over standard SOI n-channel devices. The mobility enhancement is very significant, almost 40%. In the long channel regime mobility enhancement is about 80%.

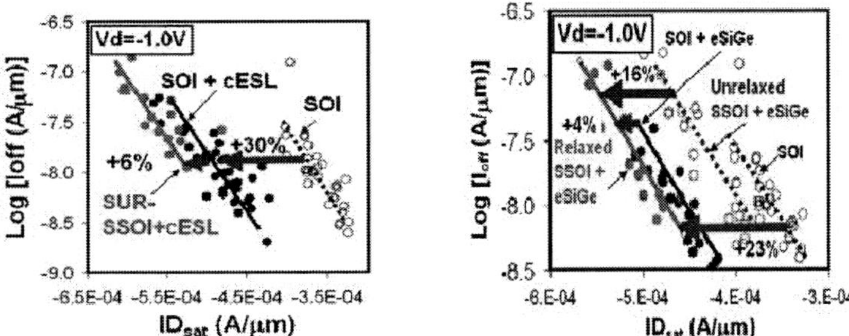

Figure 3. A significant p-channel IDsat increase is obtained for both compressive ESL and eSiGe in combination with sSOI substrates. From A. Thean et al. [15].

For a fully depleted architecture with metal gate and high k dielectric the IDsat increase through sSOI is of 16% at Lg=25nm. For the p-channels a uniaxial relaxation of the sSOI strain is induced by selective amorphization coupled to the active area layout [15]. As a result the p-channel strain can be engineered very effectively with a compressive component along the channel and while keeping the tensile component across the channel. A 23% IDsat enhancement is obtained over unstrained SOI p-channels. Figure 3 and 4 show the results obtained for p- and n-channel devices on sSOI in combination with ESL and eSiGe stressors, respectively.

The scalability of sSOI has been thoroughly investigated for ultrathin body sSOI with the fabrication of short Lg=25nm and narrow channel Wg=25nm devices [16], as well as the benefits of 40 to 50nm thick super critical strained SOI (SC- sSOI), for partially depleted (PD) MOSFETs [15], and FinFET structures [17]. Recent results also show that 2.5GPa strained Si sSOI can be manufactured [18] opening further the process window for sSOI based stressors.

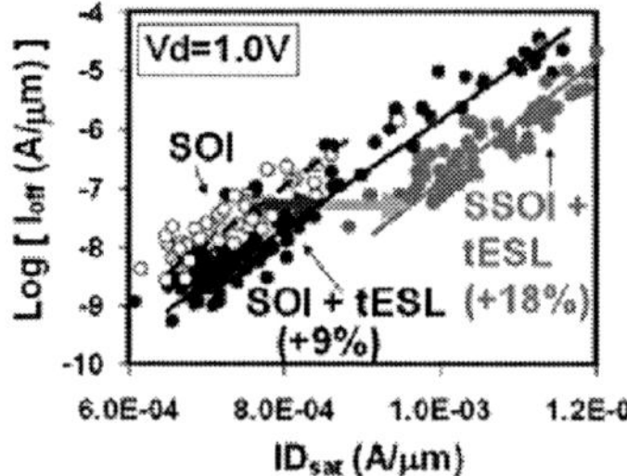

Figure 4. Almost an additional 20% current drive increase can be obtained by combining tensile ESL with sSOI substrates. From A. Thean et al. [15].

<u>Dual Crystal Orientation Surface</u>

The fabrication of hybrid orientation substrates has led to the development of CMOS technologies in which n-channel transistors are fabricated on a (100) Si surface and p-channels on a (110) Si surface. The driver for this approach is the fact that the hole mobility doubles for the (110) surface along the <110> direction as compared to the widely used (100) surface. Hybrid orientation composite SOI (HOT) is fabricated by transferring a (110) Si layer onto a (100) handle wafer. A (100) film on a (110) is another variation of a hybrid substrate [19]. For 40nm long pMOSFETs fabricated on a (110) surface, a current drive increase of 45% is achieved, but in contrast, the n-MOSFET on the (110) plane is degraded by 35% [19, 20]. To overcome the n-channel degradation, windows corresponding to the n channel regions are etched in the substrate through the buried oxide down to the (100) handle substrate. Then Si selective epitaxial growth follows after a spacer formation. A strong overgrowth is required to drive defects out of the active area with facet formation above the substrate plane. A final CMP steps planarizes the topography and a composite wafer with (110) and (100) regions embedded in the same surface is obtained [19]. The obvious drawback of this approach is that one of the devices is fabricated on bulk. It is a mixed bulk-SOI CMOS design.

An alternative approach proposed recently [21] for making wafers by Direct Silicon Bonding Si (DSB) is based on the idea that it is possible to change the orientation of selected Si regions by an amorphization templated recrystallization process. Here the selected top regions are amorphized by ion implant, and solid phase epitaxial (SPE) regrowth is initiated from bulk Si of the desired orientation as shown in figure 5. This new approach is fully compatible with conventional bulk CMOS technology and 35% improvement in PMOS performance has been demonstrated [7]. The DSB wafers with top (110) Si seem to provide more straightforward CMOS integration because the SPE of (100) Si results in fewer crystal defects than that for (110) Si.

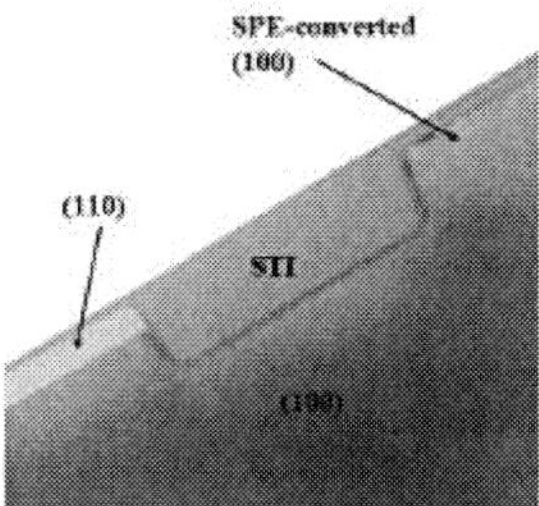

Figure 5. DSB cross section of a DSB substrate with STI separating the p-channel region on a (110) surface on the left hand side and the SPE converted (100) region on the right hand side. From C.Y. Sung et al. [7].

The DSB substrate fabrication represents a major challenge. The Smart Cut conditions have to be re-developed to take into account the absence of buried oxide and in particular the bonding becomes critical [22]. The contact between the top and substrate Si must be ohmic, oxide precipitates at the bonding interface must be minimized during Smart Cut processing and residual SiO_2 eventually dissolved in a high temperature anneal. Bonding two different crystal orientations introduces a two-dimensional dislocation network as a result of the misorientation of (100) and (110) surfaces. Results at present are very encouraging as shown in figure 6, although further work is required to improve the crystal quality of DSB as well as the interaction of these substrates with the subsequent CMOS processing and specially the SPE regrowth.

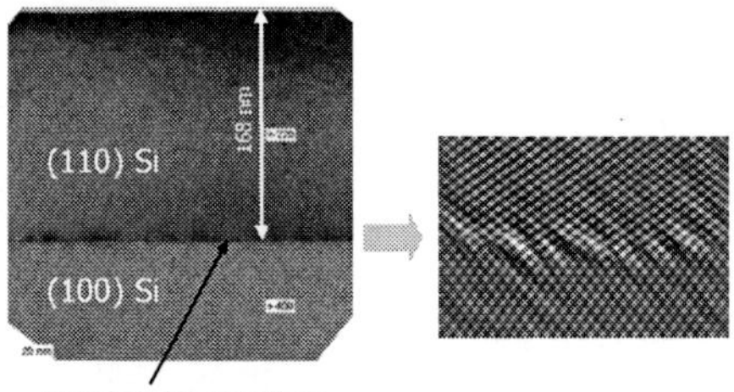

Figure 6. TEM of the DSB wafer and a HRTEM image of the bonding interface. The interface is atomically closed with no SiO2 precipitates and no threading dislocations generating at the interface. From K. Bourdelle et al. [22].

Fully Depleted Substrates

<u>Ultra thin SOI</u>

Full depletion happens when the depletion region covers the whole transistor body. The coupling between gate bias and inversion channel results in improved subthreshold characteristics and weaker short channel effects as compared to Si bulk or partially depleted device architecture. The drawback is that the FD device design is very sensitive to Si channel thickness variations. Thickness variations translate into threshold voltage V_T changes. This has been the reason for the strong domination of the partially depleted architecture. With the coming of age of Smart Cut SOI manufacturing, thickness control for both the buried oxide and the top Si have ceased to be an issue for FD devices. The top Si layer total thickness variation (LTTV) is within ±1nm for all wafers and all sites [23]. This is obtained through wafer-to-wafer control on both the mean silicon and buried oxide thickness within 0.5nm. The within wafer range (max − min) and the wafer to wafer range are shown in figure 7. Similar layer thickness control is obtained for Si films down to 200Å.

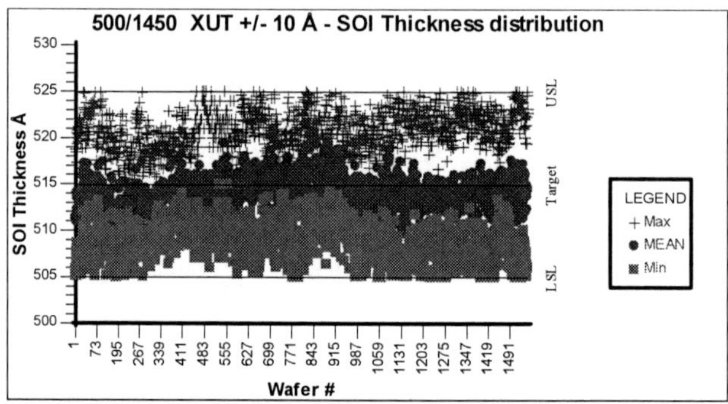

Figure 7. Thin SOI product capability: pattern of 41 points/wafer, 3 mm edge exclusion. The statistic shown corresponds to 1500 Wafers. From C. Maleville [23].

The choice of FD SOI combined to high k and metal gate minimizes the need for channel doping, thus strongly reducing the dopant dependent V_T variability and simplifying the metal gate processing. The reduction of V_T scatter translates into a much more stable SRAM cell which is otherwise a serious issue beyond the 45 nm technology node for PD SOI and Si bulk based CMOS. Another significant advantage of a weakly doped channel is a much higher mobility, which in turn amplifies the stress induced mobility enhancement via local stressors or sSOI. Due to its SOI nature and improved gate control the FD device exhibits a very low leakage even at 125°C, thus also significantly lowering static power consumption by at least an order of magnitude compared to PD SOI or bulk CMOS.

Multi Gate FETs

The Multi Gate FET (MuGFET) or FinFETs are a vertical device approach that improves electrostatic control of a short transistor channel, while further increasing IC density. These devices are in the operation regime of 1V, and for fin thicknesses ranging from 10 to 40nm fully depleted devices. While the height of the fin is set by the Si thickness, the width control as well as the roughness of the fins becomes as critical as the thickness control for the planar FD devices. The manufacturability of MuGFETs is strongly dependent on the etch selectivity of the Si fins to the underlying buried oxide. Depending on the MuGFET design an undercut may be desired for gate underlap and improved gate control. In order to add robustness to the fin formation with a highly reproducible underlap a thin buried silicon nitride can be added to the buried oxide in the fabrication of the starting SOI wafer. A further benefit of the composite dielectric nitride/oxide is to avoid the fin undercut during hydrogen sidewall smoothing of the fin's sidewall lateral roughness [24]. Fins have been found to be very stable on the buried oxide much more than poly-silicon for equivalent film widths and thicknesses.

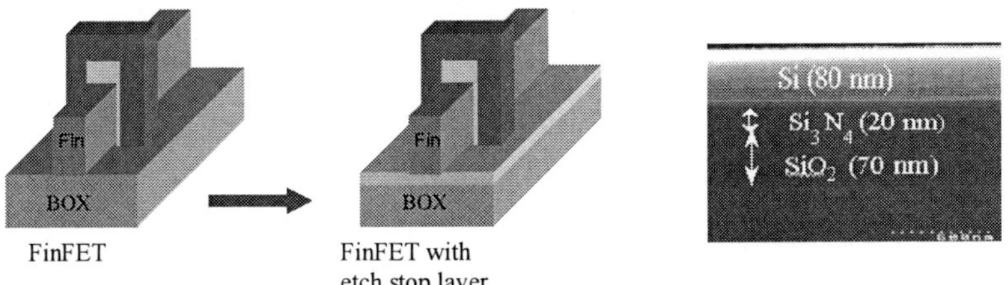

Figure 8. Schematic drawing of a FinFET with and without a composite dielectric nitride/oxide as an etch stop. The SEM micrograph shows an example of a composite nitride/oxide buried dielectric.

Advanced SOI Substrates

Capacitor less DRAMs

Circuit design has evolved to take advantage of the SOI properties and to optimize the area required per logic function. Capacitor-less one transistor DRAM cells are an important development which takes advantage of the floating body effect in SOI devices [25, 26]. The generation of excess negative or positive charge in the body can be used to store data states. In a n-channel device an excess of positive charges leads to an increase of the current drive, defining one state. The removal of positive charges from the body decreases the channel current, defining the opposite state. The industrial potential of the floating body cell (FBC) is important considering that it eliminates the process complexity associated to the stack or trench capacitor in a DRAM process. Very dense embedded memory blocks can be realized with standard SOI process with footprints as small as $4F^2$, where F is the minimal feature size [25]. The development of the FBC technology is being attentively followed by the IC community. Recently a 128Mb FBC was reported [27] and its scalability to 32nm has been proven [28]. Further, the FBC concept has been proven to work as well for FinFET architecture [29]. The development of a capacitor less DRAM cell constitutes a breakthrough that can considerably reduce the cost of ownership of SOI technology with respect to its Si bulk equivalent and could enable a significant process simplification of today's very complex DRAM ICs.

Ultra thin buried oxide SOI

Typical buried oxide (BOX) layer thicknesses for SOI are about 150nm or higher. The BOX reduction to 50nm or below enhances the parasitic device layer coupling with the substrate, but for particular applications it may be the desired effect. For example for very low power applications, ultra-thin BOX offers the possibility to easily form buried n and p regions in the handle substrate, which can be used as back gates. By applying a back bias, the off current can be reduced by an order of magnitude or more, while in forward bias mode it lowers the device V_T, resulting in a current drive increase [30]. It is a particularly interesting concept for short channel effect control in a fully depleted MOSFET architecture and in low power SOC applications .

For ICs where local MOSFET self-heating can be a major concern, ultra thin BOX SOI may be an advantageous solution. A factor of 3 improvement in thermal conductance can be achieved by reducing the BOX thickness from 150nm to 20nm [31]. The tradeoff is, of course, increasing the parasitic junction capactances.

For very low power SOCs, SOI with ultra thin buried oxide (<50nm) will enable IC architectures where n and p regions are defined in the handle substrate for back bias generation through the buried oxide. Since attaining the highest performance is not the focus here, these SOI CMOS solutions will target the lowest power consumption and longest battery lifetime. Low standby and low operating power devices will be built by taking full advantage of dielectric isolation.

<u>High Impedance SOI</u>

SOI offers design advantages over traditional bipolar devices resulting in a considerable reduction of crosstalk between RF analogue and digital logic elements plus easy integration with passive elements. High impedance SOI substrates enhance these advantages for mixed analogue/digital circuits combined to RF circuitry, voltage controlled oscillator (VCO) and low noise amplifiers (LNA), making them more robust to process variations [32].

High impedance SOI is an evolutionary SOI approach where the handle or base wafer is a high resistivity (HR) substrate (>1 kOhm-cm) [29]. SOI technology provides complete oxide isolation, cutting off direct paths of substrate injection noise and a high resistivity substrate reduces the capacitive coupling, further minimizing the substrate related RF loss. Because of the SOI nature of the high impedance substrate, latchup is of no concern, and ground shields can be avoided. Thus, higher density, smaller footprint mixed RF/analogue/digital IC can be achieved. Elimination of substrate crosstalk improves integrated passive components performance, becoming comparable to what can be achieved on InP [32].

CMOS SOI constitutes a cost effective alternative to GaAs and BiCMOS technologies [33]. Moreover, layer transfer offers the unique possibility to engineer SOI HR, high impedance substrates, to optimize RF gain, while reducing noise with no significant change in the fabrication process. SOI on high resistivity substrates (SOI-HR) [32] offers a solution tailored to RF applications that will substantially improve performance of passive components.

Conclusion

Engineered substrates offer disruptive, competitive advantages that enable heterogeneous integration and bridges towards new substrate functionality. In the field of microelectronics, substrate engineering is a proven path to reduce power consumption through enhanced mobility, dual surface orientation, fully depleted devices, high impedance substrates, etc. It opens new possibilities through added substrate functionality via buried dielectrics, band gap engineering, for improvement of the n- and p-channel devices adding robustness to the device fabrication. FD SOI or MuGFET architectures enables a further IC density while improving the device performance at the same time. The integration of embedded one transistor floating body memories on SOI will result in ICs that will not only have speed and power advantages, but should also be significantly lower cost.

Aknowledgments

The work reviewed here is the result of the effort of many teams of different companies, institutes and universities. Special thanks in particular to K. Bourdelle, I. Cayrefourcq, G. Celler, B. Ghyselen, F. Letertre, C. Maleville and M. Yoshimi from Soitec; C. Raynaud from CEA / STMicroelectronics; B.-Y. Nguyen from Freescale.

References

1. J.-P. Colinge, *Silicon-on-Insulator Technology: Materials for VLSI*, 2nd edition Kluwer Press, Boston (1997).
2. M. Bruel, *Nuclear Instr. and Methods in Physics Research B*, **108**, p. 313 (1996).
3. M. Bruel, Electron Device Letters, **31**, p. 1201 (1995).
4. C. Maleville and C. Mazure, Solid State Elect., Vol 48-6, p. 855 (2004).
5. C. Mazure and A. J. Auberton-Hervé; Proc. ESSDERC Conf., p. 29 (2005).
6. C. Mazure & I. Cayrefourcq; Proc. of IEEE Int. SOI Conf., p. 1 (2005).
7. C.-Y. Sung, H. Yin, H. Y. Ng, K. L. Saenger, V. Chan, S. W. Crowder, J. Li, J. A. Ott, R. Bendernagel, J. J. Kempisty, V. Ku, H. K. Lee, Z. Luo, A. Madan, R. T. Mo, P. Y. Nguyen, G. Pfeiffer, M. Raccioppo, N. Rovedo, D. Sadana, J. P. de Souza, R. Zhang, Z. Ren, and C. H. Wann, *IEDM Tech Dig.*, p. 236 (2005).
8. I. Cayrefourcq, A. Boussagol and G. Celler, in *SiGe and Ge: Materials, Processing, and Devices*, ECS Trans. **3**, (7), p. 399 (2006).
9. Andy Wei, T Kammler, J. Höntschel, A. Mowry, H. Bierstedt, A. Hellmich, K. Hempel, J. Rinderknecht, B. Trui, R. Otterbach, and M. Horstmann, Ian Cayrefourcq, F. Metral, Mark Kennard, and Eric Guiot, in *SiGe and Ge: Materials, Processing, and Devices*, ECS Trans. **3**, (7), p. 719 (2006).
10. M. Yang, M. Ieong, L. Shi, K. Chan, V. Chan, A. Chou, E. Gusev, K. Jenkins, D. Boyd, Y. Ninomiya, D. Pendlton, Y. Surpris, D. Heenan, J. Ott, K. Guarini, C. D'Emic, M. Cobb, P. Mooney, B. To, N. Rovedo, J. Benedict, R. Mo, H. Ng, IEDM Tech. Dig., p. 453 (2003).
11. J. Jung, M.L. Lee, S. Yu, E.A. Fitzgerald and D. Antoniadis, IEEE EDL, **24**, p. 460 (2003).
12. F. Andrieu, T. Ernst, O. Faynot, Y. Bogumilowicz, J.-M. Hartmann, J. Eymery, D. Lafond,Y.-M. Levaillant, C. Dupré, R. Powers, F. Fournel, C. Fenouillet-

Beranger, A. Vandooren, B. Ghyselen, C. Mazure, N. Kernevez, G. Ghibaudo and S. Deleonibus, IEEE Int. SOI Conf, p. 223 (2005).
13. T. Ghani, M. Armstrong, C. Auth, M. Bost, P. Charvat , G. Glass, T. Hoffmann, K. Johnson, C. Kenyon, J. Klaus, B. McIntyre, K. Mistry, A; Murthy, J. Sandford, M. Silberstein, S. Sivakumar, P. Smith, K. Zawadzki, S. Thompson, and M. Bohr, IEDM Tech. Dig., p. 978 (2003).
14. H. S. Yang, R. Malik, S. Narasimha, Y. Li, R. Divakaruni, P. Agnello, S. Allen, A. Antreasyan, J. C. Arnold, K. Bandy, M. Belyansky, A. Bonnoit, G. Bronner, V. Chan, X. Chen, Z. Chen, D. Chidambarrao, A. Chou, W. Clark, S. W. Crowder, B. Engel, H. Harifuchi, S. F. Huang, R. Jagannathan, F. F. Jamin, Y. Kohyama, H. Kuroda, C. W. Lai, H. K. Lee, W. Lee, E. H. Lim, W. Lai, A. Mallikarjunan, K. Matsumoto, A. McKnight, J. Nayak, H. Y. Ng, S. Panda, R. Rengarajan, M. Steigerwalt, S. Subbanna, K. Subramanian, J. Sudijono, G. Sudo, S. Sun, B. Tessier, Y. Toyoshima, P. Tran, R. Wise, R. Wong, I. Y. Yang, C. H. Wann, L. T. Su, M. Horstmann, T. Feudel, A. Wei, K. Frohberg, G. Burbach, M. Gerhardt, M. Lenski, R. Stephan, K. Wieczorek, M. Schaller, H. Salz, J. Hohage, H. Ruelke, J. Klais, P. Huebler, S. Luning, R. van Bentum, G. Grasshoff, C. Schwan, E. Ehrichs, S. Goad, J. Buller, S. Krishnan, D. Greenlaw, M. Raab, and N. Kepler, IEDM Tech. Dig., p. 1075 (2004).
15. A. V-Y. Thean D. Zhang, V. Vartanian, V. Adams, J. Conner, M.Canonico, H. Desjardin, P. Grudowski, B. Gu, Z.-H. Shi, S. Murphy, G. Spencer, S. Filipiak, D. Goedeke, X-D. Wang, B. Goolsby, V. Dhandapani, L.Prabhu, S. Backer, L-B. La, D. Burnett, T. White, B.-Y. Nguyen, B. E. White, S. Venkatesan, J. Mogab, I. Cayrefourcq, and C. Mazuré, VLSI Tech. Dig., p. 164 (2006).
16. Andrieu, C. Dupre, F. Rochette, O. Faynot, L. Tosti, C. Buj, E. Rouchouze, M. Casse, B. Ghyselen, I. Cayrefourcq, L. Brevard, F. Allain, J. C. Barbe, J. Cluzel, A. Vandooren, S. Denorme, T. Ernst, C. Fenouillet-Beranger, C. Jahan, D. Lafond, H. Dansas, B. Previtali, J. P. Colonna, H. Grampeix, P. Gaud, C. Mazure, and S. Deleonibus in VLSI Tech. Dig., p. 168 (2006).
17. W. Xiong, C. Rinn Cleavelin, P. Kohli, C. Huffman, T. Schulz, K. Schruefer, G. Gebara, K. Mathews, P. Patruno, Yves-Matthieu Le Vaillant, I. Cayrefourcq, M. Kennard, C. Mazure, K. Shin, T.-J. King Liu, IEEE Electron Device Letters **27** (7), p. 612 (2006).
18. T. Akatsu, J.M. Hartmann, A. Abbadie, C. Aulnette, Y.-M. Le Vaillant, D. Rouchon, Y. Bogumilowicz, L. Portigliatti, C. Colnat, N. Boudou, F. Lallement, F. Triolet, Ch. Figuet, M. Martinez, P. Nguyen, C. Delattre, K. Tsyganenko, C. Berne, F. Allibert, Ch. Deguet, M. Kennard, E. Guiot, F. Metral and I. Cayrefourcq, in *Semiconductor Wafer Bonding IX*, ECS Trans. **3**, (6), p. 107 (2006).
19. M. Yang, M. Ieong, L. Shi, K. Chan, V. Chan, A. Chou, E. Gusev, K. Jenkins, D. Boyd, Y. Ninomiya, D. Pendlton, Y. Surpris, D. Heenan, J. Ott, K. Guarini, C. D'Emic, M. Cobb, P. Mooney, B. To, N. Rovedo, J. Benedict, R. Mo, H. Ng, IEDM Tech. Dig., p. 453 (2003).
20. M. Yang, V. Chan, S. H. Ku, L. Shi, K. Chan, C. S. Murphy, R.T. Mo, H.S. Yang, E. A. Lehner,Y. Surpris, F.F. Jamin, P. Oldiges, Y. Zhang, B.N. To, J. R. Holt, S.E. Steen, M.P. Chudzik, D.M. Fried, K. Bernstein, H. Zhu, C.Y. Sung, J. Ott, D. Boyd, and N. Rovedo, Proc. Symp. VLSI Tech, p.160 (2004).
21. K. L. Saenger, J. P. de Souza, K. E. Fogel, J. A. Ott, A. Reznicek, C. Y. Sung, D. K. Sadana, and H. Yin, *Appl. Phys. Lett.*, **87**, 221911 (2005).

22. K. K. Bourdelle, O. Rayssac, A. Lambert, F. Fournel, X. Hebras, F. Allibert, C. Figuet, A. Boussagol, C. Berne, K. Tsyganenko, F. Letertre, and C. Mazuré, in *High Purity Si Symposium*, ECS Trans. **3**, (4), p. 409 (2006).
23. C. Maleville, in *High Purity Si Symposium*, ECS Trans. **3**, (4), p. 397 (2006).
24. R. Zaman, W. Xiong, R. Quintanilla, T. Schulz, C. R. Cleavelin, R. Wise, M. Pas, P. Patruno, K. Schruefer and S.K. Banerjee; MRS Symp. E on Semiconductor Defect Engineering, March (2005).
25. S. Okhonin, M. Nagoga, JM Sallese, and P. Fazan, IEEE Int. SOI Conf., p.153 (2001).
26. K. Inoh, T. Shino, H. Yamada, H. Nakajima, Y. Minami, T. Yamada, T. Ohsawa, T. Higashi, K. Fujita, T. Ikehashi, T. Kajiyama, Y. Fukuzumi, T. Hamamoto and H. Ishiuchi, Proc. Symp. VLSI Tech, p. 63 (2003).
27. Y. Minami, T. Shino, A. Sakamoto, T. Higashi, N. Kusunoki, K. Fujita, K. Hatsuda, T. Ohsawa, N. Aoki, H. Tanimoto, M. Morikado, H. Nakajima, K. Inoh, T. Hamamoto and A. Nitayama, IEDM Tech. Dig., p.317 (2005).
28. T. Shino, N. Kusunoki, T. Higashi, T. Ohsawa, K. Fujita, K. Hatsuda, N. Ikumi, F. Matsuoka, Y. Kajitani, R. Fukuda, Y. Watanabe, Y. Minami, A. Sakamoto, J. Nishimura, H. Nakajima, M. Morikado, K. Inoh, T. Hamamoto and A. Nitayama, IEDM Tech. Dig., p. 569 (2006).
29. I. Ban, U. Avci, U. Shah, C. Barns, D. Kencke, and P. Chang, IEDM Tech. Dig., p. 573 (2006).
30. R. Tsuchiya, M. Horiuchi, S. Kimura, M. Yamaoka, T. Kawahara, S. Maegawa, T. Ipposhi, Y. Ohji, and H. Matsuoka, IEDM Tech Dig., p. 631 (2004).
31. N. Bresson, S. Cristoloveanu, K. Oshima, C. Mazuré, F. Letertre, H. Iwai, IEEE Int. SOI Conf., p. 62 (2004).
32. C. Raynaud, F. Gianesello, C. Tinella, P. Flatresse, R. Gwoziecki, P. Touret, G. Avenier, S. Haendler, O. Gonnard, G. Gouget, G. Labourey, J. Pretet, M. Marin, R. Di Frenza, D. Axelrad, P. delatte, G. Provins, J. Roux, E. Balossier, JC. Vildeuil, S. Boret, B. Van Haaren, P. Chevalier, L. Boissonnet, T. Schwartzmann, A. Chantre, D. Gloria, E. de Foucauld, P. Scheer, C. Pavageau, G. Dambrine; *Silicon-On-Insulator Technology and Devices XII*, ECS Proc., vol **2005-03**, edited by G.K. Celler, p. 331 (2005).
33. T. Matsumoto, S. Maeda, K. Ota, Y. Hirano, K. Eikyu, H. Sayama, T. Iwamatsu, K. Yamamoto, T. Katoh, Y. Yamaguchi, T. Ipposhi, H. Oda, S. Maegawa, Y. Inoue and Inuishi, IEDM Tech. Digest, p. 219 (2001).

ECS Transactions, 6 (4) 15-26 (2007)
10.1149/1.2728836, ©The Electrochemical Society

Physics and Integration Of Fully-Depleted Silicon-On-Insulator Devices

A. Vandooren
Freescale Semiconductor, Inc.
850 Rue Jean Monnet, 38926 Crolles, France.
Email: anne.vandooren@freescale.com, phone : +33 438923579

Abstract. This paper gives an overview of the device physics and the process integration challenges of fully-depleted (FD) silicon-on-insulator (SOI) devices for advanced technology nodes. In the first part of the presentation, the physics related to the operation of ultra-thin film devices is described as well as how device scaling is impacted by the device parameters. Next, major fabrication challenges faced by fully-depleted SOI devices are reviewed along with their most recent progress. The technology modules reviewed include isolation, substrate engineering (limits of ultra-thin layers, strained layers, buried insulator materials,...), gate stack (high k and gate workfunction engineering techniques), methods for reducing SD parasitic resistance, such as elevated source/drain, as well as stress techniques.

I. INTRODUCTION

Fully-Depleted devices are considered among the best candidates to increase the control of short-channel effects and to allow for a better scalability of CMOS devices while still maintaining compatibility with standard CMOS planar processing techniques. Extremely thin films are widely recognized for their potential for the end-of-the-roadmap transistors. The SOI substrate fabrication based on the Smart Cut® technique and others have been developed to provide commercial availability of high quality substrates. Development is on-going to fabricate films as thin as 10nm and below with good uniformity and low defectivity.

Even though thin film single gate devices do not have as good scalability as double-gate devices, they offer the major advantage of full compatibility to planar CMOS processing with some fabrication process modifications. Specific modules, however, need to be introduced. Like in other CMOS devices, a high permittivity gate dielectric is required to continue device scaling along with a metal gate for compatibility with high k and workfunction adjustment. Undoped channels are desired in order to reduce dopant fluctuation effects and improve mobility. While vertical isolation from the substrate is provided by the Buried oxide (BOX), various lateral isolation schemes may be possible due to the thin silicon film and inherent small topography. Another major modification compared to existing planar devices is the implementation of a raised source/drain architecture. This is typically carried out using selective epitaxial growth to locally increase the thickness of the silicon film allowing for optimized silicidation and reduced series resistance. The applicability of existing techniques for mobility enhancement used in planar bulk or PDSOI technologies needs to be evaluated for the specific case of thin film devices. Currently, most of these techniques are based on process-induced strain obtained from contact etch stop layer (CESL), embedded SiGe in the SD regions, or Stress Memorization Technique (SMT). Substrate-induced strain techniques, not yet widely used in today's technologies, can also provide additional boost in thin film devices. This paper ends with a description of a proposed fabrication flow for FDSOI devices. Their electrical properties, along with their impact on basic circuit performance, in SRAM bit cells and ring oscillators, are presented. The device applicability at the platform level is also briefly discussed.

II. DEVICE PHYSICS OF THIN FILM DEVICES

A fully-depleted device operates under full depletion of the silicon channel. This provides specific properties of conduction, modified coupling between the interfaces, as well as other unique advantages. As a result, parameters such as the thickness of the silicon film or the buried oxide affect the scalability of the devices, unlike in PDSOI devices. Additional effects specific to thin film devices caused by scaling include mobility reduction, quantum confinement, gate induced floating body effects, dopant fluctuations, and self-heating. These also need to be taken into account in order to fully understand the resulting device's electrical behaviour.

II.1 Device Scaling

Scaling rules for thin film devices are quite different from bulk or PDSOI devices. This results from the device operating in full depletion mode and the front and back interface potential coupling. A natural scale length, λ, has been defined as a parameter representative of the ability of a device to scale. λ is equal to $\sqrt{\dfrac{\varepsilon_{si}}{\varepsilon_{ox1}}t_{si}t_{ox1}}$ in single gate fully-depleted device and depends on silicon film thickness, front gate oxide thickness and front gate dielectric permittivity. In general, the gate length of the device needs to be approximately 5 to 10 times longer than the natural scale length. One can easily observe that the reduction of the silicon film reduces λ and improves the scaling of the device (Fig. 1). While the doping concentration, doping profile, and front gate dielectric thickness in bulk devices are the main parameters for controlling short channel effects, FDSOI devices provide an additional control knob through silicon film adjustment. The constraint on the silicon film thickness is however extremely tight as 20nm-long devices will require film thickness as low as 5nm.

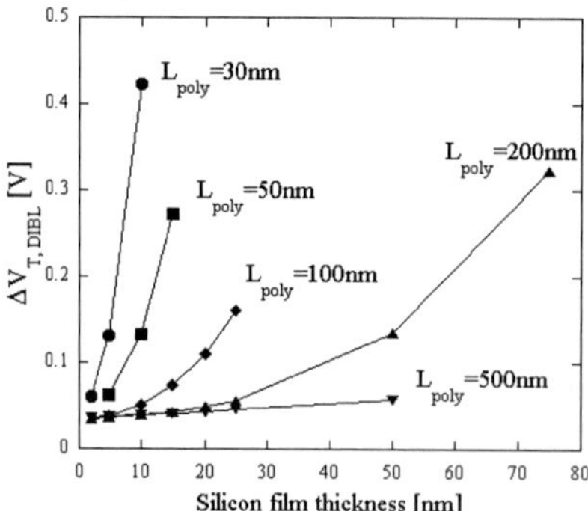

Figure 1 Variation of simulated DIBL versus silicon film thickness for gate length from 0.5µm down to 30nm. [1]

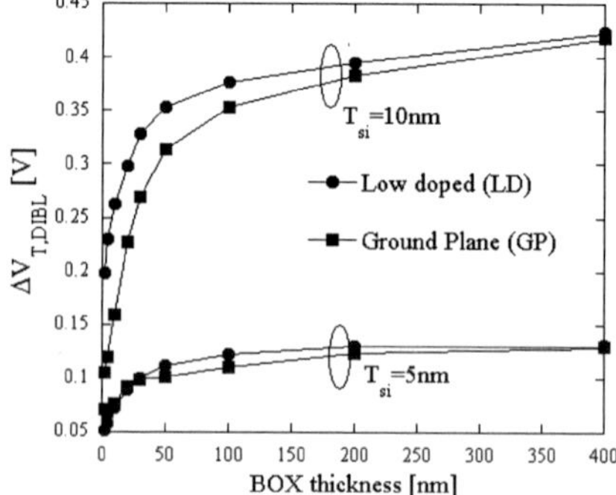

Figure 2 Variation of simulated DIBL with BOX thickness in FDSOI devices with T_{si}=10nm and 5nm in low doped substrates or ground plane [1]

Another scaling effect specific to FDSOI devices is caused by the electric field lines penetrating from the drain side into the buried oxide and the underlying substrate. This results in an increase of the potential at the back interface called Drain Induced Virtual Substrate Biasing (DIVSB) and degrades the device performance through interface coupling. This effect can be completely suppressed by using a thin BOX combined with a ground plane architecture (Figs. 2 and 3).

I_{on} vs. I_{off} simulated data are shown in Fig. 4 for various silicon film thickness and BOX thickness in a low doped (LD) or ground plane (GP) architecture.

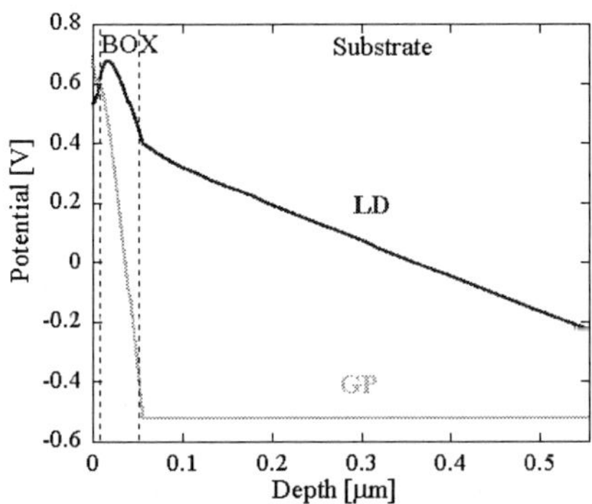

Figure 3 Simulated potential lines at mid-channel in LD and GP architectures [1]

Figure 4 Simulated I_{on} vs. I_{off} for various T_{si} and T_{BOX} for LD and GP architectures.[1]

II.2 Additional Scaling Effects

<u>Quantum confinement</u>: Very thin silicon layers induce a vertical confinement of the carriers in the film, which results in a quantization of the energy levels of the carriers inversely proportional to the SOI thickness. Typically, for films thinner than 5nm, the threshold voltage (V_T) increases due to the increase of the ground state energy level. This V_T increase also degrades its sensitivity to film thickness variation according to the

equation, $\sigma_{V_T} = -\left(\dfrac{\hbar^2 \pi^2}{q m^* t_{si}^2}\right) \dfrac{\sigma_{t_{si}}}{t_{si}}$.

<u>Gate induced floating body effect</u>: Thinning of the gate oxide results in increased gate leakage. In partially-depleted SOI devices and in FDSOI devices biased in back accumulation, charging of the body occurs due to the injection of majority carriers from the gate. This results in the occurrence of a kink effect in the I_{DS} vs. V_{GS} characteristic.

<u>Mobility</u>: The mobility evolution with film thickness is shown in Figure 5. As the silicon film thickness decreases, the mobility first decreases but rebounds around 5nm film thickness and then drops again for a film thickness below 3nm. This can be explained as follows (Fig. 6): as the silicon film is reduced, the reduction of the quantum well in the 4-fold valley increases phonon scattering and reduces the mobility. For film thinner than 5nm, the 4-fold valley energy is lifted up and the occupancy of the 2-fold valley is increased, which increases the mobility. Finally for films thinner than 3nm, the reduction of the quantum well in the 2-fold valley increases phonon scattering and reduces again the mobility.

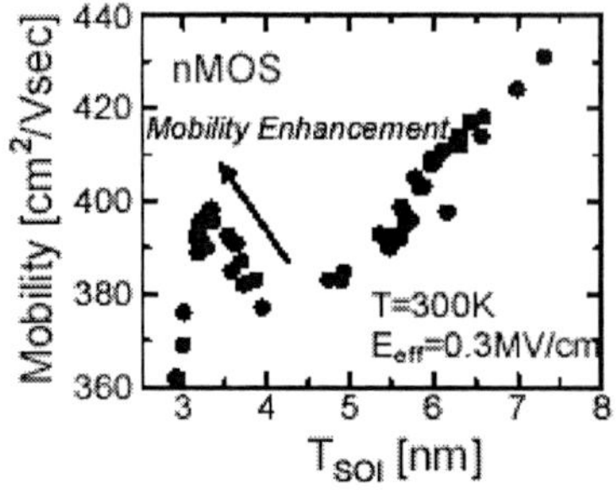

Figure 5 Experimental T_{SI} dependence of electron mobility at 300 K. [2]

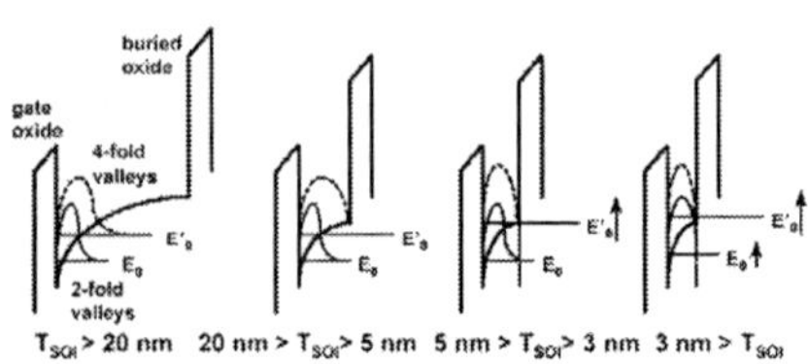

Figure 6 Conduction band diagram evolution with T_{SI}.

III. TECHNOLOGY MODULES

Specific technology modules are required for the fabrication of FDSOI devices. Requirements and evolution of the starting SOI substrates are first described. Next, transistor isolation techniques are compared. The gate stack module options which include high k materials for control of the transistor gate leakage and implementation of single or dual metal gates for adjusting the workfunction are discussed in detail. While selective epitaxy performed in the Source/Drain regions can reduce the excessive series resistance, potential problems and process requirements need to be considered. Techniques for improving mobility by using process-induced strain techniques, such as contact etch stop layers, or by using substrate-induced strain techniques, such as SSOI, are presented. Finally, integration of multiple substrate orientation as well as alternate channel materials like germanium for improving device performance are introduced.

III.1 SOI Substrates

Continuous improvements in the fabrication techniques of SOI substrates (SIMOX, wafer bonding and Smart Cut®, …) have led to today's low defect, high quality SOI materials used in production. Fully-depleted devices, however, require an extremely thin top silicon film with low defectivity and good uniformity. Thin silicon SOI substrates down to +/- 10Å uniformity have already been achieved by SOI vendors.

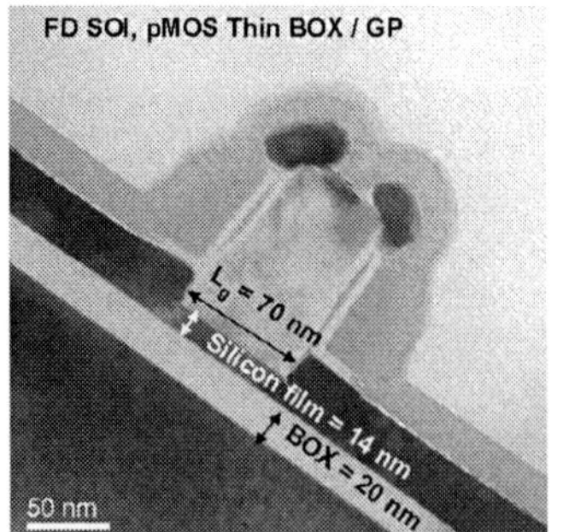

Figure 7 TEM cross-section of thin BOX/ ground plane device [3].

Short channel	S Exp. L_g=70 nm	DIBL Exp. L_g=70 nm
$T_{BOX}=20$ nm/ GP	71 mV/dec	28 mV
$T_{BOX}=145$ nm	69 mV/dec	66 mV

Table 1 Comparison of SCE between ground plane/thin BOX and thick BOX FDSOI architectures [3].

As discussed in the previous section, thin BOX devices are needed to suppress the DIVSB effect and improve control of DIBL. SOI substrates with a thin BOX obtained through the Smart Cut® technique are becoming available and have been used to build FDSOI devices. For the thin BOX to be efficient, it needs to be combined to a ground plane. The ground plane can be formed by dopant implantation through the BOX into the underlying substrate. A TEM cross-section of the device is shown in Fig. 7. The SCE improvement obtained is listed in Table 1. [3]

Today, engineered SOI substrates are becoming more and more attractive. They include modification of the device layer to improve mobility using strained silicon, different crystal orientation, or other material such as germanium. They also involve alternative buried insulator materials, such as Si_3N_4, Al_2O_3 or diamond to lower the self-heating effect by increasing the thermal conductivity.

Another approach to build thin film SOI devices is the Silicon-On-Nothing (SON) approach where thin silicon devices are built on bulk substrates using selective epitaxy of thin layers of SiGe and Si. After the SiGe is selectively removed, an empty tunnel underneath the top silicon layer is created, which is then filled by insulating materials. The advantages of this approach are the low cost of the starting bulk substrate and the good uniformity of the SiGe and Si layers from the epitaxy process [4].

III.2 Isolation Module

Three different isolation schemes are compared. Lateral isolation can be obtained by MESA, Shallow Trench Isolation (STI) or LOCOS isolation techniques. The **MESA isolation** is simple and is based on etching active silicon islands and stopping selectively on the buried oxide layer. Special care needs to be taken due to corners effect for parasitic corner conduction and reduced gate dielectric reliability. In addition, careful monitoring and reduction of the BOX consumption during the fabrication flow is necessary to avoid excessive topography between the field oxide and the active silicon regions. The **STI isolation** technique, widely used in today's technologies, is more complex and more costly. But because the trench is so shallow, due to the thin silicon film, this technique may no longer be justified. The **LOCOS isolation** has been widely used in less aggressive planar CMOS technologies but was abandoned mostly because of the bird's beak formation requiring additional local doping and causing reduction of the active silicon width. In FDSOI devices, because the amount of silicon to oxidize is very limited, the encroachment issue is also significantly reduced.

III.3 Gate Stack Module

The reduction of the gate oxide thickness causes the gate leakage to become unacceptably high due to direct tunnelling current. It is therefore necessary to move away from the currently used nitride oxides to gate dielectric materials with higher permittivity. Many high k materials have been studied in the literature in the past years. It was shown that high k materials have many issues. They cause V_T instability, Fermi pinning effect which modifies the gate workfunction, degrade the carrier mobility, and exhibit poor reliability. These are especially strong in the case of polySi gate material which tends to lead to the introduction of high k and metal gate electrodes simultaneously. Metal gates have a better potential to adjust to the proper workfunction, but band edge values needed for bulk devices are still difficult to achieve.

The use of metal gates is also very attractive for several other reasons. They fully suppress the poly depletion effect and lower the inversion capacitance equivalent thickness (CET) of the device. They have lower resistance and eliminate boron penetration. The requirement on the gate material workfunction is quite different in FDSOI (Fig. 7). If the channel is doped and band edges metals are used, a very high doping concentration will be necessary. Channel doping allows for V_T modulation but results in large V_T dispersion issue and mobility degradation. On the other hand, if the channel is kept undoped, a single gate with a midgap workfunction gives symmetrical V_Ts for nMOS and pMOS devices. This is definitely simpler to integrate. However, the V_Ts obtained are relatively high. To reduce the V_T, distinct metal gates for nMOS and pMOS are needed, with a workfunction slightly below and above midgap. In any case, modulation of the V_T is mandatory for circuit applications, and techniques to vary the metal workfunction, at least slightly, need to be implemented. Possible techniques are modulation by implantation of nitrogen or total silicidation of a polySi gate with various dopants and/or silicide phases.

Undoped channel have the additional advantage of improved mobility thanks to the low effective electric field in normal operation. Therefore, mobility degradation due to the use of high k can be compensated and mobility values similar to bulk devices with SiON gate dielectric can be obtained (Fig. 8).

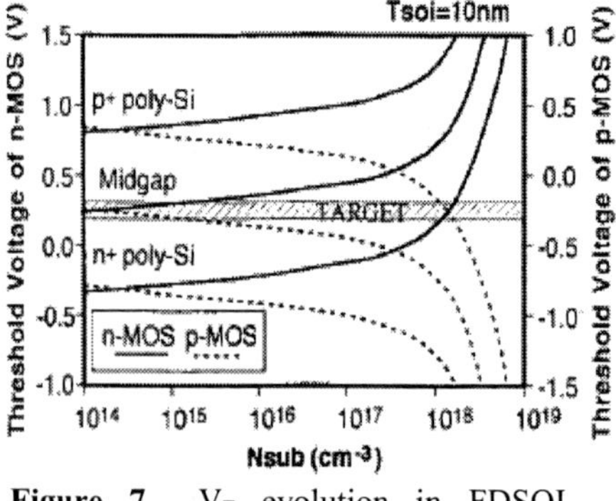

Figure 7 V_T evolution in FDSOI devices with channel doping concentration for N+ like, P+ like and midgap gate workfunctions [5] (T_{SOI}=10nm, T_{ox1}=5nm).

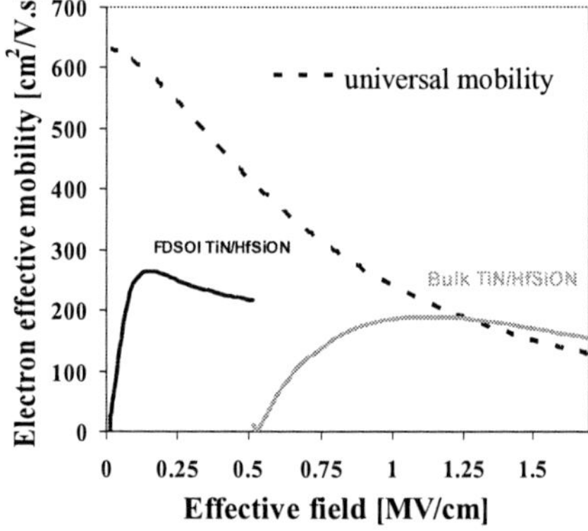

Figure 8 Effective mobility vs. Effective electrical field comparison in FDSOI with high k and bulk devices with SiON.

The process integration of metal gate/high k involves modification of the gate etch process and selective removal of the high k material from the source/drain areas. For thin film devices, it is critical to minimize the consumption of the silicon during these two steps. The high k material is a good gate etch stop material, preventing silicon recess. Different processes exist to remove the high k selectively to the silicon, depending on the high k material used. To avoid any contamination of the metal gate material in the front-end line, the gate needs to be fully encapsulated to resume processing.

Dual metal gate integration is very challenging. Various integration schemes have been proposed. In one of them [6], a first metal is deposited, then removed from the active area where unnecessary, before a second metal is deposited. This requires one to selectively remove the first metal exposing the high k material. The simultaneous gate etch of a single and dual metal stack is yet another issue. Another integration approach consists of performing full silicidation of a polysilicon gate. Due to the snowplow effect, the dopants in the polySi are pushed toward the gate dielectric interface which can modify the overall workfunction [7]. Issues related to this approach are the control of the full silicidation and related uniformity requirements, and the compatibility with high k material for which the workfunction modulation could not be demonstrated so far.

III.4 Source/Drain Module

All ultra-thin film devices have to face the main challenge of high series resistance in the Source/Drain regions, due to the thinness of the film. It was shown that once the silicide thickness reaches the silicon film thickness, the SD junction resistance increases dramatically due to the reduction of the effective contact area. In addition, full silicidation of the thin silicon film can cause the formation of voids due to the diffusion of silicon atoms from underneath the gate once the entire Si film is consumed. To reduce the series resistance, thicker source/drain regions are necessary as well as thinner silicides. Recessed channel techniques have been used to locally thin down the silicon in the channel region, but their implementation in manufacturing is doubtful. Another technique consists of elevating the Source/Drain regions by selectively growing silicon on those regions and is known as selective epitaxy growth (SEG). There are, however, some general issues related to any SEG process: faceting, micro- and macro-loading effects, pre-epi surface preparation, thermal budget impact on CMOS processing,... In addition, very thin Si films are quasi-stable and their shape changes after thermal treatment. They tend to agglomerate into silicon islands to reduce their interface energy. It is unclear for crystalline silicon where the agglomeration starts and what causes it. This results in active silicon shrink (Fig. 9), which is already visible after performing the H_2 pre-bake, and in silicon agglomeration (or islanding) (Fig. 10), which depends on the temperature of both the pre-bake and the SEG. By properly tuning the process conditions, good quality SEG was performed on silicon films as thin as 3 or 4nm.

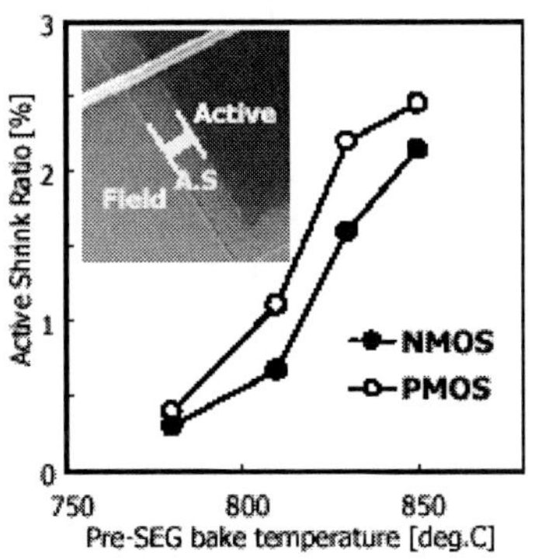

Figure 9 Active shrink evolution with SEG pre-bake temperature [8]

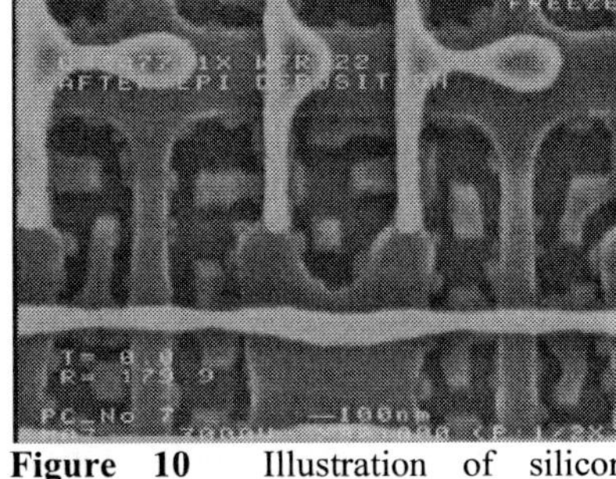

Figure 10 Illustration of silicon agglomeration phenomenon

Selective epitaxy is usually implemented in the SD regions of the device. However, selective epitaxy can also be carried in the extension regions of the device to further reduce the series resistance. In this case, two SEG steps are performed. A trade-off exists between reducing the series resistance and increasing the Miller capacitance.

III.5 Mobility Enhancement Techniques

Because scaling down the devices is becoming increasingly difficult, techniques to boost the carrier mobility to increase the drive current are being implemented in CMOS technologies. As of today, most techniques are based on process-induced strain. Global channel strain can also be induced by the substrate, but is not yet implemented in manufacturing.

<u>Process-induced techniques</u>: Most widely used process-induced techniques are either based on stress induced by layers such as contact etch stop nitride (CESL), by replacement of the SD regions with a different material such as SiGe (SiC), or by SMT of the gate. Dual stress liners, i.e. compressive (for pMOS) and tensile (for nMOS), are implemented on both bulk and PDSOI technologies. In FDSOI devices, the effectiveness of strained CESL has also been assessed [9]. The amount of strain transferred into the channel increases as the silicon film thickness is reduced and decreases when the raised SD regions become thicker (Fig. 11). Implementation of SiGe in the SD regions of pMOSFETs has been reported on FDSOI devices using the Ge condensation technique and elimination of the SD recess step [10]. Finally, SMT has also been successfully applied to FDSOI devices showing a more effective strain transfer as the channel thickness is decreased [11].

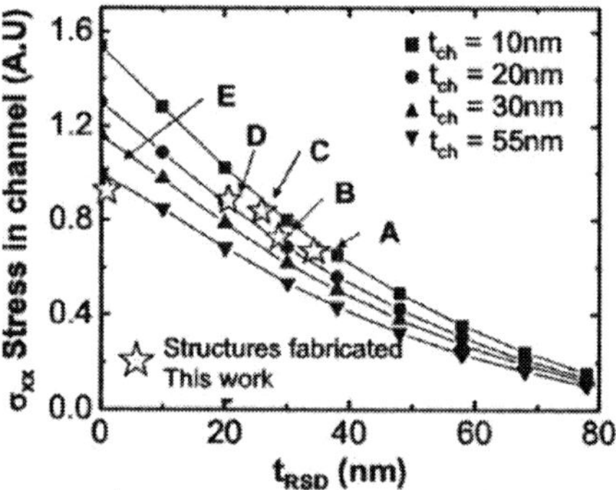

Figure 11 Evolution of stress-induced in the channel with channel thickness and raised SD thickness [9]

<u>Substrate-induced techniques</u>: Bi-axial strain can also be induced globally at the wafer level. A bi-axial tensile strain results from silicon epitaxially grown on a relaxed SiGe substrate. Both electron and hole mobilities can be improved but the improvement for holes only occurs for Ge content larger than 30-40%. SiGe can be transferred onto an SOI wafer and then have a strained silicon layer grown on top of it (SGOI). Even though device improvement was demonstrated on SGOI, the presence of the underlying SiGe layer greatly complicates the device integration and alters performance due to defectivity, difference in dopant diffusion in SiGe vs. Si, Ge up-diffusion, lower SiGe band-gap,... Another approach consists of transferring the strain Si layer directly onto the BOX/substrate (SSOI). The bi-axial strain can be maintained through device processing. In addition, a silicon thickness larger than the critical thickness can be used before relaxation occurs, which also renders this technology attractive for PDSOI applications.

(Multiple) Substrate orientation: In (100) substrates, the electron mobility is maximized compared to alternate crystal orientation. However, the hole mobility is larger in substrates with the (110) orientation. It is therefore desirable to be able to build devices on two distinct orientations on the same wafer. This was demonstrated in [12].

Germanium channel: The large carrier mobility in germanium is drawing wide interest. In addition, high k gate materials are compatible with the Ge channel. GeOI substrates have been fabricated using either the Smart Cut® process or the condensation method. However, fabrication of such devices has not yet shown the expected performance improvement.

IV. FABRICATION FLOW & DEVICE ELECTRICAL PROPERTIES

A fabrication flow of the FDSOI devices integrating metal gate, high k, and raised SD is illustrated in fig. 12. The starting SOI wafers are first adjusted to the correct film thickness by thermal oxidation. Then isolation is performed followed by the gate stack deposition. TiN or TaSiN gate materials are deposited on HfO_2, HfSiON or HfZrO high k dielectrics. The metal is capped with polySi for topography and silicidation compatibility. The metal gate is etched stopping on the high k material, and a nitride offset spacer is formed to encapsulate the metal gate. The high k material is then removed selectively from the active SD regions. It is beneficial to keep the high k material as an etch stop layer when creating the offset spacer to avoid recessing the silicon film. Next, selective epitaxy is performed in the extension regions followed by extensions and pocket implants, if necessary. D-shape SD spacers are formed followed by a second selective epitaxy step, if needed. The SD regions are implanted and silicided with NiPt. A copper backend process completes the process flow. A TEM cross-section of devices with TiN gate is shown in Fig.13. Capping of the gate will be necessary for aggressive technology nodes to prevent epitaxy on the gate and mushroom creation, which may cause implant shadowing effects and creates risks of shorts with the SD contact plugs. Capping can be implemented through an oxide or a nitride layer deposition on top of the gate and removed after performing the SEG steps.

Figure 12 Process fabrication flow

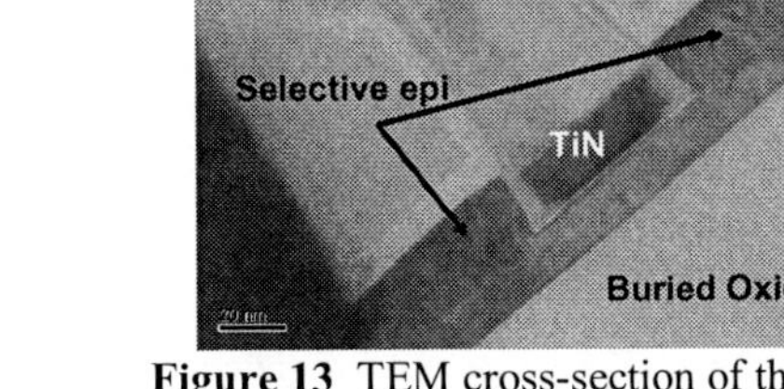

Figure 13 TEM cross-section of the fabricated FDSOI device. [13]

The device electrical characteristics (I_{DS} vs. V_{GS} and I_{DS} vs. V_{DS}) are illustrated in Figs. 14 and 15 for 45nm devices. The short channel effects are relatively well controlled by the channel thickness and thermal budget (Figs.16 and 17). Further control of SCE can be achieved by adding mild pocket implants.

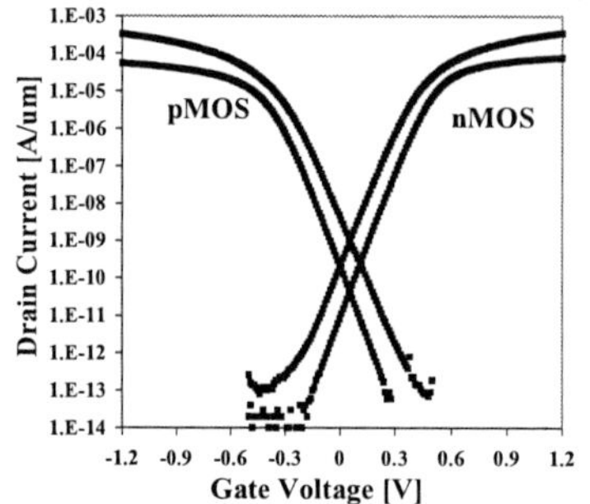

Figure 14 I_{DS} vs. V_{GS} characteristics [13]

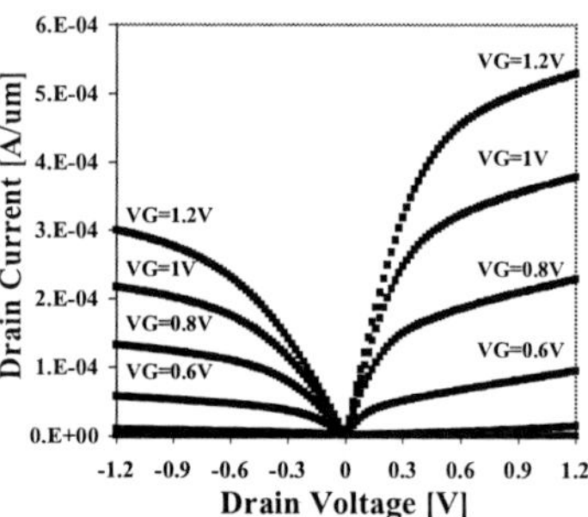

Figure 15 I_{DS} vs. V_{DS} characteristics [13]

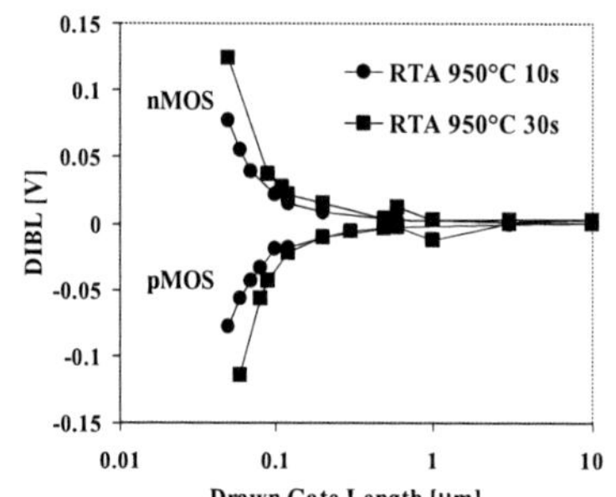

Figure 16 DIBL evolution with gate length [13]

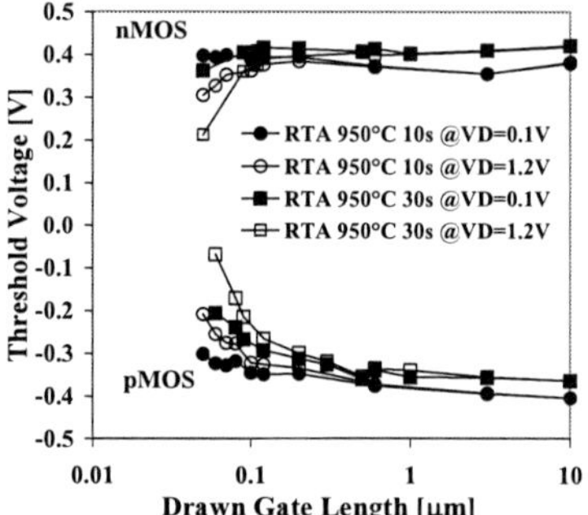

Figure 17 Threshold voltage evolution with gate length [13].

V. ELECTRICAL CIRCUIT PERFORMANCE

V.1 SRAM Bitcell

Functional SRAM bitcells were demonstrated. The butterfly curves of the $0.525\mu m^2$ cell are illustrated in Fig.17. Thanks to the intrinsic robustness of FDSOI devices at high temperature, the cell still functions correctly at 125°C with good characteristics. At 125°C, the SNM remains larger than 15% of V_{DD} down to 0.7V (Fig. 18). The stand-by current is also well controlled at this temperature thanks to the use of high k and the intrinsic isolation from the substrate which reduces the junction leakage.

In addition, better matching properties are expected by using an undoped channel and a metal gate. The matching is typically a function of the doping concentration in the channel ($N_c^{1/4}$) and of the grain size in the polysilicon gate, which can result in local fluctuation of the gate workfunction. Matching values down to 2.6mV.um to 3mV.um have been extracted from the high k/metal gate FDSOI devices (Fig. 19 and Table 1).

The advantages of FDSOI on the scaling of SRAM cells are obvious. Because of the limited depth of the trench isolation compared to that of bulk technology, the trench aspect ratio is much smaller and the filling is performed with more ease. In addition, the absence of wells allows for a more compact cell, with the N+-P+ spacing becoming equivalent to standard active-to-active spacing.

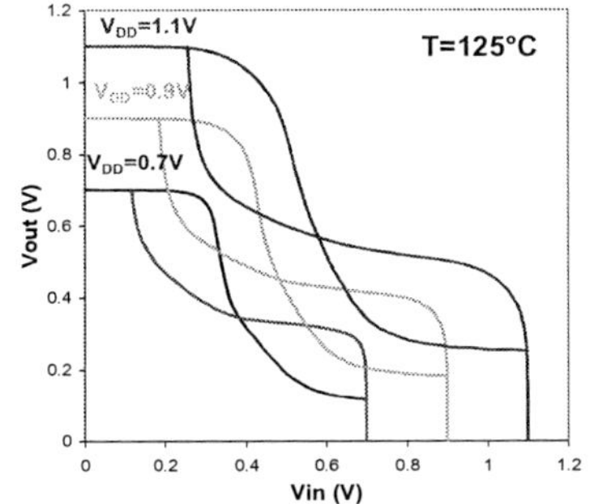

Figure 17 Butterfly curve of FDSOI SRAM bitcell

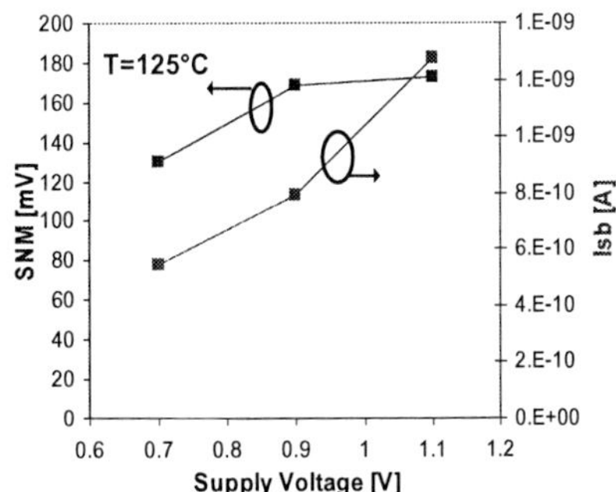

Figure 18 Signal to Noise Margin and stand-by current versus supply voltage

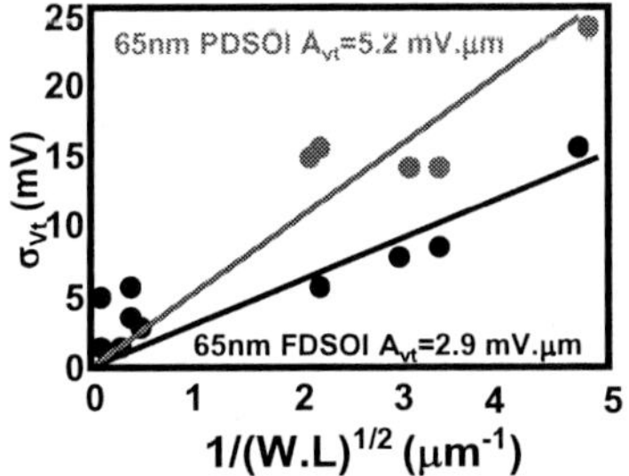

Figure 19 Matching comparison between FDSOI and PDSOI technologies

Avt (mV.μm) comparison

Technology	nmos	pmos
FDSOI	2.9	2.6
PDSOI	6.1	4.2
Low-Power Bulk	8.2	7.5

Table 2 Comparison of technology matching of V_T [14].

V.2 Ring Oscillators

Ring oscillator performance of stage delay versus dynamic power consumption is illustrated in Fig.20. It is compared to a standard PDSOI technology. While the stage delay is increased by 10ps due to lower drive current (larger CET), the power consumption is reduced by half due to their reduced junction capacitance, which makes this technology very appealing for low power applications.

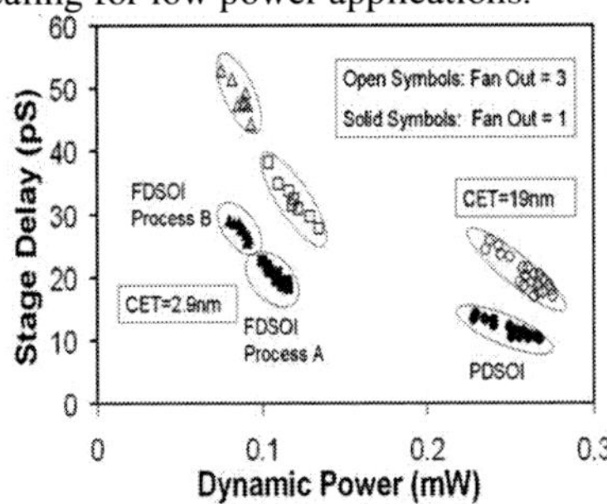

Figure 20 Ring oscillator stage delay evolution vs. dynamic power in FDSOI and PDSOI technologies (L=65nm W=0.42μm) [14]

V.3 I/O Co-Integration

For circuit compatibility, I/O devices are needed. GO1/GO2 device co-integration was achieved using high k dielectric and the process flow illustrated in Fig. 21. A thick bottom oxide is first grown over the entire active silicon and etched away from the GO1 active areas. The high k layer is then deposited everywhere, on top of the thick bottom oxide in the GO2 active areas and on top of a chemical oxide in the GO1 active areas. Both GO1 and GO2 gates are etched simultaneously stopping on the high k layer.

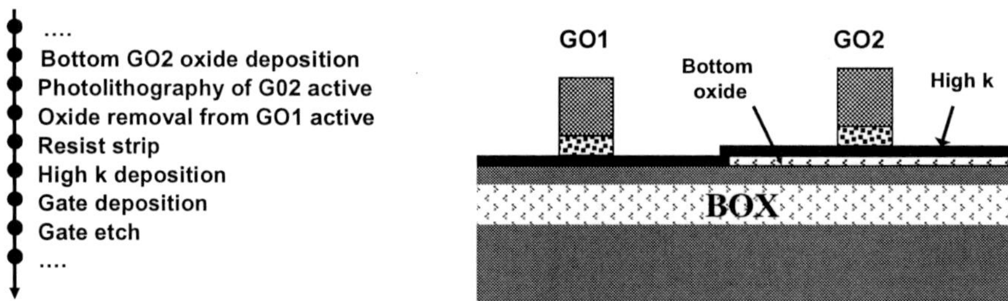

Figure 21 Co-integration flow of GO1 and GO2 devices and final devices

VI. CONCLUSIONS

FDSOI devices are attractive for next technology nodes due to their improved ability to scale and their compatibility to standard CMOS planar processing. However, SOI substrates with very thin film and very good uniformity will be required. In addition, selective epitaxy in the SD regions will be mandatory when using such thin films. Metal gates and workfunction tuning will be likely required for compatibility to high k gate dielectrics and for V_T adjustment. However, V_T adjustments within only 150mV to 200mV from midgap are sufficient. This is less stringent than for bulk devices. Additional attractive advantages of this technology when keeping an undoped channel are the reduced dopant fluctuation effect, improved matching properties, and increased mobility due to the operation at lower electric field.

REFERENCES

[1] A. Vandooren et al., *IEEE Intl. SOI Conf.*, pp. 25-27, 2002.
[2] K. Uchida et al., *IEDM Tech. Digest*, pp. 47-50, 2002.
[3] C. Gallon et al., *IEEE SOI Intl. Conf.*, pp. 17-18, 2006.
[4] S. Monfray et al., *IEDM Tech. Digest*, p.645, 2001.
[5] H. Shimada et al, *IEEE Trans. Electron Dev.*, 44, 11, pp. 1903-1907, 1997.
[6] S. Samavedam et al., *IEDM Tech. Digest,* pp. 433-436, 2002.
[7] C. Fenouillet-Beranger, *ESSDERC*, pp. 158-161, 2006
[8] C. B. Oh et al., *IEDM Tech. Digest*, pp., 423-426, 2003.
[9] D.V. Singh et al., *IEEE SOI Intl. Conf.*, pp. 178-179, 2005.
[10] K. J. Chui et al., *IEDM Tech. Digest,* 2005.
[11] D. V. Singh et al., *IEDM Tech. Digest,* pp. 505-508., 2005
[12] Yang et al., *Symp. on VLSI Technology*, pp. 160-161, 2004
[13] A. Vandooren et al., *IEEE Intl. SOI Conf.*, pp. 221-222, 2005
[14] A. Vandooren et al., *IEDM Tech. Digest,* pp. 11.5.1, 2003

ECS Transactions, 6 (4) 27-32 (2007)
10.1149/1.2728837, ©The Electrochemical Society

IMPACT OF METAL SILICIDE LAYOUT COVERING SOURCE/DRAIN DIFFUSIONS ON PARASITIC RESISTANCE OF TRIPLE-GATE SOI MOSFET

K. Yoshimoto[1], Y. Omura[1] and H. Wakabayashi[2]

[1]ORDIST/Grad. School of Eng., Kansai University, 3-3-35, Yamate-cho, Suita, Osaka, 564-8680 Japan
[2]Atsugi Tech. Ctr, Sony Corp., 4-14-1, Asahi-cho, Atsugi, Kanagawa, 243-0014 Japan

Abstract. This paper describes an impact of silicide layout in the S/D region on parasitic resistance of the multiple-fin triple-gate (TG) SOI MOSFET. For devices with narrow S/D region, it is demonstrated that Π-shape layout with a thin silicide film results in the lowest parasitic resistance. On the other hand, for devices with wide S/D region, it is shown that a deep '*localized*-silicide' layout results in the lowest parasitic resistance. The silicide/Si contact area should be as large as possible. It is also strongly suggested that a new technique is needed to reduce the parasitic resistance in sub-10-nm channel TG SOI MOSFET's.

INTRODUCTION

Silicon-on-insulator (SOI) MOSFET's are very promising for effectively suppressing the short-channel effects and achieving high drivability and low standby power consumption [1]. For aggressive scaling, it is anticipated that fin-type SOI MOSFET's are essential since they offer footprint reduction, high drivability, and short-channel effect suppression. However, parasitic resistance, such as source/drain (S/D) diffusion resistance and contact resistance, has been discovered to be a serious problem in triple-gate (TG) SOI FinFETs [2-4]. For the purpose of introducing silicide film, the elevated S/D structure using selective epitaxial growth (SEG) technique is being widely considered [3]. However, no silicide film structure has been successfully introduced [3] because the silicidation technique often yields additional difficulties.

This paper examines how to introduce the silicide film to the S/D diffusion region so as to produce high drivability TG SOI MOSFET's. We propose many configurations of the silicide film for n^+ S/D diffusion that can reduce the parasitic resistance of the S/D diffusion region.

DEVICE STRUCTURE AND SIMULATIONS

3-D device simulations were conducted using TCAD-*DESSIS* (Synopsys) [5]; this study focuses on the multiple-fin TG SOI MOSFET having narrow source and drain regions (see Fig. 1(a)); Fig. 1(a) also details source diffusion (10^{20} cm^{-3}) with a 10-nm-thick NiSi film with the resistivity of 1×10^{-5} Ωcm [6] as an example. It is assumed that the contact resistivity (R_C) between the silicide and the n^+ diffusion region is typically 10^{-8} Ωcm^2. It is assumed that the device has an unsilicided length (L_{un}) of 10-nm; L_{un}

represents the length of the region not covered by the gate electrode and the S/D contact. L_{SC} is the length of the electrode contact on the 100-nm-deep S/D diffusion region and the length of silicided region: L_{SC} = 30nm. Si-fin height (h) and width (W_S) are 100 nm and 40 nm, respectively. In this paper, we change the width of the silicided region (W_{SD}) from 40 nm to 80 nm to examine the impact of the multiple-fin structure with wide source and drain regions on the parasitic resistance of S/D diffusion. The simulations also examine the impact of silicide depth (H_S) on the resistance of S/D diffusion.

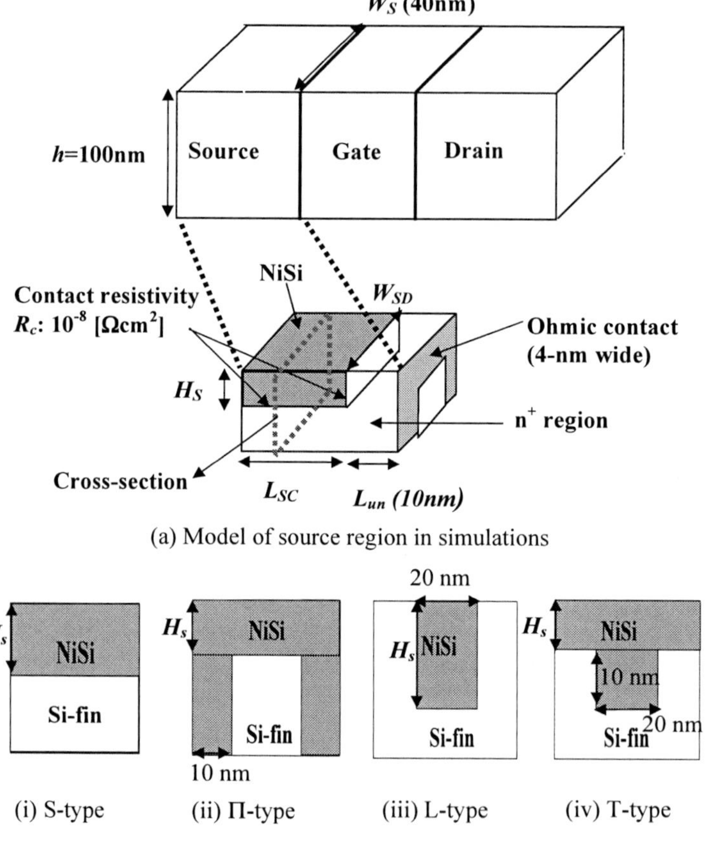

(a) Model of source region in simulations

(i) S-type   (ii) Π-type   (iii) L-type   (iv) T-type

(b) Layout models for silicide region

Figure 1. Model assumed for source diffusion of fin-type triple-gate (TG) SOI MOSFET

In addition, we consider the layout of the silicide film that covers the n$^+$ region and/or is buried into the n$^+$ region. Cross-sectional layouts of the source diffusion region (denoted by broken line) assumed here are shown in Fig. 1(b). First is the 'S-type', where the silicide film covers only the top surface of the n$^+$ region. Second is the 'Π-type', where the silicide film covers the top and both side surfaces of the n$^+$ region. Third is the 'L-type', where the silicide film is localized near the center of the top surface. Last is the 'T-type', where the silicide film covers the top surface and a part of the silicide film is buried in the n$^+$ region. The simulations assess 10-nm-thick unsilicided silicon regions with all of the layouts shown in Fig. 1(b).

In estimating the effective resistance of the source diffusion region, we assume that electrons enter the 4-nm-thick inversion layer under the gate electrode [7]. We consider that the effective inversion layer is thicker than that (~2 nm) of conventional bulk devices [7] because the volume inversion effect can not be neglected in such a thin-fin-type device.

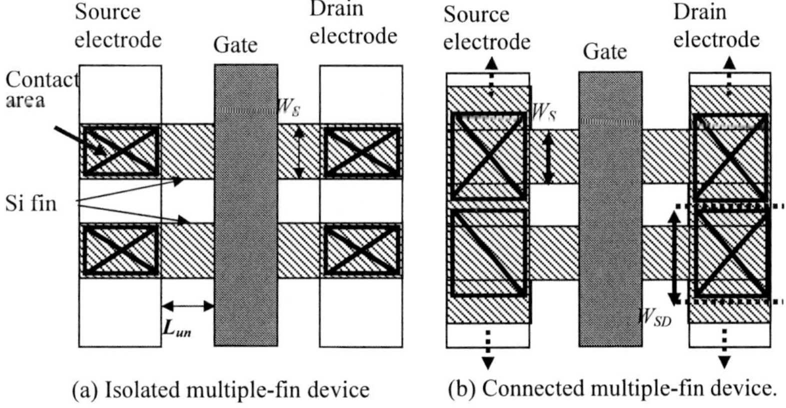

(a) Isolated multiple-fin device   (b) Connected multiple-fin device.

Figure 2. Top views of multiple-fin TG-SOI MOSFET.

RESULTS AND DISCUSSION

We compare effective source resistances of an isolated multiple-fin TG SOI MOSFET and a connected multiple-fin TG SOI MOSFET. Top views of the assumed TG SOI MOSFET structures are shown in Fig. 2; Fig. 2(a) shows the isolated multiple-fin TG SOI MOSFET with a narrow S/D region, while Fig. 2(b) shows the connected multiple-fin TG SOI MOSFET with a wide S/D region. Figure 3 plots simulated source resistance (R_S) versus L_{SC} for the multiple-fin SOI MOSFET having the narrow or wide S/D region. For the multiple-fin device with the narrow S/D region, we can see the following. The parasitic resistance decreases as L_{SC} increases, which is anticipated from the result given by L. T. Su et al. [7]. With the 'S-type' structure, a larger H_S yields a lower resistance as expected. With the 'Π-type', a smaller H_S yields the lowest resistance [8]; increasing H_S increases the resistance. This is due to the reduction in silicide/Si contact area and Si volume according to [7]. The 'L-type' structure yields a surprising result; a larger H_S yields almost the same result as the 'Π-type' with a smaller H_S. It can be considered that the increase in silicide/Si contact area results in lower parasitic resistance. With the 'T-type' structure, a larger H_S yields a resistance that is low, but not the lowest value. From this consideration, we can recommend either the 'Π-type' structure with a small H_S value or the 'L-type' structure with a large H_S value in designing multiple-fin TG SOI MOSFET's with the narrow S/D region.

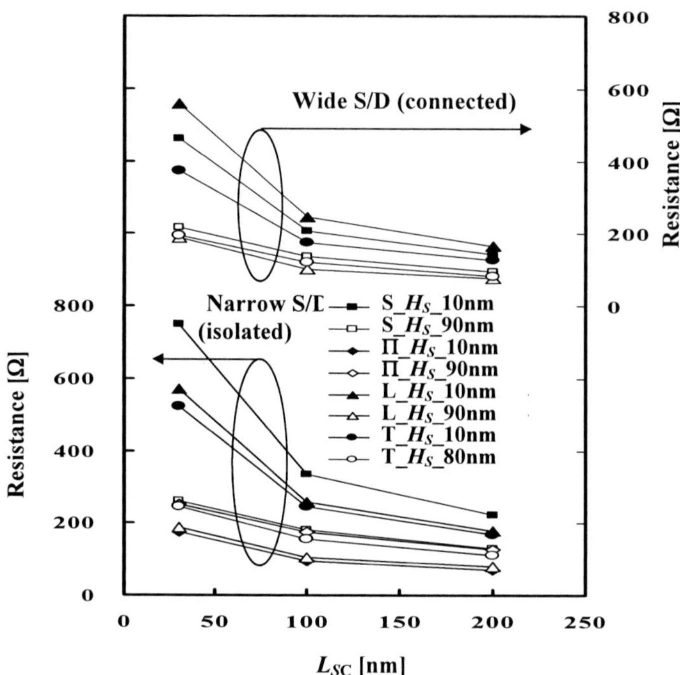

Figure 3. Simulated parasitic resistance of source region of multiple-fin device with a narrow S/D region and multiple-fin device with a wide S/D region.

For the multiple-fin SOI MOSFET with the wide S/D region, it is also seen that the parasitic resistance decreases as L_{SC} increases. The features of the multiple-fin device with the wide S/D region are as follows.

(i) The minimal resistance is gained with the 'L-type' structure with large H_S value.

(ii) 'T-type' with a large H_S value and 'S-type' with a large H_S value result in low resistance.

We also evaluated the impact of contact resistivity (R_C) between the silicide film and the n^+ diffusion region on parasitic source resistance (R_S); simulation results are shown in Fig. 4 for different L_{SC} values. It is seen that R_S value strongly decreases as L_{SC} increases. Assuming the standard process technology, we have $R_C=10^{-7}$ Ωcm^2. In addition, from device simulations, we also have channel resistance (R_{ch}) values of TG SOI MOSFET; R_{ch} value of a 30-nm-channel device is 680 Ω for $W_S=40$ nm, $h=100$ nm, and $t_{ox}=2$ nm at $V_g=1$ V, $V_{th}=0.25$ V and $V_d=0.1$ V [4], R_{ch} value of a 30-nm-channel device is 360 Ω for W_S=40 nm, h=100 nm, and t_{ox} =1 nm at V_g=1 V, V_{th}=0.25 V and V_d=0.1 V, and R_{ch} value of a 20-nm-channel device is 310 Ω for W_S=40 nm, h=100 nm, and t_{ox}= 1 nm at V_g= 1 V, V_{th}=0.25 V and V_d=0.1 V; these channel resistance values were carefully extracted from channel potential profiles so that the parasitic resistance holds almost constant regardless of device geometry examined here. When R_C=10^{-7} Ωcm^2, R_S values of $\Pi_H_S_$10nm (devices with the narrow S/D region) and $L_H_S_$90nm (devices with the wide S/D region) are 500 Ω and 514 Ω, respectively, at L_{SC}= 100 nm. Since R_{ch} of the 30-nm-long channel device and 20-nm-long-channel device is less than 1 kΩ, it can be suggested that R_C should be as low as 10^{-8} Ωcm^2 and L_{SC} should be at least ~100 nm. It is clearly seen that the conventional 'S-type' silicide is not suitable for future TG SOI MOSFET's. In addition, it can be anticipated that present S/D engineering will face a serious technical barrier given the device structure of a sub-10-nm channel TG SOI MOSFET with a very small parasitic resistance. Therefore, we have to propose a new technique to reduce the parasitic resistance.

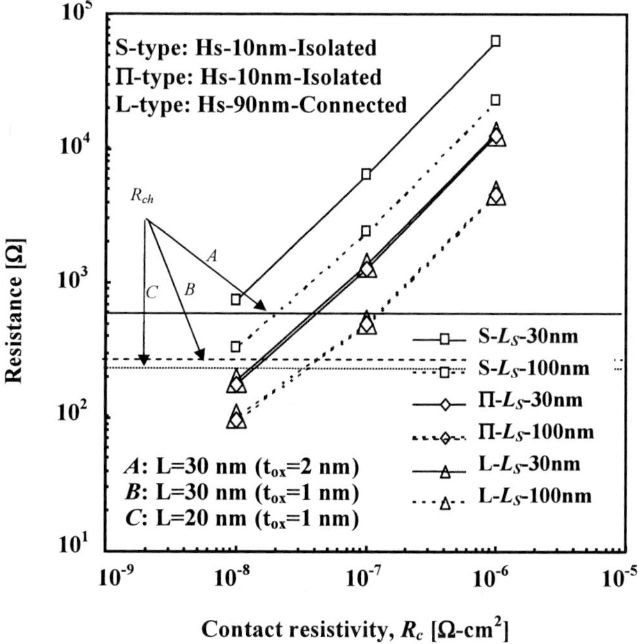

Figure 4. Simulated parasitic resistance of source region versus contact resistivity for various silicide film structures. L denotes the channel length.

Practical device design demands that we evaluate the entire performance of the TG SOI MOSFET, including short-channel effects [4, 9]. When a very thin fin is used to suppress the short-channel effect, the simulations should take account of the quantum transport effects [10]. These are the issues that should be considered in the next research step.

CONCLUSION

This paper has examined the impact of silicide layout in the S/D region on parasitic resistance. For multiple-fin TG SOI MOSFET's with a narrow S/D region , it has been discovered that the 'Π-type' structure with a thin silicide film results in the lowest parasitic resistance, and that the 'L-type' structure with deep, 'localized' silicide also yields low parasitic resistance. On the other hand, for multiple-fin TG SOI MOSFET's with a wide S/D region, it has been discovered that the 'L-type' structure with deep, 'localized' silicide film results in the lowest parasitic resistance, and that 'S-type' and 'T-type' structures with deep silicide film also achieve low parasitic resistance. Generally speaking, it is strongly suggested that the silicide/Si contact area should be as large as possible. However, this guideline will lose its meaning when designing a sub-10-channel TG SOI MOSFET.

ACKNOWJEDGMENT

This study is financially supported by a joint research program of Kansai University and Sony Corp., Japan. The authors wish to express their thanks to Drs. N. Nagashima, H. Ansai, Y. Tagawa, and S. Yamakawa with SONY Corp. for their critical discussion and encouragement.

REFERENCES

1. J.-P. Colinge, *Silicon-On-Insulator Technology: Materials to VLSI*, 3rd ed., Kluwer Academic Pub.(2004).
2. H. Kam, L. Chang, and Tsu-Jae King, in Abstr. *IEEE 2004 Silicon Nanoelectron. Workshop* (Hawaii, June, 2004) p. 9.
3. A. Dixit, A. Kottantharayil, N. Collaert, M. Goodwin, M. Jurczak, and Kristin De Meyer, IEEE Trans. Electron Devices, **52**, 1132 (2005).
4. H. Konishi and Y. Omura, IEEE Electron Devices Lett., **27**, 472 (2006).
5. TCAD *DESSIS/GENESISe*, ver. 8.0 Operations Man. (Synopsys Inc., Mountain View, CA, USA).
6. K. Maex and M. van Rossum, *Properties of Metal Silicides*, IEE (1995) p. 192.
7. L. T. Su, M. J. Sherony, Hang Hu, J. E. Chung and Dimitri A. Antoniadis, IEEE Electron Devices Lett., **15**, 363 (1994).
8. H. Shang, L. Chang, X. Wang, M. Rooks, Y. Zhang, B. To, K. Babich, G. Toir, Y. Sun, E. Kiewra, M. Ieong, W. Haensch, in Dig. Tech. Papers, 2006 Symp. VLSI Technol. (Hawaii, June, 2006) p. 66.
9. K. Yoshimoto, H. Konishi, and Y. Omura, submitted to IEEE Electron Devices.
10. Y. Omura, H. Konishi, and S. Sato, IEEE Trans. Electron Devices, **53**, 677 (2006).

ECS Transactions, 6 (4) 33-38 (2007)
10.1149/1.2728838, ©The Electrochemical Society

High Dose Implantation Impact on the Carrier Mobility in Ultra-Thin Unstrained and Strained SOI Films

C. Dupré[1-2], P. F. Fazzini[3], T. Ernst[1], F. Cristiano[4], J.-M. Hartmann[1],
A. Claverie[3], F. Andrieu[1], O. Faynot[1], P. Rivallin[1], F. Laugier[1],
G. Ghibaudo[2], S. Cristoloveanu[2] and S. Deleonibus[1]

[1]CEA-LETI, Minatec 17 Avenue des Martyrs, 38054 Grenoble Cedex 9, France
[2]IMEP-ENSERG, Minatec 3 Parvis Louis Néel BP 257, 38016 Grenoble, France
[3]CEMES-CNRS 29, Rue Jeanne Marvig, 31055 Toulouse Cedex 4 France
[4]LAAS-CNRS 7 Avenue du Colonel Roche - 31077 Toulouse Cedex 4 France

Based on electrical measurements and transmission electron microscopy imaging we propose in this paper a possible explanation for the measured mobility degradation with gate length reduction in short-channel MOSFETs. The carrier mobility in end-of-range implantation regions was investigated in Silicon-On-Insulator (SOI) and tensily-strained Silicon-On-Insulator (sSOI) substrates. Wafers with ultrathin films (from 8 to 35 nm) were Ge implanted at various concentrations, then annealed. The Pseudo-MOSFET measurements showed a mobility decrease (from 5 to 34%) as the implantation dose increased. The results are relevant for the optimization of the sources and drains regions of advanced nano-scale SOI transistors.

Introduction

Substrate-induced strained silicon is a promising candidate for transistor performance enhancement (1-2). Detailed analysis of the transport-limiting mechanisms showed a degradation of the mobility enhancement for short-channel transistors. This phenomenon was observed for both bulk and SOI, strained and unstrained devices, doped and intrinsic channels (3-6). Low temperature studies evidenced that Coulomb or neutral interactions were responsible for such degradation (3-6). Source/Drain (S/D) processing (such as gate etching or ion implantation) could be at the origin of this phenomenon. We studied here the impact of S/D ion implantation on strain relaxation and/or defects generation that would induce mobility degradation (7-8).

Our approach is detailed in Fig. 1. SOI and strained-SOI (sSOI) substrates with miscellaneous Si layer thicknesses were Ge-implanted at 3keV. Doses between 5.10^{14} and 2.10^{15} atoms.cm^{-2} were used. The samples were then annealed at 600°C. Since, the crystal was strongly damaged (defects in End-Of-Range (EOR) region), the Pseudo-MOSFET technique was used to characterize the Si/BOX interface transport properties. At appropriate substrate biasing, a conduction channel is formed at the back interface. The variable Si film thickness enabled the conduction channel to electrically probe the EOR region (Fig.1 c-d). We have implanted Ge (instead of As which is normally used in the extensions of n-MOSFETs transistors) because (i) Ge atoms are not electrically active, which makes the use of the Pseudo-MOSFET technique possible and (ii) its atomic mass is very close to the one of As (generating same kind of defects with similar defect densities, checked by Crystal TRIM (9-10)).

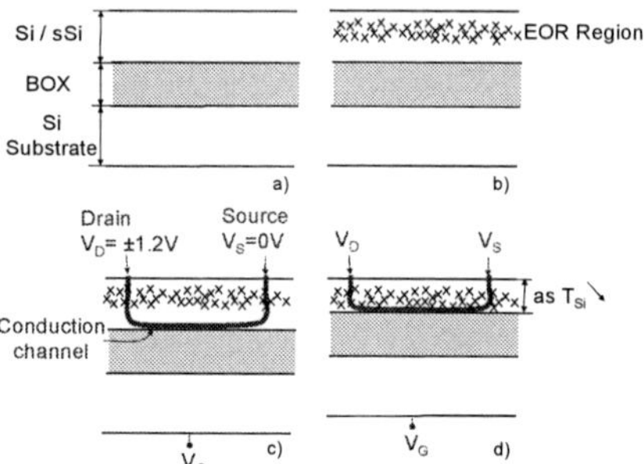

Figure 1. Detailed experimental procedure for wafer implantation and Pseudo-MOSFET evaluation.

For short channel transistors, carriers reaching the S/D zones are directly impacted by crystalline defects. This experiment is a simple way of highlighting and quantifying the impact of implantation defects on transport properties.

Sample preparation

Two types of SOI substrates were used in this study. The first family consisted of unstrained SOI substrates (prototypes with 100nm thick buried oxide, BOX), thinned down thanks to several dry oxidation / wet de-oxidation cycles. The final Si overlayer thickness (as determined by spectroscopic ellipsometry) was equal to 8, 18, 28 and 34 nm for samples A, B, C and D, respectively (Table I). The second batch consisted in very thin (5 nm) sSOI substrates, which were thickened by Reduced Pressure – Chemical Vapor Deposition. The surface preparation conditions (an "HF-last" wet cleaning followed by a 2 minutes H_2 bake at 750°C) as well as the actual growth conditions (SiH_2Cl_2 + HCl gaseous precursors, T = 750°C, P = 20 Torr) were such that no Si agglomeration occurred (11). We ended up with 15, 25 and 33 nm of sSi on top of 210 nm thick BOX (samples 1, 2 and 3, respectively).

TABLE I. Matrix of the SOI wafers

SOI sample	T_{Si}(nm)	sSOI sample	T_{sSi}(nm)
A	8		
B	18	1	15
C	28	2	25
D	34	3	33

A quarter of each wafer was saved for reference, while the other three quarters were implanted with increasing Ge doses (5.10^{14}, 1.10^{15} and 2.10^{15} atoms/cm² @ 3 keV, respectively) through a 1nm top oxide. The wafers were then submitted to one hour recrystallisation anneal under N_2 at 600°C. This sample matrix (Table 1) allowed us to quantify the impact of (i) the Si overlayer thickness, (ii) the nature of the Si film (unstrained versus tensile-strained) and (iii) the implanted dose.

Electrical characterization

The Pseudo-MOSFET technique (12) was used to study the SOI wafers prior to any CMOS processing. The typical current $I_D(V_G)$ characteristic is shown in Figure 2.

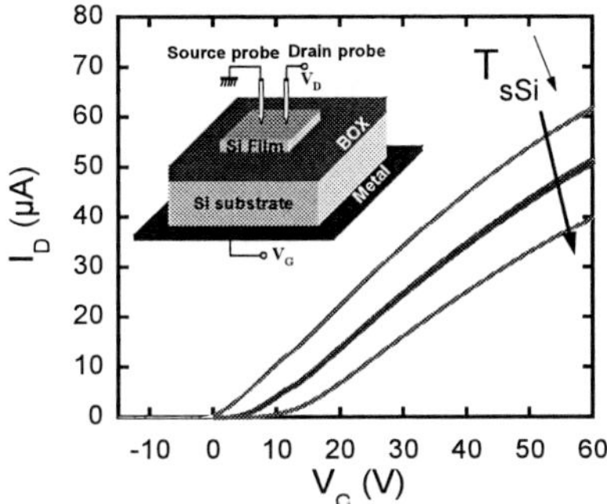

Figure 2. Drain current I_D versus gate voltage V_G for implanted sSOI substrates of different thicknesses (samples 1, 2 and 3) and schematics of the Pseudo-MOSFET.

The Pseudo-MOS technique allows quantifying the impact of the implantation on an important transport parameter: the effective mobility μ_{eff} extracted by

$$\mu_{eff} = \frac{I_D}{Q_{inv}.V_D.f_g} \qquad [1]$$

I_D being the drain current, V_D the drain bias, f_g the geometrical factor (empirical value $\approx$ 0.75) and $Q_{inv} = C_{OX}.(V_G - V_T)$ the inversion charge with C_{OX} the oxide capacitance and V_T the threshold voltage extracted using the $Y=I_D/G_m^{0.5}$ function method (13). In the following, the effective mobility will be shown for a given value of $Q_{inv}=2\times10^{12}$ cm^{-2} corresponding to the strong inversion regime. The effective field can then be deduced from $E_{eff}=\eta Q_{inv}/\varepsilon_{Si}$ in such fully depleted SOI films; we obtain $E_{eff}= 0.15$ MV.cm^{-1} for electrons. As the thickness decreases, the conduction channel penetrates the defective EOR region and so probes the EOR region's defects (see Fig. 1). Before implantation and annealing, the effective mobilities of electrons decrease with both the Si and sSi thicknesses (Fig. 3), a trend which has previously been reported (14). Higher electron mobilities are achieved in thickened sSOI than in standard SOI substrates.

Our goal is to analyze the impact of implanted Ge atoms as a function of (i) the dose and (ii) the film thickness and strain. After implantation and annealing, the mobility decay in thinner films is accentuated (Fig. 3). Indeed, the implantation-induced damaged region gradually approaches the film-BOX interface and directly affects the back-channel mobility. For a given thickness, the effective electron mobility decreases as the Ge implantation dose increases, the reduction being stronger for sSOI than for SOI substrates (34% versus respectively 5%, see Fig. 4). The measurement dispersion is strongly reduced for implanted samples (Figs. 3-4). Defects induced by implantation may have enhanced the contact efficiency of S/D probes with the Si film.

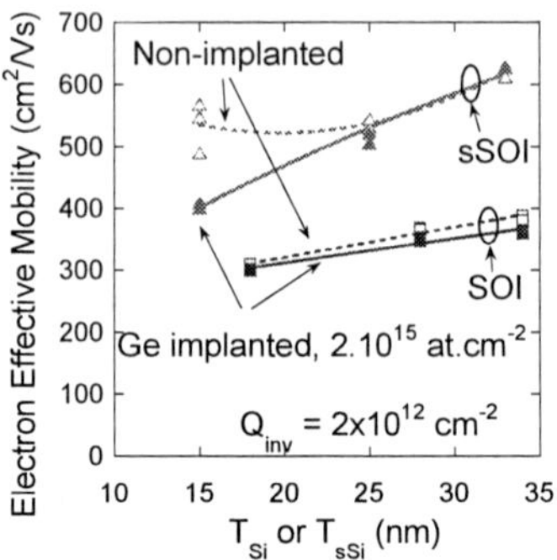

Figure 3. Measured electron effective mobility as a function of the Si (SOI) or sSi (sSOI) layer thickness either before or after implantation and annealing.

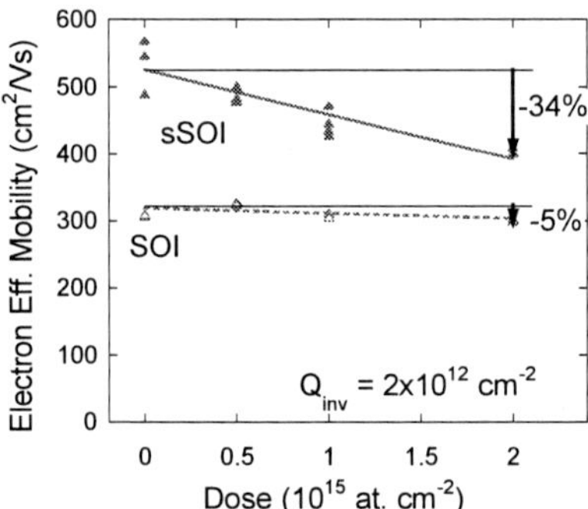

Figure 4. Measured electron effective mobility as a function of the implantation dose for the sample B (18nm) and sample 1 (15nm)

In the following, we discuss some possible causes for the mobility reduction on implanted samples:

(i) UV Raman measurements revealed that the tensile strain is maintained for Ge implanted samples (Fig. 4 in (5)). Thus, macroscopic strain relaxation can not be invoked to explain the mobility reduction.

(ii) The maximum Ge concentration in the Si film does not exceed x=0.05 (extracted from SIMS profiles, not shown here). This concentration is not sufficient to induce any significant $Si_{1-x}Ge_x$ alloy scattering phenomenon that would degrade the mobility (15).

(iii) The Si film-BOX interface trap density, extracted from the sub-threshold slope of the Pseudo-MOSFET, did not show any significant variations with the implanted dose. This rules out the possibility of a heavily damaged back interface leading to the mobility decrease.

The nature of the EOR defects in SOI and sSOI (<311> interstitial clusters, loops…) has been investigated by plan-view Transmission Electron Microscopy (TEM). Figure 5 demonstrates the presence of interstitial clusters. We have shown in Fig. 3 and 4 that mobility was sensitive to implantation. Such a conclusion is strengthened by morphological studies: interstitial clusters have indeed been observed in our samples.

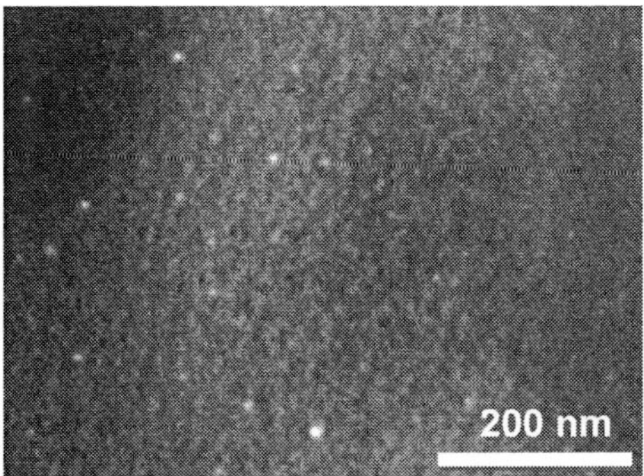

Figure 5. Plan-view TEM: evidence of extended defects in SOI substrates with Ge implantation and annealing for sample D. This TEM plan-view image have been obtained in a [422] Weak Beam DF condition with an excitation error s>0.

Conclusion

The impact of high-dose ion implantation on transport properties and crystal quality in thin SOI structures has been investigated. EOR defects have a significant impact on the back-channel mobility, which is degraded in ultra-thin films and for higher implantation doses. The degradation is approximately 5% and 34% for SOI and sSOI respectively. This electrical degradation is associated with plan-view TEM observations: interstitial clusters have been observed. This implies that the S/D implantation should be optimized in order to avoid transport penalties, in particular for ultra-thin sub-50 nm SOI MOSFETs.

Acknowledgments

This work has been carried out in the frame of CEA-Léti / ALLIANCE collaboration and was partially funded by MEDEA 2T101 SILONIS and the ATOMICS European Project (FP6, contract no. 027152).

References

1. M. Lee et al., *J. Appl. Phys.*, **97**, 011101 (2005).
2. I. Cayrefourcq et al., *in Silicon-on-Insulator Technology and Devices XII, ECS Proc Vol. 2005-03,* (The Electrochemical Society, Pennington, NJ, USA, 2005), pp.191-206.
3. K. Romanjek et al., *Proc. of 6th Workshop On Low Temperature Electronics*, 201 (2004).
4. F. Andrieu et al., *VLSI Tech.*, 176-177 (2005).
5. F. Andrieu et al., *IEEE SOI Conference*, 223 (2005).
6. F. Andrieu et al., *Solid State Electr.*, **50**, 566 (2006).
7. A. Cros et al., *IEDM Tech. Dig.*, 663 (2006).
8. G. Xia et al., *IEEE Trans. Elec. Dev.*, **51**, 2136 (2004).
9. M. Posselt et al., *J. Elecrochem. Soc.*, **144**, 1495 (1997).

10. L. Laânab et al, *Nucl. Inst. and Meth. B*, **96**, 236 (1995).
11. C. Jahan et al., *J. Crystal Growth*, **280**, 530 (2005).
12. S. Cristoloveanu et al., *IEEE Trans. Elec. Dev.*, **47**, 1018 (2000).
13. G. Ghibaudo et al., *Electron. Lett.*, **24**, 543 (1988).
14. C. Gallon et al., *Microelec. Eng.*, **80**, 241 (2005).
15. G. von Busch et al., *Helv. Pys. Acta (Switzerland)* **33**, 437 (1960).

ECS Transactions, 6 (4) 39-44 (2007)
10.1149/1.2728839, ©The Electrochemical Society

LOW TEMPERATURE CHARACTERIZATION OF HIGH-K DIELECTRIC METAL GATE FD-SOI NMOSFETS

L. Zafari[a], J. Jomaah[a], G. Ghibaudo[a], O. Faynot[b], A; VanDooren[c]

[a] IMEP, MINATEC/INPG, 38016 Grenoble, France
[b] CEA/LETI, 38054 Grenoble,France
[c] Freescale, Crolles 38921, France

Fully Depleted devices with HfO_2 dielectric and TiN gate electrode have been studied at different temperatures. Noise measurements have also been carried out on these devices. The effects specific to short channels are evidenced and studied. The precise doping density of the silicon film has also been calculated.

Introduction

In order to meet the requirements of ITRS, new device architectures and materials have been introduced. Thin film silicon-on-insulator (SOI) technology has proved to offer advantages over bulk silicon technology, specially reduced short channel effects (1) and improved subthreshold swing (2). Fully Depleted (FD) SOI MOSFETs have demonstrated even more satisfying characteristics such as elimination of the kink effect over partially depleted devices (3).

On the other hand, high permitivity dielectrics have shown promising characteristics to replace conventional SiO_2 gate oxides due to their capability to reduce gate leakage current for the same electrical capacitance (4). Currently hafnium-based dielectrics are among the most studied ones (5). To overcome the problem of polydepletion effect in poly_Si gates and also the incompatibility between high-κ materials and poly-silicon, metal gates have been introduced (6).

The combination of these two solutions (SOI CMOS with high k dielectrics) could thus be a reliable candidate for sub 45nm CMOS technologies (7). However there are still issues raised by this integration related to mobility degradation due to the use of high-κ dielectrics (8) and ultra-thin silicon film as well as short channel transport issues (9).

In this paper we have studied the characteristics of Fully Depleted nMOS devices with a high-κ dielectric (HfO_2) and a metal gate (TiN) at different temperatures from 300K down to 10K and for gate lengths from 0.05 to 10 µm.

Experimental Devices

In the studied Fully Depleted devices, Atomic Layer Deposition (ALD) method was used to deposit the gate dielectric. The gate stack includes a thin interfacial layer (IL) of 1 nm SiO_2, followed by a 3 nm HfO_2 layer. TiN is used as the gate electrode. The silicon film is 10nm thick and undoped ($\approx 10^{15}/cm^3$) and the thickness of the buried oxide (BOX) is 145nm.

Different gate lengths from 0.5 to 10 µm have been chosen so that the effects due to short channel transport could be better distinguished. A large temperature range has also been applied, which allows to study more completely the physical mechanisms influencing the transport. A fully automatic LF noise measurement system has been used for noise measurements. All experiments have been carried out in the linear regime (Vd=30mV), and at depletion mode of the back interface (Vg₂=0V).

Experimental Results

The variation of the threshold voltage Vt with temperature is plotted in Fig.1_a for different gate lengths. It could be seen in Fig. 1_b that in spite of the fact that the silicon film is thin (10nm), a slight dependence is observed with the gate length but it is negligible comparing to bulk or partially depleted SOI CMOSFETs.

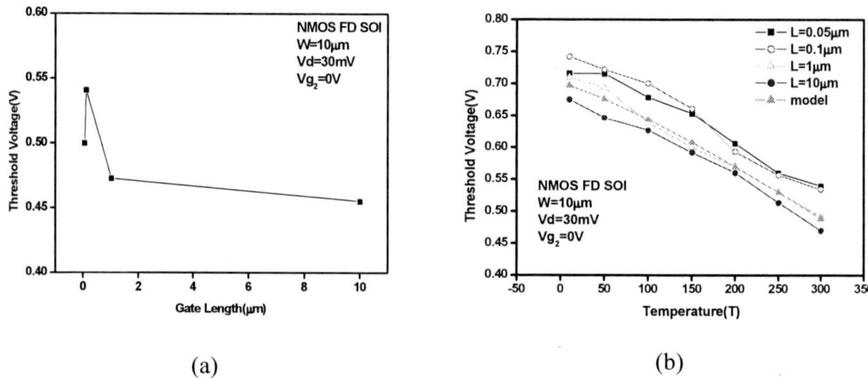

(a)   (b)

Figure 1. a_ Threshold voltage at different temperatures for L=10, 1, 0.1, 0.05μm. b_ Threshold voltage versus gate length at room temperature for the same devices

In the studied devices, since the silicon film is undoped (naturally doped), the last term in the classic definition of threshold voltage (equation 1) (10), does not significantly depend on temperature. Besides Titanium Nitride (TiN) is a midgap metal with respect to silicon (Si) and so $\Phi_{ms} = -\Phi_f$. From the above paragraph, equation 2 could be derived,

$$V_t = \varphi_{ms} + 2\varphi_f + \frac{qN_a t_{si}}{C_{ox}} \qquad (1)$$

$$\frac{dV_t}{dT} = \frac{d\varphi_f}{dT} \qquad (2)$$

where Na is the doping density, t_{si} the silicon film thickness, $q\Phi_{ms}$ the metal-semiconductor workfunction difference and $\Phi_f = kT.\ln(\frac{N_a}{n_i})$ is the Fermi potential.

Based on equation (2) and Fig.1, the doping density N_a of the Si film could be extracted, yielding Na $= 6 \times 10^{15}$/cm³ for the studied devices. Since the silicon film is not intentionally doped, this parameter is unknown during the fabrication process.

From the subthreshold slope at each measurement temperature, the interface state density is determined from equation (3) where Cox is the gate oxide capacitance, S the subthreshold swing, and k the Boltzman constant. In Fig.2_a it could be seen that for longer devices the interface state density is in the state-of-the-art range and the HfO_2 layer has not deteriorated the quality of the interface. It could also be noticed that this value increases for devices with a small gate length.

$$N_{it} = \frac{C_{ox}}{q}\left(\frac{qS}{kT}-1\right)$$
(3)

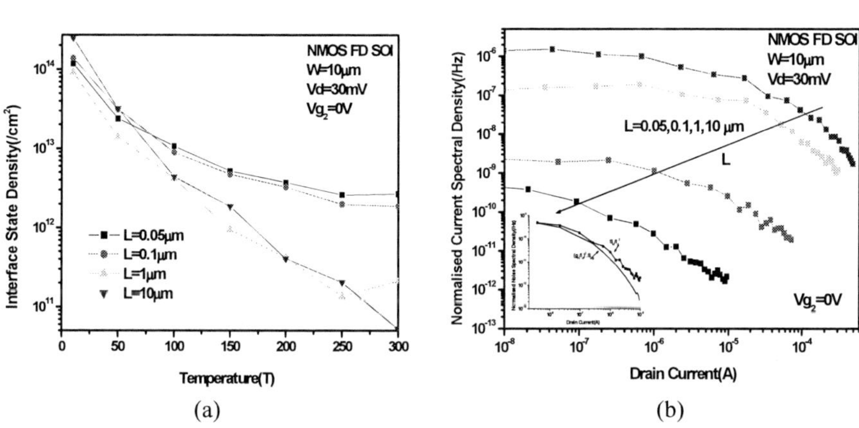

Fig.2_a_Interface state density at different temperatures calculated from the subthreshold slope; b_ low noise measurements at room temperature.

Low frequency noise measurements carried out on these devices, reveal the capture and release of carriers in traps located in the oxide near the silicon interface as the dominant noise source (Fig.2_b) (14). In this model (15), the normalized drain current noise spectral density is given by equation (4) where S_{Vfb} denotes the flat band voltage spectral density and is used to calculate the trap density from equation (5)

$$\frac{S_{Id}}{I_d^2} = \frac{g_m^2}{I_d^2}S_{Vfb}\left(1+\alpha\mu_{eff}C_{ox}\frac{I_d}{g_m}\right)^2$$
(4)

$$S_{Vfb} = \frac{q^2 kT\lambda N_t}{WLC_{ox}^2 f}$$
(5)

where, λ =0.1nm stands for the tunneling constant in the SiO_2, N_t denotes the interface trap density and f is the extraction frequency.

Using equation (5), shorter devices show a higher trap density than the longer ones. The value, calculated for devices with L=10μm is 6-10x10^{17}/eVcm^3, which is slightly higher than that corresponding to state of the art poly/SiO_2 devices (1-3x10^{17}/eVcm^3) but lower than TiN/HfO_2.bulk devices studied earlier (16).

From the above discussions, it could be concluded that the density of traps in short devices is higher both at the interface (interface state density calculated from subthreshold swing) and in the oxide (trap density calculated from noise measurements).

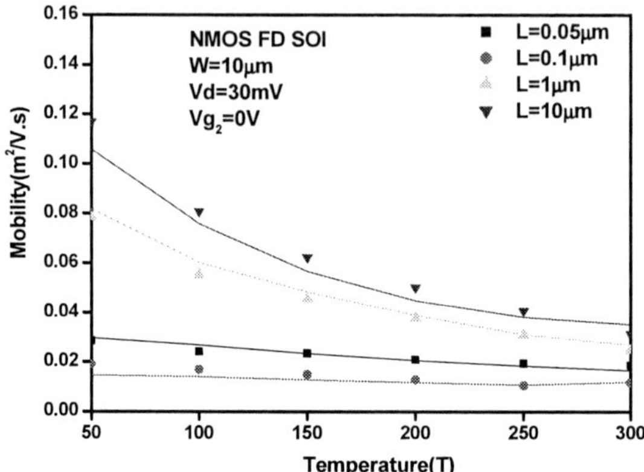

Figure 3. Experimental data (symbols) and model fit (line) for low field mobility μ_0 versus temperature for various gate lengths. μ_n =2000 for L=10 and μm, and 380 for L=0.05μm while μ_{ph} is around 220 for all gate lengths.

The $Y(Vg) = Id / \sqrt{gm}$ function technique (11) has been used for extracting the low field mobility μ_0 at different temperatures and different gate lengths as is shown in Fig.3. This technique allows to suppress the effect of series resistance and also the mobility attenuation factor.

For long channels, as expected, μ_0 is increasing at low temperature due to the phonon scattering reduction (10). In contrast, at small gate lengths (sub 0.1μm), μ_0 is nearly independent of temperature.

In order to distinguish between different scattering mechanisms and to better understand the physical origins of the mobility degradation, the mobility curves extracted at different temperatures have been fitted to the empirical model given in equation (6) (17)

$$\frac{1}{\mu_0} = \frac{T}{300 \times \mu_{ph}} + \frac{1}{T \times \mu_c} + \frac{1}{\mu_n} \qquad (6)$$

where μ_{ph}, μ_c, μ_n are phonon scattering, Coulomb scattering and neutral scattering parameters respectively.

For short gate lengths the neutral scattering parameter has a significant impact on the mobility (μ_n =380, μ_{ph} =220), while in longer devices it is negligible (μ_n =2000, μ_{ph} =220). That is why in short devices the mobility is independent of temperature. This phenomenon is characteristic of the presence of neutral defects in the silicon or at the interface near source and drain. It indicates that the transport in short channels is mainly controlled by source-drain implant-induced defects extending over 30-40nm from source and drain regions (9,12).

It could also be noted that for devices with L=10μm and 1μm the mobility values are approximately in agreement with the results obtained for devices with SiO_2 dielectrics (13) and do not show a significant difference.

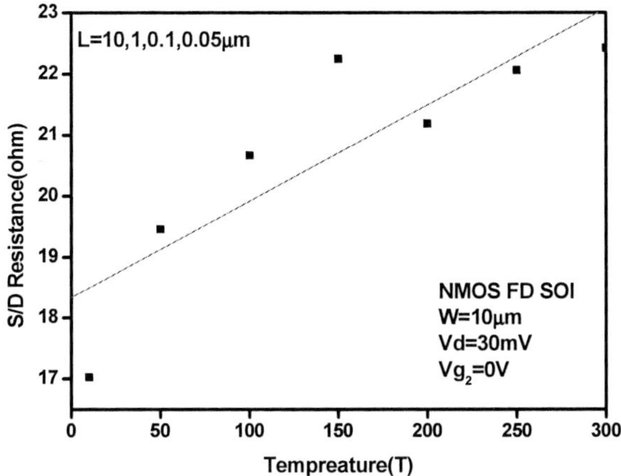

Fig.4_Series resistance variation with temperature

The series source/drain resistance Rsd extracted at different temperatures is plotted in Fig.4. For the extraction of this parameter the $X = 1/\sqrt{g_m}$ function (11) has been used to deduce the mobility attenuation parameter (θ_1) and thereby Rsd has been extracted.

An approximately linear dependence with temperature is observed in agreement with a metallic behavior of highly accessible region. The value of Rsd at room temperature is about 300 $\Omega.\mu$m, which is satisfactory for such ultra thin silicon film and approximately in line with the ITRS requirements (7).

Conclusions

Electrical characteristics of FD-SOI devices with HfO_2/TiN gate stack have been studied in a large temperature range (10-300K) and also for a large gate length range (0.05-10μm). From the evolution of threshold voltage at different temperatures, the doping density of the silicon film has been estimated, which would remain unknown otherwise. In devices with long gate lengths, the interface state density, oxide traps and the mobility are comparable with those in devices with SiO_2 dielectrics, demonstrating an overall good gate stack quality. However, sub 100nm devices exhibit higher interface and also oxide trap density, and, lower mobility values at even ambient temperature. From the fact that the mobility remains temperature-independent in these devices, it was concluded that in addition to higher interface and oxide trap densities, the neutral defects located near the source/drain region were also responsible for the mobility reduction in sub 100nm devices.

References

1. K.K.Young, *IEEE Transactions on Electron Devices*, **36**, p. 399-402 (1989).
2. J.P. Colinge, *IEEE Electron Device Letters*, , **7**, p.244-246 (1986).
3. S.P.Edwards, *IEEE Transactions on Electron Devices*, **35**, p.1012-1020 (1988).
4. M.Fischetti, *Journal Comput.*, **2**, p. 73-79, (2003).
5. Y.Kim et al., *IEDM Tech. Dig.*, p. 455-458 (2001).

6. V.Misra, in *High dielectric constant materials – VLSI MOSFET applications*, H. R. Huff and D. C. Gilmer Editors, p. 415-434, Springer , Berlin (2005).
7. International Technology Roadmap for Semiconductors, Semiconductor Industry Assoc., ,San Jose, CA (2005).
8. S.Saito, *IEDM Tech. Dig.*, p. 33.3.1-4 (2003).
9. A. Cros et al, *IEDM Tech. Dig.*, (2006).
10. S.Sze, in *Physics of semiconductor devices* , John Wiley, New York (1981).
11. G. Ghibaudo, *IEE Electronic Letters.*, 24, p. 543-545 (1988).
12. K. Romanjek et al, *Solid State Electron*, **49**, p. 721-726 (2005).
13. K. Uchida et al, *IEDM Tech. Dig.*, p.47-50 (2002).
14. L.Zafari et al, *SOI Confernce*, p.35-36 (2006).
15. G. Ghibaudo, et al, *Phys Stat Sol (a)*; **124**, p.571-581 (1991).
16. B.Guillaumot et al, *IEDM Tech. Dig,*.p.355-358 (2002).
17. D.Jeon et al, *IEEE Transactions on Electron Devices,* **36**, p. 1456-1463 (1989).

ECS Transactions, 6 (4) 45-50 (2007)
10.1149/1.2728840, ©The Electrochemical Society

UNIAXIAL STRAIN ENHANCEMENT IN SOI TECHNOLOGY FOR THE IMPROVEMENT OF CHANNEL AND EXTERIOR RESISTANCES

Da Zhang[a], D. Goedeke, V. Dhandapani, J. Hildreth, C.C. Fu, T. Kropewnicki, A. Lu, M. Jahanbani, H. Martinez, R. Noble, D. Eades, N. Liu, L. Kang, B.Y. Nguyen, V. Kolagunta, M. Hall, J. Cheek, S. Venkatesan

Freescale Semiconductor Inc., Technology Solutions Organization, 3501 Ed Bluestein Blvd., MD K-10, Austin, TX 78721, USA.
a) Electronic mail: da.zhang@freescale.com.

Strain enhancement has been proven a viable enabler for transistor technology scaling. The implementation and performance of source/drain embedded SiGe (eSiGe) for devices on silicon-on-insulator (SOI) substrate are presented. With proper process controls to mitigate substrate challenges, significant PFET drive current improvement is obtained with SOI eSiGe. The impact of dimensional scaling factors on device enhancement is explored, and results demonstrate a good extendibility for eSiGe. It is also demonstrated that an optimized integration can lead to gate dielectric reliability improvement. The eSiGe contributions to the reductions of channel and exterior resistances are quantitatively depicted. Hole mobility at high and low vertical fields has been calculated, and results show that the eSiGe enhancement of mobility changes little with vertical field variation.

Introduction

With the conventional scaling methodology arriving at the cliff for a continuous transistor performance boosting in the deep sub-micron technology regime, channel stressing approaches, such as source/drain embedded SiGe (eSiGe), have emerged as a viable enhancement enablement for keeping the device performance in line with the technology roadmap requirements [1-4]. In a typical eSiGe technique, a layer of SiGe is epitaxially grown in a pre-recessed Si source/drain (S/D) region. The lattice mismatch between the epitaxial and substrate layers induces an effective strain to the channel in a preferred direction to appreciably improve carrier mobility, resulting in a drive current enhancement. For example, according to results from piezoresistance studies, for PFET with (100) surface and <110> channel orientation, every 100 MPa strain in the channel direction leads to ~7% carrier mobility increase [5]. S/D stressor keeps the channel material intact, thereby obviating potential channel leakage degradation or gate stack re-engineering. Starting at 90nm node, product manufacturing incorporating eSiGe on bulk substrate has been reported [1].

Silicon-on-insulator (SOI) substrate technology has been widely adopted for high performance, low power applications due to its unique advantages such as parasitic capacitance reduction and superb local isolation. Coupling eSiGe with SOI is therefore of critical importance for the industry. SOI substrate induces some intrinsic challenges for embedded S/D stressor. The S/D depth is limited on SOI, therefore eSiGe recessing and stressor defectivity control needs to be well controlled to realize the performance enhancement. This is to be discussed in this article. An eSiGe device on SOI with ~500 angstroms body thickness and ~35 nm gate length is demonstrated here to provide appreciable drive current enhancement.

The lifetime of S/D stressor technology relies largely on its sensitivity to transistor dimensional scaling. The impact of critical geometric factors, such as gate pitch, is investigated. Results verified reasonably robust eSiGe enhancement scalability to 65nm node and beyond. In addition, it has been demonstrated that eSiGe can impose an improvement on device reliability with optimized integration. This work has also quantitatively explored the eSiGe contributions to the improvements of both device channel and exterior resistances. The mobility improvement at both high and low vertical electrical fields has been derived based on characterizations of devices with a series of gate dimensions. It is concluded that the mobility enhancement does not change appreciably with vertical field variation.

SOI Substrate Impact

SOI substrate imposes an intrinsic limit to the magnitude of SD recessing that is applicable. As an extreme example, an over-etch reaching the buried oxide (BOX) definitely induces process failure as no semiconductor seed layer exists for a subsequent epitaxial SiGe growth (Fig. 1). Less apparent is that when the remaining S/D Si after recessing is very thin, epitaxial growth can still fail. Our study shows that the issue is mainly caused by silicon migration under the in situ hydrogen prebake step for epitaxy. (The prebake eliminates native oxide on the surface to enable high-quality epi growth.) For two wafers with the same recessing process profile as illustrated in Fig. 2a, one is

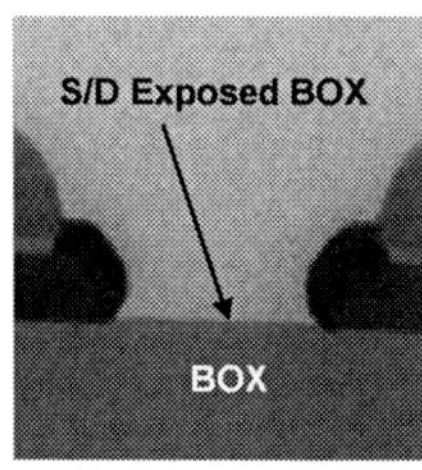

Fig. 1. TEM showing that with BOX exposed after S/D recessing, stressor epitaxy in S/D is not able to proceed.

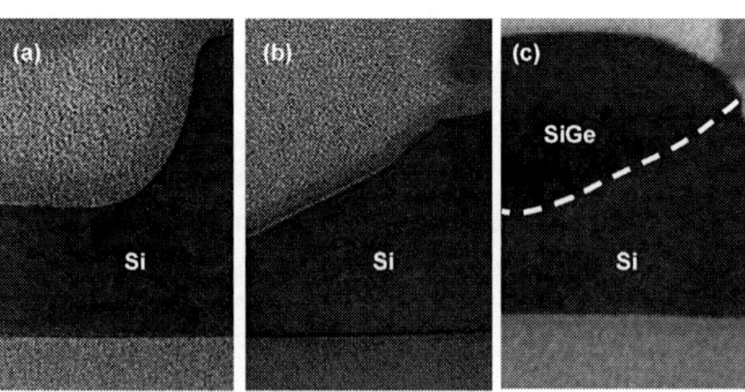

Fig. 2. TEM of S/D profiles at different stages: a) after S/D recessing; b) after hydrogen prebake; c) after epitaxy.

processed through prebake only (Fig. 2b), while the other was processed through prebake and epitaxy (Fig. 2c). Results show that once prebake is performed, the recessing profile is changed due to silicon migration. It is due to increased silicon migration mobility under hydrogen presence at elevated temperature. Additional epi does not change the interface profile much. It is therefore critical to maintain sufficient remaining silicon after prebake to ensure good S/D stressor growth. Fig. 2 also indicates that under severe silicon migration, the recessing slope near the channel can be significantly flattened, leading to an eventually much reduced strain generation. Therefore it is also important to minimize silicon migration through mainly optimization of the prebake step.

eSiGe on SOI: Structure and Performance

Illustrated in Fig. 3a is an eSiGe PFET on SOI substrate. The SOI body thickness is 500 angstroms, and the gate dimension is 35nm. Recessing is conducted after finishing S/D spacer. Nickel silicidation is applied after eSiGe. Dual etch stop layers (dESL) are applied. For PFETs, a mild compressive etch stop layer (cESL) material is deposited on top of the device. HR-TEM and Fourier Transform Diffractogram (Fig. 3b) confirmed that the epitaxial stressor is defect-free.

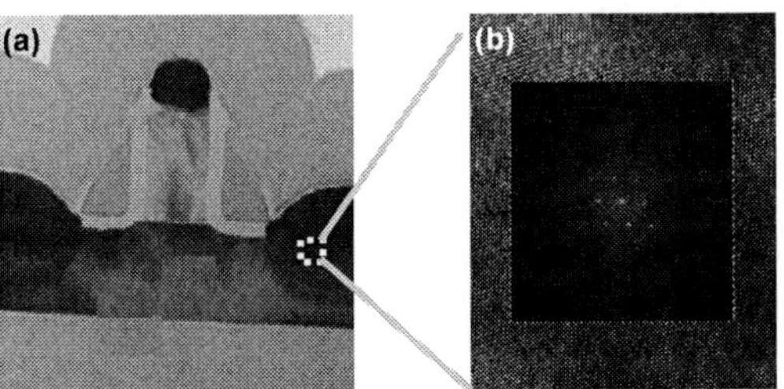

Fig. 3. SOI eSiGe PFET: a) device structure; b) HR-TEM
and Fourier transform diffractogram from S/D SiGe region.

Fig. 4 shows the Ion-Ioff characteristics for eSiGe and Si reference devices. At 100 nA/µm off state current, a significant drive current enhancement is realized. The linear state drain current and transconductance (Gm) as a function of gate overdrive current have also been characterized. A Gm increase confirms strain induced mobility enhancement from eSiGe.

As device dimensions continue to shrink, impact on eSiGe enhancement from critical geometric factors, such as gate pitch (GP) and active width, needs to be studied to evaluate its extendibility to future technology nodes. Devices with varying GP from 0.26 to 1.04 µm are first examined. The active width is 1 um. As shown in Fig. 5, the drive current enhancement from the uniaxial stressor shows a maximum at large GP of 1.04 µm. There is a trend of decreasing enhancement with decreasing GP. However, the drive current improvement is still significant even at the minimum pitch of 0.26 µm. This shows the important value of uniaxial stressor for 65nm+ nodes. The GP

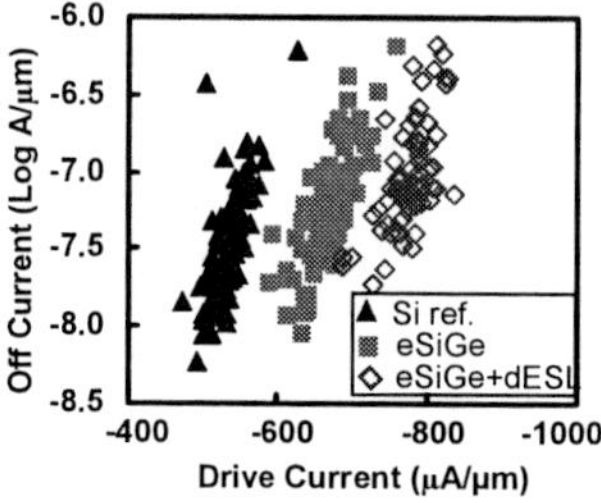

Fig. 4. Off current as a function of drive current for strained and unstrained PMOS devices.

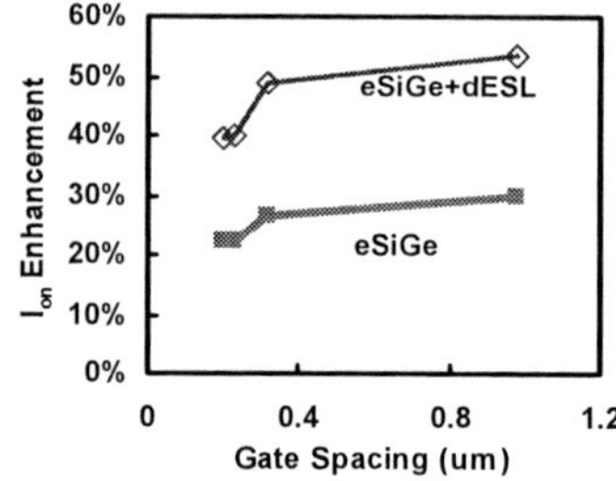

Fig. 5. I_{on} enhancement as a function of gate spacing for eSiGe and eSiGe+dESL strained devices.

dependence of uniaxial enhancement also indicates that in circuit design, relaxing GP in critical paths may be considered to fully realize the stressor benefit. The impact from device width is also studied for a range of dimension from 0.12 to 1 µm. The GP is at 0.26 µm. Relative to the performance of a Si reference with no stressor, the Ion enhancement as a function of device width for three stressor conditions: eSiGe, cESL, and eSiGe+cESL, is presented in Fig. 6. It depicts that eSiGe has a consistent enhancement level through the range of widths studied. Although cESL enhancement degrades at narrow width, eSiGe+cESL produces an improvement with much less width dependence.

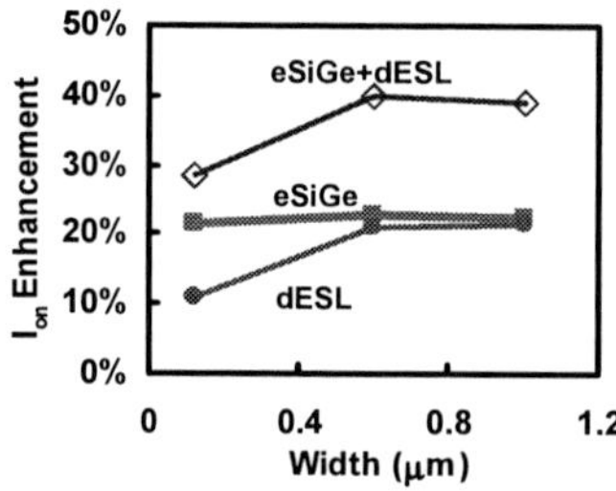

Fig. 6. I_{on} enhancement as a function of device width for different strained devices.

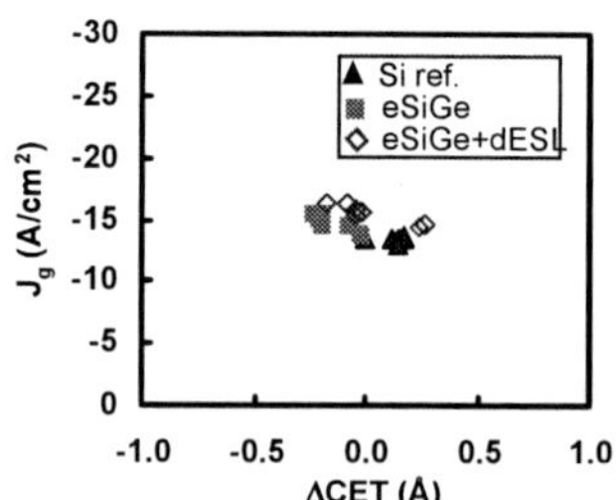

Fig. 7. Jg at -1V Vg as a function of CET for strained and unstrained devices.

eSiGe processes do not directly change any material in the gate stack, however there is thermal impact. Under careful minimization of the thermal budget related to epitaxy, the thermal impact can be reduced to mitigate any potential gate stack degradation. Evaluation of gate leakage and gate dielectric capacitance equivalent thickness (CET) shows that the performance is comparable for devices with and without eSiGe (Fig. 7). Characteristics of time dependent dielectric breakdown (TDDB) for eSiGe and Si reference devices are also studied for this work, and results are shown in Fig. 8. With a proper integration sequence optimization, an even improved TDDB has been demonstrated for eSiGe devices.

In addition to mobility enhancement, eSiGe also contributes to the improvement of contact properties [3]. The reason is that SiGe material has a reduced band gap relative to Si. The Schottky barrier height at the silicide-semiconductor S/D contact, which approximately proportions with half the band gap, is therefore reduced for eSiGe, resulting in reduced contact resistance. To quantify the eSiGe benefits to the reduction of channel resistance and exterior resistances, a dR/dL approach is used to characterize eSiGe and Si reference devices with a close range of effective gate lengths [6]. Resulting total resistance as a function of L_{eff} is shown in Fig. 9. The total resistance is defined as Vd/Id at linear state, and L_{eff} is derived from C-V measurements and TEM characterization. The interception on the R-axis approximates the exterior resistance. In comparison, it is shown that eSiGe reduces exterior resistance by ~33%. This carries a significant meaning for the extendibility of eSiGe to future nodes where exterior resistance represents an increased impact to the overall device performance.

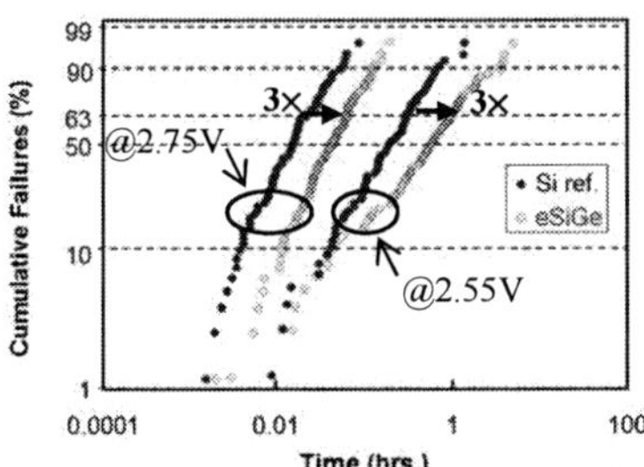

Fig. 8. TDDB for eSiGe and Si ref. devices at 2.55V and 2.75V stress voltage.

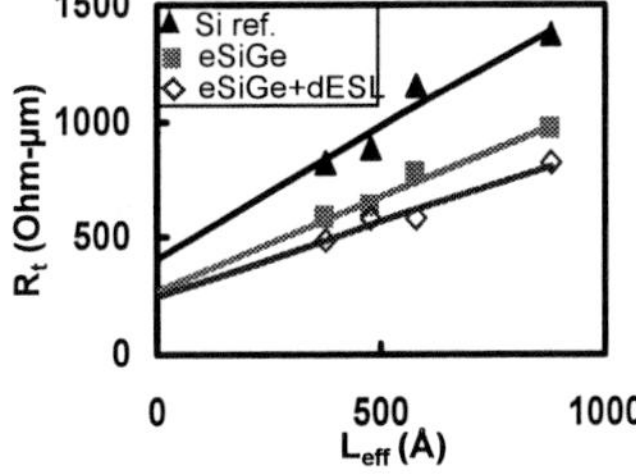

Fig. 9. R_t as a function of L_{eff} for strained and unstrained devices.

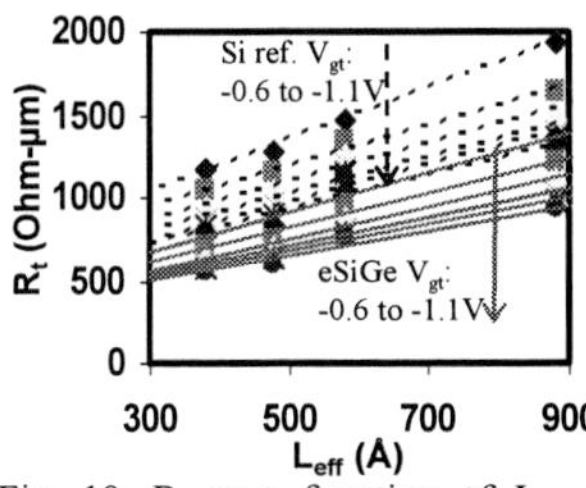

Fig. 10. R_t as a function of L_{eff} at different V_{gt} for eSiGe and Si reference devices.

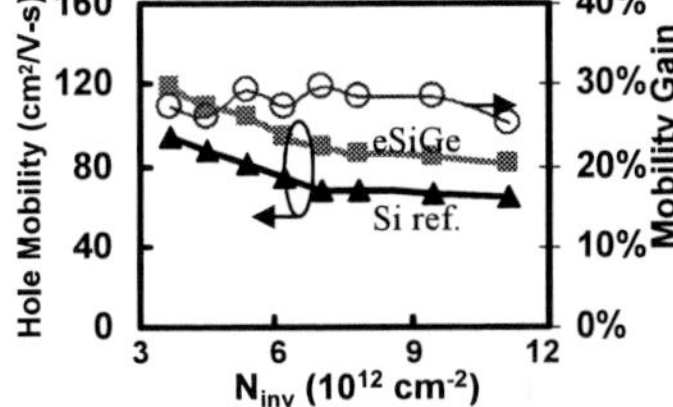

Fig. 11. Hole mobility as a function of Ninv for eSiGe and Si reference, and eSiGe mobility improvement.

To investigate the dependence of mobility improvement on vertical fields, dR/dL characterizations are done at a series of gate voltages for eSiGe and Si reference devices (Fig. 10). Mobility at different Vg is derived from the slope of the R-L_{eff} plot, and total inversion charge (N_{inv}) at different Vg is calculated from C-V characterizations. Mobility as a function of N_{inv} is thereafter created (Fig. 11). It shows that the mobility enhancement from eSiGe does not change much with the variation of N_{inv} or vertical electric field, indicating good eSiGe performance with voltage scaling.

Conclusions

S/D eSiGe stressor on SOI substrate has been demonstrated to effectively enhance PFET drive current. While the Ion enhancement effect decreases with gate pitch, it is valuable at 65nm node and beyond. Reducing active width has a less impact on the enhancement performance. No gate leakage degradation is induced from eSiGe, and the gate dielectric reliability can be improved from the S/D stressor. It has been derived that eSiGe appreciably improves contact resistance in addition to its contribution to channel resistance reduction. The strain induced mobility enhancement does not change much with vertical field scaling.

ACKNOWLEDGEMENTS

The authors would like to thank process groups from Austin Technology & Manufacturing Center and characterization fellows from Physical Analysis Laboratories for their invaluable support to the work.

REFERENCES

[1] T. Ghani, *et al.*, IEDM Tech. Dig., 978 (2003).
[2] P.R. Chidambaram, *et al.*, VLSI Symp. Tech. Dig., 497 (2004).
[3] D. Zhang, *et al.*, VLSI Symp. Tech. Dig., 26 (2005).
[4] L. Smith, *et al.*, IEEE Electron Device Lett., **V26**, 652 (2005).
[5] C. S. Smith, Phys. Rev., **V84**, 42 (1954).
[6] K. Rim, S. Narasimha, M. Longstreet, A. Mocuta, and J. Cai, IEDM Tech. Dig., 43 (2002).

ECS Transactions, 6 (4) 51-56 (2007)
10.1149/1.2728841, ©The Electrochemical Society

Heavy Ion Induced Upset Cross-Sections of 150nm SOI CMOS SRAMs Fabricated In UNIBOND[TM]

S. T. Liu[a], D. K. Nelson[a], J. C. Tsang[a], and H. L. Hughes[b]

[a]Honeywell Aerospace, Plymouth, MN, 55441, USA
[b]Naval Research Laboratory, Washington, DC, 20375, USA

Abstract. This paper discusses the heavy ion performance of the 150nm SOI CMOS SRAMs with a hardened active delay element in the feedback path. The limiting upset cross-section is found to be related to the average resistance of the delay element.

INTRODUCTION

SOI SRAMs have been developed for commercial and space applications because of their low power, high speed and good single event upset (SEU) performance. However, conventional 6T SRAMs are sensitive to SEU. Insertion of resistors in the feedback path has been proposed to mitigate SEU sensitivity at the expense of write time, which increases rapidly with the path resistance. To further improve single event hardness, an active delay element (ADE) with very large resistance ($>1x10^6$ ohms) has been used [1]. The ADE is essentially a switched resistor that shortens the write cycle when the switch is turned on. When the switch is off, as a result of the high ADE resistance ($>1x10^6$ ohms), good proton SEU performance is realized and explained with a double hit (DH) mechanism [1]. The DH mechanism relates the observed limiting proton upset cross-section ($\sigma_{PL} < 10^{-17}$ cm^2/bit) of SOI SRAM/ADE with the physical cross-sections of the two sensitive volumes in the DH path and varies as the square of SOI thickness. The DH path consists of the ADE and any one of the three off-transistors in the SRAM cell. This observation makes SOI even more advantageous in the case of proton SEU performance.

Recently very low limiting heavy ion upset cross-section ($\sigma_{HL} \approx 2x10^{-12}$ cm^2/bit) in these SOI SRAMs was discussed and also attributed tentatively to a DH mechanism [2]. The purpose of this paper is to measure the ADE resistance used in the SOI SRAM cell and to correlate them to the observed limiting heavy ion upset cross-sections for the worst-case heavy ion performance at 125C temperature. Consequently an empirical functional relationship between the two is realized.

EXPERIMENTAL PROCEDURES

Experimental 150nm SOI SRAM/ADEs are fabricated with partially depleted (PD) silicon-on-insulator (SOI) CMOS technology in UNIBOND substrates obtained from SOITEC [3]. The buried oxide of UNIBOND is 145 nm and the top SOI thickness is 150 nm. A 3nm gate oxide grown at low temperature is used in the 1.8V core devices, and a 7nm gate oxide is used in the 3.3V input output (I/O) devices. Both gate oxides are nitrided. The SRAM/ADE cell circuit diagram including three small parasitic nodal capacitances is shown in Figure 1. The gate capacitances and drain capacitances of the MOSFETs are included in the SPICE models. Shown in Figure 1 is the ADE, consisting of a switch shunted by a resistor, Rsh. The switch is controlled by a write word-line (WWL). The memory cell in Figure 1 becomes the conventional 6T SRAM when Rsh is shorted. The resistor can be passive or active in general. We use an active element here to achieve high resistance. When the switch is open (Rsh=Roff), the effective resistance is basically the resistance of Rsh, and thus it is high. When the switch is closed (Rsh=Ron), the effective ADE resistance is the on-resistance of the switch, and thus it is low.

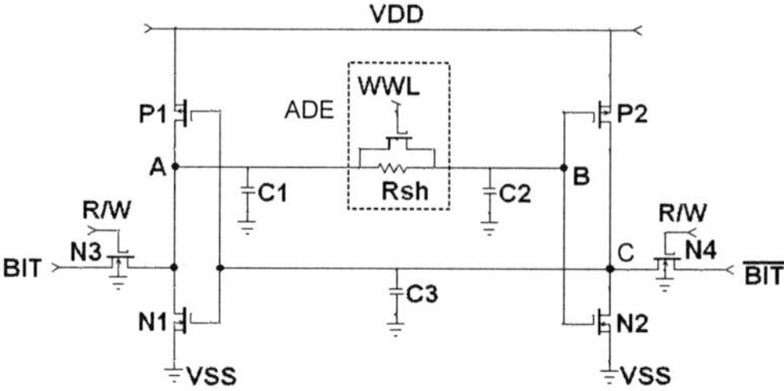

Figure 1. Schematic diagram of an SOI SRAM cell with an active delay element.

We characterized the ADE resistances on the sister wafers from the same lot used for the SRAM/ADEs in the heavy ion test. The ADE resistance as a function of the ADE voltage (V_{BA}) with temperature as a parameter at -55, 25, and 125C were measured. Since the worst SEU performance occurs at the lowest resistance (125C), only the 125C ADE resistance will be discussed here

For the heavy ion test at 125C, the SRAM under test was thermally contacted by a Zener diode heater and the chip temperature was measured by a junction diode on chip. The SEU test was conducted at Lawrence Berkeley National Laboratories in California using six heavy ion species N, Ne, Ar, Kr, Xe, Bi with energies of 67, 90, 180, 387, 612, and 941 MeV, and effective linear energy transfers (LET) at normal incidence are 3.5, 5.8, 15.1, 40.3, 63.2, and 93.8 MeV-cm^2/mg, respectively. The fluence of these heavy ions was carefully selected so as not to exceed 3×10^7 cm^{-2}, especially for the Bismuth ions, to avoid excess vacancy generation, which may produce high leakage current and mask the real heavy ion induced upsets [4].

RESULTS

The ADE resistance as a function of voltage (V_{BA}) at 125C is shown in Figure 2. In standby, we assume transistor N1 is off, and V_A and V_B are equal to V_{DD}. In the event when N1 is hit, V_A goes to 0V, and V_B starts to drop from V_{DD}. As V_B reaches 0.69V (the switch point of the N2/P2 inverter), the output of N2/P2 goes from 0V toward V_{DD}. If V_B is <0.69V, the memory cell flips. This sequence of events is analyzed by SPICE simulation shown in Figure 3. The average value of the ADE resistor in the event from 1.65V to 0.69V is measured to be 1.9×10^6 ohms.

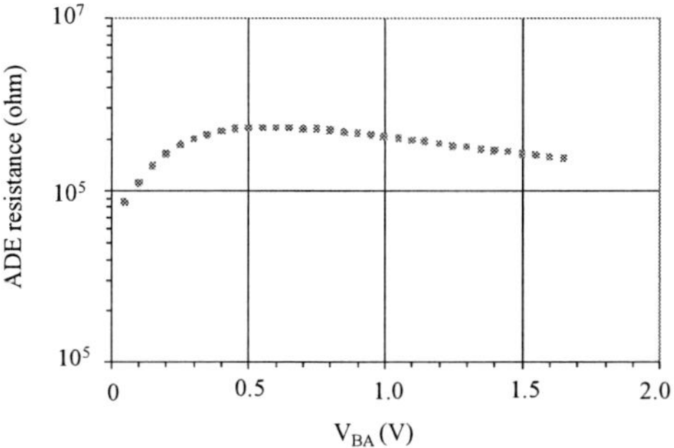

Figure 2. The ADE resistance as a function of voltage across it when the off-transistor N1 is hit and the switch is open (off state) .

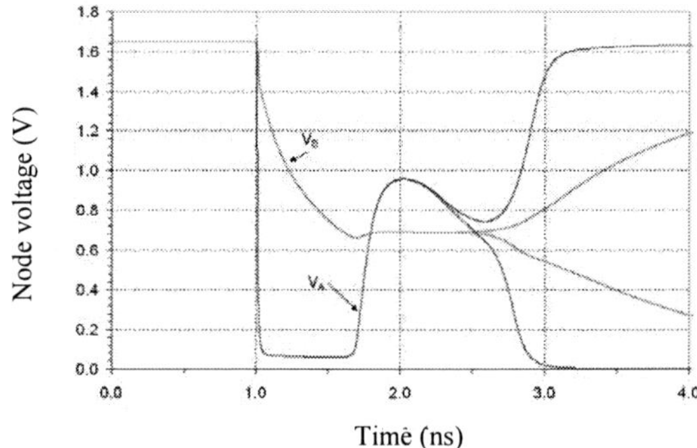

Figure 3. Transient analysis of the node voltages (V_A and V_B in Figure 1) using SPICE simulation of the SOI SRAM/ADE, when the switch is open and N1 is hit. Upset occurs at a critical charge of 76.89fC (lower V_A and V_B curves) corresponds to an onset LET of 51MeV-cm^2 /bit. The charge at upper V_A and V_B curves is 76.88fC.

The upset cross-section (σ) as a function of effective LET (x) without geometrical correction is shown in Figure 4. The data show two distinct regions. Each one is fitted to a four parameter Weibull analysis given by [5]

$$\sigma = \sigma_{HL} [1 - \exp \{ - (\frac{x - x_0}{W})^S \}], \tag{1}$$

where σ_{HL} is the limiting cross-section, x_0 is the onset LET (MeV-cm^2/mg), and W (MeV-cm^2/mg) and S (dimensionless) are fitting parameters. We use the largest value of 1.2×10^{-12} cm^2/bit as the limiting upset cross-section from the high Weibull branch. In Figure 3 the onset of a hit, as simulated in SPICE, corresponds to an LET of 51MeV-cm^2/mg, which is very close to the x_0 value from the Weibull analysis here, confirming the limiting upset mechanism is most likely a single hit.

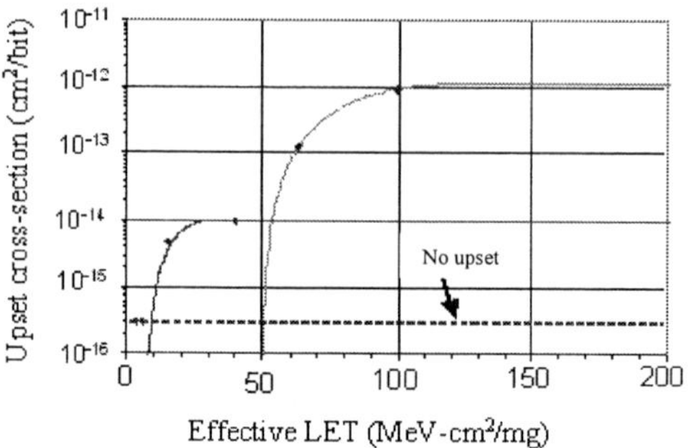

Figure 4. Heavy ion induced upset of an SOI SRAM/ADE when the switch is open.

Similarly, we measure the ADE resistance when the switch is closed. This is shown in Figure 5. The average value of the ADE resistor in the event from 1.65V to 0.69V is measured to be 1.6×10^4 ohms in this case.

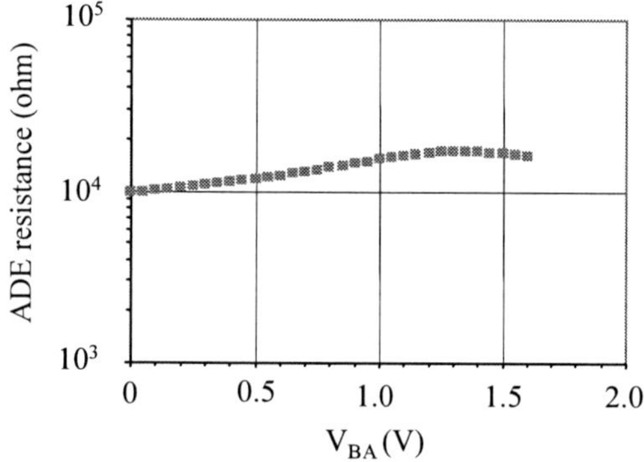

Figure 5. The ADE resistance as a function of the ADE voltage across it when off-transistor N1 was hit, while the switch is closed (on state).

The corresponding heavy ion induced upset cross-section in this case when the switch is closed, is shown Figure 6.   A solid line in Figure 6 is a Weibull fit with four parameters. A much higher value of σ_{HL} is observed at 6.0×10^{-9} cm^2/bit for the case when the switch is open (1.2×10^{-12} cm^2/bit). Thus, the high value of Rsh is advantageous when a switch can be implemented in the memory cell. Note that the onset at the lower Weibull branch in Figure 4 corresponds to a double hit of 5.5 MeV-cm^2/bit as predicted by SPICE.

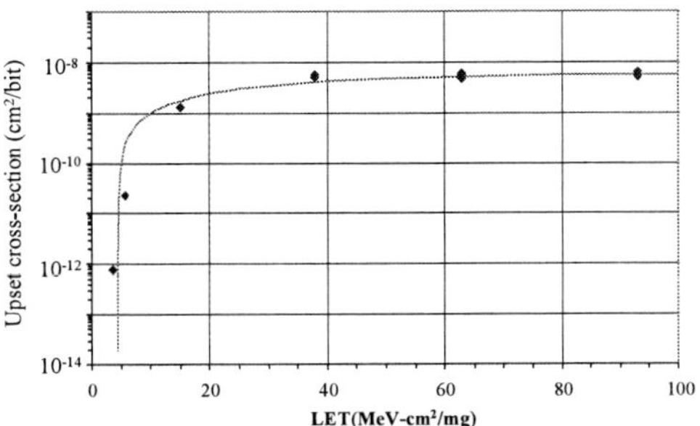

Figure 6. Heavy ion induced upset of an SOI SRAM/ADE when the switch is closed.

DISCUSSION

Results of the limiting upset cross-sections (σ_{HL}) and the ADE resistances are summarized in TABLE I. Also included is the value at Rsh=0 of 6T SRAM (in a separate chip built with same device geometry) at 1×10^{-8} cm²/bit [7]. The value can also be obtained if we consider that the sensitive volumes of 6T in SOI SRAM are those of three off-transistors: N1, P2, and N4 (Figure 1). We assume hits by the primary ions and secondary recoiled ions from all angles are all collected, and the sum of the surface areas of the three off-transistors is equal to 1.068×10^{-8} cm² from the layout of the memory cell (1 bit).

TABLE I. Results of typical 150nm SOI SRAMs (125C).

SRAM type	Rsh (ohms)	σ_{HL} (cm²/bit)
6T	0	1×10^{-8} [7]
ADE switch closed	1.6×10^{4}	6.0×10^{-9}
ADE switch open	1.9×10^{6}	1.2×10^{-12}

Using the data given in Table 1, we can fit the three data points to a function for the normalized limiting upset cross-section ($y = \sigma/\sigma_0$) as a function of the normalized ADE resistance ($r=R/R_0$) given by

$$y = \frac{1}{r^2+1} \ . \tag{2}$$

This function is shown as a solid curve in Figure 7 for which R_0 is a fitting parameter and is equal to 1.6×10^{4} ohms, and σ_0 is the value at $r=0$ and is equal to 1×10^{-8} cm²/bit. The data are shown in open triangles. The fit is reasonable. When the normalized resistance is large, the normalized upset cross-sections can be described simply by r^{-2}. We are preparing different ADE samples with lower resistance to further validate this functional relationship and shall present the findings elsewhere [8].

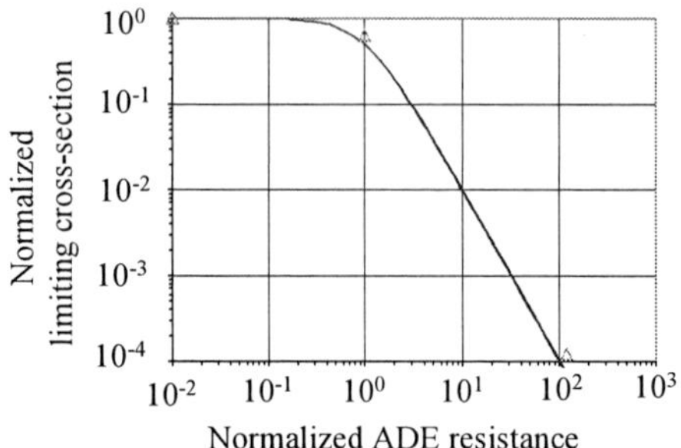

Figure 7. The normalized limiting upset cross section as a function of the normalized ADE resistance for the single port SOI SRAMs.

CONCLUSION

In conclusion, we have shown that the normalized limiting heavy ion upset cross-section of SOI SRAM/ADE is determined by the normalized ADE resistance. The functional relation between the two suggests that the limiting upset cross-section is produced by a single hit.

ACKNOWLEDGEMENTS

We would like to acknowledge the funding support of the Defense Threat Reduction Agency and the Naval Research Laboratory for this work. We would also like to thank Dave Roehler and Terry Fabian of Honeywell for technical assistance.

REFERENCES

1. S. E. Diehl, A. Ochoa, P. V. Dressendorfer, R. Koga, W. A. Kolasinski, *IEEE Trans. Nucl. Sci. vol. 29(6)*, pp/ 2031-2039 (1982).
2. S. T. Liu, H.Y. Liu, D. Nelson, N. Brewster, K. Golke, H. L. Hughes, and J. F. Ziegler, *IEEE Trans. Nucl. Sci.* vol. 53(6), pp. 3487-3493 (2006)
3. S. T. Liu, H.Y. Liu, D. Nelson, N. Brewster, K. Golke, H. L. Hughes, and J. F. Ziegler, *IEEE Trans. Nucl. Sci.* vol. 53(6), pp. 3487-3493 (2006).
4. SOITEC SA, Parc technologique des Fontaines, Brenin, Crollex Cedex, 38926, France.
5. J. F. Ziegler, SRIM-2003 version, www.SRIM.org.
6. E. L. Petersen, J. C. Pickel, J. H. Adams, Jr., and E. C. Smith, *IEEE Trans. Nucl. Sci.* vol. 39 (6), pp. 1577-1599 (1992).
7. P. J. McMarr, M. E. Nelson, S. T. Liu, D. Nelson, K. J. Delikat, P. Gooker, B. Tyrrell, and H. Hughes, *IEEE Trans. Nucl. Sci.,* vol. 52 (6), pp.2481-2486 (2005).
8. S. T. Liu, D. Nelson, J. C. Tsang, H. Y. Liu, P. J. McMarr, and H. L. Hughes, to be submitted to 2007 Nuclear Space Radiation Effects Conference, July 23-27.

SESSION 2

FinFETS AND OTHER DEVICES

ECS Transactions, 6 (4) 59-69 (2007)
10.1149/1.2728842, ©The Electrochemical Society

Intrinsic Advantages of SOI Multiple-Gate MOSFET (MuGFET) for Low Power Applications

Weize Xiong[a], C. Rinn Cleavelin[a], Che-Hua Hsu[b], Mike Ma[b], K. Schruefer[c], Klaus Von Armin[c], T. Schulz[c], I. Cayrefourcq[d], C. Mazure[d], P. Patruno[d], M. Kennard[d], Kyoungsub Shin[e], Xin Sun[e], Tsu-Jae King Liu[e], K. Cherkaoui[f] and J.P Colinge[f,g]

[a] Texas Instruments Inc., SiTD, Dallas, TX USA
[b] Central R&D, United Microelectronics Corporation (UMC), Hsin-Chu City, Taiwan
[c] Infineon Technologies, Neubiberg, Germany
[d] SOITEC S.A., Bernin, France
[e] University of California, Berkeley, CA USA
[f] Tyndall National Institute, Cork, Ireland
[g] University of California, Davis, CA USA

Abstract

MuGFET structure improves local transistor mismatch compared to planar bulk MOSFET. This enables further SRAM cell size reduction. GIDL current is well controlled even with a mid-gap metal gate. MuGFETs have low subthreshold leakage if L_g/W_{si} ratio is kept above 1.5. The advantage of MuGFET subthreshold leakage suppression is even more pronounced at higher temperatures. Furthermore MuGFETs are compatible with local strain techniques to improve carrier mobility. The aforementioned qualities, along with low manufacturing cost of single mid-band-gap metal gate, make MuGFET a good candidate to replace planar bulk MOSFET for Low-Power Applications.

Introduction

Despite recent breakthroughs in high-K and metal gate technology ([1-3]), Multiple-Gate MOSFETs (MuGFETs) are still good candidates to replace planar CMOS for low-power applications. By using an undoped thin body, MuGFETs provide a single mid-band-gap metal gate solution for low-power devices ([4,5]). Single-metal gate MuGFETs reduce manufacturing costs compared to dual work-function metal gates. Low-power electronics is particularly sensitive to cost. MuGFETs also have superior random dopant fluctuation immunity, due to the use of an undoped body. Lower transistor mismatch enables the continuation of SRAM cell size reduction ([6]). Furthermore, MuGFETs have good Short Channel Effect (SCE) control and high degree of process compatibility with planar CMOS ([7-9]).

In this paper, we present experimental and theoretical analysis of single mid-gap metal gate MuGFETs in terms of subthreshold leakage, SCE control, GIDL leakage, transistor mismatch and strain engineering. The purpose of this paper is to demonstrate the feasibility of MuGFETs for low-power applications.

Device Fabrication

Standard Unibond® SOI and strained SOI (sSOI) substrates were used as starting materials. The thickness of the silicon top layer, T_{si}, is 60nm and the buried oxide layer is

145nm thick. T_{si} defines the eventual fin height, H_{si}. The sSOI strained silicon layer had an initial bi-axial strain of 1.5 GPa. Our previous study demonstrated the strain in sub-20nm fins is 1.1Gpa after processing ([10]). (110) sidewall fins were laid out at 0 or 90 degrees rotation with respect to <110> notch, while (100) sidewall surface fins were laid out at 45 or 135 degree rotation. The silicon film was doped p-type (2×10^{15}/cm^3) for both SOI and sSOI. No other channel implants were used. Fins down to 11nm in width were patterned with 193nm lithography and RIE. H_2 annealing was used to round the fin corner and to smooth the fin sidewalls. The gate dielectric was 1.8nm SiO_2, grown by *in-situ* steam oxidation to ensure uniform oxide thickness on fin corners and surfaces. The gate electrode was 7nm-thick, 3GPa-tensile LPCVD TiSiN. It was capped with 200nm poly-Si (Figure 1). After a planarization step, gate lengths, L_g, down to 50nm were formed. Source and drain implants were activated by 1000°C, 10s RTA.

To study the local strain effects, 50nm PECVD nitride Contact Etch Stop Liners (CESL) were used. Tensile CESL, *tCESL* was 1GPa and compressive CESL, *cCESL* was 1.2GPa. The film had a low compressive CESL of 0.1GPa.

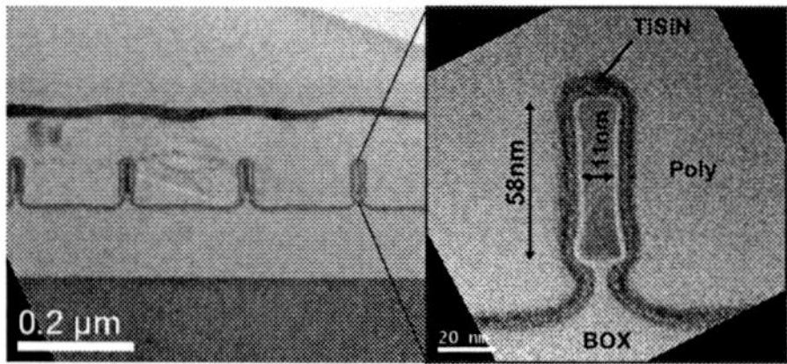

Figure 1: Cross sectional TEM of fins under the TiSiN metal gate electrode

Analysis of Transistor Leakage Components

<u>Subthreshold Drain Leakage</u>

To control subthreshold leakage and reduce Short Channel Effects (SCE), bulk CMOS scaling relies on 1) gate oxide (T_{ox}) thickness reduction to increase gate capacitive coupling to the channel, and 2) an increase of substrate doping (N_A) concentration to reduce the penetration of the electric field from the drain in the channel region ([11]). However, T_{ox} scaling is limited by direct tunneling current while increasing N_A reduces mobility and increases junction leakage.

Subthreshold leakage in MuGFETs is controlled by the thin body ([8]). Numerical simulation in ([11]) suggests a reasonable subthreshold leakage can be achieved if L_g is 5-10 times larger than the natural length, λ ([11]). For double gate devices, λ is given by

$$\lambda = \sqrt{\frac{\varepsilon_{si}}{2\varepsilon_{ox}}\left(1 + \frac{\varepsilon_{ox}}{4\varepsilon_{si}}\frac{W_{si}}{T_{ox}}\right)W_{si}T_{ox}} \qquad [1]$$

λ is derived from the 3D Poisson equation ([12]). It represents the penetration distance of the drain electrical field into the channel. λ is a function of fin width W_{si} and gate dielectric thickness T_{ox}. A small λ is desirable to minimize short-channel effects, which translate into smaller W_{si} or thinner T_{ox}. Equation [1] suggests that one can trade T_{ox} reduction with fin width reduction. This is not possible for single gate planar bulk or PDSOI MOSFET, since the increasing channel doping without reduction of T_{ox} would results in unacceptable threshold voltage ([11]). Thus, MuGFETs can prolong the life of plasma nitrided gate oxide (PNO) and delay the introduction of high-K. This option further reduces the manufacturing cost of MuGFETs compared to planar CMOS with high-K and dual workfunction metal gate.

Figure 2 shows the experimental data of Drain Induced Barrier Lowering (DIBL) versus L_g/W_{si}. The figure contains 70 devices with different L_g and W_{si} combinations. 80mV/V *DIBL* is attainable if L_g/W_{si} is >1.5 ([13]). Figure 3 shows the *Ion vs. Ioff* plot of NMOS MuGFETs with three different W_{si}. Steeper slope is observed with wider fins, indicating degraded SCE control. When W_{si} equals 40nm subthreshold leakage becomes too high for low-power applications at Lg=50nm.

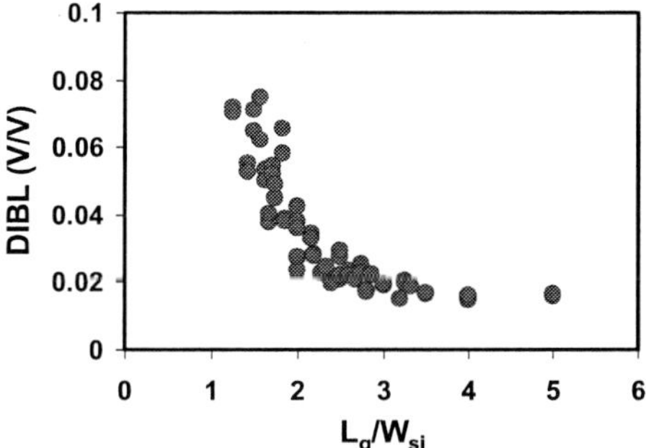

Figure 2: *DIBL* as a function of physical gate length, L_g, to fin width, W_{si} ratio. Data set contains 70 devices with gate length ranging from 50nm to 100nm and fin width ranging from 20nm to 40nm. 80mV/V *DIBL* is attainable for L_g / W_{si} >1.5. T_{ox} = 1.8nm.

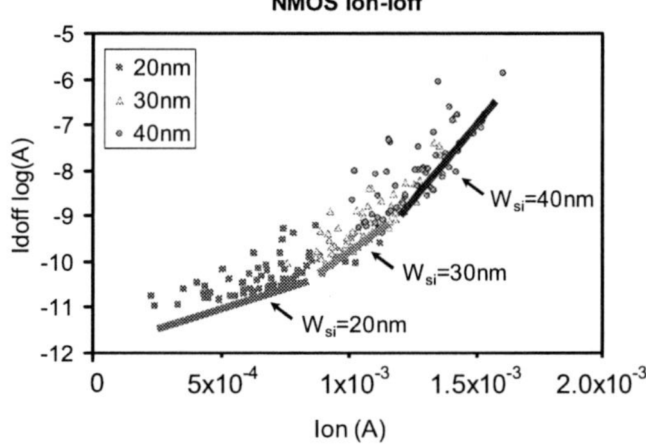

Figure 3: Ion vs. Ioff plot of W_{si} = 20nm, 30nm and 40nm.

Today's consumer electronics operate at temperatures up to 125°C. Thus, the temperature effects on devices are important. The leakage current of a thin-film SOI MuGFET is mostly due to diffusion, rather than to recombination/generation. Its dependence on temperature is equal to: ([14])

$$I_{off} = \frac{qH_{si}}{L_n}\frac{n_i^2}{N_A}\int_0^{W_{si}} e^{\frac{q\phi(x,T)}{kT}}\,dx \qquad [2]$$

where, L_n and $\Phi(x,T)$ are the electron diffusion length in P-type fins and the potential distribution in the body between the two gates, respectively. The narrower the fin width, the lower the value of the potential $\Phi(x,T)$. As a result, extremely low leakage currents are found in narrow MuGFETs at high temperature. Figure 4 shows the evolution of the leakage current with temperature. The leakage current in a 350nm-long device at T=200°C and V_{DS}=1V is only 10pA per fin.

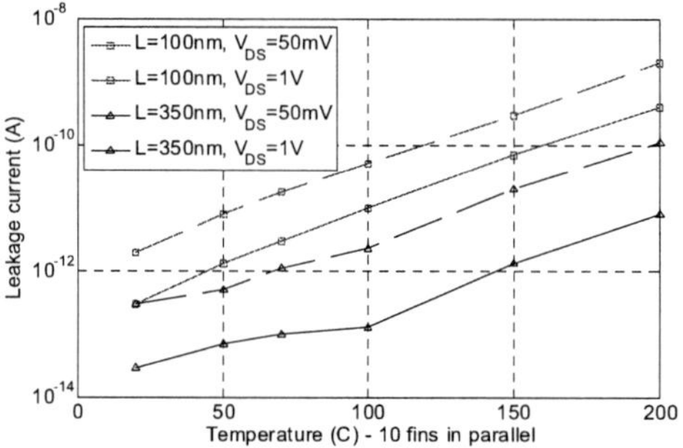

Figure 4: Leakage current in MuGFETs as a function of temperature.

In any MOSFET, the increase of leakage current with temperature results in an increase of subthreshold slope (SS). The very small leakage current of MuGFETs allows the SS to remain closer to the ideal value kT/q ln(10) at higher temperatures than in bulk devices (Figure 5). SS increases above the ideal value because the widening of the channel depth due to the reduction of the surface electric field in weak inversion. When this effect is taken into account, the subthreshold slope can be written ([15]):

$$S = \frac{nkT}{q} \ln(10) \left(1 + \frac{dQ_{tot}/d\phi_s}{C_{ox}} \right) \qquad [3]$$

with

$$n = \left[\frac{kT}{q} \frac{d\left(\ln(Q_{inv})\right)}{d\phi_s} \right]^{-1} \qquad [4]$$

Figure 5: Sub-threshold slope as a function temperature. MuGFET W_{si} =30nm

Gate Induced Drain Leakage (GIDL)

As planar CMOS continues to scale, thinner gate oxide, higher extension and HALO doping are needed to reduce short-channel effects and parasitic series resistance. Unfortunately GIDL current increases with junction abruptness as well as reduction in gate oxide thickness ([16]). The origin of GIDL is band-to-band tunneling in the drain extension under the gate. With a large drain-to-gate bias, enough band-bending occurs for valence-band electrons to tunnel into the conduction band ([17]). GIDL current is a function of average electrical field in the gate to drain overlap region modeled by: ([18])

$$J = AE_{TOT} \exp(\frac{-B}{E_{TOT}}) \qquad [5]$$

Where A is a pre-exponential parameter; B is a physically based exponential parameter. E_{TOT} in [5] is the electrical field at the point of maximum band-to-band tunneling. For simplicity, if we ignore the doping-gradient-induced electrical field; E_{TOT} can be estimated using the surface electrical field ([19])

$$E_s = \frac{V_{dg} - V_{FB} - 1.2}{3T_{ox}} \qquad [6]$$

Thin body MuGFETs offer lower E_s, which reduces GIDL. GIDL current also decreases with fin width ([18]). Furthermore, since the reduction of fin width allows relaxed gate oxide scaling, GIDL could be further reduced using MuGFETs.

Several components in the MuGFET structure could potentially increase GIDL. One of them is the high electrical field at the fin corners ([19]). Corner rounding and the use of an undoped body can relieve this problem ([20]). Fin corner-rounding has been demonstrated with oxidation and H_2 annealing ([21, 22]). Another potential GIDL problem is the gate work function. GIDL is expected to increase when the gate work function is moved away from the band edges ([23]). The impact of gate workfunction on GIDL manifests itself through V_{FB} in equation [6]. In undoped fully depleted MuGFETs, the drain-body potential is set predominantly by effective gate work function. Figure 6 shows measured I_dV_g curves of a MuGFET with two different gate electrodes. As the work function increases, the entire I_dV_g curve shifts to the left, increasing the effective GIDL current at any given gate voltage. Extrapolating the GIDL current for TiSiN gate device at Vg=0V, the GIDL level is under $1x10^{-12}$A/μm, negligible compared to the subthreshold leakage.

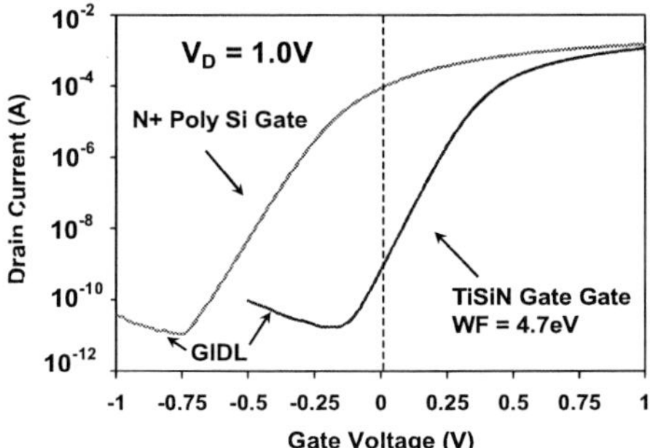

Figure 6: Effect of gate Workfunction on GIDL

<u>Other Leakage Components</u>

When MuGFETs are made on SOI, diode-related junction leakage is reduced or eliminated ([24]). That leaves gate leakage as the only other major leakage component in MuGFET devices. Gate leakage near 1V is due to intrinsic direct tunneling. It is exponentially dependent in gate dielectric thickness ([25]). As mentioned earlier, SCE control in MuGFETs can be achieved through fin width reduction without aggressively scaling the gate oxide thickness. Keeping a relative thicker gate dielectric reduces gate leakage.

Transistor Mismatch

Embedded SRAMs occupy a significant portion of modern-day VLSI chips. As SRAM cell size continues to scale, cell transistors local mismatch increases due primarily to Random Doping Fluctuation (RDF). As a result, SRAM cell stability is reduced and yield becomes problematic ([26]). For planar bulk CMOS with uniformly doped channels, the effect of random doping fluctuation can be modeled by ([27-29])

$$\sigma_{Vt} = \frac{1}{\sqrt{W * L_g}} \frac{T_{ox}}{\varepsilon_{ox}} \left(\frac{\sqrt[4]{4q^3 \varepsilon_{si} \phi_B}}{2} \right) \sqrt[4]{N_A} \qquad [11]$$

σ_{Vt} is the standard deviation of V_t. W and L_g are the transistor width and gate length and N_A is the channel doping concentration, respectively. σ_{Vt} becomes worse as W and L_g become smaller and channel doping is increased.

MuGFETs with metal gates and undoped bodies can significantly reduce random doping related transistor mismatch. Figure 7 shows that the V_t mismatch is reduced by >50% using metal gate MuGFETs, compared to typical 65nm node planar bulk devices. Light HALO implants do not significantly affect the V_t mismatch. Polysilicon-gate MuGFETs with heavily doped fins ($5 \times 10^{17}/cm^3$) have a V_t mismatch similar to that of planar bulk devices.

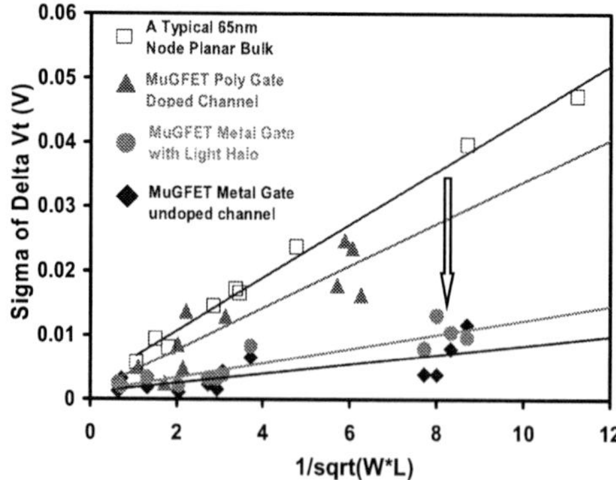

Figure 7: Measured linear threshold voltage local mismatch

In future MuGFETs, W_{si} and L_g in the range of 10nm are expected. Concerns about RDF in such small fins have been raised ([30]). The argument is: even a single dopant in such small volume can change the overall dopant concentration dramatically. For example, one dopant atom in a 10nmx10nmx10nm volume results in an effective $1x10^{18}/cm^3$ doping level. The probability of having n dopants in this volume for $1x10^{15}/cm^3$ doped Si is 10^{-3n}, where n is an integer.

To explore the scaling potential of MuGFETs in terms of RDF, we performed two-dimensional simulation of n-channel MuGFETs by solving the Poisson equation numerically using Comsol Multiphysics^{TM} [31]. The gate oxide thickness is 2nm and the silicon in the channel is nominally intrinsic. A single P-type doping atom is represented by a small sphere with a diameter of 0.5nm containing an equivalent doping concentration $N_A=1/V$, where V is the volume of the sphere. The doping atom can be moved around and placed at any location (x, y) in the silicon. Mid-gap gate electrode was used for all devices.

Figure 6 shows the electron concentration in an undoped trigate device and with one donor atom near threshold. Fin width = fin height = 10nm.

The doping atom was then moved to different positions. Tall devices with H_{si}=60nm and W_{si}=10nm were also simulated. The electrical characteristics of devices with or without a doping atom are shown in Figure 7. The increase of V_t brought about by the presence of a doping atom depends on the position of the atom in the device. In general, V_t is the maximum if the atom is near the center of the device cross section and it decreases as fin height or fin width increase. V_t spread of a device caused by a single dopant atom randomly located in the channel 10nm x 10 nm cross section is 36mV. This is in agreement with the numerical simulation done in [32], which showed V_t is insensitive to doping concentration below $1x10^{19}/cm^3$.

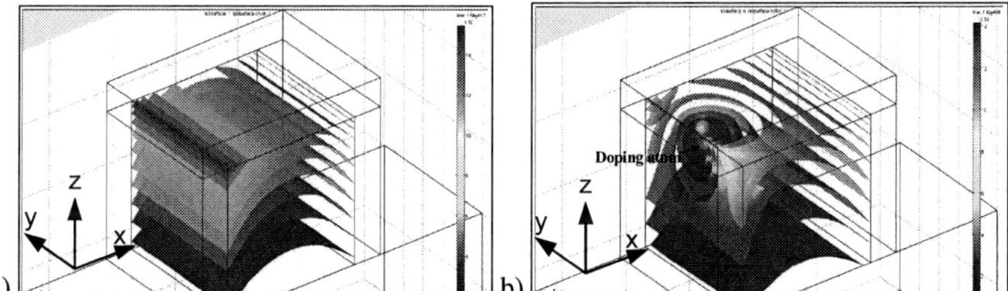

Figure 6: a) Electron concentration in an undoped trigate device with *fin height =fin width* =10nm and V_G=0.2V. b) Concentration in the same device as a) with a single acceptor doping atom located at (x,y,z) = (5nm, 5nm, 9nm) (at the top left corner).

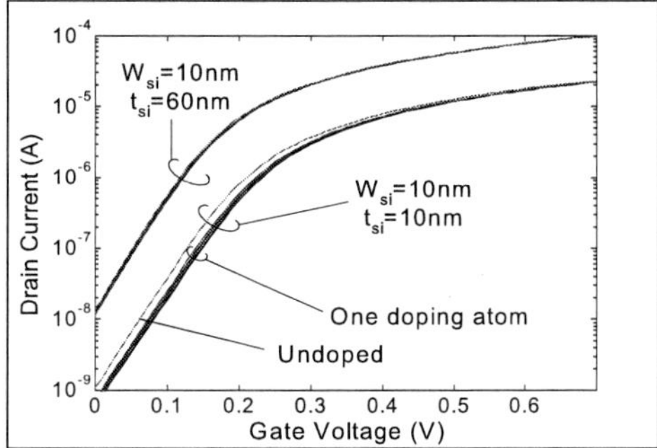

Figure 7: Drain current *vs.* gate voltage for trigate devices with L_g=10nm and V_{DS}=50mV.

Strain Engineering and Mobility

Mobility enhancement through strain engineering has enabled planar CMOS to increase performance without significant L_g and T_{ox} scaling ([33]). To be competitive, MuGFETs must be compatible with known strain techniques. Significant progress has been made in this arena. Various strain techniques to boost the carrier mobility have been reported on MuGFETs in references ([34 – 43]).

One of the advantages of the MuGFET is the high hole mobility (110) surface is readily accessible. The draw back of the (110) surface is low electron mobility. The focus of this work is to bring the (110) electron surface mobility in par or exceed (100) surface mobility.

<u>Tensile Metal Gate and sSOI</u>

Figure 8 and Figure 9 plot the measured mobility versus electrical field. By applying a high tensile metal gate (3GPa), mobility in the (110) surface was enhanced by ~100% compared to the (110) surface universal curve. It is 85% of the (100) universal curve. We would like to point out the electrical field at V_{DD} = 1.0V in the MuGFET is only 0.2MV/cm. This is much lower than the electrical field in planar bulk CMOS which is typically at 1MV/cm due to high channel doping. The electron mobility of the MuGFET at this point is already twice that of planar bulk devices on (100) surface.

The impact of tensile metal gate on (100) electron surface mobility is minimal. Hole mobility in (110) was also unaffected, but the (100) hole mobility was improved by 15%.

When the tensile metal gate is applied on sSOI substrate, the electron mobility for both (110) and (100) devices is improved and goes beyond the (100) electron universal curve. A peak mobility of 765cm^2/V-S was observed. Combining sSOI substrate and tensile metal gate technology, the electron mobility difference between (110) and (100) is minimized. Hole mobility in (100) surface is improved by 40%, but the (110) direction is degraded by sSOI substrate by 30%.

There are several advantages of using strain techniques with gate electrode and sSOI substrate over CESL liner and embedded SiGe: 1) mobility enhancement is not limited to short-channel devices (long channel device also benefit from strain induced current gain) and 2) strain techniques are not pitch limited and are highly scalable.

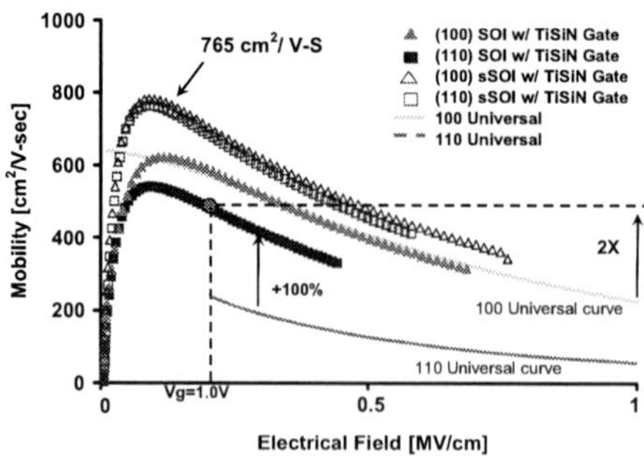

Figure 8: Electron mobility vs. transverse electric field. sSOI and tensile gate pushes (110) electron mobility 20% above the (100) universal curve.

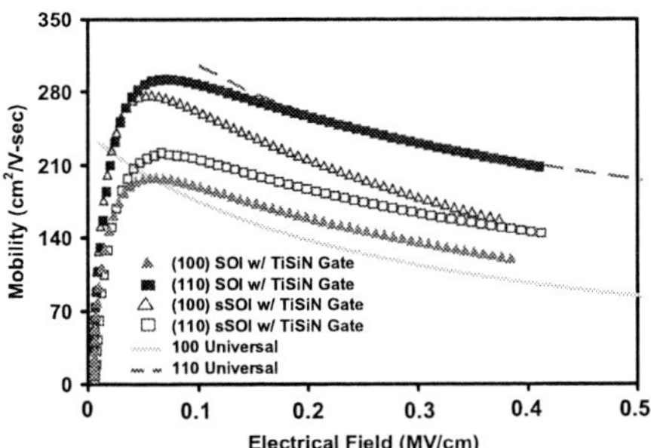

Figure 9: Hole mobility vs. transverse electric field. (110) hole mobility is not affected by the tensile gate, but (100) hole mobility is improved by 15% over the (100) universal curve. sSOI degrades (110) hole mobility, but improves (100) hole mobility.

Combining CESL Stress with Tensile Metal Gate and sSOI

Combining tCESL (1GPa) with tensile metal gate, we observed a 12.5% improvement in NMOS short-channel I_{on} *vs.* I_{off} (Figure 10). But we did not see any improvement for long-channel devices. Figure 10 also indicates sSOI substrates improve I_{on} in both long-channel and short-channel devices, and tCESL is additive to the sSOI and metal gate strain.

Figure 10 shows cCESL strain degrades NMOS I_{on} by 8%. Interestingly, we did not observe a significant improvement for (110) PMOS with our cCESL stress liner (Figure 11), most likely due to the relative low stress level of the cCESL film. But this cCESL film was enough to recover the hole mobility loss associated with the use of a sSOI substrate for short-channel devices.

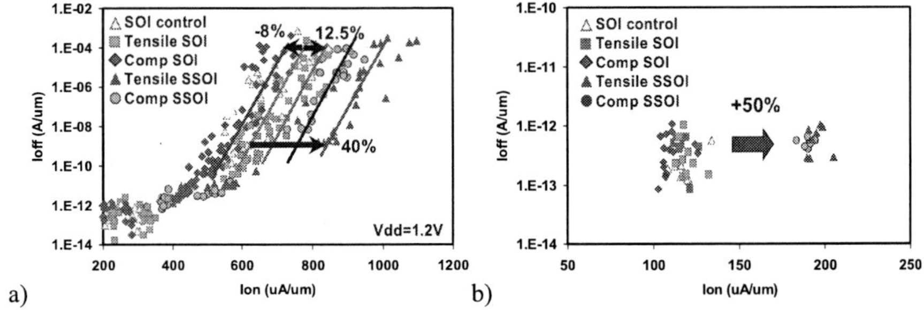

Figure 10: Impact of tCESL and sSOI on a) short-channel NMOS I_{on} *vs.* I_{off}; b) long-channel (1μm) NMOS I_{on}

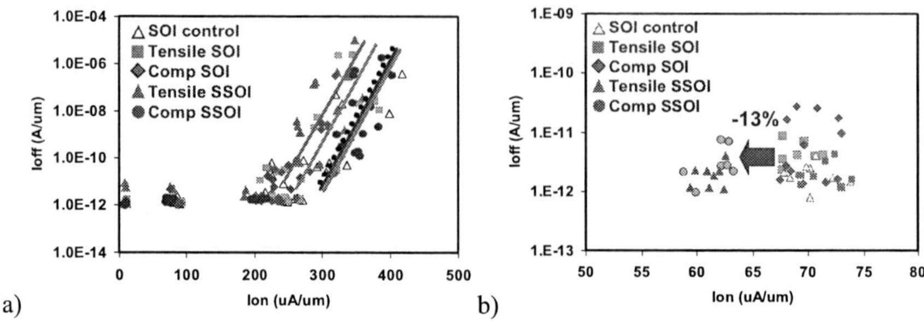

Figure 11: Impact of strain engineering on a) short-channel PMOS I_{on} *vs.* I_{off}; b) long-channel (1 μm) Ion vs. Ioff

Summary

We have demonstrated that the MuGFET structure improves local transistor mismatch compared to planar bulk MOSFET. This enables further SRAM cell size reduction. GIDL current in a MuGFET is well controlled even with a mid-gap metal gate. MuGFETs have low subthreshold leakage if L_g/W_{si} ratio is kept above 1.5. The advantage of MuGFET subthreshold leakage suppression is even more pronounced at higher temperatures. Furthermore MuGFETs are compatible with local strain techniques to improve carrier mobility. The aforementioned qualities, along with low manufacturing cost of single mid-band-gap metal gate, make MuGFET a good candidate to replace planar bulk MOSFET for Low-Power Applications.

Acknowledgments

The authors would like to thank ATDF and SEMATECH staffs for wafer processing and characterization support.

References

1. www.intel.com
2. www.ibm.com
3. www.sematech.com
4. Leland Chang et al., IEDM Technical Digest, Dec. 2000 Page(s):719 - 722
5. W. Xiong et al., 2006 NATO International Advanced Research Workshop Nano-Scaled Semiconductor-on-Insulator Structures and Devices Proceedings
6. A. Thean et al, IEDM Technical Digest 2006, 34.2
7. D. Hisamoto et al., IEDM Technical Digest, Page(s):833 – 836, Dec., 1989
8. B. Doyle et al., Symp. VLSI Tech. Dig., 2003, pp. 133 134.
9. J. P. Colinge, Solid-State Electronics, Vol. 48, pp. 897-905, 2004
10. W. Xiong et al, IEEE Electron Device Letters, Vol. 27 No. 7, page(s) 612 -614
11. Ran-Hong Yan et al., IEEE Transaction on Electron Devices, vol. 39. no. 7, 1992 pp. 1704 -1710
12. K. Suzuki et al., IEEE Trans. Electron Devices, Vol. 40, December 1993, pp. 2326 – 2329
13. Chi-Woo Lee et al., Solid-State Electronics (in press, Feb. 2007)

14. T. Rudenko et al., The Third European Conference on High Temperature Electronics, pp.83-86, 1999
15. T. Rudenko et al., IEEE Electron Device Letters, Vol. 23, No. 3, pp. 148-150 , 2002
16. S.A Parke et al, IEEE Transaction on Electron Devices, vol. 39. no. 7, July 1992, pp. 1694 -1703
17. J. Chen et al., IEEE Electron Device Letters, Vol. 8, No. 11, 1987, pp. 515 -517
18. Yang-Kyu Choi et al., Japanese Journal of Applied Phys. Vol. 42, 2003, pp. 2073-2076
19. Hyun-Sook Byun et al., International Conference on Simulation of Semiconductor Processes and Devices, 2006 pp. 267 -270
20. W. Xiong et al., IEEE International SOI Conference, 2003, pp. 111 -113
21. E. Landgraf et al., Solid State Electronics, Vol. 50., 2006, pp. 38 -43
22. W. Xiong et al., IEEE Electron Device Letters, Vol. 25, no. 8, 2004, pp. 541-543
23. N. Lindert et al., IEEE Electron Device Letters, Vol. 17, No. 6, 1996, pp. 285 – 287
24. J. P. Colinge, Kluwer Academic Publishers, Third Edition.
25. Yijie Zhao et al., Solid State Electronics, Vol. 48, 2004, pp. 1801 -1807
26. D. Burnett et al., Tech. Digest VLSI Symp. 1994, pp. 15–16.
27. T. Mizuno et al., IEEE Transaction on Electron Devies, Vol. 41, No. 11, pp. 2216 -2221
28. T. Mizuno et al., Janpanes Journal of Applied Physics, Vol. 35, 1996, pp. 842 -848
29. P.A Stolk, et al., IEEE Transaction on Electron Device. Vol. 45, no 8 pp 1960-1971, 1998
30. C. Gustin et al., IEEE Electron Device Letters, Vol. 27, No. 10, 2006, pp 846 – 848
31. www.comsol.com
32. S. Xiong et al., IEEE Transaction on Electron Device, Vol. 50, No. 11, 2003, pp. 2255 – 2261
33. S.E. Thompson et al., IEEE Transaction on Electron Device, Vol 54, no. 5, pp 1010-1021, 2006
34. N. Collaert et al., IEEE Electron Device Letters, Vol. 26, pp. 820-822, 2005.
35. P. Verheyen et al., Symp. VLSI Technology Digest, Pages:194 - 195, June, 2005
36. W. Xiong et al., IEEE Electron Device Letters Vol. 27, No. 7, 2006 pp. 612-614
37. Weize Xiong et al., 2006 Device Research Conference Proceeding, pages 39-40
38. N. Collaert et al., Symp. VLSI Technology Digest, Pages:64-65, June, 2006
39. J. Kavalieros et al., Symp. VLSI Technology Digest, Pages:62-63, June, 2006
40 Tsung-Yang et al., Symp. VLSI Technology Digest, Pages:68-69, June, 2006
41. K-M Tan et al., IEEE Electron Device Letters, Vol. 27, No. 9, 2006, pp. 769 -771
42. C.Y. Kang et al., IEDM Technical Digest., Page(s): 833 – 836, Dec., 2006
43. C. H. Hsu et al., Accepted to 2007 VLSI TSA Conference

ECS Transactions, 6 (4) 71-82 (2007)
10.1149/1.2728843, ©The Electrochemical Society

Fabrication and Power-Management Demonstration of Four-Terminal FinFETs

K. Endo, Y. Liu, M. Masahara, T. Matsukawa, S. O'uchi and E. Suzuki

National Institute of Advanced Industrial Science and Technology
1-1-1 Umezono, Tsukuba, Ibaraki 305-8578, Japan

A high-aspect-ratio and damage-free vertical ultrathin channel (UTC) for a vertical-type double-gate (DG) MOSFET has been fabricated by using low-energy neutral-beam (NBE) etching. Also, dynamically power-controllable CMOS technology has been investigated using separated-gate four-terminal (4T) FinFETs. We demonstrate that the power consumption of the CMOS inverter can dynamically be controlled using the 4T-FinFET.

Introduction

The decreased feature size of metal-oxide-semiconductor (MOS) devices in ultra-large-scale-integrated circuits (ULSIs) requires the nano-scale complementary MOS (CMOS) fabrication technology. Recently, the non-planar double-gate (DG) MOSFET has been recognized as the most promising candidate for nano-scale CMOS technology thanks to its intrinsic strength against short-channel effects and its ability to control the leakage current while maintaining a high drive current [1]. Among the DG MOSFETs, the fin-type FET (FinFET) with a vertical-ultrathin-channel (UTC) has attached much attention due to the compatibility of its fabrication process with the conventional CMOS process [2]. In the usual three-terminal (3T) FinFET as shown in Fig. 1(a), the gate electrodes are under the same potential and the threshold voltage (V_{th}) is fixed depending on the work-function of the gate material. In contrast to the 3T-FinFET, a V_{th} controllable four-terminal (4T) FinFET as shown in Fig. 1(b) has been proposed [3]. For the future ultra-low power circuits design, the flexible control of the V_{th} will inevitably be required. The V_{th} of the 4T-FinFET can flexibly be controlled by adding the bias voltage to the second gate electrode.

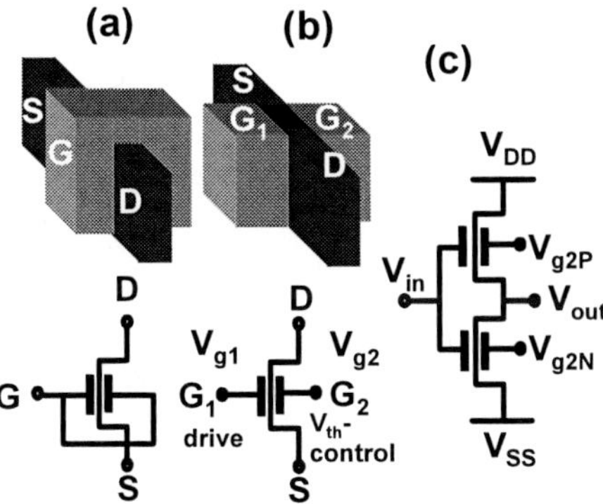

Fig. 1. Schematic illustrations and their circuit diagram of the (a) conventional 3T-FinFET, (b) separated-gate 4T-FinFET and (c) CMOS inverter using 4T-FinFETs.

However, there are still many unsolved issues to introduce FinFETs into the ULSI circuits. Fabrication of 4T-FinFETs and their co-integration with the conventional 3T-FinFETs requires additional process development. Also, fabrication of upstanding silicon-fins formed by the etching process from the silicon substrate still faces several challenges. By using conventional reactive-ion etching (RIE), generation of defects due to irradiation of charged particles and vacuum ultraviolet (UV) photons during processing gives rise to serious issues concerning device performance and reliability of the UTC [4,5]. Recently, a damage-free neutral-beam etching (NBE) system has been proposed to overcome the disadvantages of the plasma etching [6-8]. In particular, an NBE system utilizing neutralization of negative ions provides simple and highly efficient NBs for ULSI fabrication [9].

In this study, to find a potential solution for a damage-free nano-scale CMOS fabrication, we evaluated the effect of the NBE on the performance of n-channel 3T-FinFETs [10]. We also investigated fabrication and co-integration of 4T-FinFETs with conventional 3T-FinFETs by a newly developed fabrication process, and successfully demonstrated the power-controllable CMOS inverter operation using the flexible V_{th} 4T-FinFETs [11].

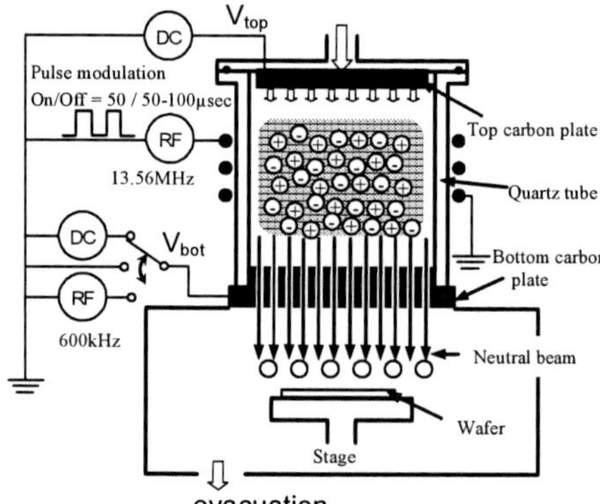

Fig. 2. Schematic diagram of the neutral-beam etching system. Negative ions in the plasma can be efficiently neutralized by passing through the carbon aperture.

Neutral Beam Etching

The NBE system used in this study is shown in Fig. 2 [12]. The diameter and the length of the quartz tube are 10 and 25 cm, respectively. Chlorine (Cl₂)-based gas chemistry is used and a pulse-time modulated rf (13.56 MHz, on-time/off-time=50 μs/100 μs) power is applied to generate the plasma with a large amount of negative ions in the afterglow during the period when the plasma was off [12]. The plasma thus consists of negative and positive ions. The process chamber is separated from the plasma chamber by a carbon plate that is fitted at the bottom of the chamber. Numerous apertures in the carbon plate extract NBs from the plasma. In this system, for generating NBs, only negative ions are accelerated from the plasma. Namely, the negative ions can be efficiently neutralized by passing through the carbon aperture. Collimated and high-flux Cl NBs with a neutralization efficiency of almost 100% can be generated by using Cl-ions in any negative acceleration voltages, whereas low neutralization efficiency and low flux is observed by using the positive ions. In this condition, the estimated peak energy of Cl NBs is around 5 eV, that is, much lower than the conventional RIE. The NBE can perfectly eliminate the charge build-up and photon-radiation damage from the plasma. A conventional RIE with a 13.56-MHz inductively coupled plasma (ICP) and Cl_2-based chemistry is also used for comparison of the device characteristics fabricated by the NBE and the RIE.

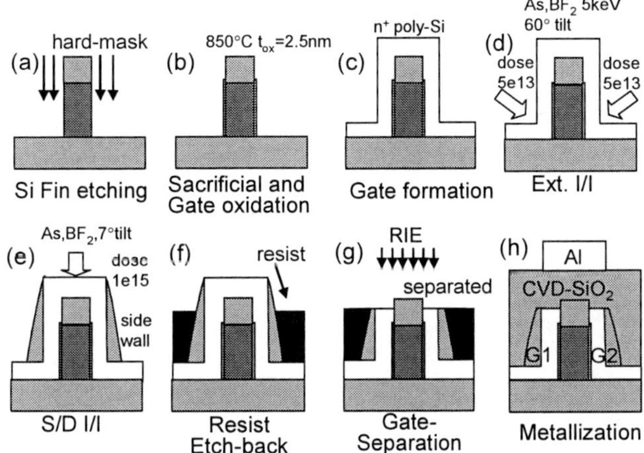

Fig. 3. Fabrication process flow for the 4T-FinFET. After the two-step ion implantation (d)-(e), the gate separation process as shown in (g) is carried out.

Device Fabrication

The starting materials for the MOSFET fabrication were (100) and (110) oriented silicon-on-insulator (SOI) wafers; thus, the channel-orientations of the fabricated FinFETs were (110) and (111), respectively. The process flow for the FinFETs is summarized in Fig. 3. An electron-beam (EB) resist (SAL-601) was used to pattern Si-fins or to transfer the patterns to a SiO_2 hard-mask. After the 100-nm-high Si-fin etching using NBE or conventional RIE, the 2.5-nm-thick gate oxide was formed at 850°C and then covered with the n^+ poly-Si. The gate electrode was then patterned by the same EB resist mask. Due to the three-dimensionally shaped Si-fin, the thickness of the spin-coated EB resist was thinner at the top of the Si-fin than that at the other planar portion. Consequently, the EB resist could not endure the gate etching, which resulted in excess removal of the poly-Si. We overcame this problem by using a 50-nm-thick SiO_2 hard-mask for the gate-electrode formation. Firstly, the gate pattern was transferred to the hard-mask using a fluorocarbon based RIE; a poly-Si gate was then formed using Cl_2+HBr-based gas chemistry.

After the gate electrode was formed, a shallow implantation into the extension of the source/drain (S/D) was performed. To distribute impurity atoms uniformly into the vertical channel, 60-degree tilted implantation was carried out at an acceleration energy of 5 keV and a dose of 5×10^{13} cm^{-2} in each side [13]. A 1-nm-thick screening oxide was used to suppress the significant dopant loss [14-15]. A S/D implantation was performed at a dose of 1x 10^{15} cm^{-2} after a 150-nm-thick gate-sidewall was formed by using a chemical vapor deposited (CVD) SiO_2.

Then, the poly-Si gate for the 4T-FinFET was selectively separated by using a newly developed etch-back process as shown in Fig. 3(g), whereas the 3T-FinFET was protected by the thicker photoresit. After Al electrode metallization, the devices were sintered in 3% H_2 ambient at 450°C for 30 min.

Performance Improvements using NBE

Figure 4 shows cross-sectional scanning transmission electron microscope (STEM) images of the fabricated Si-fins. Anisotropic Si-fins with an ideally rectangular shape were successfully fabricated using the SiO_2 hard mask and the collimated low-energy NBs. For comparison of the electrical characteristics, we also fabricated similar rectangular-shaped Si-fin with the conventional RIE (Fig. 4b). We measured the current-voltage (I-V) characteristics of the fabricated n-channel 3T-FinFET with either NBE-processed or RIE-processed UTCs on the (100) SOI substrate. Figure 5 shows the I-V characteristics of the fabricated FinFETs. The FinFETs with fin width (t_{fin}) = 50 nm and gate length (L_g) = 110 nm fabricated by both the NBE and the RIE were found to have superior sub-threshold characteristics of around 80 mV/decade and the same V_{th}. Thus, the influence of ion and UV radiation during the RIE process on V_{th} is negligible. The sub-threshold slope of the NBE processed FinFET is slightly lower than that of the RIE processed FinFET probably due to the difference in the interfacial trap density revealed by the capacitance-voltage (C-V) measurement described in the following paragraph. Moreover, it is noteworthy that a large difference exists on the drain on-current and that of the NBE processed FinFET is significantly improved. The superior on-currents for the NBE processed devices are demonstrated regardless of the various t_{fin}. The drain current of the RIE-processed FinFET, on the other hand, significantly degraded the current reduction became noteworthy with decreasing t_{fin}.

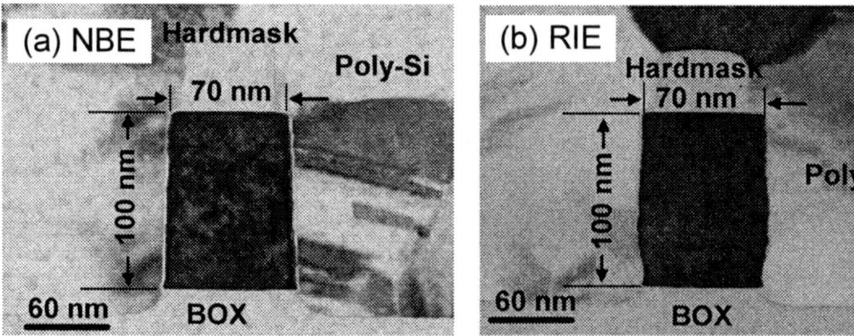

Fig. 4. Cross-sectional scanning-TEM (STEM) images of the Si-fins by using (a) the neutral beam etching and (b) the conventional reactive ion etching.

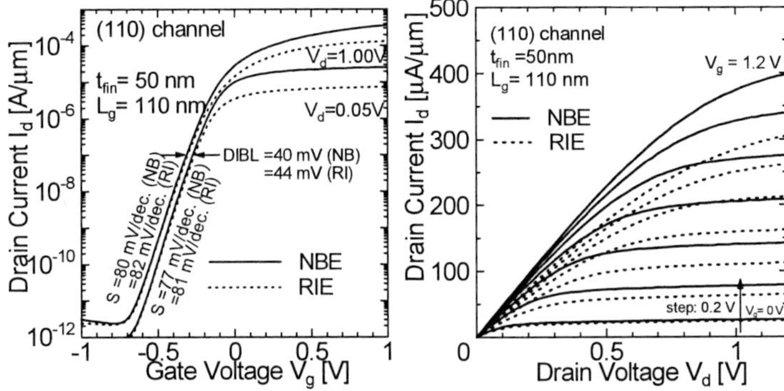

Fig. 5. Comparison of the electrical characteristics of the fabricated FinFETs using the NBE and the RIE with t_{fin}=50 nm and L_g=110 nm: (a) I_d-V_g characteristics and (b) I_d-V_d characteristics.

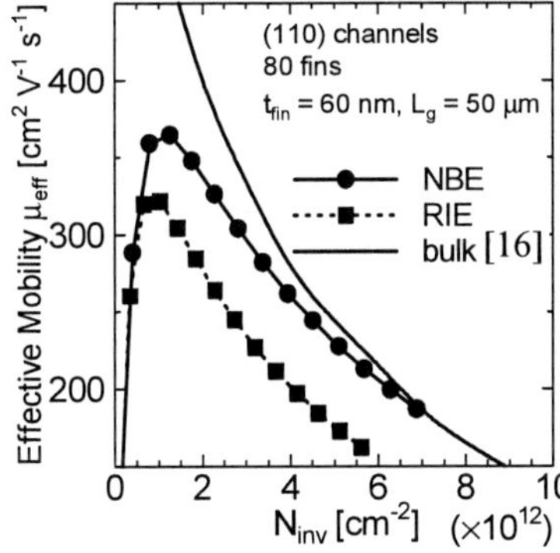

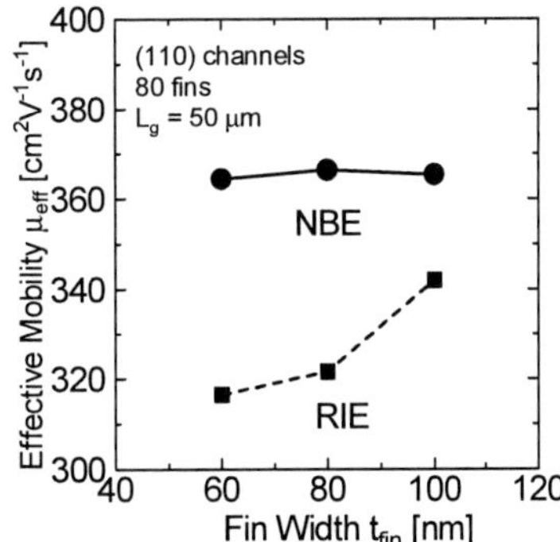

Fig. 6. Comparison of the effective mobility of the (110) channel between the NBE and RIE processed FinFETs.

Fig. 7. Comparison of peak effective mobility as a function of fin width (t_{fin}).

To investigate such a difference in detail, we fabricated multiple-channel FinFETs with 80 channels and L_g=50 µm and examined the carrier mobility. The electrical characteristics of the multiple FinFET also showed the supreme sub-threshold characteristics of 63 mV/decade. The electron mobility of the FinFETs as a function of the induced inversion charge density (N_{inv}) was calculated from the drain conductance and inversion charge density and is shown in Fig. 6. It is apparent that the electron mobility of the NBE processed channel is superior to that of the RIE processed one and is almost equivalent to the bulk planar (110) MOSFET [16]. We measured the mobility of the channel with the different t_{fin} and found that the superiority of the NBE processed channel was unchanged as shown in Fig. 7. On the other hand, the remarkable mobility reduction of the RIE processed channels were observed as t_{fin} decreased.

We also fabricated the n-channel 3T-FinFETs on (110) SOIs with and without an additional 2.5-nm-thick sacrificial oxidation. As shown in Fig. 8, NBE processed mobility also increased even on the (111) channel. In addition, we found that the sacrificial oxidation of the (111) channel slightly increase the mobility. However, it did not recover the distinguished mobility difference between the NBE and the RIE. This result is consistent with the previous report, which also indicated the difficulty in the mobility improvement in spite of various surface treatments [5]. This result also suggests that the sacrificial oxide is too thin to remove the degraded layer of the Si-fin due to the plasma irradiation.

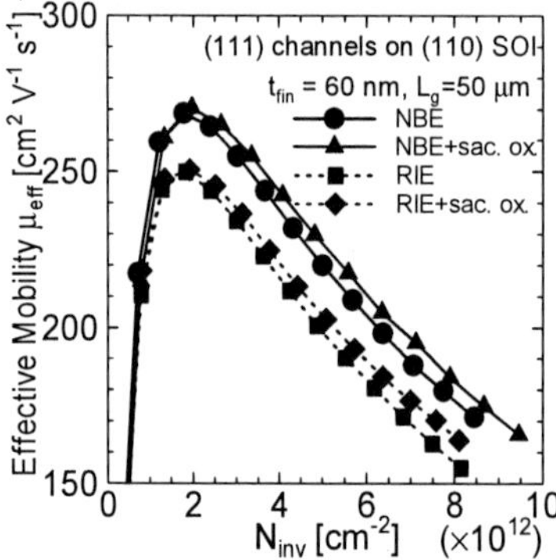

Fig. 8. Comparison of the effective mobility of the (111) channel fabricated on a (110) SOI. Sacrificial oxidation did not improve the mobility significantly.

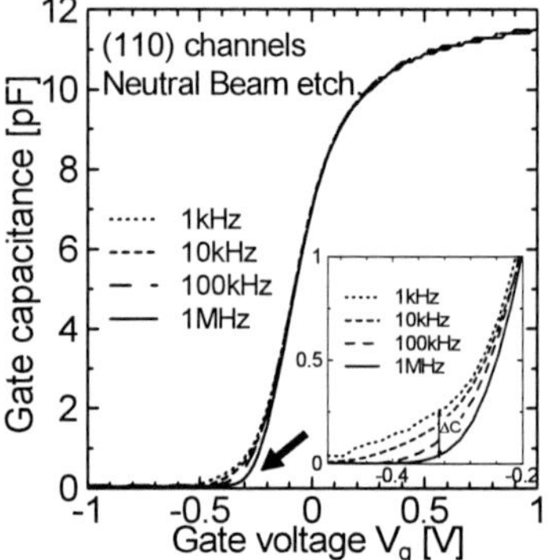

Fig. 9. C-V characteristics of the multiple-channel FinFETs with different operating frequencies. The offsets in the curves correspond to the interface traps.

To explain why mobility differences took place between NBE and RIE as shown in Figs. 6 and 8, we measured the C-V characteristics of the FinFETs to check the interfacial trap density. The C-V characteristics with different frequencies of both NBE and RIE processed devices in Fig. 9 show slight offset in the capacitance (ΔC), which corresponds to the interface traps. The calculated interface trap density (D_{it}) of the fabricated devices form ΔC [17] are below 10^{11} cm^{-2}. The interfacial trap density is slightly higher for device with the RIE process than that with the NBE process. At low N_{inv} below 1×10^{12} cm^{-2} in Figs. 6 and 8 where the observed mobility is almost comparable between the RIE and the NBE processed devices, the mobility has $N_{inv}^{1/2}$ dependence, which indicates that the mobility degradation is caused by the Coulomb scattering [18]. This implies that the mobility difference between the RIE and the NBE process caused by the Coulomb scattering due to the interfacial trap

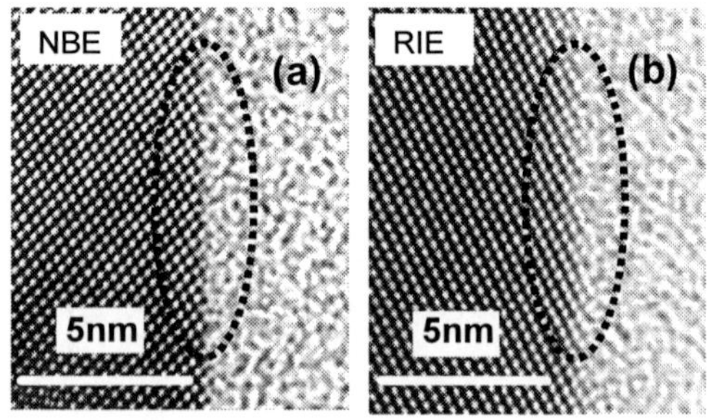

Fig. 10. Cross-sectional TEM images of the Si-fins sidewalls fabricated using (a) NBE and (b) RIE. The sidewall of the NBE processed Si-Fin appears more abrupt than that of the RIE processed one.

is negligible because the trap density remained below 10^{11} cm^{-2} regardless of the fabrication method used.

To investigate the mobility difference further, we next measured the surface roughness, which triggers the roughness scattering. Because direct measurement of the surface roughness of the sidewall is very difficult, we tried to visualize the roughness by taking the high-resolution TEM images. Figure 10 compares the cross-sectional TEM images for the Si-fins fabricated by the NBE and RIE. We found that the NB process can produce a defect-free and atomically smooth sidewall surface, whereas the conventional plasma etching processes causes crystal defects and large roughness. The remarkable contrast of the electric characteristics between the NBE and the RIE processed devices can thus be explained by the difference in the atomic-level roughness and damages of the etched channels. The previous report on the surface-roughness scattering of double-gate SOI inversion layers [19] supports that the nanometer-level surface roughness may cause the significant mobility fluctuation revealed also in this study. Because N_{inv} is proportional to the effective electric field (E_{eff}), we investigated the dependence of the mobility on N_{inv} at the high N_{inv} region. It was found that all the observed mobility had weak N_{inv} dependence of $N_{inv}^{-0.6}$ than the theoretical N_{inv}^{-2} dependence [20]. This is consistent with the previous report that revealed $E_{eff}^{-0.6}$ dependence due to the surface roughness [5]. Because the slope of the curve is determined by the surface roughness correlation length L [18,20], the higher L value as compared to the electron wave length results in observed weak dependence of N_{inv} and the difference in the mobility may be caused by the difference in scattering cross-section between the NBE and the RIE process due to the different roughness amplitude [18].

The roughness may be created through high-energy ion bombardment and UV irradiation (which de-bonds the Si-Si network). Once the surface roughness is created, the atomic level roughness may be preserved throughout the fabrication process as indicated by the inefficiency of the additional sacrificial oxidation or previously reported various surface treatments [5]. The almost unchanged mobility indicated that a thicker oxidation or the other treatment such as high-temperature hydrogen annealing is required to remove the damaged layer after RIE. On the other hand, the neutral-beam etching effectively suppresses the surface roughness due to the removal of the high-energy ions and UV irradiation. Since the feature size of the CMOS device is reaching to the nano-meter scale, the damage-free NBE technology is found to be one of the effective ways for the fabrication of the future devices.

Fabrication of 4T-FinFET

For the future ultra-low power circuits design, the flexible control of the V_{th} will inevitably be required. We then investigated fabrication of the V_{th}-controllable 4T-FinFET by using a newly developed gate-separation method. Figure 11(a) shows a scanning electron microscope (SEM) plan-view of the fabricated 3T-FinFET after the gate etching. The Si-fin is fully surrounded by the gate poly-Si. After the gate side-wall spacer formation using a CVD-SiO$_2$ as shown in Fig. 11(b), As and BF$_2$ ions were implanted into the source and drain region. After the ion implantation, the EB resist was spin-coated on the whole chip for the gate separation etching. The EB resist was then partially removed so that the top of the poly-Si for the 4T-FinFET region was selectively revealed as shown Fig. 11(c). Figure 11 (d) shows a SEM image of the 4T-FinFET after the gate separation etching. The poly-Si gate was successfully separated by the gate separation etching and the fin-top was revealed through a resist opening. In contrast, the

3T-FinFET was fully protected by the thicker photo-resist during this gate separation etching and thus the 3T- and 4T-FinFETs were successfully co-integrated. Figure 12 shows the cross-sectional STEM view of the co-fabricated 3T- and 4T-FinFETs. 18-nm-thick rectangular fin was successfully fabricated. On the contrary, the resist opening position was selectively etched during the gate-separation process and the gate-separated 4T-FinFET was successfully fabricated.

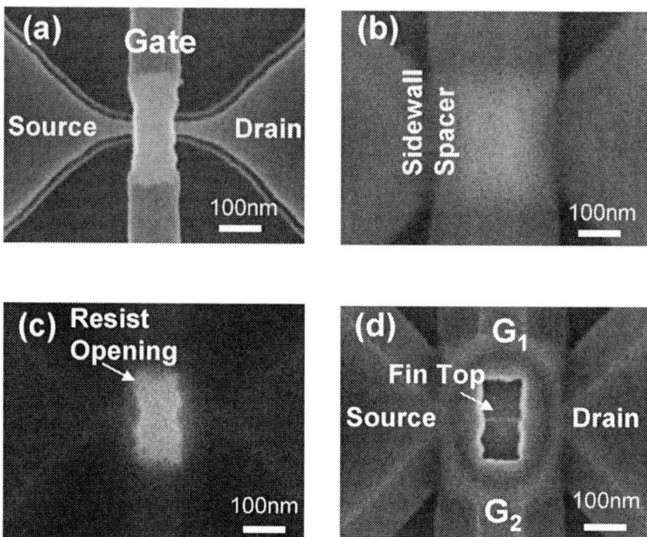

Fig. 11. SEM plan-view of the FinFETs (a) after the gate poly-Si etching, (b) after the side-wall spacer formation, (c) after the EB resist coating and partial ashing, (d) after the gate separation etching.

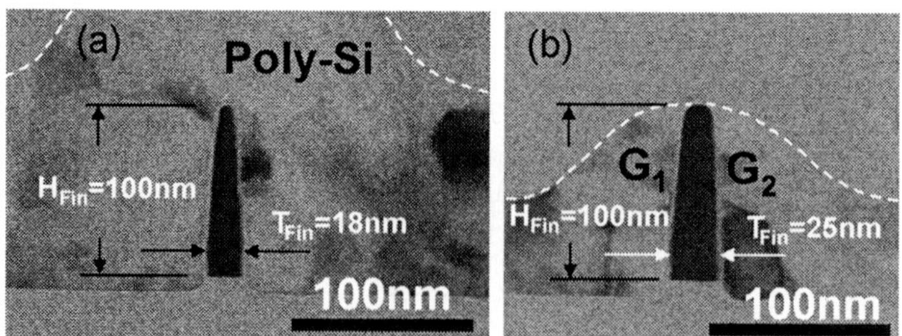

Fig. 12. Cross-sectional STEM images of (a) 3T-FinFET and (b) 4T-FinFET. They were successfully co-integrated using the newly developed etch-back process.

CMOS inverter using 4T-FinFETs

The I_d-V_{g1} characteristics of 4T-FinFETs are shown in Fig. 13. The flexible V_{th} shift is clearly seen by applying the bias voltage to the control-gate (G$_2$). Figures 13(a) and 13(c) show the I_d-V_{g1} curves for the single drive (SD) mode where the V_{g2} is fixed. In this case, the V_{th} shift rate γ defined by $-\delta V_{th}/\delta V_{g2}$ increases with the reducing fin width (t_{fin}). However, the subthreshold-slopes are deteriorated in this static biasing mode because the second gate (G$_2$) side is changed from the depletion to inversion state [21]. On the contrary, in the synchronously driving double gates (DD) operation where both V_{g2} and V_{g1} are simultaneously driven with an offset voltage, the I_d-V_{g1} characteristics can be controlled with no degradation in the subthreshold-slope as shown in Figs. 13(b) and 13(d). This result indicates that the DD mode operation provided by appropriate circuit geometry is promising for the future integration of 4T-FinFETs. Moreover, it should be noted that all the values of γ are higher than 0.2. This indicates the higher controllability of the V_{th} as compared with the conventional body-bias control of the planar MOSFETs.

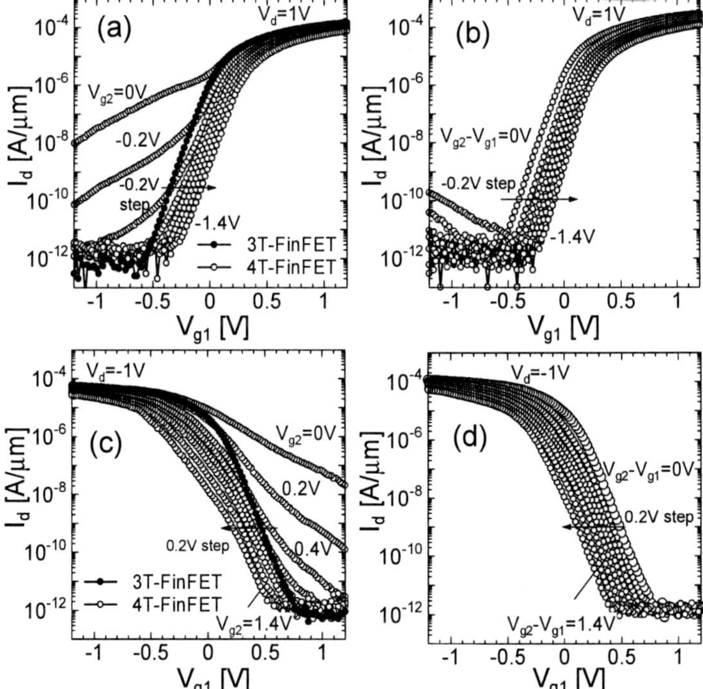

Fig. 13. I_d-V_{g1} characteristics of (a) single drive (SD) mode nMOS, (b) synchronized double gate drive (DD) mode nMOS, (c) SD mode pMOS, (b) DD mode pMOS operations for the 4T-FinFET with t_{fin}= 50 nm and L_g= 110 nm.

The 4T-FinFETs were successfully integrated and Fig. 14 shows the fabricated CMOS inverter composed by 4T-FinFETs. Figures 15 and 16 show the input-output characteristics of the CMOS inverter. To control the characteristics of the CMOS inverter, both the second-gates of the nMOS and pMOS need to be controlled. Figure 15 shows the inverter characteristics when one V_{g2} is fixed and the other is varied. In this case, the logical threshold voltage can be flexibly changed. On the contrary, by changing both the V_{g2n} and V_{g2p} synchronously towards the opposite direction, the short circuit current of the CMOS inverter can be dynamically controlled with keeping the logical threshold as shown in Fig. 16. This implies that the power-managed CMOS operation can be accomplished. The lower logical threshold voltage below 0.5 V is the issue for the further optimization. We found that the peak short circuit current can exponentially be reduced with decreasing V_{g2} as shown in Fig. 17. This indicates that the CMOS circuits are tunable for both the high- and low-power operations, which strongly suggests the advantage of the power-managed CMOS circuit using 4T-FinFETs.

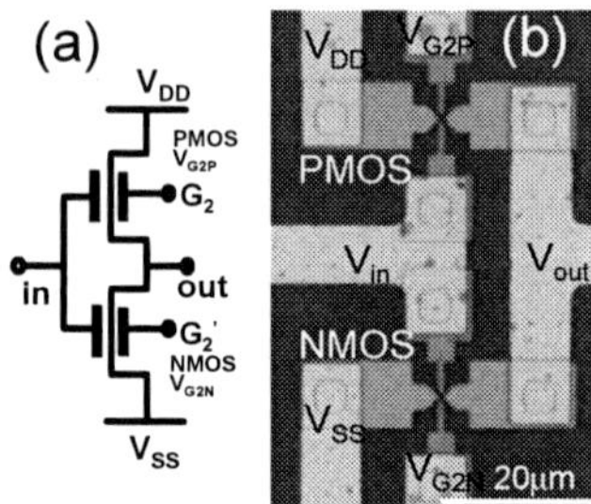

Fig. 14. (a) Circuit diagram and (b) optical microscopic view of the CMOS inverter composed by 4T-FinFETs.

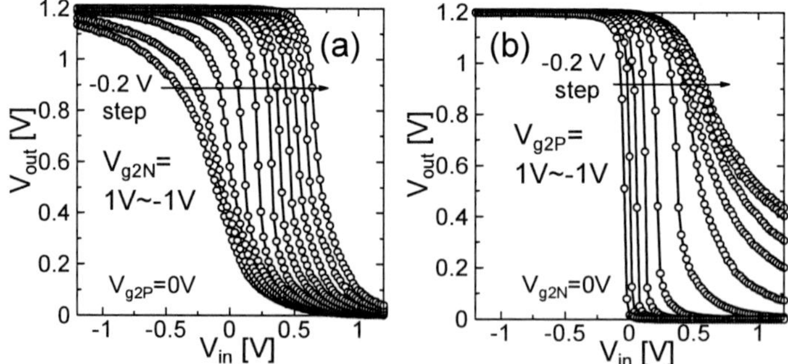

Fig. 15. Input-Output characteristics of the CMOS inverter composed by 4T-FinFETs with $t_{fin}=$ 50 nm and $L_g=$ 110 nm; (a) $V_{g2p}=$ 0V and V_{g2n} is varied, (b) $V_{g2n}=$ 0V and V_{g2p} is varied. The logical threshold voltages of the CMOS inverter can be flexibly controlled.

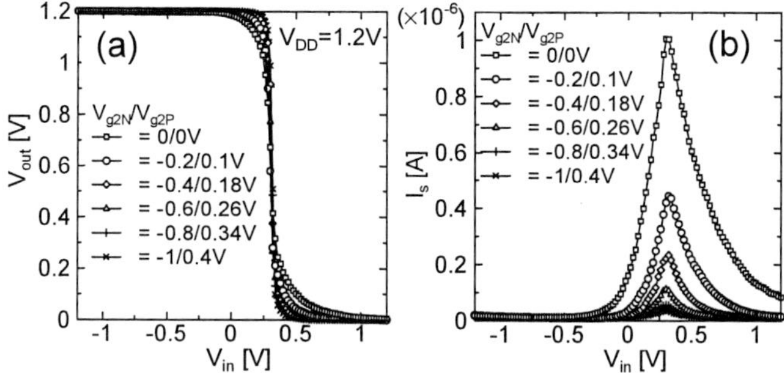

Fig. 16. (a) Input-Output characteristics of the CMOS inverter composed by 4T-FinFETs (t_{fin}= 50 nm and L_g= 110 nm) with various control-gate biasing conditions and (b) short circuit currents. Drastic reduction of the short circuit currents was demonstrated.

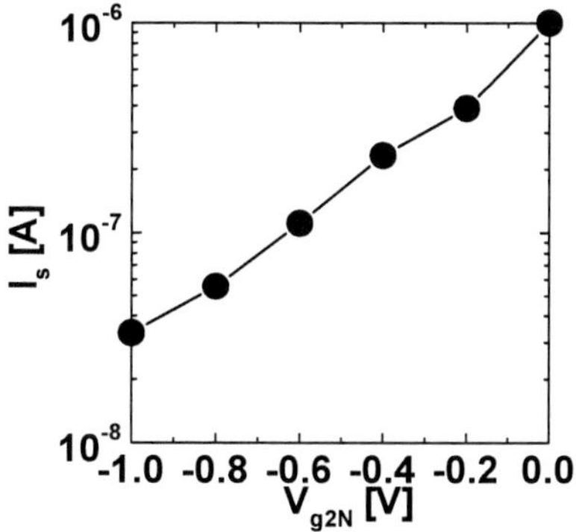

Fig. 17. Peak short circuit current of the CMOS inverter as a function of V_{g2}. We revealed an exponential reduction of the short circuit current with decreasing control-gate bias voltage for both the high-power and low-power operations.

Conclusion

Damage-free fabrication of high-aspect-ratio rectangular Si-fins using the newly developed NBE has been investigated. The fabricated FinFETs provide higher electron mobility than that obtained using a conventional etching process. The improved electron mobility of the NBE-processed FinFETs is explained by the atomically-flat surface of the NB processed channels. In addition, dynamically power-controllable CMOS inverter has been successfully demonstrated using the newly-integrated separated-gate 4T-FinFETs. Exponential reduction of the CMOS short circuit current has been revealed by applying the bias voltage to the control-gate. These results strongly suggest the advantage of the power-managed CMOS circuits using 4T-FinFETs.

Acknowledgement

The authors would like to thank Dr. S. Samukawa and Dr. S. Noda for the technical support of the neutral beam etching and helpful discussions. They also would like to thank Dr. H. Koike and Dr. T. Sekigawa for their useful discussions on the power-managed inverter circuit.

References

1. T. Sekigawa and Y. Hayashi, Solid State Electron., **27**, 827 (1984).
2. D. Hisamoto, W. Lee, J. Kedzierski, H. Takeuchi, K. Asano, C. Kuo, E. Anderson, T. King, J. Bokor, and C. Hu, IEEE Trans. Electron Devices, **47**, 2320 (2000).
3. Y. Liu, M. Masahara, K. Ishii, T. Sekigawa, H. Takashima, H. Yamauchi, and E. Suzuki, IEEE Electron Device Lett., **25**, 510 (2004).
4. G. S. Oehrlein, R. M. Tromp, Y. H. Lee, and E. J. Petrillo, Appl. Phys. Lett., **45**, 420 (1984).
5. C. J. Petti, J. P. McVitte, and J. D. Plummer, IEDM Tech. Dig., 104 (1988).
6. F. Shimokawa, Y. Ishikawa, and M. Okigawa, Jpn. J. Appl. Phys., **40**, L1346 (2001).
7. F. Shimokawa, J. Vac. Sci. Technol., **A 10**, 1352 (1992).
8. T. Yunogami, K. Yokogawa, and T. Mizutani, J. Vac. Sci. Technol., **A13**, 952, (1992).
9. S. Samukawa, K. Sakamoto, and K. Ichiki, J. Vac. Sci. Technol., **A20**, 1566 (2002).
10. K. Endo, S. Noda, M. Masahara, T. Kubota, T. Ozaki, S. Samukawa, Y. Liu, K. Ishii, Y. Ishikawa, E. Sugimata, T. Matsukawa, H. Takashima, H. Yamauchi, and E. Suzuki, IEEE Trans. Electron Devices, **53**,1826 (2006).
11. K. Endo, Y. Ishikawa, Y. X. Liu, T. Matsukawa, S. O'uchi, K. Ishii, M. Masahara, J. Tsukada, H. Yamauchi, T. Sekigawa, H. Koike, and E. Suzuki, Proc. IEEE SOI Conf., 81 (2006).
12. S. Noda, H. Nishimori, T. Ida, T. Arikado, K. Ichiki, T. Ozaki and S. Samukawa, J. Vac. Sci. Technol., **A22**, 1506 (2004).
13. K. Endo, M. Masahara, Y. Liu, T. Matsukawa, K. Ishii, E. Sugimata, H. Takashima, H. Yamauchi and E. Suzuki, Jpn. J. Appl. Phys., **45**, 309 (2006).
14. M. Masahara, S. Hosokawa, T. Matsukawa, K. Endo, Y. Naitou, H. Tanoue, and E. Suzuki, Appl. Phys. Lett., **85**, 4139 (2004).
15. M. Koh-Masahara, K. Esuga, H. Furumoto, T. Shirahata, E. Seo, K. Shibahara, S. Yokoyama, and M. Hirose, Jpn. J. Appl. Phys., **38**, 2324 (1999).
16. H. Irie, K. Kita, K. Kyuno and A. Toriumi, IEDM Tech. Dig., 225 (2004).
17. D. K. Schroder, Semiconductor material and device characterization, Wiley, New York, (1998).
18. J. Koga, S. Takagi, and A. Toriumi, IEDM Tech. Dig., 475 (1994).
19. F. Gámiz, J. B. Roldán, P. Cartujo-Cassinello, J. A. López-Villanueva, and P. Cartujo, J. Appl. Phys. **89**, 1764 (2001).
20. S. Takagi, A. Toriumi, M. Iwase, and H. Tango, IEEE Trans. Electron Devices, **41**, 2363 (1994).
21. M. Masahara, Y. Liu, S. Hosokawa, T. Matsukawa, K. Ishii, H. Tanoue, K. Sakamoto, T. Sekigawa, H. Yamauchi, S. Kanemaru, and E. Suzuki, IEEE Trans. Electron Devices, **51**, 2078-2083 (2004).

ECS Transactions, 6 (4) 83-87 (2007)
10.1149/1.2728844, ©The Electrochemical Society

Parasitic Effects Depending on Shape of Spacer Region on FinFETs

Y. Kobayashi[a], K. Tsutsui[a], K. Kakushima[a], V. Hariharan[b], V. R. Rao[b], P. Ahmet[c],
and H. Iwai[c]

[a] Department of Electronics and Applied Physics, Tokyo Institute of Technology, J2-69,
4259, Nagatsuta, Midoriku, Yokohama, 226-8502, Japan
[b] Department of Electrical Engineering, Indian Institute of Technology Bombay, Powai,
Mumbai-400076, Maharashtra, India
[c] Frontier Collaborative Research Center, Tokyo Institute of Technology, J2-69, 4259,
Nagatsuta, Midoriku, Yokohama, 226-8502, Japan

Parasitic resistance and capacitance relating to spacer region of
FinFETs were investigated by changing shape of the spacer region.
The trade-off relationship between these two parasitic elements
was demonstrated on the expansion of the fin width in the spacer
region. The gate delay characteristic of the FinFETs was optimized
by gradual expansion with triangular shape. It was indicated that
not only parasitic resistance but also parasitic capacitance on the
spacer region was significant for transistor performance.

Introduction

FinFETs are candidates of ultra short channel transistors for the future generation [1]-
[3]. But parasitic resistance of spacer region on FinFETs has been recognized to be a
significant problem [4]-[5]. A simple method to reduce the parasitic resistance is to
expand the fin width of spacer regions and bring the source/drain (S/D) contacts close to
the gate [6]-[8]. However such a geometrical change would cause an increase in the
fringing capacitance between the S/D and gate. This means that there exists a trade off
between the parasitic resistance and the parasitic capacitance. Therefore, optimization of
geometrical structure of the spacer regions must be considered on the design of high
performance FinFETs. In this paper, change of the parasitic effects depending on shape
of the fin in the spacer regions is investigated through the simulation.

Device Structure and its Variation for Simulation

A planer cross section of n-channel FinFET geometry used for the 2D simulation is
illustrated in Figure 1 (a). Poly Si gate length of 30nm and gate oxide (SiO$_2$) thickness of
1.2nm were commonly used in the simulations. The overlap of the gate and S/D doped
region (1E20cm^{-3}) is 3nm for each and impurity profile at the edge of S/D doped region is
abrupt down to the channel doping (1E15cm^{-3}). The spacer region is covered with
passivation oxide (SiO$_2$). Figure 1 (b) shows step by step shape change of the spacer
region of the FinFET corresponding to the region indicated by broken line in Fig.1 (a).
Starting from the shape of the left side, which is identical to the corresponding region in
Fig.1 (a), the fin is expanded outward with triangular shape. The degree of the expansion
is represented by vertex angle of α, and the vertex position is fixed at the point separated
with *Lspn*=5nm form the gate edge. After the expansion (*i.e.* increase in α) is reached the
shape of middle in Fig.1 (b) in which the base of triangular is identical to *Wsp*, the

expansion continues in the shape of trapezoid so that width of the expanded region is within *Wsp*. The expansion is ended at $\alpha=90^{\circ}$ as indicated by shape of right side in Fig.1 (b). Using the 2D device simulator, parasitic resistance, that is resistance between channel edge and silicide contact, total gate capacitance which includes fringing capacitance, and gate delay of the transistor were evaluated as a function of α, for various fin width, *Wfin*, and size of spacer region before the expansion ($\alpha=0^{\circ}$), *Lsp/Wsp*.

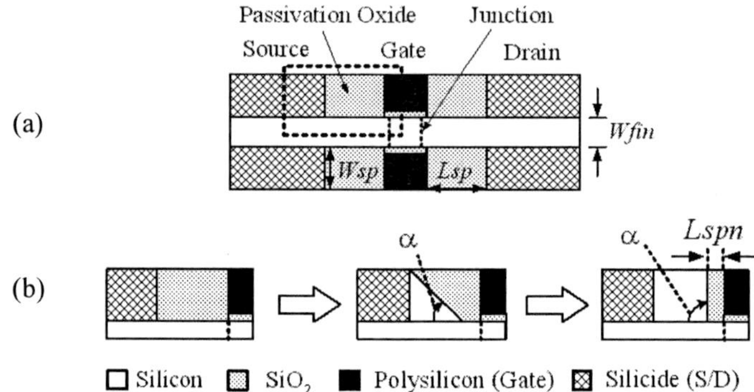

Figure 1, Simulated FinFET. (a) Planner schematic of basic structure. (b) Step by step shape change of spacer region represented by expansion angle of α.

Results and Discussion

Figure 2 shows the parasitic resistance and the total gate capacitance, which are normalized by fin height, and the gate delay as a function of expansion angle of α, for *Wfin*=5 nm, 10 nm or 20 nm. It was found that the parasitic resistance decreased while total gate capacitance increased with increase of α. Since intrinsic capacitance relating to the gate is considered to be independent of the spacer region shape, the increase in total gate capacitance indicates increase in parasitic capacitance due to fringing capacitance. Thus, this relation indicates a trade–off between parasitic resistance and parasitic capacitance. The trade-off is reflected on the delay depending on α as shown in the same figure. As increase of α, the delay decreased with α up to 20°, but it increased again with further expansion; $\alpha>40^{\circ}$. The improvement in the region of small α is due to decrease of parasitic resistance, while the degradation in the region of large α is due to increase in parasitic capacitance.

Variation of parasitic resistance by varying the *Wfin* is apparent, while total gate capacitance is insensitive to *Wfin* variation. Especially, increase in parasitic resistance for small α and narrow *Wfin* is significant. However, increase in gate delay is not so large even in this condition, *i.e.,* small α and narrow *Wfin*. This indicates that the gate delay is governed by parasitic capacitance dominantly rather than parasitic resistance.

Figure 3 shows results in the same type plots in which dependence on size of spacer region, *Lsp/Wsp*, is focused. As the spacer size becomes large, increase in parasitic resistance at small α and increase in parasitic capacitance at large α are more significant. As a result, variation of the gate delay becomes large for the spacer shape change.

Although range of the change of shape and size of the spacer region is limited in this work, the optimum value of α was fond to be 20-30° independent of *Wfin, Lsp* and *Wsp*. Such a gradual change of fin width through the spacer region is considered to be desirable to suppress the parasitic effects involving the trade-off relationship.

Conclusion

The trade-off relationship between the parasitic resistance and the parasitic capacitance was demonstrated for the variation of shape and feature size of spacer region on FinFETs. The gate delay, which is noticed as a figure of merit of the FinFET performance, is affected by both parasitic elements, and effect of the parasitic capacitance was found to be significant rather than that of parasitic resistance. The optimum shape of the spacer region was fond to be the triangular like expansion of fin width with vortex angle, α, of 20-30°. The optimum value of α was insensitive to the fin width and the spacer size as far as the varied range employed in this work.

Acknowledgments

This work was partially supported by Grant-in-Aid for Scientific Research on Priority Areas by the Ministry of Education, Culture, Sports, Science and Technology, Japan.

References

1. D. Hisamoto, W. -C. Lee, J. Kedzierski, E. Anderson, H. Takeuchi, K. Asano, T. -J. King, J. Bokor and C. Hu, *IEDM Tech. Dig.*, Dec. 1032(1998)
2. X. Huang, W. -C. Lee, C. Kuo, D. Hisamoto, L. Chang, J. Kedzierski, E. Anderson, H. Takeuchi, Y. -K. Choi, K. Asano, V. Subramanian, T. -J. King, J. Bokor, C. Hu, *IEDM Tech. Dig.*, Dec., 67(1999).
3. D. Hisamoto, W. -C. Lee, S. Ahimed and C. Hu, *IEEE Trans. Electron Devices*, **47**, 2320(2000).
4. Y. K. Choi, N. Lindert, P. Xuan, S. Tang, D. Ha, E. Anderson, T. -J. King, J. Bokor and C. Hu, *IEDM Tech. Dig.*, Dec. 421(2001).
5. A. Dixit, A. Kottantharayil, N. Collaert, M. Goodwin, M. Jurczak, and K. D. Meyer, *IEEE Trans. Electron Devices*, **52**, 1152(2005).
6. D. -S. Woo, B. Y. Choi, W. Y. Choi, M. W. Lee, J. D. Lee and B. -G. Park, Electronics Lett., **39**, 1154(2003).
7. J. Kedzierski, M. Ieong. E. Nowak, T. S. Kanarsky, Y. Zhang, R. Roy, D. Boyd and H. -S. P. Wong, *IEEE Trans. Electron Devices*, **50**, 952(2003)
8. G. Chen, R. Huang, X. Zhang, L. Yang, D. Zhao and Y. Wang, *Proc. 7th International Conf. on, Solid-State and integrated Circuits Technology*, Oct., **1**, p. 122(2004).

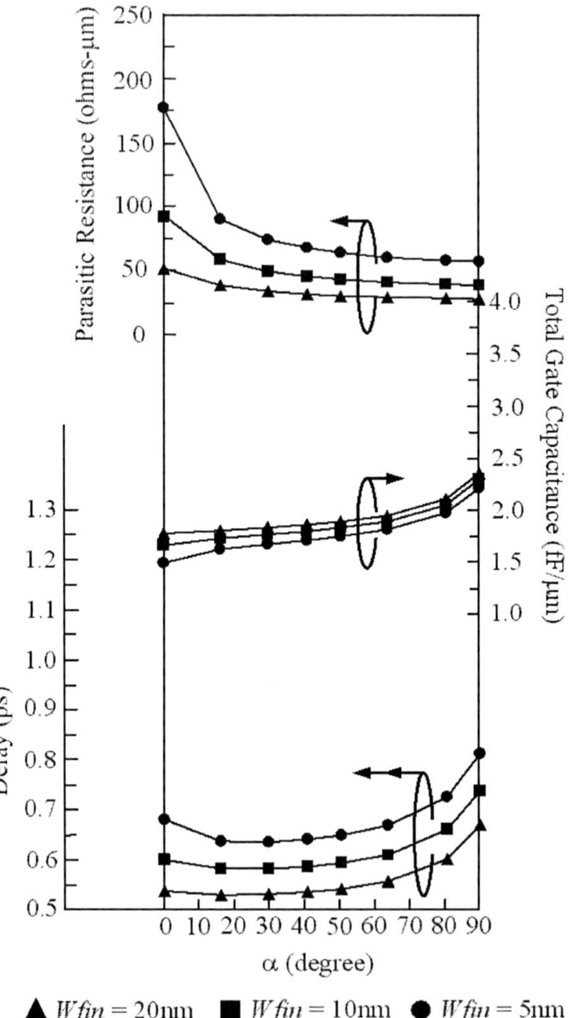

Figure 2, Parasitic resistance, total gate capacitance and gate delay depending on α for various *Wfin*. *LsD/WsD*=40 nm/30 nm are constant.

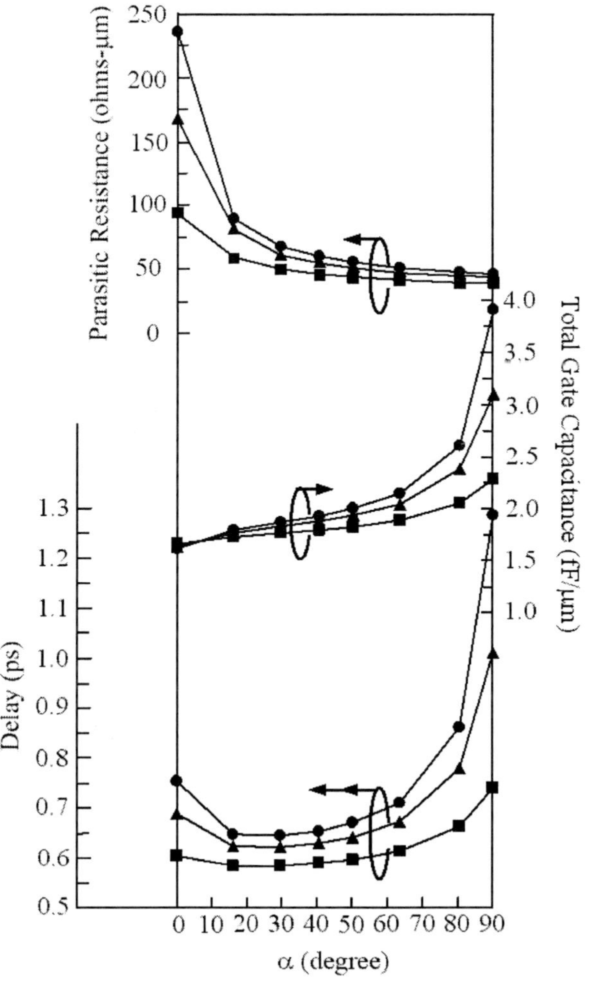

Figure 3, Parasitic resistance, total gate capacitance and gate delay depending on a for various *Lsp/Wsp*. *Wfin*=10 nm is consant.

ECS Transactions, 6 (4) 89-94 (2007)
10.1149/1.2728845, ©The Electrochemical Society

Modeling the Back-Gate Coupling Effect in Triple-, Pi- and Omega-GateFETs

R. Ritzenthaler[a,c], M. Gaillardin[b,c], K. Akarvardar[c], O. Faynot[a],
C. Jahan[a], and S. Cristoloveanu[c]

[a] LETI (CEA), 17 rue des martyrs, 38054 Grenoble Cedex 09, France
[b] DIF (CEA), BP 12, 91680 Bruyères-Le-Châtel, France
[c] IMEP Minatec-INPG, 3 Parvis Louis Néel, Grenoble BP257, France

In multiple-gate devices, coupling effects due to the 3-D architecture modify the electrical characteristics of the transistor. In this paper, electrical measurements have been carried out in order to examine the role of the back-gate voltage. In agreement with numerical simulations, it is shown that the back-gate influence is reduced for narrow devices due to a strong coupling between the different faces of the main gate. In narrow Fin devices, the Pi- and Omega-shapes of the gate provide an excellent electrostatic control of the field lines and potential, both in the buried oxide and in the silicon channel. The radiation-induced charge trapping in the BOX underneath the channel can be then much limited: narrow Fin devices exhibit a higher tolerance to ionizing radiations than its single-gate counterparts. Additionally, it is shown that the 2D dimensional coupling in these structures can be analytical calculated for Triple- and Pi-gate FETs. These models are compared with numerical simulations and show a very good agreement.

Introduction

As the scaling of classical CMOS is approaching its limit, new architectures like Triple-gate FETs (Figure 1) (1) are examined. When the gate penetrates into the buried oxide (BOX), the transistor is commonly called Pi-gateFET (2) or, for lateral overetch under the silicon channel, Omega-gateFET (3) (Figure 1). These 3D structures present new and interesting behaviours like lateral coupling effects due to the gate shape.

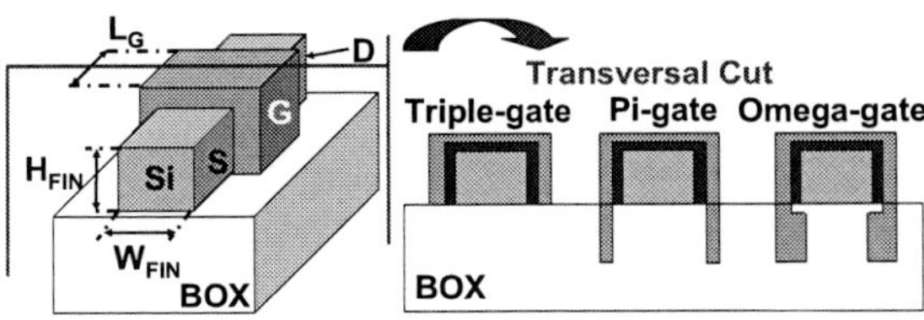

Figure 1. Schematic configurations of Triple, Pi- and Omega-gate FET transistors.

The purpose of this paper is to investigate the back-gate bias and coupling effects in these structures. Using extensive measurements and 3D numerical simulations, the first part of this paper is focused on the influence of the Fin width and gate shape on the back-gate bias sensitivity. An interesting application, the resistance to ionizing radiations, is shown in the second section. In the last section, an analytical model is presented and compared with numerical simulations in the case of the Triple-gateFETs and Pi-gateFETs.

Lateral coupling

Process description of experimental devices

Advanced FinFET-based devices have been fabricated (performance described in (4)). They feature HfO_2 high-K dielectric and TiN/poly-Si metal gate stack. Figure 2 shows the TEM cross-section of a 60 nm silicon finger array with the surrounding poly-Si/TiN/HfO_2 gate stack. The SOI film and BOX thicknesses are respectively 25 and 100 nm. Tested devices are NMOSFETs with undoped channels ($N_A \approx 10^{15}$ cm^{-3}). The gate length L_G is 10 µm. Fin height H_{FIN} is equal to SOI thickness, i.e. 25 nm. The overetch depth into the BOX is about 27 nm, while the lateral penetration underneath the channel is evaluated to 5 nm. Fin width W_{FIN} varies from 10 µm down to 30 nm and allows investigating the different channels of the main gate. A wide FinFET behaves like a Fully Depleted planar MOSFET whereas a narrow FinFET is governed by 3D effects (5).

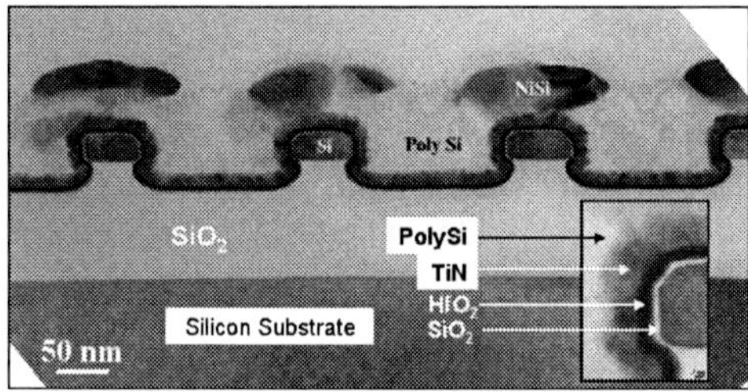

Figure 2. TEM cross section of 60 nm wide Fins array with the TiN/HfO_2 gate. Insert shows body/gate and oxide/gate region.

Experimental measurements and numerical simulations

The threshold voltage V_{T1} variation with substrate bias V_{G2} is shown in Figure 3 for various Fin widths (from W_{FIN} = 10 µm down 50 nm). The threshold voltage was extracted by taking the gate voltage for a constant current of 0.1 µA, normalized by (W_{FIN} + 2H_{FIN})/L_G. For wide devices, increasing the substrate voltage from accumulation to inversion results in a decrease of the threshold voltage. Each curve exhibits two linear regions: the initial slope reflects the 1D vertical coupling effect and is followed (for more positive V_{G2}) by a more rapid drop due to the activation of the back channel. The linear decrease in threshold voltage with increasing back-gate bias is typical for Fully Depleted planar MOSFETs and can be modeled by the classical coupling relation between the front and the back gates formulated by Lim and Fossum (6). For narrow devices, the lateral gates can screen a narrow silicon body from the back-gate influence. In Figure 3, it is seen that for devices with very small Fin width the threshold voltage is rather insensitive to back-gate voltage.

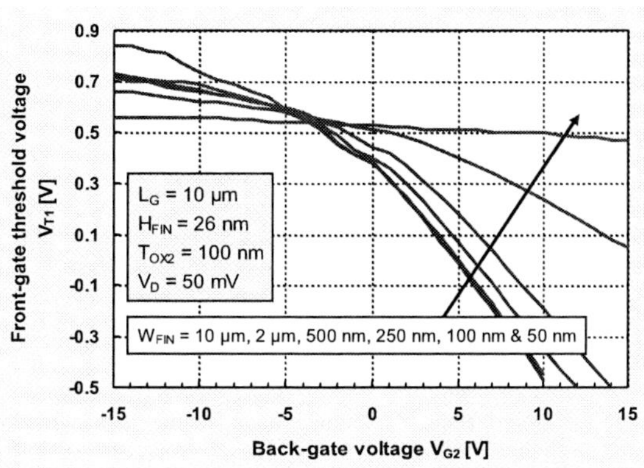

Figure 3: Threshold voltage V_{T1} versus back-gate bias V_{G2} in long N-channel Omega-gateFETs with Fin widths W_{FIN} of 10 µm, 2 µm, 500 nm, 250 nm, 100 nm and 50 nm. V_{T1} is extracted from experiments using the constant current method.

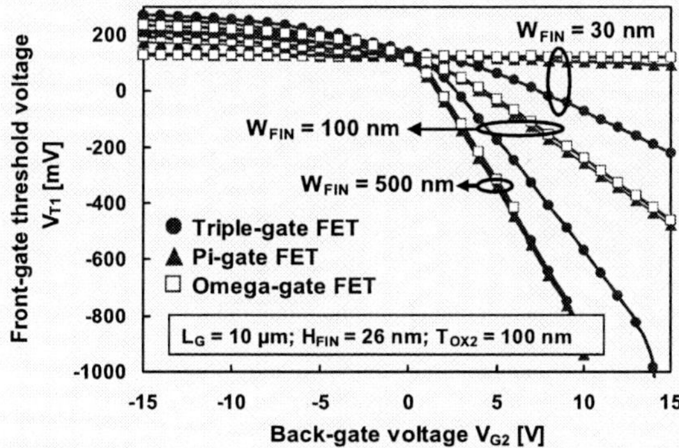

Figure 4. Numerical simulations of the front gate threshold voltage V_{T1} versus back-gate bias V_{G2} for Triple- (circles), Pi- (triangles) and Omega-gate (squares) FETs. Fin widths W_{FIN} are 30, 100 and 500 nm. Gate oxide thickness t_{OX1} is 2 nm and BOX thickness t_{OX2} is 100 nm. $V_D = 50$ mV and $L_G = 10$ µm.

In Figure 4, numerical simulations of threshold voltage versus substrate voltage curves are plotted for various Fin widths and transistor configurations (Tri-gate, Pi-gate and Omega-gate geometries). For wide devices ($W_{FIN} = 500$ nm), where the influence of the lateral gates is negligible, the characteristics of all three devices are obviously identical. As the Fin width is reduced ($W_{FIN} = 100$ nm), the lateral coupling tends to screen the threshold voltage from the substrate voltage as observed in the previous figure. The curves for the Pi- and Omega-gateFETs are roughly equivalent. However, the threshold voltage of the Triple-gateFET is far more sensitive to back-gate bias. This difference is clearly visible for very narrow channels (W_{FIN} = 30 nm). We conclude that the buried oxide zone located underneath the body has also to be well controlled in order to ensure immunity to back-gate action. Thanks to the overetch of the lateral gates, the Pi-gateFET and even better the Omega-gateFET structures act like 'quasi gate-all-around' transistors.

Resistance to radiations

In an irradiative atmosphere, charged particles generate electron/hole pairs in silicon body, gate oxide and buried oxide. In Fully Depleted structures, the radiations modify the back interface potential, and then degrade the front-gate threshold voltage via the vertical electrostatic coupling. As a consequence, irradiating the transistors leads to a front-gate threshold voltage shift similar to a positive back-gate biasing. This mechanism has experimentally been measured on the Omega-FET devices (7). In Figure 5, the front-gate threshold voltage shift (difference between irradiated and non-irradiated devices) ΔV_{T1} is shown for various channel widths (from W_{FIN} = 10 µm down to W_{FIN} = 40 nm) and different irradiation doses (from 10 krad up to 500 krad). For wide devices (similar to Fully Depleted planar transistors), the front-gate threshold voltage degradation is clearly visible and goes up to 200 mV for a 500 krad dose. However, this effect is drastically reduced for narrow Omega-gateFET devices, whatever the irradiation dose. Thanks to the lateral coupling in Omega-gateFET transistors, the narrow transistors are intrinsically very immune to radiations (7).

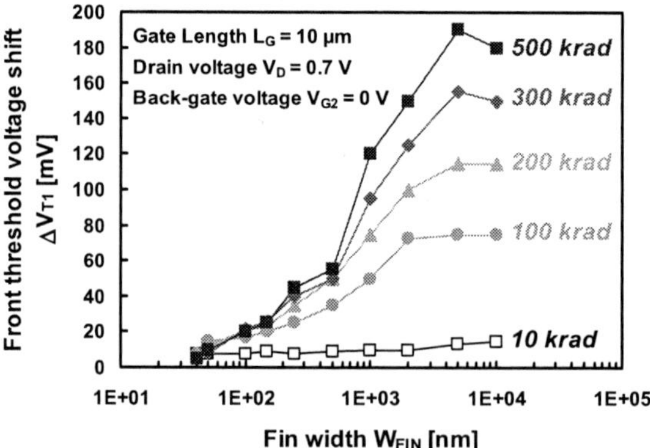

Figure 5. Front threshold voltage shift ΔV_{T1} vs. silicon Fin width W_{FIN} for several total dose steps. L_G = 10 µm, V_D = 0.7 V and V_{G2} = 0 V.

Modeling

Compared to the one-dimensional Lim and Fossum model, the electrostatic problem to be solved in a triple-gate structure is two-dimensional due to the influence of the lateral gates. Solving the 2D Poisson equation in the structure (details presented in (8)), with the correct boundary conditions, yields the following formula of the coupling coefficient α =

dV_{T1}^{DEP2}/dV_{G2} in a Triple-gateFET (with the back-gate biased into depletion) as a function of Fin width W_{FIN} and height H_{FIN}:

$$\alpha(W_{FIN}, H_{FIN}) = \frac{\partial V_{V1}^{DEP2}}{\partial V_{G2}} = \left[\frac{2\sqrt{2}}{\sinh(2\sqrt{2}\,\dfrac{H_{FIN}}{W_{FIN}})} \right] \frac{\dfrac{C_W(W_{FIN})}{C_{OX1}}}{1 + \dfrac{2\sqrt{2}}{\tanh(2\sqrt{2}\,\dfrac{H_{FIN}}{W_{FIN}})}\,(\dfrac{C_W(W_{FIN})}{C_{OX2}})} \qquad [1]$$

where $C_W = \varepsilon_{Si}/W_{FIN}$, C_{OX1} and C_{OX2} are respectively the lateral silicon film capacitance, front and back oxide capacitances.

We demonstrate that this expression can simply be extended to the case of a Pi-gateFET by considering that the silicon film capacitance C_W is replaced in [1] by the association of C_W and the overetch vertical capacitance C_{Pi} :

$$C_{Pi} = \frac{C_W C_{OV}}{C_W + C_{OV}} \qquad [2]$$

with:

$$C_{OV} = \varepsilon_{BOX} \Big/ t_{OV} \qquad [3]$$

where t_{OV} is the gate overetch in the BOX.

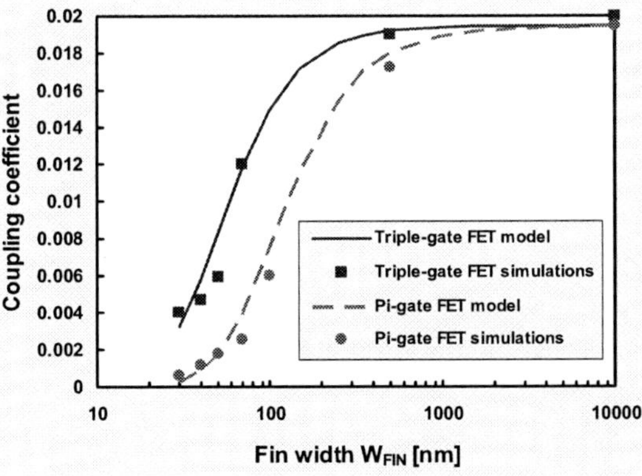

Figure 6. Comparison of the coupling coefficient α between the analytical models of Eqs. [1] and [2] and the numerical simulations for Triple- (squares) and Pi-gate (circles) FET structures. Gate oxide thickness t_{OX1} is 2 nm and BOX thickness t_{OX2} is 100 nm. $V_D = 50$ mV and $L_G = 10\ \mu$m.

These proposed analytical compact models are compared to the coupling coefficients extracted from numerical simulations in the Figure 6 and show a very good agreement.

For wide devices, the coupling coefficient tends towards t_{OX1}/t_{OX2} (t_{OX1} being the gate oxide thickness and t_{OX2} the BOX thickness), which is the value obtained for Fully Depleted planar devices. For narrow devices, the coupling coefficient tends towards zero, meaning that devices are intrinsically immune to the back-gate bias. For a given Fin width W_{FIN}, it is also clearly visible that the coupling coefficient is smaller for Pi-gateFETs (Eq.[2]) compared to Triple-gateFETs (Eq.[1]); this is due to the better control of the silicon/BOX potential by the gate overetch in Pi-gateFETs. The case of Omega-gateFET is more complex due to the boundary conditions at the silicon/BOX interface.

Conclusions

The coupling effect occurring in non-planar vertical structures (Triple-, Pi- and Omega-gateFETs) was investigated. The coupling between front- and back-gate becomes two dimensional and leads to a complete screening of the back-gate influence for narrow transistors. We demonstrated that these structures are then very immune to radiations. Pi-gate and Omega-gateFETs are even more immune than Triple-gateFETs to the influence of the back-gate and BOX defects because of the gate-induced control of the body/BOX interface. It was also shown that the 2D coupling coefficient for non-planar vertical structures can be analytically calculated for Triple- and Pi- gateFETs. These models were compared to numerical simulations and experiments, showing a very good agreement.

Acknowledgments

The author would like to thank the staff of LETI facilities for device fabrication support, IMEP characterization group, and Drs. Paillet and Ferlet.

References

1. B. Doyle et al., "Tri-Gate Fully-Depleted CMOS Transistors: Fabrication, Design and Layout", *Digest of Technical Papers, 2003 Symposium on VLSI Technology*, pp. 133-134, 2003
2. J.-T. Park, J.-P. Colinge, C.H. Diaz, "Pi-Gate SOI MOSFET", *IEEE Electron Device Letters*, vol. 22, no. 8, pp. 405-406, 2001
3. F.-L. Yang et al., "25 nm CMOS Omega FETs", *IEDM'02 Technical Digest*, pp. 255-258, 2002
4. C. Jahan et al., "10nm ΩFETs transistors with TiN metal gate and HfO$_2$", *VLSI Tech. Dig.*, pp. 112-113 ,2005
5. S. Cristoloveau, R. Ritzenthaler, A. Ohata, O. Faynot, " 3D size effects in advanced SOI devices", *Int. J. of High Speed Electronics and Systems*, vol. 16, no. 1, pp. 9–30, 2006
6. H.K. Lim and J.G. Fossum, "Threshold voltage of thin-film silicon-on-insulator (SOI) MOSFETs", *IEEE Transactions on Electron Devices*, vol. 30, p1244, 1983
7. M. Gaillardin, P. Paillet, V. Ferlet-Cavrois, S. Cristoloveanu, O. Faynot, and C. Jahan, "High tolerance to total ionizing dose of Ω-shaped gate field-effect transistors", *Appl. Phys. Lett. 88, 223511*, 2006
8. K. Akarvardar, A. Mercha, S. Cristoloveanu, P. Gentil, E. Simoen, V. Subramanian and C. Claeys, "A 2-D model for interface coupling in triple-gate transistors," *IEEE Transactions on Electron Devices*, to be published in April 2007 issue

ECS Transactions, 6 (4) 95-100 (2007)
10.1149/1.2728846, ©The Electrochemical Society

Origin of Transient Gate Current Observed in Pseudo-MOS Transistor

Shingo Sato[1], Nguyen Quang Tuan[2], Sorin Cristoloveanu[2], and Yasuhisa Omura[1]

[1]High-Technology Research Center, Kansai University, 3-3-35, Yamate-cho, Suita, Osaka, 564-8680 Japan
[2]IMEP (UMR CNRS-INPG-UJF), ENSERG, BP 257, 38016 Grenoble Cedex 1, France

We discuss the origins of the transient gate current observed in the pseudo-MOS transistor in order to extract optional semiconductor parameters and its physical mechanisms. We analyze the time evolution of the gate current; its delay time is long even in quasi-static measurements at low gate voltages. Experiments show that the transient gate current characteristics are ruled by the generation-recombination process in the SOI layer and thermionic emission currents through the source and drain contacts. We address how to evaluate the generation-recombination lifetime in the SOI layer from the transient gate current behavior.

Introduction

The pseudo-MOSFET technique has been proposed to characterize Silicon-On-Insulator (SOI) wafers efficiently and rapidly (1). This is a quite useful and simple method for measuring the electrical properties of SOI layers and buried-oxide (BOX) layers because it consists of applying two metal probes to the SOI wafer. Experimental results for various types of SOI wafers have been reported (2).

To better understand the characteristics of pseudo-MOSFETs, analyses of transient current characteristics observed at every terminal are critical. In this paper, we interpret the physics of transient phenomena appearing in the SOI layer by using various step-pulse gate voltages.

Experimental Setup

To analyze the origin of the transient gate current observed in the pseudo-MOS transistor, we measured the current as a function of the height of the gate voltage. The sample tested here is the UNIBOND SOI wafer fabricated by SOITEC. The measurement system is shown in Fig. 1 and the corresponding device parameters are summarized in Table 1. We used two probes of a four-point-probe system (produced by Jandel) and an HP 4156B precision semiconductor analyzer. The SOI layer is separated into compact islands (5 mm x 5 mm) to reduce the influence of any leakage current caused by unexpected pipes between the substrate and the SOI layer. When we measured the transient current characteristics of pseudo-MOS transistor, we used the stand-by mode of the HP 4156B to hold the specific initial voltage before applying the gate pulse to the substrate. We performed all experiments in the dark condition to eliminate the influence of external illumination.

TABLE I. Device parameters.

Parameters	Values	Units
SOI layer thickness	70	nm
Buried oxide layer thickness	145	nm
Nominal boron concentration of SOI layer	2.0×10^{15}	cm^{-3}
Nominal boron concentration of substrate	2.0×10^{15}	cm^{-3}
Island size	5x5	mm^2

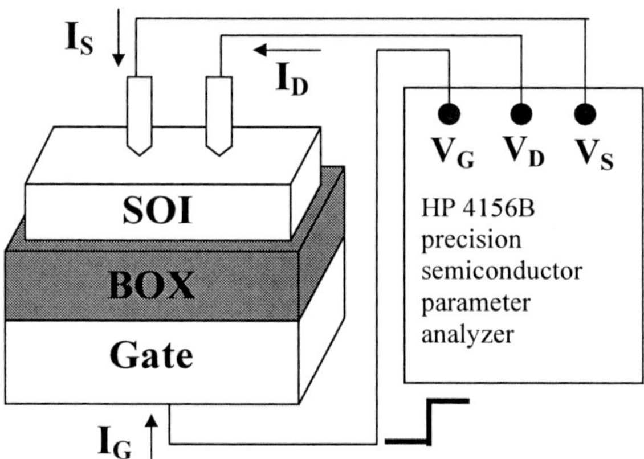

Figure 1. Pseudo-MOS transistor and measurement system.

Experimental Results

First we performed quasi-static current measurements to examine whether the previous results reported by N. Bresson et al. (2) were reproduced in our sample. Figure 2 shows drain- and gate-current characteristics as a function of gate voltage swept from negative to positive value; the machine-defined delay time is 0.5 [sec] and the drain-to-source voltage is 0.2 [V]. The drain and gate currents were measured with two-different sweep directions; i. e., the gate voltage was swept from negative to positive value and from positive to negative value. Sweep direction is denoted by the symbol of the arrow in the illustration. Figure 2 shows a distinct transient gate current depending on sweep direction. The transient gate current has a long delay time because the transient behavior can be observed even in the quasi-static measurement. Though this behavior is already reported by N. Bresson et al. (2), we focus on the fact that the transient behavior of gate current depends on sweep direction.

Since the electrostatic potential at the top surface of SOI layer except for the source and drain contacts is basically floating as strongly suggested from both experimental (2) and theoretical (3) points of view, we can anticipate that the gate current is caused by the global modulation of the electrostatic potential inside the top Si layer.

When we apply Gauss' law to the entire pseudo-MOS transistor, other than the source and drain contacts, we have the following comprehensive relationship.

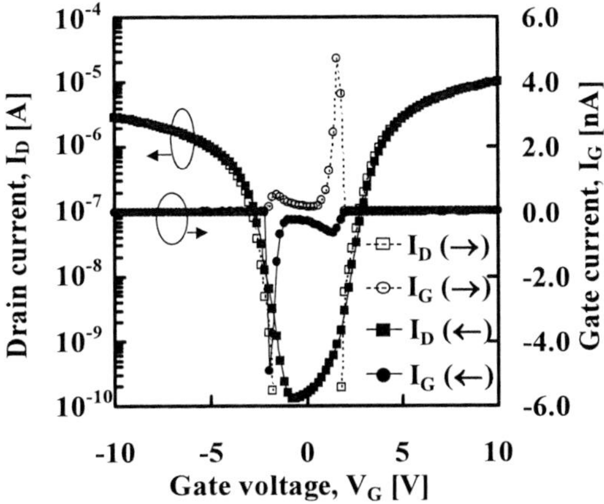

Figure 2. Drain- and gate-current characteristics for the two sweep directions.

$$Q_G + Q_{GIT} + Q_{OX} + Q_{BIT} + Q_{SOI} + Q_{TIT} = \Phi_S + \Phi_D \qquad [1]$$

where Q_G, Q_{GIT}, Q_{OX}, Q_{BIT}, Q_{SOI} and Q_{TIT} stand for the charge stored in the gate (substrate) region, the interface-trap charge between BOX and the gate (substrate), the oxide charge of BOX, the interface-trap charge between the SOI layer and BOX, the space charge for the SOI layer, and the interface-trap charge at the top surface of the SOI layer, respectively. Φ_S and Φ_D are the electric fluxes through the source and the drain contacts, respectively. We have the following current equation when we take the current continuity in the substrate into account and use eq. [1].

$$I_G = \frac{\partial(Q_G + Q_{GIT})}{\partial t}$$
$$= \frac{\partial(\Phi_S + \Phi_D)}{\partial t} - \frac{\partial(Q_{OX} + Q_{BIT} + Q_{SOI} + Q_{TIT})}{\partial t} \qquad [2]$$

We can neglect the first term (displacement current) of eq. [2] because the contact size of the probes is much smaller than Si island size. We can expect from eq. [2] that the gate current consists of the generation-recombination of carriers in the SOI layer, the charging-discharging at the interface traps, and the carrier supply from the source and drain contacts.

We turn out attention to an analysis of the time evolution of the gate current. We evaluated the transient current characteristics after applying the step-pulse voltage to the gate (substrate) terminal.

Figure 3 plots the transient current characteristics obtained at each terminal. The step-pulse voltage from the base of 0 [V] to 2.4 [V] is applied to the gate (substrate) terminal. The drain voltage supplied is 0.2 [V].

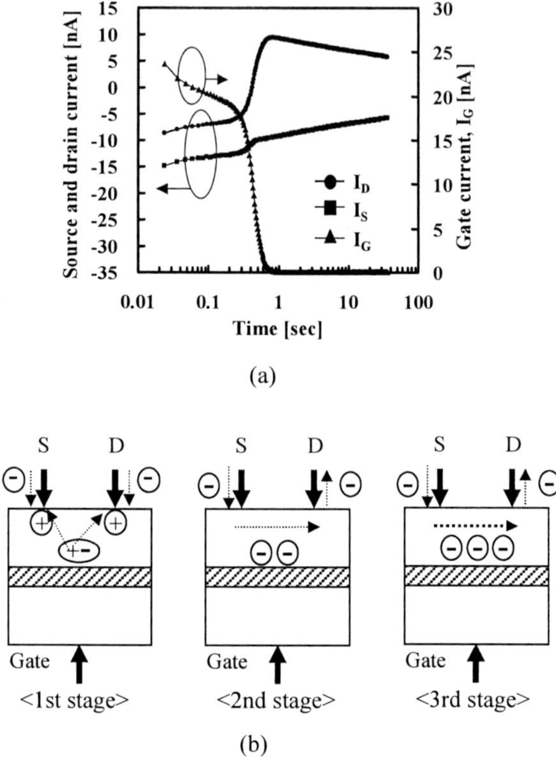

Figure 3. Transient current characteristics at each terminal. The step-pulse voltage from the base of 0 [V] to 2.4 [V] is applied to the gate terminal.

We can divide the transient current characteristics into three stages. The source and drain currents have negative polarity in the first stage. The drain current suddenly has positive polarity in the second stage. In the last stage, the drain current holds its positive polarity, while the source current holds its negative polarity in all stages. It is anticipated that the modulation of electrostatic potential in the SOI layer is primarily responsible for the electrons' behavior because the positive pulse voltage is applied to the gate (substrate) and hole depletion from the SOI layer should be very fast. The behavior of the transient gate current shown in Fig. 3 can be explained as follows. When the positive voltage is applied to the gate (substrate) terminal, the electrostatic potential in the SOI layer also shifts to the positive value as shown in (2). This generates electron-hole pairs in the depleted SOI layer due to a thermal effect; generated holes run away to the source and the drain terminals. At the same time, the thermionic emission electron current flows from the source and the drain contacts to the SOI layer. The sum of these current components is observed in the first stage at the source and the drain contacts in Fig. 3. In the second stage, some electrons are stored in the SOI layer and the electrostatic potential of the SOI layer takes a lower value than in the first stage, hence the potential difference between the metal probes and the SOI layer is reduced. This process accelerates the electronic flow from the SOI layer to the drain terminal and the electron flow from source to drain. We think that the abrupt increase in drain current in the second stage is caused by this physical mechanism.

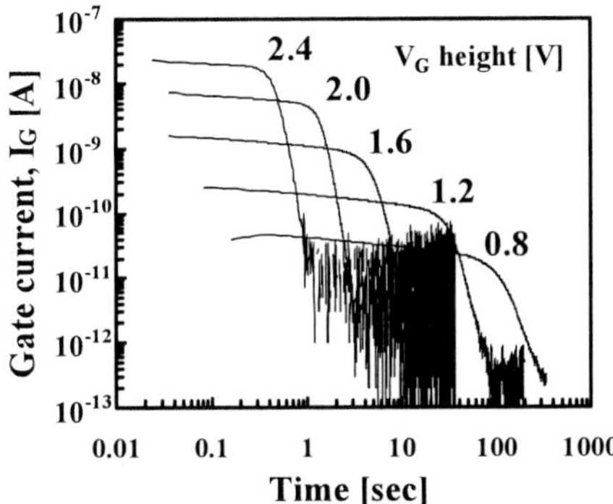

Figure 4. Transient gate current characteristics obtained at the gate terminal with parameter of the gate-voltage height.

Figure 4 shows transient gate current characteristics for various step-pulse heights of gate voltage. The drain voltage supplied is 0.2 [V]. We can see that the time evolutions of the gate current exhibit two characteristic decay processes. The gate current value evaluated in the early stage rises when the height of gate voltage rises as seen in Fig. 4. This behavior is responsible for the gate voltage dependence of the thermionic emission current through the contacts, and it occurs with the same mechanism as observed in the Schottky diode (4). In addition, the decay process in the second stage becomes more rapid as the gate voltage increases. This behavior corresponds to a reduction in effective resistance between the metal contact and the SOI layer; this idea can be validated by assuming an electrical equivalent circuit for the whole system. It can be expected that decay curve kinks represent drastic falls in effective resistance between the metal contacts and the SOI layer.

Figure 5 shows the transient characteristics of gate current with the parameter of initial gate voltage; in this experiment, the gate voltage was reduced from the initial positive voltage to 0 [V]. The drain voltage is 0.2 [V]. Initial electron density stored in the SOI layer increases with the initial gate voltage. When we apply a low initial voltage to the gate (substrate) terminal, the gate current decay is primarily characterized by a single process ($I_G \sim t^{-1/3}$). It is strongly suggested that this decay is due to the recombination process ruled by the diffusion of background carriers. The second decay process appears in the time range of 0.1 [sec] to 100 [sec] ($I_G \sim t^{-1}$) for a high initial gate voltage. This decay is the typical discharging process ruled by the self-induced field of electrons stored in the SOI layer (5) (reverse process of that shown in Fig. 3(b)), and this is supported by the electron emission from the SOI layer to the source and the drain terminals. With a low initial gate voltage, stored electrons are ready to recombine with surrounding holes because of low electron density. We think that sum of the electron emission current from the SOI layer to the source and the drain contacts and the recombination current of electrons created with a high initial gate voltage rules the phenomena.

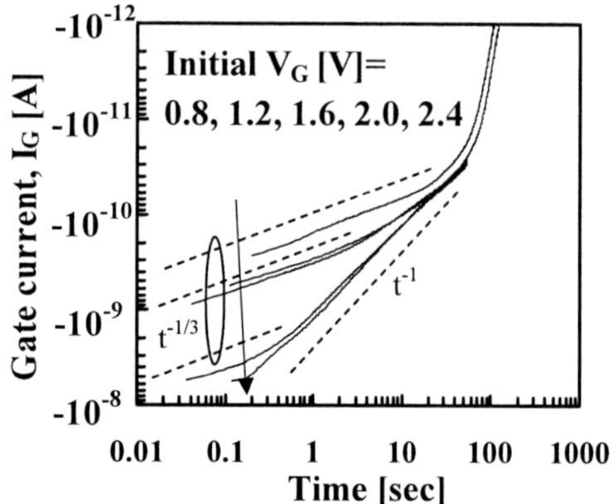

Figure 5. Transient gate current characteristics obtained at the gate terminal. The gate voltage is reduced from initial positive voltage to 0 [V].

Conclusion

We studied the origin of the transient gate current observed in the pseudo-MOS transistor. We analyzed the physical mechanisms of transient current characteristics observed at each terminal; some aspects of gate-current decay identify the generation-recombination process (including charging and discharging process of traps) in the SOI layer and the thermionic emission process of electrons through the contacts. It is also suggested that the extraction of characteristic time constants and trap density requires new mathematical models describing the transient gate-current characteristics.

References

1. S. Cristoloveanu, D. Munteanu and M. S. T. Liu, IEEE Trans. Electron Devices, **47**, 1018 (2000)
2. N. Bresson, F. Allibert and S. Cristoloveanu, Proc. Int. Symp. SOI Technol. and Dev., (ECS, Quebec, April, 2005) PV2005-03, 317 (2005)
3. S. Sato, K. Komiya, N. Bresson, Y. Omura and S. Cristoloveanu, IEEE Trans. Electron Devices, **52**, 1807 (2005)
4. D. K. Schroder, *SEMICONDUCTOR MATERIAL AND DEVICE CHARACTERIZATION*, 2nd ed, Wiley-Interscience, New York (1998)
5. C. H. Sequin and M. F. Tompsett, *CHARGE TRANSFER DEVICES*, Academic, New York (1975).

ECS Transactions, 6 (4) 101-106 (2007)
10.1149/1.2728847, ©The Electrochemical Society

Impact of Improved Mobilities and Suppressed 1/f Noise in Fully Depleted SOI MOSFETs Fabricated on Si(110) Surface

W. Cheng[a], A. Teramoto[a], C. Tye[b], P. Gaubert[a], M. Hirayama[a], S. Sugawa[b] and T. Ohmi[a]

[a] New Industry Creation Hatchery Center, Tohoku University
Aza-Aoba 6-6-10, Aramaki, Aoba-Ku, Sendai, Japan
[b] Graduate School of Engineering, Tohoku University
Aza-Aoba 6-6-10, Aramaki, Aoba-Ku, Sendai, Japan

In this study, we reveal that Si(110) surface is flatted using HF(0.5wt%)/H_2-UPW solution in N_2 ambience without light illumination. We demonstrate that the electron effective mobility on Si(110) is obviously improved by introducing this new chemical surface flattening process. Especially, the improvement of electron mobility become larger while the effective field is increased. As a result, the electron mobility is improved about 1 4 times at the high effective field equals 1MV/cm, which is mostly limited by surface roughness. We consider that the improvement of the electron mobility is originated in the good suppression of surface micro-roughness in this experiment. However, the 1/f noise level is almost the same even introducing the chemical surface flattening process. In this paper, it is observed that the 1/f noise level of MOSFETs on Si(110) is suppressed very well by introducing the novel accumulation mode device structures.

I. Introduction

To realize the high performance analog-digital mixed circuits, it is very important to improve the effective mobility and suppress the 1/f noise [1]. It was reported that the hole effective mobility in p-MOSFET on Si(110) surface is much larger than that on conventional Si(100) surface [2,3]. However, the electron effective mobility is lower and 1/f noise in MOSFETs on Si(110) surface is too large to replace the CMOS fabricated on Si(100) surface [3]. Several efforts are reported to improve the electron mobility in Si(110) devices such as strained silicon [4] and novel accumulation mode device structure [5]. However, there is almost no work reported to improve the mobility by suppressing the surface micro-roughness, which was reported as the main limitation of mobility performance at high field. On the other hand, 1/f noise dramatically increases with the continuous downscaling in device size and is a severe problem for analog and RF applications. A suppression of 1/f noise level is an urgent issue for electronic devices. However, it is not realized using this new chemical surface flattening process.

In this study, firstly we experimentally demonstrate the effect of a suppression of Si(110) surface micro-roughness using HF(0.5wt%)/H_2-UPW solution in N_2 ambience without light illumination. Secondly, we concentrate the performance of n-MOSFETs on Si(110) surface after this new chemical surface flattening process and its mechanism. Finally, the advantage of introduction accumulation mode device to 1/f noise characteristics and its mechanism is described.

II. Experimental

Both the accumulation mode (AM) and the conventional inversion mode (IM) SOI n- and p-MOSFETs on Si(110) surfaces are employed in this experiment. UNIBOND p-type SOI wafers (SOI/BOX=65/100 nm) with resistivities equal 9~18 ohm·cm were used and the SOI layer doping concentration was adjusted by ion implantation before mesa isolation. The doping concentrations in both n- and p-type SOI layers are 1×10^{16} and 2×10^{17} cm^{-3}. An increasing of surface micro-roughness is observed because that Si(110) surface is very easily attacked by the dissolved oxygen, OH⁻ ion in solution and light illumination during surface flattening process and wafer cleaning process. For this reason, a new chemical surface flattening process roughness using $HF(0.5wt\%)/H_2$-UPW solution, in which 1.6mg/L H_2 is added and dissolved oxygen is reduced less than 1ppb, in N_2 ambience without light illumination was employed to suppress the surface roughness and improved 5-steps room temperature wafer cleaning technology without light illumination was employed to avoid the increasing of surface micro-roughness during wet cleaning [6,7]. 7.5nm gate oxides were formed by microwave-excited high-density plasma oxidation (radical oxidation) at 400°C after wet cleaning to realize high-quality oxidation on Si(110) surface [8]. For IM MOSFETs, P^+ and B^+ ions (1.0×10^{16} cm^{-2}) are implanted to gate Poly-Si layer (300 nm) for n- and p-MOSFET, respectively. For AM MOSFETs, P^+ and B^+ ions (1.0×10^{16} cm^{-2}) are implanted to gate Poly-Si layer (300 nm) for p-MOSFET and n-MOSFET, respectively. As^+ and BF_2^+ (1.5×10^{15} cm^{-2}) ions are implanted to Source/Drain region for n- and p-MOSFET, respectively. The process flow is shown at fig. 1 (a) and the schematic of inversion mode and accumulation mode device structure is shown at fig. 1 (b). Figure 2 (a) shows the schematic diagram of the experimental setup and the flow of the new chemical surface flattening process and (b) shows the improved 5-steps room temperature wafer cleaning process flow.

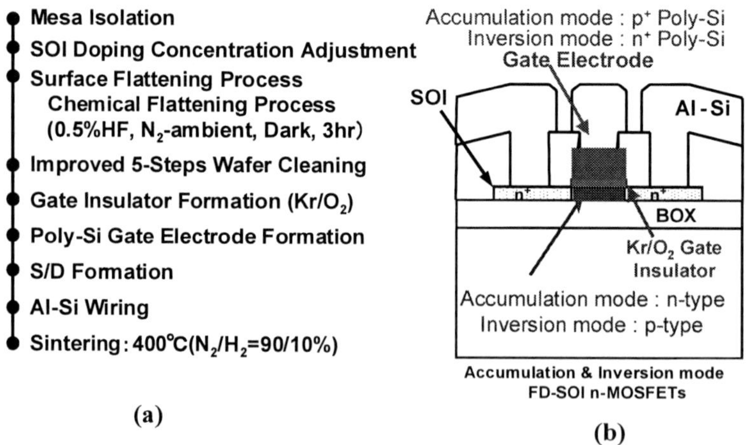

(a)

(b)

Figure 1. (a) Process flow of device fabrication in this experiment. (b) Schematic view of inversion and accumulation mode n-MOSFETs device structure.

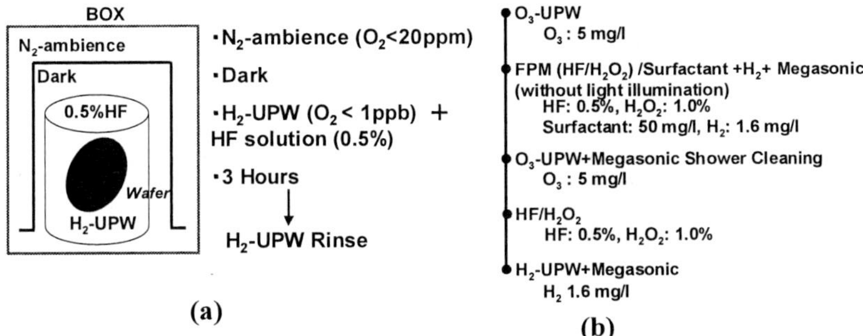

Figure 2. (a) Schematic diagram of the experimental setup and the flow of the new chemical surface flattening process and (b) Improved 5-steps room temperature wafer cleaning process flow.

III. Results and Discussions

A. Si(110) Surface Flattening Process

It has been pointed out that the micro-roughness of Si(110) surface easily increases in the alkaline or neutral solutions originating from the effect of the anisotropic etching by OH⁻ ion on the Si surface and the oxidation of silicon by dissolved oxygen in solutions [6]. It has also been reported that light illumination surprisingly affects the oxidation of silicon and increases surface roughness [7]. In this study, the effect of surface flattening using the new chemical surface flattening process roughness by HF(0.5wt%)/H₂-UPW (O₂<1ppb) solution on Si(110) surface is experimentally revealed. In this experiment, surface flatting process was carried out in N₂ ambience in order to thoroughly reduce the dissolved oxygen below 1ppb in solutions without light illumination and improved 5-steps room temperature wafer cleaning technology without light illumination was employed to avoid a surface micro-roughness induced by oxidation of surface during flattening and wafer cleaning processes.

As a result, an anisotropic etching is observed of Si(110) surface. The etching rate of (100) surface is about 1.5 times of that on (110) surface and very slow. Figure 3 shows the results of initial and the sample treated with chemical flattening process 3 hours. The Ra, rms and P-V values reduce from 0.15 nm to 0.1 nm, 0.18 nm to 0.14 nm and 1.9 nm to 1.5 nm, respectively. There is no groove observed on Si(110) surface in this experiment. We consider that the suppression of the micro-roughness on Si(110) is originated from the slight etching by the small amount of OH⁻ ion in the solution and there is no increasing of roughness caused by oxidation in solutions by adding the H₂ in this solution and the effect of light interception. Commonly, it is very difficult to keep the flatted Si(110) surface at the wafer cleaning process such as RCA cleaning because of the attack from large amount of OH⁻ ion and oxidation in solutions. To keep the flatted surface state, an improved 5-steps room wafer cleaning is employed in this experiment, in which there is only very small amount of OH⁻ ion existed and almost no oxidation attacked during the cleaning process. It is experimental demonstrated that there is not any change before and after the improved 5-steps room wafer cleaning process.

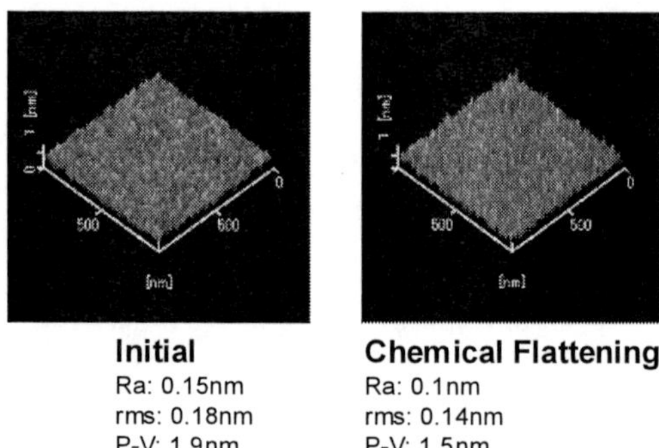

Figure 3. AFM images of initial Si(110) surface and that after 3 hours chemical flattening process.

B. Enhancement of The Device Performance on Si(110) Surface with Flattening Process

Electrical characteristics of MOSFETs on Si(110) surface were evaluated. The channel direction of <100> for n- MOSFETs was measured in this experiment.

Figure 4 (a) shows the linear I_d-V_g characteristics measured at drain voltage equals 0.05V for FD-SOI n-MOSFETs on Si(110) surface. It is obvious that the current drivability of n-MOSFET after chemical flattening process is larger than that of n-MOSFET fabricated on initial Si(110) surface. At the gate bias equals 4V, the drain current is improved about 1.3 times by introducing the chemical surface flattening process. Figure 4 (b) shows the Si(110) orientated FD-SOI n-MOSFETs transconductance characteristics. The peak value of n-MOSFET with chemical surface flattening process is about 1.1 times larger than that fabricated on the initial surface and keeps greater g_m value at higher filed.

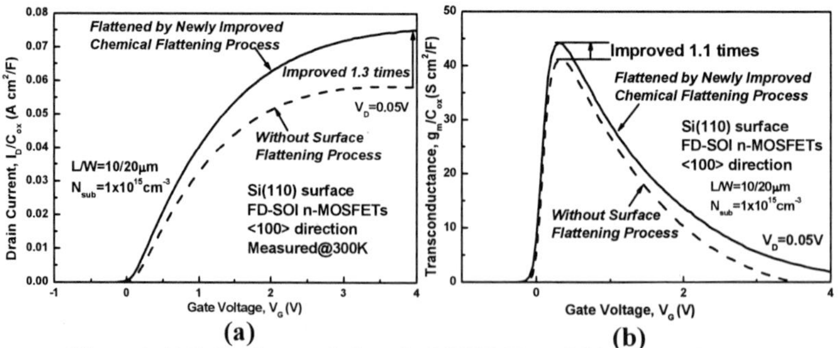

Figure 4. (a) I_d-V_g characteristics of n-MOSFETs and (b) transconductance characteristics on Si(110) surface.

Figure 5 shows the μ_{eff}-E_{eff} characteristics of n-MOSFET on initial Si(110) surface and the flatted surface. It has been reported the mobility in MOS devices at high field is limited by the surface roughness [9]. In this experiment, a suppression of the very small micro-asperity of Si(110) surface is achieved and as a result, a flatted surface is realized. The mobility curve of n-MOSFET on the initial surface agrees that reported by T. Mizuno et al., very well. On the other hand, the mobility of n-MOSFET fabricated on the flatted surface is obviously improved compared with that on the initial surface at the same effective field. The improvement of the effective electron mobility becomes larger while the effective filed increases in this experiment. At the high field equals 1MV/cm dominated by surface roughness, the effective electron mobility is improved about 1.4 times. The slope of the mobility curve is also one of the most important points. It is observed that the slope of the mobility curve of device on flatted surface is especially improved at high filed from the power of −1.2 to −1. It is demonstrated that this new chemical surface flattening process is very effective to improve the effective electron mobility at high field on Si(110) surface.

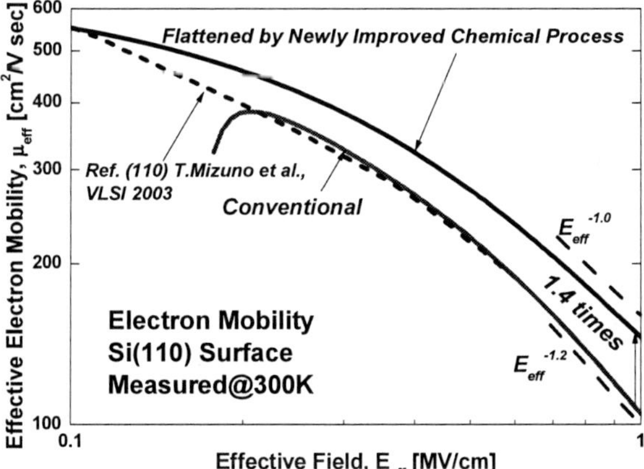

Figure 5. μ_{eff}-E_{eff} characteristics of n-MOSFETs on Si(110) surface.

C. Suppression of 1/f Noise in Accumulation Mode Devices

It is very effective to improve the mobility characteristics using this chemical surface flattening process. However, there is not an obvious improvement of 1/f noise characteristics observed. Generally, the origin of 1/f noise is explained as the fluctuation in the number of carriers caused by interface traps and the fluctuation of conductivity. It has been reported that the effective filed in accumulation mode device is smaller and bulk current in accumulation mode is existed compared with conventional inversion mode devices [5]. The approach of reducing 1/f noise level by introducing the accumulation mode device structure is carrier out in this study. Figure 6 shows 1/f noise characteristics of inversion mode and accumulation mode p-MOSFETs fabricated on Si(110) surface. It is observed that a reduction of noise level about one digit is realized using accumulation mode device. We consider that the 1/f noise suppression of p-MOSFET operated of accumulation mode is originated from the changing of interface traps and the existence of bulk current in this device structure.

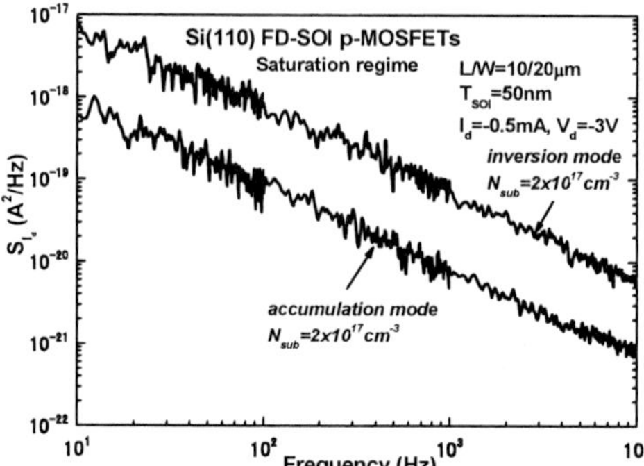

Figure 6. An suppression of 1/f noise level in accumulation mode p-MOSFET compared with conventional inversion mode on Si(110) surface.

IV. Conclusion

The new chemical surface flattening process using HF(0.5wt%)/H$_2$-UPW solution in N$_2$ ambience without light illumination is effective to suppress the micro-roughness of Si(110) surface. The performance of n-MOSFETs on Si(110) surface after this new chemical surface flattening process is obviously improved because of the improvement of the mobility caused by the reducing of surface roughness scattering. Finally, 1/f noise level is reduced about one order of magnitudes by introducing accumulation mode device structure. A high-speed and low-noise device is realized using this improved surface flatten process and novel device structure.

Acknowledgments

This work was conducted as a part of the project under Grant-in-Aid for Specially Promoted Research (project No. 18002004), supported by Japanese Ministry of Education, Culture, Sports, Science and Technology.

REFERENCES

[1] J. Colinge: *IEEE Trans. Electron Devices.* Vol. 45, No. 5, pp. 1010 (1998).
[2] T. Sato, Y. Takeishi and H. Hara: *Phys. Rev. B.* Vol. 4, No. 6, pp. 1950 (1971).
[3] A. Teramoto, T. Hamada, H. Akahori, K. Nii, T. Suwa, K. Kotani, M. Hirayama, S. Sugawa and T. Ohmi: *IEDM Tech Dig.,* pp.801 (2003).
[4] T. Mizuno, N. Sugiyama, T. Tezuka, T. Numata, T. Maeda, S. Takagi: *IEEE Trans. Electron Devices.* Vol. 51, No. 7, pp. 1114 (2004).
[5] W. Cheng, A. Teramoto, M. Hirayama, S. Sugawa and T. Ohmi: *Jpn. J. Appl. Phys.* Vol. 45, pp. 3110 (2006)
[6] T. Ohmi: *J. Electronchem. Soc.,* Vol. 143, No. 9, pp. 2957 (1996).
[7] H. Morinaga, K. Shimaoka, and T. Ohmi: *J. Electronchem. Soc.,* Vol. 153, No. 7, pp. G626 (2006).
[8] T. Ohmi, M. Hirayama and A. Teramoto: *J. Phy. D: Appl. Phys.* Vol. 39, R1 (2006).
[9] S. Takagi, A. Toriumi, M. Iwase, H. Tango: *IEEE Trans. Electron Devices.* Vol. 41, No. 12, pp. 2357 (1994)

ECS Transactions, 6 (4) 107-111 (2007)
10.1149/1.2728848, ©The Electrochemical Society

The "C" Shape Behavior of the Gate Induced Floating Body Effect in function of temperature in PD SOI nMOSFETs

Paula Ghedini Der Agopian [1*], João Antonio Martino[1], Eddy Simoen[2] and Cor Claeys[2, 3]

[1]Laboratório de Sistemas Integráveis -LSI/ PSI/USP - University of São Paulo,
Av. Prof. Luciano Gualberto, trav. 3 n° 158, 05508-900 – SP – Brazil.
*biba@lsi.usp.br
[2]IMEC, Kapeldreef 75, B-3001 Leuven, Belgium
[3] KU Leuven, Electr.Eng. Dept., B-3001 Leuven, Belgium

The temperature influence on the gate induced floating body effect in partially depleted (PD) Silicon-on-Insulator (SOI) nMOSFETs is investigated, through experimental results and two-dimensional numerical simulations. A "C" shape behavior of the transconductance second peak as a function of temperature is observed, due to the competition between two effects: the generation-recombination process and the effective electric field mobility degradation.

Introduction

As MOS devices are scaled down to submicron levels, there is a significant gate tunneling current increase (1,2), since the scaling of gate oxide thickness is not proportional to the bias reduction applied to devices. For devices with gate oxides smaller than 5nm the gate direct tunneling current is predominant (3). This current increases with a nearly exponential dependence on the gate voltage. Although the gate tunneling current is composed of several components, one of the smaller components called electron valence band (EVB) tunneling current, which flows between the gate and the body, is responsible for the appearance of a floating body effect in PD SOI MOSFETs.

With the EVB gate current increases, the floating body is charged and the potential increases exponentially. This body potential increase causes a threshold voltage reduction and results in a higher drain current. This drain current increase can be observed through the second peak in the transconductance (gm) (4).

This floating body effect is called the 'linear kink effect (LKE)' or the "gate induced floating body effect (GIFBE)" and occurs in partially depleted (PD) SOI transistors with floating body or in fully depleted ones with the back interface accumulated.

The aim of this work is to evaluate the GIFBE behavior with the temperature variation in PD SOI nMOSFETs.

Devices Characteristics

The processing characteristics of the studied devices are: gate oxide (t_{oxf}), silicon film (t_{Si}) and buried oxide thicknesses (t_{oxb}) of 2.5 nm, 100 nm and 390 nm, respectively, and a channel doping concentration of 5.5×10^{17} cm^{-3}. The channel width (W) is 10 μm for the measured devices, while for the performed simulations the curves were obtained per unit width, since the simulations are two-dimensional. The channel lengths ranged from 0.3 μm to 5 μm.

Besides fitting the experimental curves, the use of numerical simulations as an auxiliary tool in the further analysis of the experimental results was more fundamental in nature and aiming to get physical insight into the displacement of the transconductance second peak with temperature. Simulations were performed using the Atlas program.

Analysis and Discussions

It is already known that when the temperature increases, the amplitude of the transconductance second peak is attenuated (5). Hence, it tends to disappear with the temperature increase due to a generation-recombination process, resulting in a smaller body potential variation. From figure 1, one can note that the generation-recombination process is not only responsible for the second peak amplitude reduction, but also for the shift of the transconductance second peak threshold (Vt_2) toward a higher gate voltage. This displacement of the transconductance second peak toward a higher gate voltage is desirable to achieve a better control of the transistor. The Vt_2 is obtained from the derivative of the second peak of the transconductance.

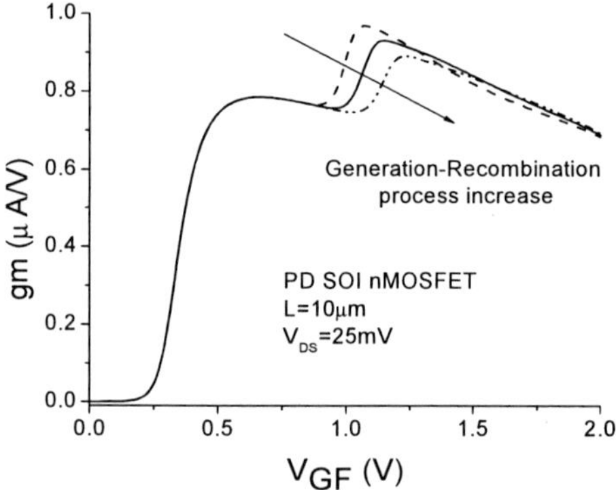

Figure 1. Simulation of the generation-recombination process influence on the GIFBE.

The global behavior of the experimental Vt_2 as a function of temperature from 100K to 450K for a transistor with 0.3 μm channel length is shown in figure 2. As already reported for high temperatures [5], the transconductance second peak amplitude is reduced for a temperature increase. Starting from the measurement data shown in figure 2, it is possible to verify by simulations that also for low temperatures the same behavior is obtained.

One can also note that for high temperatures, as previously explained, the transconductance second peak threshold occurs at a higher gate voltage. However for low temperatures the behavior of Vt_2 is inverse. When the temperature decreases the Vt_2 increases once again.

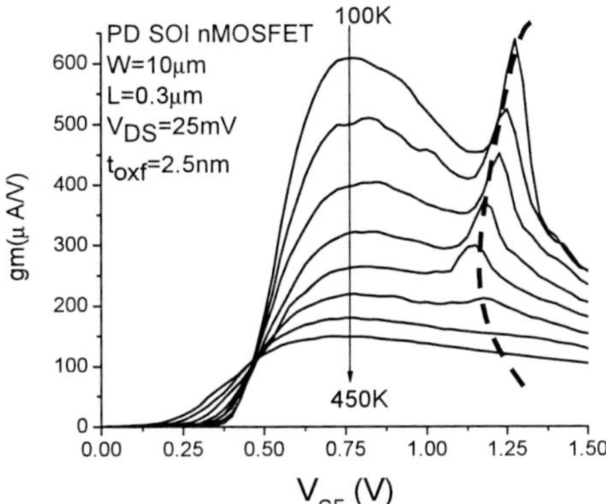

Figure 2. Experimental transconductance behavior as a function of gate voltage for temperatures varying from 100K to 450K.

Transistors from a 0.13μm SOI CMOS technology having different channel lengths, from 0.3μm to 5μm, were measured and it was observed that all devices present the same "C" shape behavior as shown in figure 3.

In order to evaluate the Vt_2 shift with temperature, numerical simulations were performed considering the low field mobility (μ_0) and the electric field mobility degradation (Θ) constants as a function of the temperature. One can note in figure 4 that, neglecting the mobility effects, Vt_2 tends to occur earlier with the temperature decrease even if the temperature is lower than 300K. This effect is only due to the generation-recombination process, which is smaller at lower temperature. The higher body potential variation causes an earlier threshold voltage reduction and consequently a lower Vt_2.

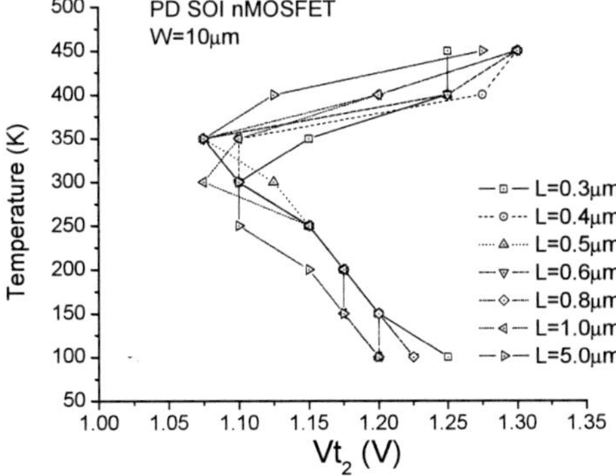

Figure 3: Experimental transconductance second peak threshold as a function of temperature for several channel lengths.

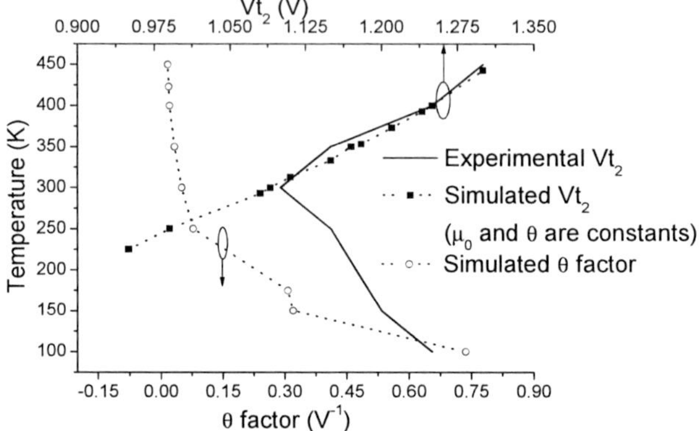

Figure 4: Competing influence of the generation-recombination process and the Θ factor on the Vt$_2$.

The effective mobility degradation with temperature was also extracted experimentally in the range from 100K to 450K. Figure 4 also shows that the Θ factor reduces when the temperature increases.

Evaluating the behavior of the experimental Vt$_2$ one can say that, as the Θ factor for high temperatures is small, the generation-recombination process is predominant, so that the behavior of the simulated Vt$_2$ neglecting the mobility effects is the same as the experimental one. When the temperature decreases, the Θ factor cannot be neglected anymore and its influence becomes higher than the generation-recombination process. As a result, the Vt$_2$ increases again due to the increment of the mobility degradation.

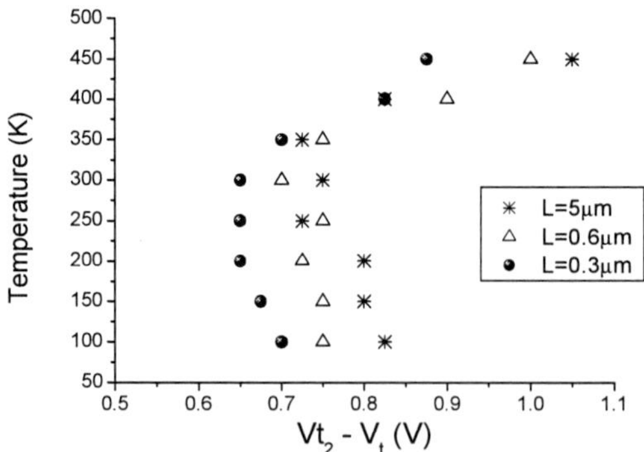

Figure 5: The simulated floating body effect behavior for several temperatures without taking into account the threshold voltage influence.

Due to the threshold voltage (V$_t$) variation with temperature, a study of the GIFBE without the threshold voltage influence was performed. Figure 5 shows the GIFBE threshold overdrive (Vt$_2$-V$_t$) for different temperatures. Although the threshold voltage increases with temperature reduction, one can see that Vt$_2$ − V$_t$ still presents a "C" shape behavior although it is less pronounced at low temperatures.

In order to evaluate the halo influence on the GIFBE, measurements were performed for different structures, with halo and without halo. These measurements were done in a temperature range from 100K to 300K. Table I shows that both structures, with and without halo, present the same behavior, suggesting that the halo has no influence on Vt_2. The results for the longer devices are not shown because the halo does not influence these devices and consequently the results are the same.

TABLE I. Comparison of Vt_2 for structures with and without halo.

	Channel Length	100K	150K	200K	250K	300K
Halo	L=0.3μm	1.25	1.2	1.175	1.15	1.15
	L=0.4μm	1.2	1.2	1.175	1.15	1.1
	L=0.5μm	1.2	1.2	1.175	1.15	1.075
	L=0.6μm	1.225	1.2	1.175	1.15	1.075
Without Halo	L=0.3μm	1.2	1.175	1.175	1.15	1.1
	L=0.4μm	1.2	1.2	1.175	1.125	1.1
	L=0.5μm	1.2	1.2	1.175	1.125	1.1
	L=0.6μm	1.2	1.2	1.175	1.1	1.1

Conclusion

The carrier generation- recombination process increases as the temperature increases, resulting in a transconductance second peak amplitude reduction and the shift of this second peak toward a higher gate voltage. On the other hand, with a temperature decrease, the effective electric field mobility degradation factor increases exponentially. Therefore, the observed "C" shape behavior, that is revealed both experimentally and by simulation, results from a competition between the generation-recombination process and the effective electric field mobility degradation factor (Θ).

The same "C" shape behavior was obtained in case when the temperature variation of the threshold voltage was not taken into consideration.

The halo implantation has no influence on the GIFBE in the temperature range from 100K to 300K.

Acknowledgments

Paula Ghedini Der Agopian and João Antonio Martino thank CNPq for the financial support for execution of this work.

References

1. A. Mercha, J.M. Rafi, E. Simoen, E. Augendre, C. Claeys, *IEEE Transactions on Electron Devices*, vol. 50, p.1675- 1682, no. 7, July (2003).
2. J. Pretet, T. Matsumoto, T. Poiroux, S. Cristoloveanu, R. Gwoziecki, C. Raynaud, A. Roveda and H. Brut, *ESSDERC*. p. 515, (2002).
3. K.F Schuegraf. and Chenming Hu, *IEEE Transaction on Electron Devices*, Vol.41. p.761- 767, no. 5, May (1994).
4. P.G.D.Agopian, J. A. Martino, E. Simoen, C. Claeys, *Microelectronics Journal*, vol. 38, p.114-119, (2007).
5. L.Vancaillie, V. Kilchytska, P.Delatte, L.Demeus, H. Matsuhashi, F.Ichikawa, D.Flandre, *SOI Conference, 2003. IEEE International*, p.78-79, (2003).

ECS Transactions, 6 (4) 113-118 (2007)
10.1149/1.2728849, ©The Electrochemical Society

Hot Carrier Instability Mechanism in Accumulation-Mode Normally-off SOI nMOSFETs and Their Reliability Advantage

R. Kuroda[a,c], A. Teramoto[b], W. Cheng[b], S. Sugawa[a] and T. Ohmi[b]

[a] Graduate School of Engineering, Tohoku University,
6-6-10 Aza-Aoba, Aramaki, Aoba-ku, Sendai, Japan
[b] New Industry Creation Hatchery Center, Tohoku University,
6-6-10 Aza-Aoba, Aramaki, Aoba-ku, Sendai, Japan
[c] JSPS Research Fellow (DC1)

This work studies the degradation mechanism of the hot carrier instability in normally-off Accumulatin-Mode SOI nMOSFETs. From the experimental examination of degradations in transistor-transfer characteristics and low frequency noise level at various transistor operation bias regions, it is proposed that the major degradation in the Accumulation-Mode nMOSFETs is the positive charge and interface state generations at the front gate insulator/Si interface at the Drain side due to the injection of hot holes that are generated by impact ionizations caused by channel hot electrons. It is experimentally shown that the Accumulation-Mode nMOSFETs have higher hot carrier immunity than the Inversion-Mode FD-SOI nMOSFETs. Also, effects of the hot carrier stress to the transistor-transfer characteristics are shown to be smaller for devices with higher dopant concentration of SOI layer (N_{SOI}), which originates from the smaller influence of the quality of the front gate insulator/Si interface to the current drivability in higher N_{SOI} devices.

I. Introduction

The Accumulation-Mode SOI MOSFETs, in comparison to the Inversion-Mode FD SOI MOSFETs, have several advantageous characteristics, such as higher current drivability, lower low frequency noise and suppressed kink effect, and it has been proposed that this device gives analog, digital and their mixed circuits better performances (1-3). These advantages become larger as dopant concentration of SOI layer (N_{SOI}) is increased (2,3). However, there have been only small amount of works reported on the reliability issues of this device (4,5). In this paper, we study the degradation mechanism of hot carrier instability in Accumulation-Mode nMOSFET. Low frequency noise analysis is incorporated in addition to the transistor transfer characteristics degradation analysis to determine the major degradation and its mechanism caused by hot carrier stress. Devices with different N_{SOI} are studied to quantitatively analyze the correlation between N_{SOI} of the device and the effect of the degradations. Finally, the comparison of hot carrier immunities between Accumulation-Mode and Inversion-Mode nMOSFETs is made.

II. Experimental

Devices used in this work are Accumulation-Mode and Inversion-Mode FD SOI nMOSFETs on bonded SOI wafer. Gate geometry is W/L=20.0/0.6 μm. Thickness of the SOI and BOX layers are 50 nm and 100 nm, respectively. The front gate insulator film is 7.5 nm-thick radical oxidized SiO_2 film (6). Devices with two N_{SOI} are prepared for both cases, 1×10^{16} and 2×10^{17} cm^{-3} n-type for Accumulation-Mode, and 1×10^{16} and 2×10^{17} cm^{-3} p-type for Inversion-Mode nMOSFETs, respectively. The devices do not contain the LDD structure. Schematic views of the Accumulation-Mode and Inversion-Mode nMOSFETs operating at the saturation region are shown in Figs.1 (a) and (b), respectively. In the Accumulation-Mode MOSFETs, the voltage drop at the Drain terminal is due to MOS capacitive coupling at the Drain side instead of the pn-junction capacitive coupling in the case of the Inversion-Mode MOSFETs (1). Hot carrier stress is applied for every device at (V_G-V_{th})=0.5 V and V_D=3.5 V for 10^4 second. The back substrate is grounded during the stress and following all the measurements.

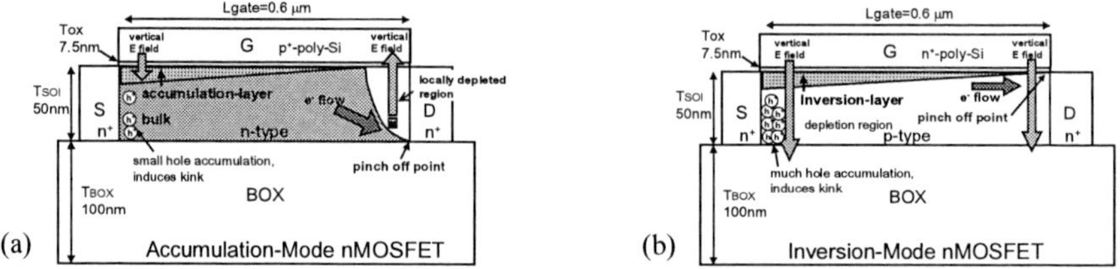

Figure 1. Schematic views of (a)Accumulation-Mode and (b) Inversion-Mode nMOSFETs operating at the saturation region.

III. Results and Discussions

A. Hot Carrier Degradation Mechanism

Figure 2 shows the reduction of drain current (I_D) in linear region for the stressed four devices. Firstly, Accumulation-Mode devices show smaller degradation than that of Inversion-Mode devices. For the Inversion-Mode devices, degradation is larger in the higher N_{SOI} device, which originates from the higher lateral and vertical electric field at the pinch-off point in the higher N_{SOI} device. On the contrary, for Accumulation-Mode devices, the degradation is slightly smaller in the higher N_{SOI} device.

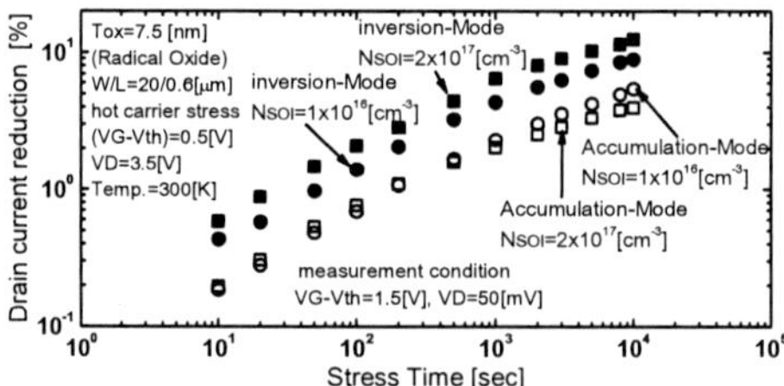

Figure 2. Reduction in I_D (in linear region) as functions of stress time for Inversion-Mode and Accumulation-Mode nMOSFETs.

Figure 3 shows the I_D-V_D characteristics for Accumulation-Mode nMOSFET with N_{SOI} equals to 2×10^{17} cm⁻³ before and after stress, where (a) shows the results when Source and Drain terminals during the measurement are same as these during the stress, and (b) shows the results when Source and Drain terminals during the measurement are reversed form these during the stress. There is a clear difference in the measured characteristics after stress between (a) and (b) at the saturation region where V_D > 2.5 V. It is the region where the kink effect starts to appear due to the floating body effect induced by hole accumulation at the Source side during the measurement. In (a), the kink effects before and after stress are the same, however in (b), kink effect is suppressed after stress compared to before stress. This result indicates that there are stress induced defects at the Drain side during the stress that works as the recombination centers for holes during the measurement in (b).

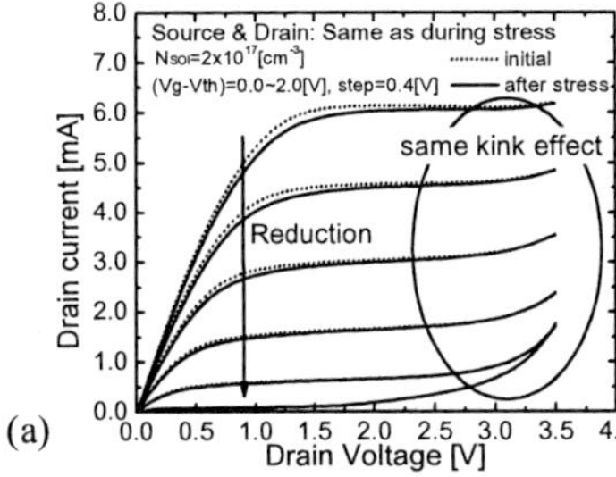
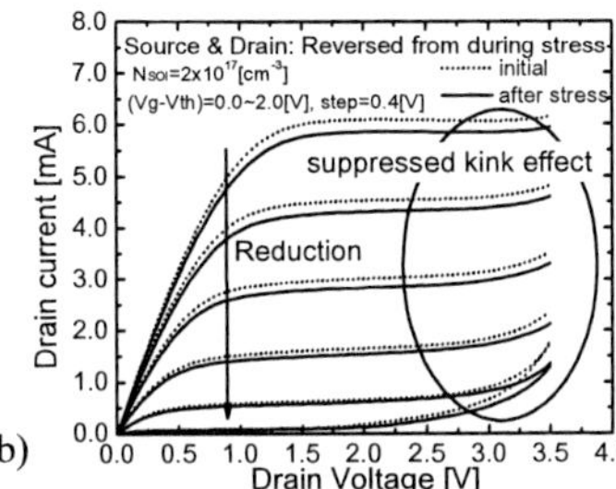

(a)   (b)

Figure 3. I_D-V_D characteristics before and after the hot carrier stress for 10^4 second. The measured device is Accumulation-Mode nMOSFET with N_{SOI}=2x10^{17} cm^{-3}. (a) when Source and Drain are the same as these during the stress, and (b) when Source and Drain are reversed form these during the stress.

Then to investigate the exact part where the defects are induced by the hot carrier stress, low frequency noise measurement is incorporated. And the correlation between the carrier conduction path and the stress induced low frequency noise increase is clarified as follows. Figure 4 shows the low frequency noise measurement results before and after the hot carrier stress, where (a) and (b) are the results measured in the linear region and (c) and (d) are the results measured in the saturation region. (a) and (c) are measured when Source and Drain are the same as these during the stress, and (b) and (d) are measured when Source and Drain are reversed from these during the stress. Schematic views of the carrier conduction are also shown in each figure.

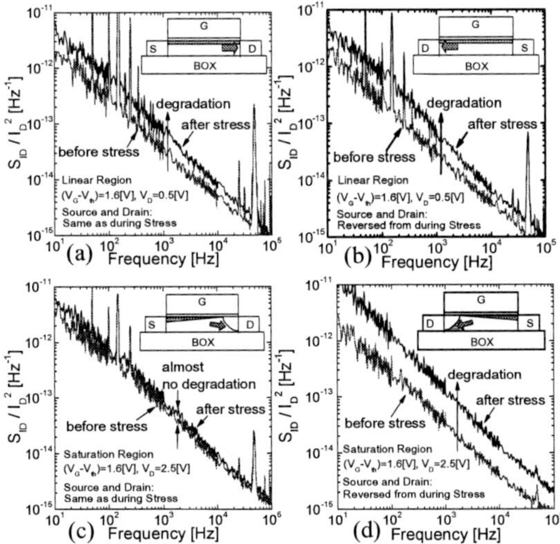

Figure 4. Low frequency noise measurement results, where (a) and (b) are measured in a linear region, and (c) and (d) are measured in a saturation region. In (a) and (c), Source and Drain are same as these during stress, and in (b) and (d), they are reversed from during stress. The electron flow path for each case is also shown in each figure.

In Fig. 4 (a) and (b) (linear region), stress induced low frequency noise increases is clearly observed and the results are symmetrical between (a) and (b). However in Fig. 4 (c) and (d) (saturation region), the low frequency noise increases are different, where almost no increase is observed in (c) although clear increase is observed in (d). Figure 5 is the summary of the low frequency noise increase at 10 Hz measured in the various bias points. Knowing that the low frequency noise characteristic is very sensitive to the interface states on the carrier conduction path, these results indicate that the hot carrier stress in the Accumulation-Mode nMOSFET induces the interface states at the front insulator/Si interface at the Drain side during the stress.

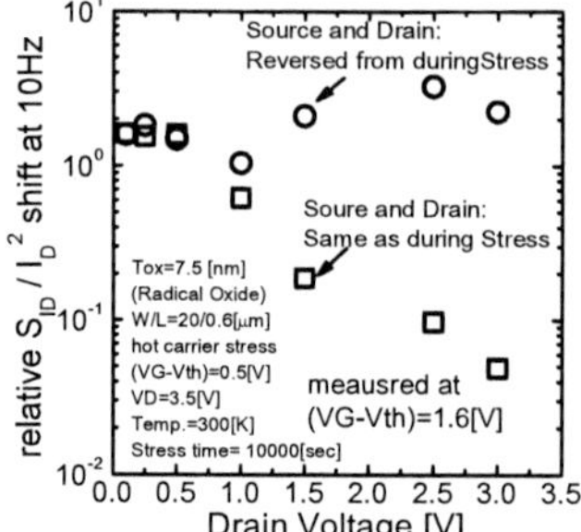

Figure 5. Summary of the low frequency noise shift after stress measured when Source and Drain are same or reversed from these during stress. The result clearly shows the difference at the saturation working region, indicating that the unsymmetrical degradation occurs at the front interface at the Drain side during the stress.

Based on the experimental results, hot carrier induced degradation mechanism in Accumulation-Mode nMOSFETs is proposed as follows, which is schematically explained in Fig. 6.

1. At pinch-off point, channel hot electrons appear.
2. Impact ionizations are caused by channel hot electrons.
3. Hot holes are generated by the impact ionization.
4. Generated hot holes are drained toward the front gate insulator/Si interface due to the vertical electric field in the locally depleted region at the Drain side, and create states.

Since the generated hot holes are the secondary product of the channel hot electrons, the number of the hot holes is much smaller than the channel hot electrons. Consequently, as it is shown in Fig. 2, the hot carrier degradation is smaller for Accumulation-Mode nMOSFETs than for Inversion-Mode nMOSFETs in which degradation is mainly caused by the channel hot electrons.

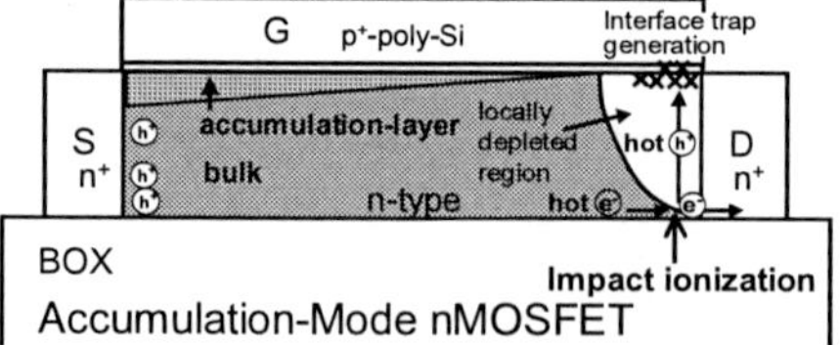

Figure 6. Schematic view of the major hot carrier degradation mechanism in Accumulation-Mode nMOSFETs. Hot holes generated by the impact ionization by channel hot electron create interface states and positive fixed charges at the front insulator/Si interface at the Drain side.

B. Effect of the SOI Dopant Concentration

Figure 7 shows the measured I_D-V_G characteristics and their 2nd derivatives of the Accumulation-Mode nMOSFETs with two N_{SOI}. In Fig. 7 (a), in which N_{SOI} is 2×10^{17} cm^{-3}, dopant concentration is so high that the subthreshold current is composed only of the bulk current that is controlled by the front gate terminal. In this device, bulk current still occupies more than 20 % of the total current drivability after the accumulation-channel is created. However in (b), in which N_{SOI} is 1×10^{16} cm^{-3}, subthreshold and the total ON currents is mostly composed of the accumulation-channel current.

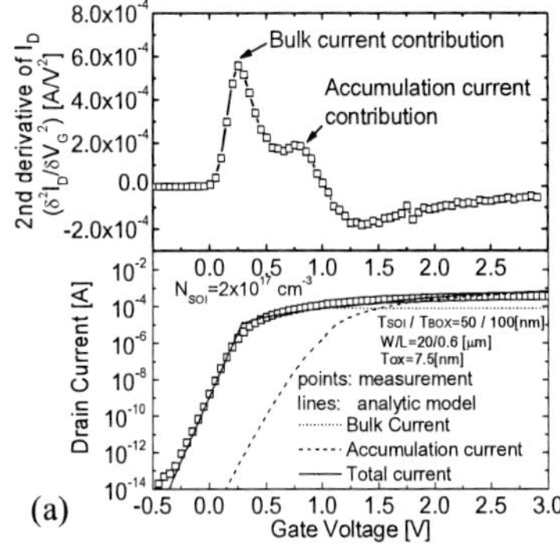

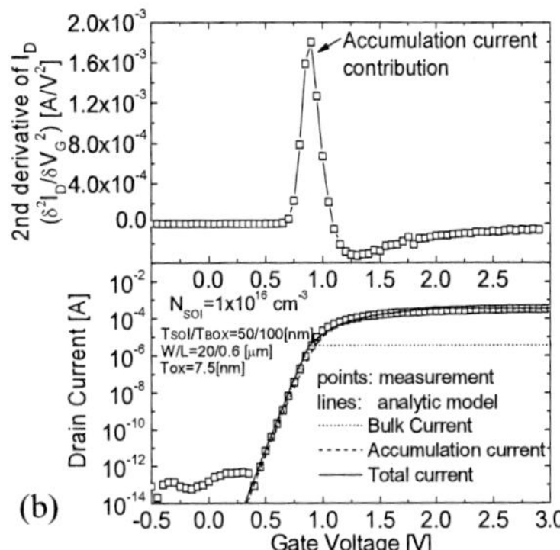

Figure 7. Measured subthreshold current characteristics and their 2nd derivatives of the Accumulation-Mode nMOSFETs with (a) N_{SOI} is 2×10^{17} cm^{-3}, and (b) N_{SOI} is 1×10^{16} cm^{-3}.

Figure 8 shows the linear I_D reduction of the stressed nMOSFETs as functions of the various gate bias conditions. Accumulation-Mode nMOSFETs show lower degradation in current drivability than Inversion-Mode nMOSFETs as discussed in the previous subsection. For Accumulation-Mode device with higher N_{SOI}, reduction is lower than that with lower N_{SOI} especially at the low gate bias region where the bulk current dominates the total current conduction for higher N_{SOI} device. This result indicates that for the Accumulation-Mode devices, the effect of the hot carrier stress induced degradation, which is revealed as interface state creation at the front insulator/Si interface, becomes smaller for higher N_{SOI} device, although the amount of the interface state creation should be higher for the higher N_{SOI} device because of the higher lateral and vertical electric field at the Drain side.

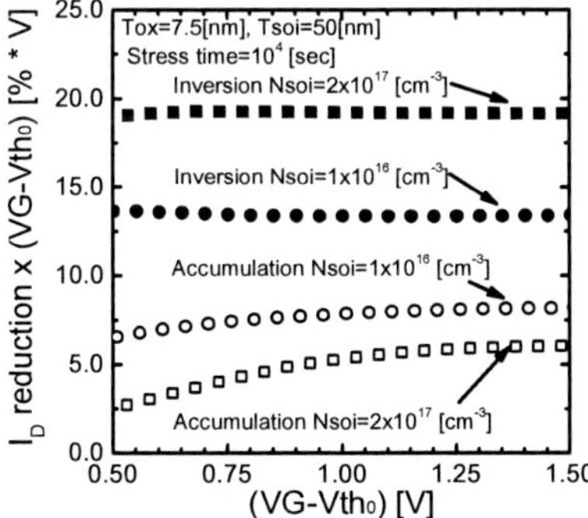

Figure 8. Linear I_D reductions for stressed devices at various gate bias points. Accumulation-Mode nMOSFET with higher N_{SOI} shows lower reduction than that with lower N_{SOI}.

Figure 9 shows the threshold voltage shift (ΔV_{th}) and subthreshold swing factor degradation (ΔS-factor) by the stress for Accumulation-Mode devices. For higher N_{SOI} device, it is shown that subthreshold characteristics do not change much by the stress, indicating that there is only small amount of the degradation at the back insulator/Si interface even though this region is near the pinch-off point. However, this result is reasonable because the vertical electric field that drains the channel hot electrons toward the back interface is almost zero at the Drain side. However in the lower N_{SOI} device, interface generation at the front interface significantly degrades the subthreshold characteristics. Consequently, the effect of the hot carrier stress to the transistor transfer characteristics is smaller for higher N_{SOI} devices in the Accumulation-Mode nMOSFETs.

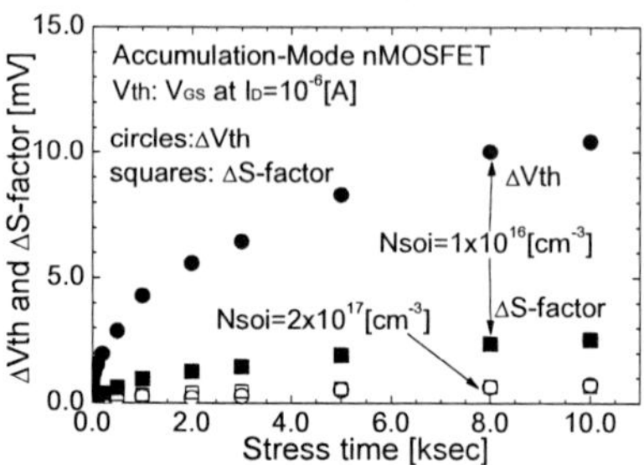

Figure 9. V_{th} shift and S-factor degradation by hot carrier stress in Accumulation-Mode nMOSFETs. Higher N_{SOI} device shows very small degradation because subthreshold is dominated by the bulk current.

IV. Conclusion

In the Accumulation-Mode nMOSFETs, unlike in the Inversion-Mode SOI nMOSFETs, the major degradation induced by hot carrier stress is the positive charge and interface state generations at the front insulator/Si interface at the Drain terminal side caused by hot holes. Since the hot holes are the secondary product of the impact ionization induced by the channel hot electrons, the magnitude of the degradation is smaller for Accumulation-Mode nMOSFETs than that of Inversion-Mode nMOSFETs as experimental results show. Furthermore, since the magnitude of the degradations in the transistor transfer characteristics is smaller for devices with higher dopant concentration of the SOI layer, trade-offs between the transistor performances improvements by increasing dopant concentration of the SOI layer and the hot carrier reliability concern do not arise.

Acknowledgments

This work was conducted as a part of the project under Grant-in-Aid for Specially Promoted Research (project No. 18002004), supported by Japanese Ministry of Education, Culture, Sports, Science and Technology.

REFERENCES

1. J. Colinge, *IEEE Trans. Electron Devices*, **37**, 718 (1990).
2. W. Cheng, A. Teramoto, M. Hirayama, S. Sugawa and T. Ohmi, *Jpn. J. Applied Physics*, **45**, 3110 (2006).
3. W. Cheng, A. Teramoto, P. Gaubert, M. Hirayama and T. Ohmi, *Intl. Conf. Solid State and Integrated Circuits Technology*, 65 (2006).
4. A. Acovic, L. K. Wang, F. Brady, and N. Haddad, *SOI Conference proceedings*, 134 (1992).
5. F. L. Duan, S. P. Sinha, D. E. Ioannou and F.T. Brady, *IEEE Trans. Electron Devices*, **44**, 972 (1997).
6. T. Ohmi, M. Hirayama and A. Teramoto, *J. Physics D. Applied Physics*, **39**, R1 (2006).
7. M. Toita, S. Sugawa, A. Teramoto, T. Akaboshi, H. Imai and T. Ohmi, *Intl. Reliability Physics Symposium proceedings*, 313 (2003).

SESSION 3

TFTs, PHOTONS, PHONONS, AND ELECTRONS

ECS Transactions, 6 (4) 121-132 (2007)
10.1149/1.2728850, ©The Electrochemical Society

Thin Film Transistor and ULSIC Technologies - Parallel or Crossing?

Yue Kuo

Thin Film Nano and Microelectronics Research Laboratory
Texas A&M University, College Station, TX 77843-3122

In the last 2 decades, the thin film transistor (TFT) technology has advanced dramatically accompanied with extraordinary growth of market. In this paper, the author compared the history of TFTs and ULSIC as well as their markets. The unique material, process, and device properties of the a-Si:H TFT was responsible for its current success and established the foundation for developments beyond liquid crystal displays (LCDs). Due to different device characters and limits, TFTs and ULSIC have separate application areas. However, they can be properly integrated to create new functions and products. The knowledge obtained from the large-area TFT array studies can help ULSIC migration to 15-inch wafers.

Comparison of Development History of ULSIC and TFTs

Figure 1 shows a short summary of some important developments in the history of IC and TFTs (1-3). TFT is a field-effect transistor (FET) usually fabricated on glass or an insulating material. Therefore, it is a SOI device. Although the concept of the MISFET was developed in 1925, it took about 20 years to demonstrate the feasibility of a solid state transistor and another 20 years to develop into a tremendous industry. Similarly, approximately 20 years after the demonstration of the first CdS TFT, the amorphous silicon (a-Si:H) TFT was invented, which is the foundation of today's immense active matrix (AM) LCD industry. Today, MOSFETs have been used in varieties of logic, memory, sensing, imaging, etc. products. However, TFTs are exclusively used in displays except a small number of the medical imagers (3).

Figure 2 shows the evolution of the sizes of the IC and TFT LCDs (3,4). Both industries share the same thought of enlarging the substrate size to increase the production throughput and to lower the cost. For the IC industry, it takes 10 to 12 years to switch to a new generation of wafers. The next generation of 15" wafers will probably be introduced into production around 2011-2013. However, the first generation of glass substrates for a-Si:H TFT LCDs was introduced in 1990 with the size around 12"x16". Therefore, from the substrate size point-of-view, the a-Si:H TFT industry is 20 years ahead of ULSIC. While the wafer size in the IC industry is standardized, the glass substrate size in the TFT LCD industry has never been standardized. Each major manufacturer has its own specification. The newest generation of glass substrate size is about 2.1 m by 2.4 m, which will probably be introduced into production by 2008.

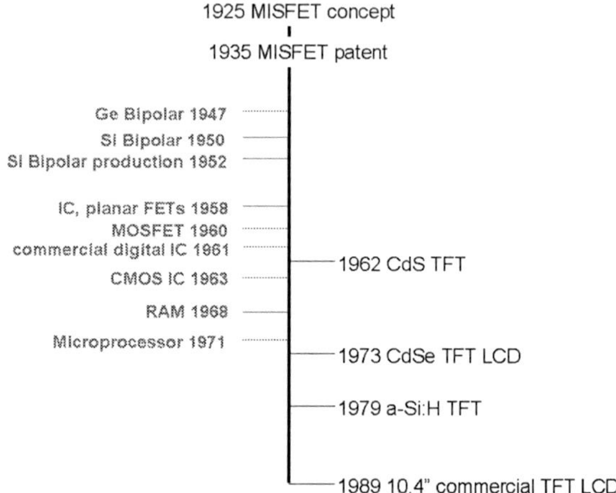

Fig. 1. Progresses of ULSIC and TFT Technologies (1,2,3).

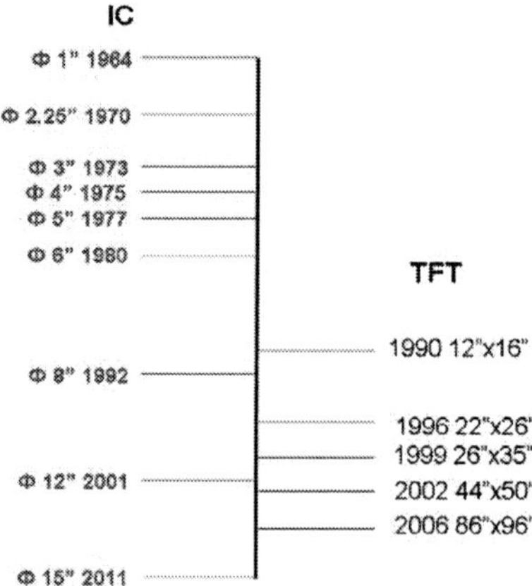

Fig. 2. Substrate size evolutions of IC and TFT LCD (3,4).

As shown in Figure 3 (5,6), in spite of the difference in technology considerations, IC and LCD industries experience dramatic growth starting in 1970 and 1990, separately. The former time coincides with the production of microprocessors and memories. The latter time is contributed by the including of a-Si:H TFTs into panels. Since the cost of the TFT array fabrication is about 20% that of the whole TFT LCD, strictly speaking, the 2005 market value of TFT alone should be about $10B, which is about 5% of IC market. Judged from the IC vs. TFT production history, i.e., 35 years vs. 15 years, and the range of product applications, i.e., multi- vs. single-purpose, the TFT industry is a great success.

Discussions in this paper are focused on a-Si:H TFTs and polycrystalline silicon (poly-Si) TFTs. TFTs can also be made from other types of semiconductor materials, such as organic molecules, transparent oxides, or compounds, with various characteristics. The discussion of these devices is beyond the scope of this paper.

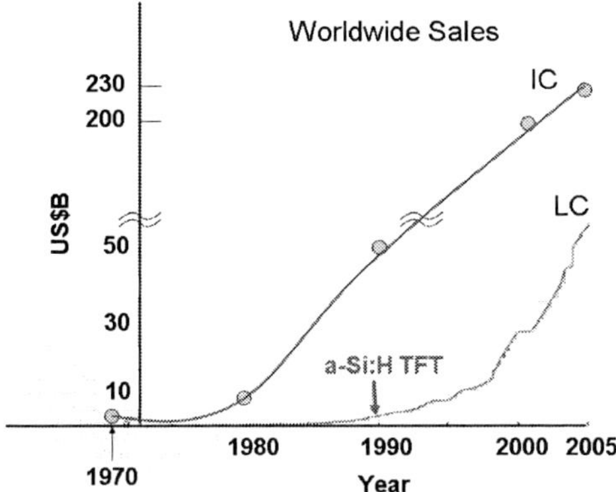

Fig. 3. Worldwide sales of IC and TFT LCD (5,6).

Technology Uniqueness of a-Si:H TFTs

Since the early stage of the TFT development, its application was clearly focused on flat panel displays, specific as the pixel driving device for LCDs. Therefore, the device performance was geared toward this goal. Among all available solid state devices, the a-Si:H TFT was chosen because its fabrication process is adequate for

1) large-area substrates,
2) low process temperatures, and
3) high throughputs.

There are many unique scientific and engineering issues in the large-area TFT array fabrication, which are difficult to extrapolate from the conventional IC fabrication data. For example, since all thin film layers of the TFT are prepared by deposition, there is great flexibility in selecting the composing materials, e.g., organic, inorganic, high-k, or low-k films. Almost all types of transistor structures have been tested on TFTs (3). Furthermore, the plasma process provides freedoms in modifying bulk, surface, and interface properties of the film. There are a large number of literatures discussing materials, processes, and devices of TFTs (3,7). In spite of its higher field-effect mobility, the poly-Si TFT is available only to limited number of large-area LCDs because its fabrication process is difficult to satisfy all above three requirements. However, due to the lack of the substrate limitation, i.e., no single crystal silicon wafer is required, the poly-Si TFT can be prepared with different structures, materials, and modification methods.

Table 1 list some of our research results on the a-Si:H TFT area, which are just examples to show the technology uniqueness.

Table 1 Examples of unique a-Si:H TFT technology issues

Tech Area	Topics
Materials	• Low-defect thin a-Si:H channel layer with a high H content, low SiH_2 component, and low surface roughness. A dual- or multi-layer a-Si:H film, which contains separate interface and bulk layers, may be used to achieve good device performance and high throughputs (8-11). • Slightly nitrogen-rich SiN_x gate dielectrics with a high H content, e.g., 30%. A dual-layer dielectric structure, which contains a SiN_x interface layer and a bulk dielectric with a different composition or material, is often used for better device characteristics and yield (12,13,14). • Highly conductive heavily doped n^+ amorphous or microcrystalline films (15). • Highly conductive and transparent conductor for pixel electrode, such as ITO (16).
Processes	• A generalized deposition-etching mechanism in PECVD a-Si:H, SiN_x, and n^+ deposition processes (17-20). • Interface damages due to excessive etching mechanism in a-Si:H deposition process (21). • Large-area PECVD SiN_x thin film deposition process, i.e., rate, thickness uniformity, composition, stress, and particles, related to plasma power due to competitive deposition and etching mechanisms (19,22,23). • n^+ nanocrystalline phase formation due to hydrogen enhanced etching mechanism (14,24, 25). • RIE sloped interconnect line etch, which is plasma phase chemistry and physics dependent, critical to subsequent thin film step coverage and yield (26,27). • RIE of ITO based on balanced removal of In and Sn components (28,29,30). • Copper, which can be etched with a plasma-based, room-temperature process, for gate, source, and drain electrodes of the TFT with minimum concern of diffusion into gate SiN_x under ordinary operation conditions (31-34).
Devices	• A generalized relationship between the threshold voltage of the a-Si:H TFT and its gate SiN_x dielectric property, which greatly reduces the number of experiments to obtain optimized TFT characteristics (19,22,35). • a-Si:H/gate SiN_x interface damages related to dangling bond formation (21). • Plasma generated short wavelength radiation damages to bulk, and interface of an a-Si:H TFT and repairing method (36,37). • CMOS-type a-Si:H TFTs fabricated at $\leq 300°C$ (38).
Structures	• Complete self-aligned, inverted, staggered tri-layer TFT fabricated using 2 photomasks with large process windows and excellent device characteristics (39,40). • Vertically redundant, horizontally redundant, or multi-channel TFT

<table>
<tr><td></td><td>structures with improved on-current or negligible photosensitivity demonstrated (41,42,43).
• Floating gate a-Si:H TFT nonvolatile memories including an embedded thin a-Si:H film in the gate dielectric structure demonstrated (44).
• Bipolar a-Si:H TFTs with a large current gain successfully prepared at 300°C (45).</td></tr>
</table>

TFT Technology Development Trends

The current success of the TFT technology is due to the large-area AM LCD product. In the near future, the value of this product will increase continuously because the tremendous demand of the consumer market. The industry a-Si:H TFT research and development activities are focused on productivity enhancement, such as increasing the throughput or decreasing the cost. The most common topics include mask number reduction, device and interconnect dimension shrinkage, high-conductivity interconnect line preparation, pixel aperture ratio enlargement, yield and reliability enhancement, etc.. The poly-Si TFT activities have been focused on decreasing the amorphous-to-polycrystalline silicon crystallization temperature, increasing the large-area fabrication capability and throughput, extending device and panel reliabilities, including pixel TFTs and periphery driving circuit on the same substrate, etc.. It is desirable to fabricate TFTs on the flexible substrate for easy transport and storage.

There are many new TFT research activities toward non-LCD applications. One successful example is the medical x-ray imager that is based on the principle of using the a-Si:H TFT to address the pixel composed of a pin diode underneath an x-ray scintillation layer (46). TFTs have tremendous advantages over other solid state devices in selecting materials, structures, fabrication methods, and substrates. For example, Figure 4 shows two types of TFT structures: the top one is exclusively used in a-Si:H TFT and the bottom one is used in poly-Si TFT. By controlling the individual layer's composing material, structure, or contact method, new functions can be created with the TFT alone or including the attached device. Table 2 lists examples of some of these novel devices. Most applications are in sensing a wide range of chemical (gases and liquid), biological, medical, optical, and magnetic properties. TFTs have also been made into solid state memories.

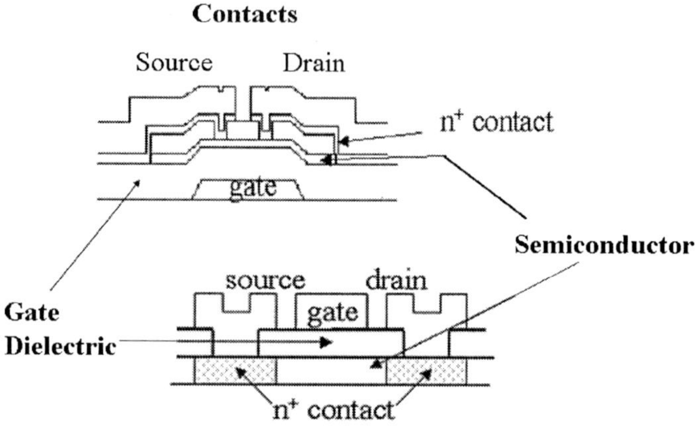

Fig. 4. (a) Inverted, staggered tri-layer and (b) coplanar TFT structures.

Table 2 Non-LCD TFT applications with respect to unique materials or structures (47)

Area	Function	Principles of operation
Gate dielectric materials and structures	pH sensing	H^+ adsorption in suspended gate dielectric structure
	Memory	PZT gate dielectric in poly-Si, SnO_2, or organic TFT
Semiconductor material	Gas sensing	Humidity, alcohols, N_2O adsorption on organic semiconductor layer
	IR detection	I_d sensitive to temperature in a-Si:H TFT
Gate electrode	Gas sensing	H_2 decomposition on Pd gate electrode
	Bio sensing	Biomolecule reaction with specific penicillinase or ssDNA on gate electrode
S/D electrodes	Protein/DNA sensing	Contact resistance increase (due to biomolecule adsorption) caused transfer characteristics change
	Artificial retina	Photo conductivity of a-Si:H layer attached to a-Si:H TFT
	X-ray imaging	Scintillation layer on diode attached to S/D electrode
TFT structures	Photo sensing	I_{ph}/I_{dark} ratio enhanced by split or offset gate TFTs
	Memory	floating gate a-Si:H or poly-Si TFT
	Magnetic sensing	Hall effects due to additional electrodes

Will the TFT Technology Compete or Compliment with ULSIC Technology?

Since the early development stage of the large-area TFT LCD, it has been desirable to fabricate circuit drivers with the same TFTs used to drive the pixel LC. This kind of approach has many advantages over the conventional IC drivers in areas such as yield, cost, reliability, size, and process simplicity.

Although poly-Si TFTs can be fabricated with many high- or low-temperature methods, currently, only the low-temperature processes are compatible with the low-cost glass substrate (3). Conventionally, the low-temperature poly-Si TFTs suffer from the low field-effect mobility, large leakage current, unstable threshold voltage, and kinks. These problems have been solved through the proper control of crystallization process, grain locations, grain boundary passivation, transistor structure, driving methods, etc. (3). Among all low-temperature poly-Si TFT fabrication processes, the laser crystallization method gives the best result due to its high field-effect mobility and excellent grain structure. With the laser crystallization method, the low-voltage TFT driver circuits were successfully made at the periphery of the display while the high-voltage TFTs were used in the pixel area (48). Furthermore, a "Full-Functional Panel," which includes a CPU, an audio circuit, an image-processing circuit, the image data program ROMs, audio ROMs, various RAMs, an electronic power supply circuit, a clock generation circuit, and the LCD on the same piece of glass, was also demonstrated (49). In another report (50), the "panel-size drivers," which contains driver circuits on a piece of panel-length or –width glass, was fabricated. In principle, the laser crystallized low-temperature poly-Si TFT circuit can be fabricated on a flexible plastic substrate.

In the past few decades, ULSIC and TFTs were developed separately with each of them focused on specific types of applications. This resulted in many unique devices and processes. However, with the rapid progress of the TFT technology, it is natural to raise the question on whether the TFT technology is competitive or complimentary to the ULSI technology.

Table 3 lists the material, process, and device properties of the a-Si:H TFT, poly-Si TFT, and single crystal Si MOSFET.

Table 3 Comparison of Materials, Process and Device Properties of TFTs and MOSFETs

	a-Si:H TFT	Low-Temp. Poly-Si TFT	MOSFETs
Semiconductor material	Amorphous film full of defects	Grain and grain boundary defectives	Single crystal with almost no defects
Gate dielectric	High-k with SiN_x interface	SiO_2	SiO_2
Gate	Refractory metals, clad Al or Cu, etc.	Poly-Si or metals	Poly-Si or polycide
Doping method	PECVD n^+ film	Ion implantation or laser	Ion implantation
Process temp.	350°C	500°C	1100°C
Substrate	Glass, plastic, etc. almost no size limit	Glass with high temperature resistance	Si wafer with limited size
Mobility	1 cm^2/Vs	10-200 cm^2/Vs	500 cm^2/Vs
Operation	High V	Low or high V	Low V

ULSIC is a well studied technology with established manufacturing processes. Although the individual poly-Si TFT fabricated from the laser crystallization method can have a field-effect mobility as high as that of the MOSFET, the process is complicated and less mature than that of ULSIC. Therefore, for the high-speed, high-performance applications, MOSFETs will be the dominate device. However, for applications where the substrate material, size, or cost prohibits the utilization of wafers, the low-temperature poly-Si TFT is a preferred choice. The a-Si:H TFT is also a well investigated, mass-production technology. In spite of its low field-effect mobility, the a-Si:H TFT technology has advantages of low process temperatures, flexible structures, and lack of substrate restrictions. It is an ideal device used in combination with other devices for non-conventional applications, as shown in Table 2.

ULSIC and TFT technologies can be complimentary in many applications. The large-area TFT array fabrication experience is especially helpful for ULSIC fabrication on 15-inch wafers. For example,

- Applications. There are many applications that can take advantages of unique characteristics of MOSFETs and TFTs. For example, it has been proven that poly-Si TFTs could replace highly resistive resistors in SRAMs to reduce the leakage current and noise (51). Another possible application is to embed poly-Si TFTs into the interconnect structure of ULSIC to achieve the 3D IC structure, as shown in Figure 5 (52). A major advantage of this kind of 3D ULSIC is that the number of transistors in a chip can be increased drastically using the existing technologies without waiting for the availability of certain critical tools, such as the next generation lithography stepper. Since the three electrodes of the TFT can be made of the same material as the interconnect line, only extra Si, doped Si, and gate dielectric layers need to be added to complete the TFT. The TFT process needs to be compatible with the back-end-of-the-line, e.g., the process temperature should decrease with the increase of the number of the layer. For example, a high-temperature process is used for the bottom layer TFTs, a low-temperature crystallization process is used for the higher level TFTs, and a milder process, e.g., metal-induced crystallization method, is used for even higher level TFTs. Before this kind of chip is operable, there are many engineering problems to solve, such as the compatibility of the TFT process and the interlevel dielectric materials, the reliability of the device, and the dissipation of heat.

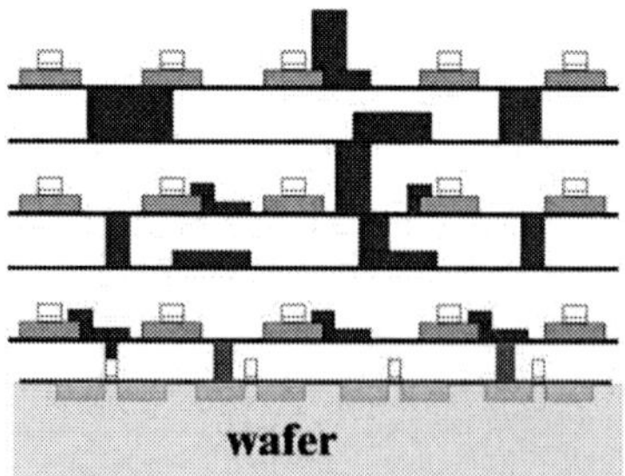

Fig. 5. 3D IC including TFTs in the interconnect structure of ULSIC (52).

Another example is to combine ULSIC with an a-Si:H photoresistor, such as a gate offset a-Si:H TFT, located on the top layer of the chip. The a-Si:H photoresistor can sense the light intensity at the surrounding environment. The logic, memory, or calculation function of the ULSIC part of the chip can be programmed with respect to the change of the ambient light to save energy or for specific operations. There are many possibilities of fabricating such kind of "smart" chips by making the attached a-Si:H TFT into a chemical, biological, photonic, magnetic, etc. sensor.

- Fabrication processes. The ULSIC fabrication processes can benefit greatly from the large-area TFT array fabrication experience. For example, we previous reported a generalized relationship on a large-area PECVD SiN_x thin film deposition process that correlated the plasma power to deposition rate, uniformity, refractive index, stress, and plasma phase particle generation (17,18,19). This kind of relationship, which can be explained by the plasma deposition-etching mechanism, is unknown to the ULSIC field because there was no need of preparing circuits over a wafer larger than 12 inches.

Another example is the multi-layer PECVD a-Si:H/SiN_x thin film deposition process and related damages. In order to warrant the consistent performance of all a-Si:H TFTs across the display panel, thin films, e.g., 50 nm for a-Si:H and 300 nm for SiN_x, need to be deposited uniformly across the whole glass substrate, e.g., > 1 m x 1m. The interface between the a-Si:H and SiN_x can be damaged by a physical or chemical mechanism if the deposition condition is improper, which causes TFT's electrical deficiencies (21). Hydrogen released in the plasma can contribute to the damage mechanism and can even influence the composition and properties of a polyimide low-k film (53). There is lack of such kind of information in conventional ULSIC processes. With the shrinking of the device geometry and the enlargement of the wafer size, plasma processes will probably be frequently used in both front- and back-end-of-the-lines to prepare new materials or structures. The above knowledge is especially useful.

Plasma etching has been used in IC production since 1970s. Most thin films etched by plasma processes are deposited by LPCVD or sputtering. The preferred etch profile is vertical. Plasma damage is an important issue of concern. On the other hand, plasma etching is also critical to the TFT array fabrication. Since the plasma etch rate of the highly hydrogenated PECVD film, e.g., a-Si:H, SiN_x, or n^+, is much higher than that of the low hydrogen-content LPCVD film, a high etch selectivity between two PECVD films is more difficult to achieve than that between two LPCVD films. Influences of plasma chemistry and ion bombardment to the PECVD film etch rate and selectivity have been investigated (54,55). This kind of problem was rarely investigated in ULSIC. In addition, the plasma etch-induced damage in a MOSFET usually shows as the SiO_2 gate dielectric breakdown caused by neutralization of accumulated charges. However, for the a-Si:H TFT, another important damage mechanism, i.e., the short wavelength light radiation, has been identified (36). This kind of damage mechanism may be critical to the nano size MOSFET that includes the high-k gate dielectric instead of SiO_2.

As the MOSFET size approaches 45 nm, a high-k gate dielectric with an equivalent oxide thickness (EOT) of 1 nm in combination with a metal gate is necessary to obtain a low leakage current and a high speed. One of the biggest challenges of the high-k gate dielectric is that it forms a poor interface layer with silicon, which has a low k value and a high interface density of states. An effective method to solve the problem is to insert an artificial SiO_2 or SiON interface layer before the deposition of the high-k film (56). This kind of gate structure, i.e., metal gate/high-k/interface layer, is common in a-Si:H TFT structure for many years (3). Although the non-stoichiometric SiN_x gate dielectric of the a-Si:H TFT is prepared using a different process, i.e., PECVD, from that of the high-k of a MOSFET, e.g., ALD, CVD, or sputtering following by rapid thermal annealing, both processes involve some critical steps that are operated in the range of non-steady state or non-equilibrium thermodynamic conditions. For example, for the inverted, staggered a-Si:H TFT, a slightly change of the PECVD a-Si:H deposition condition can affect its interface composition and defect density with the underneath gate SiN_x layer (17,21). For the sub 1 nm EOT high-k process, a minor change of the deposition or rapid thermal annealing condition can drastically alter the interface layer's composition and thickness. The interface modification knowledge obtained from the a-Si:H TFT may be applied to improve the high-k gate dielectric interface. Furthermore, in the long run, when the high-k dielectric is applied to the ultra small poly-Si TFT, e.g., with the channel built within the bulk grain, the MOSFET experience can be of a great value.

Summary

The ULSI and TFT development history and market trends have been compared and summarized. Some unique a-Si:H TFT technology issues, which were primarily developed toward the large-area LCDs, have been reviewed. The trend of the TFT technology development is clearly beyond LCDs. Due to its flexibility in selecting materials, processes, and device structures, TFTs have been fabricated into new electronic, photonic, biological, magnetic, and chemical products.

The recent demonstrations of the poly-Si TFT circuits and "System-on-Panels" stimulated the interest of the scientific community in the future development direction of TFTs and any possibility of competing with the ULSIC technology. Judged from individual device characteristics and fabrication limits, it appears that they will be successful in different markets. However, there are many opportunities to integrate TFTs with ULSIC into a "smart chip", which shows novel functions or improved circuit performance. The large-area TFT process knowledge can be helpful for the migration of the 12-inch ULSIC process into the 15-inch process.

Acknowledgments

This paper is prepared for the 2007 ECS Electronics and Photonics Division award speech. The author would like to thank his graduate students and colleagues throughout the world for continuous discussions and collaborations in the past 25 years.

References

1. R. G. Arns, *Eng. Sci. Edu. J.,* 233-240 (1998).
2. F. Stassen, Electroconference 1997, E-gruppen, Soc. Danish Eng., Oct. 1, 1997.
3. Y. Kuo, Editor, "Amorphous Silicon Thin Film Transistors", and "Polycrystalline Silicon Thin Film Transistors," Kluwer Academic Publishers, Norwell, MA, 2004.
4. IRTS Factory Integration TWG, "450mm Wafers Size Conversion," 2007/1/21.
5. Y. Ishii, Digest of Technol. Papers, AM-FPD 06, *Jpn. Soc. Appl. Phys.,* 1-2 (2006).
6. http://www.icknowledge.com/our_products/Chapter%202.pdf Chap.2, Economics Report, IC knowledge.
7. For example, ECS Proceedings of TFTT I-VII and Transactions TFT 8; JSAP AMLCD conference digests 1-12 and AMFPD 06; MRS Proceedings of Amorphous and Microcrystalline; SID conference proceedings.

8. Y. Kuo, M. Okajima, and M. Takeichi, *IBM J. Research and Development*, 43(1/2), 73-88 (1999).
9. Y. Kuo, *Vacuum*, 59, 484-491 (1999).
10. Y. Kuo, *Appl. Phys. Lett.*, 67(15), 2173-2175 (1995).
11. Y. Kuo, *MRS Proc. Amorphous Silicon*, 377, 701-706 (1995).
12. E. C. Paloura, Y. Kuo, and W. Braun, *Physica B*, 208/209, 562-564 (1995).
13. Ch. B. Lioutas, N. Vouroutzis, E. C. Paloura, and Y. Kuo, *Thin Solid Films*, 297, 28-31 (1997).
14. Y. Kuo, *J. Electrochem. Soc.*, 141(4), 1061-1065 (1994).
15. Y. Kuo, *Appl. Phys. Lett.*, 71(19), 2821-2823 (1997).
16. Y. Kuo, *Jpn. J. Appl. Phys.*, 29(10), 2243-2246 (1990).
17. Y. Kuo, *Appl. Phys. Lett.*, 63(2), 144-146 (1993).
18. Y. Kuo, *J. Electrochem. Soci.*, 142(7), 2486-2506 (1995).
19. Y. Kuo, *J. Electrochem. Soc.*, 142(1), 186-190 (1995).
20. Y. Kuo, *Appl. of Particle and Laser Beams in Materials Technology*, NATO ASI series, 283, 581-593 (1995).
21. Y. Kuo, *Vacuum*, 59, 484-491 (1999).
22. Y. Kuo, *MRS Proc. Chemical Perspectives of Microelectronic Materials III*, 282, 623-629 (1992).
23. Y. Kuo and H. H. Lee, *Vacuum*, 66(3-4), 299-303 (2002).
24. Y. Kuo and K. Latzko, *MRS Proc., Amorphous and Microcrystalline Silicon Technology Symp.*, 507, 891-896 (1998).
25. Y. Kuo, *Appl. Phys. Lett.*, 71(19), 2821-2823 (1997).
26. Y. Kuo, *J. Electrochem. Soc.*, 137(6), 1907-1911 (1990).
27. Y. Kuo and J. R. Crowe, "Slope Control of Molybdenum Lines Etched with RIE," *J. Vac. Sci. and Technol.*, A8(3), 1529-1532 (1990).
28. Y. Kuo, "Reactive Ion Etching of Indium Tin Oxide by $SiCl_4$-based Plasma – Substrate Temperature Effect," *Vacuum*, 51(4), 777-779 (1998).
29. Y. Kuo, *J. Electrochem. Soc.*, 144(4), 1411-1416 (1997).
30. Y. Kuo, *Jpn. J. Appl. Phys. Part 2*, 36(5B), L629-L631 (1997).
31. Y. Kuo and S. Lee, *Jpn. J. Appl. Phys.* 39(3AB), L188-L190 (2000).
32. Y. Kuo and S. Lee, *Appl. Phys. Lett.*, 78(7), 1002-1004 (2001).
33. S. Lee and Y. Kuo, *Thin Solid Films*, 457, 326-332 (2004).
34. S. Lee and Y. Kuo, *J. Electrochem. Soc.*, 148(9), G524-G529 (2001).
35. Y. Kuo, M. Okajima, and M. Takeichi, *IBM J. R&D*, 43(1/2), 73-88 (1999).
36. Y. Kuo, *Appl. Phys. Lett.*, 61(23), 2790-2792 (1992).
37. Y. Kuo and M. Crowder, *J. Electrochem. Soc.*, 139(2), 548- 552 (1992).
38. Y. Kuo, H. Nominanda, and G. Liu, *J. Korean Physical Soc.*, 48, S92-S97 (2006).
39. Y. Kuo, *J. Electrochem. Soc.*, 139(4), 1199-1204 (1992).
40. Y. Kuo, *J. Electrochem. Soc.*, 138(2), 637-638 (1991).
41. Y. Kuo, *J. Electrochem. Soc.*, 143(8), 2680-2682 (1996).
42. Y. Kuo, *J. Electrochem. Soc.*, 143(4), 1469-1471 (1996).
43. Y. Kuo, *Appl. Phys. Lett.*, 67(21), 3174-3176 (1995).
44. Y. Kuo and H. Nominanda, *Appl. Phys. Letts.*, 89, 173503 (2006).
45. Y. Kuo, Y. Lei, and H. Nominanda, *MRS Symp. Procs. Amorphous and Nanocrystalline Silicon Science and Technology*, 808, 709-714 (2004).
46. Y. Izumi and Y. Yamane, *MRS Bulletin*, 27(11), 889-893 (2002).
47. Y. Kuo, *Digest of Technical Papers, AM-FPD 06*, 77-80 (2006).

48. M. Hatano, H. Hamamura, M. Matsumura, Y. Toyota, M. Tai, M. Ohkura, and T. Miyazawa, ECS Trans. 3(8), 35-41 (2006).
49. Y. Yamamoto and A. T. Voutsas, *ECS Trans.* 3(8), 11-22 (2006).
50. H. Hayama, *ECS Trans.* 3(8), 3-9 (2006).
51. Y. Uemoto, E. Fujii, A. Nakamura, K. Senda, H. Takagi, *IEEE Transactions on Electron Devices,* 39, 2359 (1992).
52. Y. Kuo, *Proc. ECS 2003 ULSI Integration III Symp.,* 2003-06, 322-329 (2003).
53. T. Chung, Y. Kuo, and H. Nominanda, *ECS Proc. Copper Interconnects, New Contact Metallurgies, and Low k Interlevel Dielectric Symp.,* 22, 162-169 (2002).
54. Y. Kuo, *J. Vac. Sci. and Technol.,* A8(3), 1702-1705 (1990).
55. Y. Kuo, *IBM J. R&D,* 36(1), 69-75 (1992).
56. J. J. Peterson, C. D. Young, J. Barnett, S. Gopalan, J. Gutt, C.-H. Lee, H.-J. Li, T.-H. Hou, Y. Kim, C. Lim, N. Chaudhary, N. Moumen, B.-H. Lee, G. Bersuker, G. A. Brown, P. M. Zeitzoff, M. I. Gardner, R. W. Murto, and H. R. Huff, *Eletrochem. Solid-State Letts.,* 7(8), G164-G167 (2004).

ECS Transactions, 6 (4) 133-138 (2007)
10.1149/1.2728851, ©The Electrochemical Society

Integration of SOI-based 2D- and Si-based 3D photonic crystals

G. Kocher[a], W. Khunsin[a], J. Romero Vivas[a], K. Vynck[b], S. Arpiainen[c], S. G. Romanov[a], B. Lange[d], M. Mulot[c], F. Jonsson[a,f], T. Charvolin[e], E. Hajdi[e], D. Cassagne[b], R. Zentel[d], J. Ahopelto[c], C. M. Sotomayor Torres[a]

[a] Tyndall National Institute, University College Cork, Lee Malting, Cork, Ireland
[b] GES UMR-CNRS 5650 Université Montpellier II, Place E. Bataillon, CC074, 34095 Montpellier, France
[c] Micro and Nanoelectronics, VTT Technical Research Centre of Finland, P.O. Box 1000, FI-02044 VTT, Finland
[d] Department of Chemistry, University of Mainz, Duesbergerweg 10-14, D-55099 Mainz, Germany
[e] DRFMC/SPMM, CEA Grenoble, SiNaPS, 17 rue des Martyrs, 38054 Grenoble Cedex 9, France
[f] Current address: School of Physics and Astronomy, University of Southampton, SO17 1BJ, United Kingdom

We present progress towards a generic approach to integrate 2D photonic components embedded in 3D PhCs and couple them to single-mode ridge waveguides. 3D photonic crystals are fabricated by spatially selective self-assembly of silica and PMMA beads into an fcc lattice on a SOI patterned substrate. 2D defect layers are fabricated on 3D PMMA opals by controlled removal of specific beads through electron beam lithography. Simulations of mode matching and coupling efficiencies are presented, which provide guidelines for the fabrication of experimental structures on inverted silicon opals.

The distinct characteristics of photonic crystals (PhCs) provide the possibility to fabricate ultra-compact optoelectronic components in SOI-based structures. There are, however, multiple challenges on the way to the realization of PhC based photonic components and their integration onto a chip. For the actual fabrication of integrated components and eventually an all-optical microchip, a sound knowledge of and control of several physical principles, e.g., mode matching, waveguiding and coupling processes are necessary, as well as feasible technological concepts, such as, spatial selectivity and compatible fabrication methods. Once obstacles are removed, the approach proposed by Chutinan et.al. (1) offers particular interesting possibilities to fabricate all-optical devices with 2D PhC layers based on dielectric pillars sandwiched between two 3D PhCs.

Following on this approach, we present progress towards integration of 2D photonic components embedded in 3D PhCs and couple them to single-mode ridge waveguides. The 3D PhC "base" is fabricated by spatially selective self-assembly of silica or polymer beads into a face-centred cubic (fcc) lattice on a SOI patterned substrate. The 2D defect layer is inscribed by electron beam (e-beam) lithography either onto the top layer of the polymer opal or on a layer of resist deposited on top of the silica opal to form a template for passive or active devices (2). After processing the defect layer, a second 3D PhC is assembled on top and the whole structure can be subsequently inverted with silicon to give a full photonic bandgap. When matched to the incoming single mode ridge waveguide, these structures enable the control of emission and/or propagation directionality. We present simulations of mode matching and coupling efficiencies as well as first results on real structures. (3).

DESIGN

Our design is based on the layout suggested by Chutinan et.al. (1) containing a 2D PhC embedded in an 3D inverted opal structure. The main functionality of this design is in the 2D PhC layer, with the 3D PhC providing the essential field confinement in the vertical direction replacing total internal reflection.

In order for this combined system to work, the eigenmodes of the 2D PhC layer have to match the ones at the interface of the 3D PhC. The interface is defined by cutting the 3D PhC at a specific height. We choose to have a 2D PhC made of an array of holes with each hole located at the vertical axis of a sphere in the cut layer, instead of pillars as in Ref. (1). FDTD simulations show, that the 3D PhC has to be cut at a distance of c/12, where c is equal to the distance between similar layers (see Fig. 1).

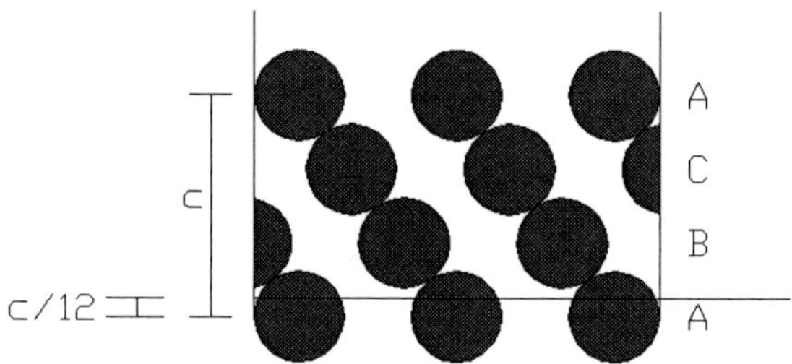

Fig. 1. The stacking layers of an fcc PhC. The distance between similar layers is given by c. A cut at a distance c/12 will expose an interface consisting of an array of holes in dielectric material.

An array of holes for the 2D PhC is most useful to incorporate both passive (waveguide) and active (light emitter) devices. Such a configuration is illustrated in Fig. 2.

EXPERIMENTAL

The opal photonic crystals were self-assembled either on pre-patterned or flat SOI substrates. Monodisperse poly(methyl methacrylate) (PMMA) and silica spheres were dispersed in de-ionised water to form a 2 mass % dispersion. The PMMA spheres were fabricated with a diameter of a = 370 nm using the modified surfactant free emulsion polymerisation technique (4, 5). The silica beads were purchased from Bangs Laboratories and had a mean diameter of 810 nm. The pattern on the SOI substrate was produced by UV lithography and etched all through the SOI top layer in an inductively coupled plasma (ICP) reactor using fluorine chemistry. The structures on the pre-patterned substrates are completely filled (3), so the thickness of the PhC corresponds to the etch depth of the structures, ranging from 5 to 10 μm.

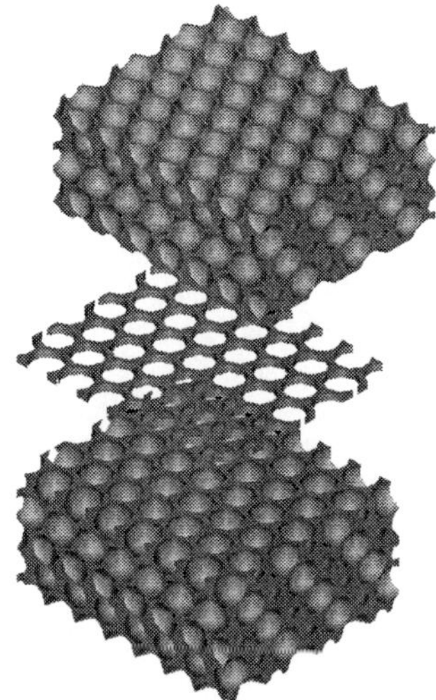

Fig. 2. Exploded view of the 3 main components of our system. The main functionality is incorporated by patterning the 2D PhC layer. The top and bottom 3D PhC provide field confinement in the vertical direction. For that to occur, mode matching between the 2D and 3D PhC components is essential.

The self-assembly technique used is a variation of the vertical deposition technique (VDT) (6) with an acoustic field applied to increase the crystalline quality. The resulting structures on flat Si substrates show an impressive fcc arrangement free of distortion throughout the whole sample in all 3 dimensions (Figure 3b) (7, 8, 9). As in VDT, the formation of the fcc lattice of the opal takes place at the meniscus formed at the suspension-air interface. The form and size of the meniscus is highly dependent on the substrate surface's topology and hydrophilicity (10).

Figure 3. SEM micrographs showing the highly regular PMMA opal with 370 nm diameter beads. a) Top view and b) Side view on flat substrate. c) Silica opal with beads of 810 nm diameter on patterned substrate.

For the pre-patterned substrates on SOI wafers, the best crystalline quality was achieved on a hydrophilic surface at the bottom of the sedimentation trench, made up by the top of the SiO_2 box layer, and hydrophobic silicon side wall surfaces.

2D PHC DEFECT LAYERS

As a first approximation, e-beam lithography is used to inscribe a 2D hexagonal superlattice on a colloidal (PMMA) 3D photonic crystal. Since PMMA is an e-beam resist, the inscription is achieved by selective exposure of single beads and subsequent development. Superlattice configurations in opals have been also simulated showing that they exhibit a PBG in higher index materials (11). Although the refractive index of the PMMA crystal is too small to open the band gap, these structures can serve as templates for double inversion processes (12) to achieve the necessary contrast in refractive index. The method can also be used to fabricate buried defects (inset to Fig. 4). Starting with a first sedimentation and sintering of the crystal (light grey layer), the uppermost layer of beads is inscribed with a lattice of defects (black beads). The second layer is sedimented on top of it (dark grey), sintered and subsequently developed to obtain a buried 2D lattice of defects (2).

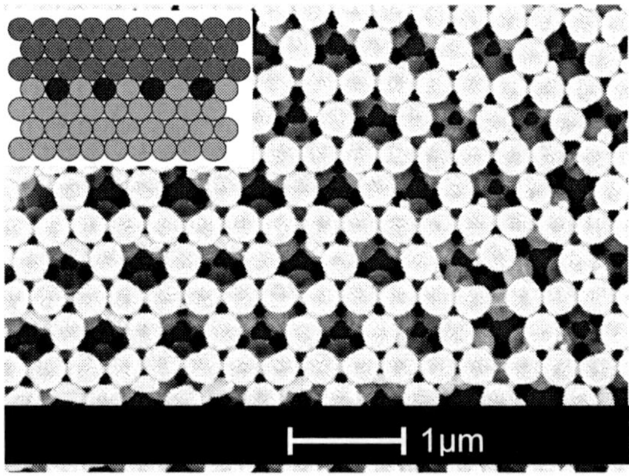

Fig. 4: A 2D defect layer inscribed into an PMMA opal with bead size 370 nm by e-beam lithography. The inset depicts the means of fabricating a buried defect layer by this method.

When patterning polymer opals with e-beam lithography, the writing field acquires a third dimension since the exposure depth becomes a crucial parameter while normally one is concerned only about the exposure of the uppermost layer of the opal. The translation and rotational symmetry of the <111> surface of the film are of course equally important and are both defined by the bead diameter. A more detailed description of the lithography procedure has been given elsewhere (3). Fig. 4 shows an example of a triangular superlattice on a PMMA opal with bead diameters of 370 nm. The second layer of the fcc lattice is clearly visible at the sites of the defects.

SILICA OPALS

To fabricate inverted silicon opals with a 2D defect layer, pre-patterned substrates for spatially selective sedimentation of template opals in wells connected to ridge waveguides were fabricated. This also enable a two-step sedimentation process (see Fig. 5). The 2D defect layer is created by e-beam lithography on a special negative resist after the first sedimentation and inversion step. After exposure and development, a second opal layer is sedimented on top and the structure is inversed again by chemical vapor deposition.

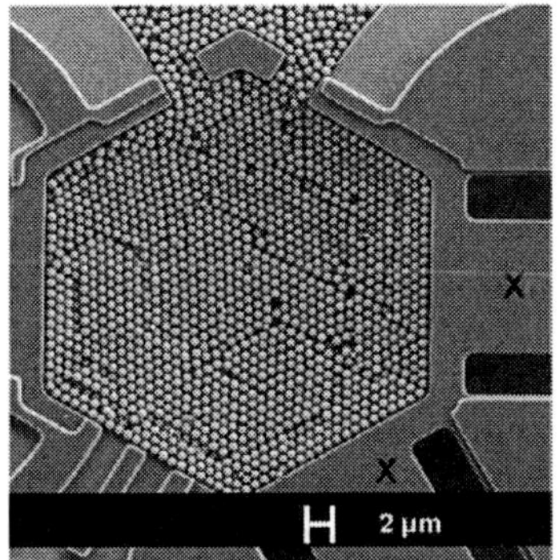

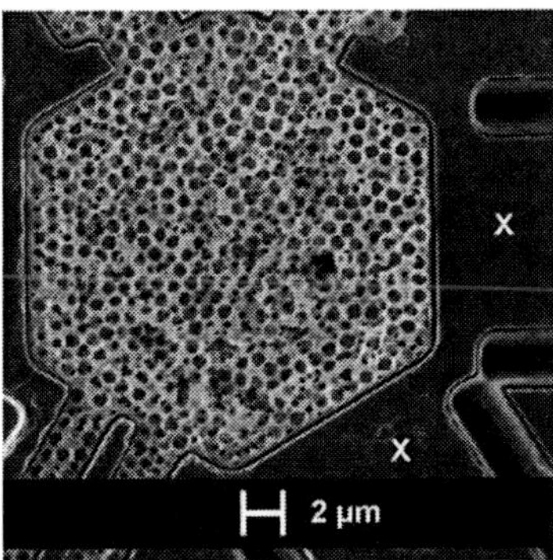

Fig. 5: Left: Selectively sedimented silica opal with bead size of 810 nm connected to ridge waveguides (denoted by "x"). Right: inverted silicon opal on similar substrate; the bead size in the template opal was 810 nm.

CONCLUSION

Opals with a high crystalline quality have been fabricated on flat and on site-selective pre-patterned substrates. 2D defect layers have been fabricated on 3D PMMA opals by e-beam lithography. FDTD simulations are used to understand and determine mode structures for emitters imbedded in 3D PhCs and to find a design template most useful to light emitter structures and coupling light to mode-matched rib waveguides. This approach, if successful, will open the door to compact silicon-based photonic circuits.

ACKNOWLEDGEMENTS

This work is supported by the Science Foundation Ireland (Grant No. 02/IN.1/172), the EU IST FET project No. 510162 (PHAT), the Academy of Finland project No. 53942, the EU NoE 511616 (PHOREMOST) and the German Research Society.

REFERENCES

1. A. Chutinan and S. John, Physical Review E **71(2)**, 026605 (2005).

2. F. Jonsson, C. M. Sotomayor Torres, J. Seekamp, M. Schniedergers, A. Tiedemann, J. Ye and R. Zentel, Microelectronic Engineering **48**, 429 (2005).
3. G. Kocher, W. Khunsin, S. Arpiainen, J. Romero Vivas, S. G. Romanov, J. Ye, B. Lange, F Jonsson, R. Zentel, J. Ahopelto, C. M. Sotomayor Torres, Solid State Electronics, doi:10.1016/j.sse.2007.01.010, (2007) in print.
4. M. Müller, R. Zentel, T. Maka, S. G. Romanov, C. M. Sotomayor Torres. Chem. Mater. **12**, 2508 (2000).
5. M. Egen, R. Zentel. Chem. Mater. **14,** 2176 (2000).
6. Z. Z. Gu. A- Fujishima and O. Sato. Chem. Mater. **14**, 760 (2002).
7. E. Von Rhein, A. Bielawny, S. Greulich-Weber, MRS Symp. Proc. 901 (2006)
8. T. Shinbrot, F. J. Muzzlo, Nature **410**, 251 (2001).
9. A. Amann, W. Khunsin, G. Kocher, E. O'Reilly, C. M. Sotomayor Torres. To be published.
10. J. Ye, R. Zentel, S. Arpianen, J. Ahopelto, F. Jonsson, S. Romanov, C.M. Sotomayor Torres, Integration of self-assembled three-dimensional photonic crystals onto structured silicon wafers, Langmuir **22**, 7378 (2006)
11. K. Vynck, D. Cassagne and E. Centeno, Optics Express 14 (15), 6668, 2006
12. G. M. Gratson, F. García-Santamaría, V. Lousse, M. Xu, S. Fan, J. A. Lewis, P. V. Braun, Adv. Mat. **18 (3)**, 461 (2005).

ECS Transactions, 6 (4) 139-144 (2007)
10.1149/1.2728852, ©The Electrochemical Society

Complementary Single-Crystal Silicon TFTs on Plastic

H.-C. Yuan[a], Z. Ma[a,*], C. S. Ritz[b], D. E. Savage[b], M. G. Lagally[b], and G. K. Celler[c]

[a] Department of Electrical and Computer Engineering, [b] Department of Materials Science and Engineering, University of Wisconsin-Madison, Madison, WI 53706, USA
[c] Soitec, Peabody, MA 01960, USA
* mazq@engr.wisc.edu

This paper reports the first demonstration of both n- and p-type thin-film transistors (TFTs) on flexible plastic substrate using single-crystal Si (110) surface-oriented nanomembrane as active layer. The Si (110) nanomembrane of 190-nm thickness is derived from UNIBOND® silicon-on-insulator (SOI) fabricated by hybrid orientation technology (HOT) and transferred to flexible plastic substrate followed by a low-temperature (< 120 °C) device process. Good device performance has been achieved under so far not-perfected device processing conditions, which includes field-effect mobility of around 95 and 45 cm^2/V-sec on the linear region for the n- and p-type Si (110) TFTs, respectively, and device I_{ON}/I_{OFF} better than 10^4 for both types of TFTs.

Introduction

Flexible microwave-frequency circuitry finds its critical applications on remote sensing, communication, and large-area steerable antenna systems (1). It is highly desirable to realize such systems on flexible plastic substrate for the advantages of low-cost, light-weight, easy stowage and transport, and robustness against fracture and deformation. However, it has been a severe challenge to realize the operation at microwave frequencies for flexible electronics that uses the widely investigated organic semiconductors, amorphous-silicon (α-Si) and poly-crystalline Si, due to their limited carrier mobility posed by the inherent material property and processing temperature constraint on plastic substrate (2). On the other hand, the high mobility values of single-crystal Si still cannot be transformed into high-speed device operation without achieving low-resistance electrode connections (3).

The recent demonstrations of flexible, high-frequency n-type thin-film transistors (TFTs) on single-crystal Si (001) surface-orientated nanomembrane that was transferred from silicon-on-insulator (SOI) to plastic substrates, with the record device speed of cutoff frequency (f_T) reaching 2 GHz and maximum oscillation frequency (f_{max}) over 3 GHz (4) have opened the possibility to realize such flexible microwave systems in the near future. However, while the Si (001) surface orientation has the highest electron mobility, which is suitable for n-channel field-effect transistors (n-FETs), the Si (110) orientation is more suitable for high-performance p-channel FETs (p-FETs) since Si (110) possesses the highest hole mobility among all the three major orientations ((001), (110), and (111)) of single-crystal Si (5)-(7). For this reason Si (110) is promising for both high-performance complementary metal-oxide-semiconductor (CMOS) digital and RF analog circuits (8). Nevertheless, whether Si (110) is suitable for high-performance flexible TFTs is still unknown since as of today neither n- or p-type TFTs on Si (110) was demonstrated. This paper describes the first demonstration of *both* n- and p-type

TFTs on low-temperature polyethylene terephthalate (PET) substrate using single-crystal Si (110) surface-oriented nanomembranes. Good device performance on both types of TFTs is demonstrated.

Device Fabrication

The starting substrate was a UNIBOND® silicon-on-insulator (SOI) bulk substrate. From top to bottom, the SOI has a 190-nm lightly-doped p-type Si (110) surface-oriented template, 200-nm buried oxide (BOX) layer and Si (001) handling substrate. The SOI substrates manufactured by this hybrid-orientation technology (HOT) (the template and the handling substrate have different surface orientations) have been shown to acquire high mobility for both n-FETs on (001) surface and p-FETs on (110) surface (9)-(11). In this report, however, only Si (110) surface-oriented template derived from SOI is used as the active device layer for the flexible TFTs. In order to realize both n- and p-type TFTs on the same active layer, n-wells were formed by phosphorus ion implantation with a dose of 3×10^{11} cm^{-2} at 50 keV through photolithographically defined regions on the bulk SOI substrate. Previous studies (4) have shown that achieving low source/drain contact resistance is critical to realize the high-frequency operation. As a result, heavily doped source/drain regions were photolithographically defined and ion-implanted with conditions of 40 keV, 4×10^{15} cm^{-2} phosphorus and 25 keV, 4×10^{15} cm^{-2} boron for the n- and the p-type TFTs, respectively, followed by a rapid-thermal annealing (RTA) at 700 °C for 20 seconds in N_2 ambient. Compared with the dopant-diffusion process used previously (12), ion implantation has the advantages of being able to precisely control the dosage and dopant depth, and has more choices on dopant species for easier optimization of doping profiles.

The "pre-doped" Si (110) surface-oriented template layer (still part of the SOI substrate) was then patterned by optical photolithography and etched into strips by reactive ion etching (RIE). After removing the photoresist, the sample was immersed in aqueous 49% HF to remove BOX. As the undercut of BOX layer progressed in HF, the "free-standing" part of the Si strips settled down and registered their positions on the Si handling substrate via weak van der Waal's force (13). When the BOX was completely removed, all the Si strips (called nanomembranes now) kept their spatial correlation as they were just patterned.

The Si (110) nanomembranes were then transferred to a PET substrate via a one-step dry-printing technique depicted elsewhere (14). A RIE step was followed to define the active device regions (to isolate the devices). Gate stack consisting of 200/20/150 nm SiO/Ti/Au was then formed on the channel regions for both n- and p-type TFTs by e-beam evaporation and liftoff. Finally, source/drain metal of 20/150 nm Ti/Au was formed on the heavily doped source/drain regions by e-beam evaporation and liftoff. Figure 1(a) shows the schematic cross-section of the finished n- and p-type TFTs on PET substrate. As described in (4), SiO can be deposited at room-temperature. In comparison with SiO_x (x=1-2) deposited by plasma-enhanced chemical vapor deposition (PECVD) (12), the process temperature applied on the plastic substrate in this report was much lowered with the highest process temperature below 120 °C. Figure 1(b) shows the optical-microscope images of a finished n-type Si (110) TFT on PET substrate. Although it is known that hole has the highest mobility along <110>, in order to retain high electron mobility the current flow for all the devices in this study is along <001>.

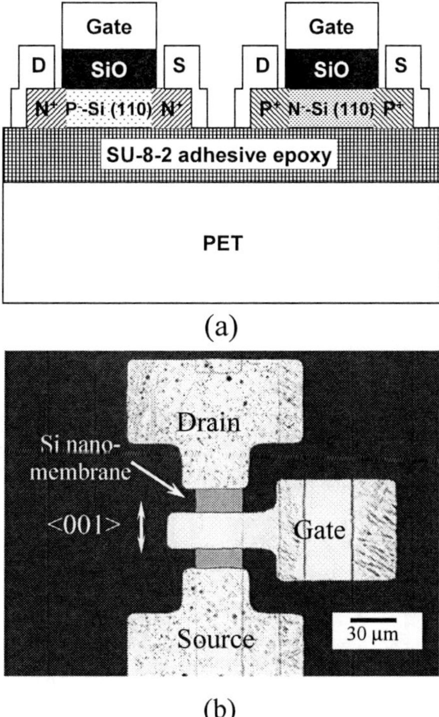

Figure 1. (a) Schematic cross-section of single-crystal Si (110) surface-oriented n- and p-TFTs o PET substrate. (b) Optical-microscope image of a Si (110) n-type TFT on PET substrate with L_G=20 μm and W_G= 30 μm.

Results and Discussions

Figure 2 shows the typical normalized current-voltage (I-V) curves and transfer characteristics of the fabricated single-crystal Si (110) n- and p-TFTs on PET substrate with both channel length (L_G) and width (W_G) of 30 μm. The threshold voltages (V_{th}) were determined by plotting the square root of drain current (I_{DS}) in saturation region versus gate voltage (V_{GS}) shown in Figure 3, and are extrapolated to be around 1 V and - 7 V for n- and p-TFTs, respectively. V_{th} can be readily adjusted and optimized for complementary operation during the well formation step or by adding extra V_{th} adjustment using ion implantation. I_{ON}/I_{OFF} ratio better than 10^4 for both n- and p-TFT is demonstrated at both low and high V_{DS} bias conditions. Gate leakage current is lower than 10 nA/mm^2 under all measured bias conditions. Subthreshold slope for both types of TFTs is around 1.25 V/decade, which is relatively large but is compatible to that on the Si (001) TFT counterpart fabricated at the same time using the identical process.

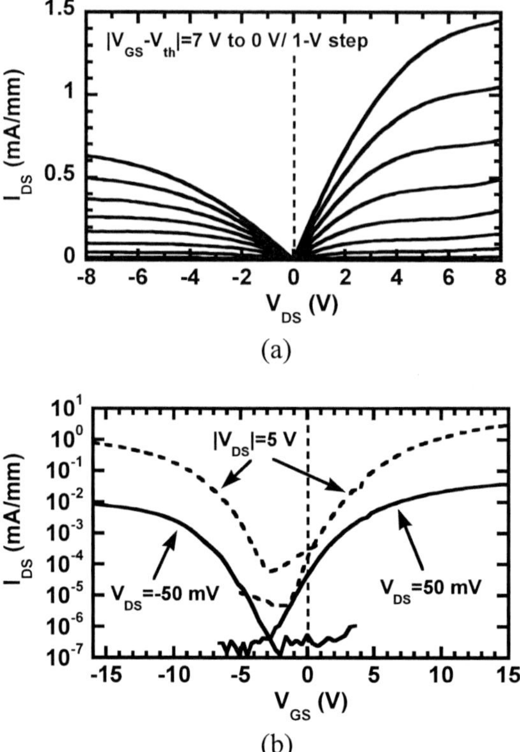

Figure 2. (a) Typical normalized I-V curves and (b) transfer characteristics of single-crystal Si (110) surface-oriented n- and p-TFTs on PET substrate. L_G=30 µm and W_G= 30 µm for both types of TFTs.

Field-effect mobility (μ) values were evaluated from the measured highest transconductance (g_m) using the following equation (15).

$$\mu = \frac{L_G g_m}{W_G C_{ox} V_{DS}}.$$

(1)

The oxide capacitance per unit area (C_{ox}) was calculated from the thickness and dielectric constant of the evaporated SiO layer. The extracted mobility values in the linear (saturation) regions at V_{DS}=±50 mV (±5 V) for the n- and the p-type TFTs are 93±10 (93±10) cm^2/V·s and 46±5 (36±4) cm^2/V·s, respectively. The hole mobility in the flexible Si (110) p-TFT is around 50% lower than that of the Si (110) n-TFT, which agrees well with the reported mobility ratio (between electron and hole) of bulk Si (110) surface-oriented substrate (11). The lower-than-expected mobility of both electrons and holes is

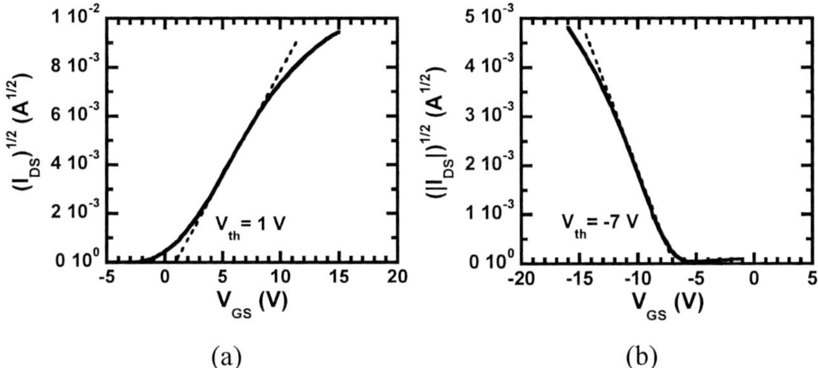

(a)   (b)

Figure 3. Threshold voltage determination on Si (110) surface-oriented (a) n-TFT and (b) p-TFT on PET substrate using the saturation-current extrapolation technique.

suspected to be attributed from high source/drain contact resistance, which is further ascribed from the non-optimized doping profiles for the heavily doped source and drain regions of these TFTs.

Conclusion

Flexible n- and p-type TFTs using single-crystal Si (110) surface-oriented nanomembranes on low-temperature plastic substrate are demonstrated for the first time. By innovatively combining "hot" (ion implantation and high-temperature annealing) and "cold" steps in the process, high-performance TFTs of different channel types are fabricated on the same low-temperature PET substrate, with field-effect mobility of around 95 and 45 cm^2/V-sec achieved for the n- and p-type Si (110) TFTs, respectively, and device I_{ON}/I_{OFF} better than 10^4 for both types of TFTs. The demonstrations of monolithic integration of both the n- and p-channel TFTs on plastic substrate show the promise for future digital applications on flexible substrate.

Acknowledgments

This work was partially supported by NSF under grant DMR0520527 and AFOSR. The Program Officer at AFOSR is Dr. Gernot Pomrenke.

References

1. R. H. Reuss, B. R. Chalamala, A. Moussessian, M. G. Kane, A. Kumar, D C. Zhang, J. A. Rogers, M. Hatalis, D. Temple, G. Moddel, B. J. Eliasson, M. J. Estes, J. Kunze, E. S. Handy, E. S. Harmon, D. B. Salzman, J. M. Woodall, M. Ashraf Alam, J. Y. Murthy, S. C. Jacobsen, M. Olivier, D. Markusm P. M. Campbell, and E. Snow, "Macroelectronics: Perspectives on technology and applications," *Proc. IEEE*, **93**, 1239 (2005).

2. S. Wagner, H. Gleskova, I.-C. Cheng, and M. Wu, "Silicon for thin-film transistors," *Thin Solid Films*, **430**, 15 (2003).
3. E. Menard, R. G. Nuzzo, and J. A. Rogers, "Bendable single crystal silicon thin film transistors formed by printing on plastic substrates," *Appl. Phys. Lett.*, **86**, 093507 (2005).
4. H.-C. Yuan, and Z. Ma, "Microwave thin-film transistors using Si nanomembranes on flexible polymer substrate," *Appl. Phys. Lett.*, **89**, 212105 (2006).
5. M. V. Fischetti, Z. Ren, P. M. Solomon, M. Yang, and K. Rim, "Six-band k·p calculation of the hole mobility in silicon inversion layers: Dependence on surface orientation, strain, and silicon thickness," *J. Appl. Phys.*, **94**, 1079 (2003).
6. H. Irie, K. Kita, K. Kyuno, and A. Toriumi, "In-plane mobility anisotropy and universality under uni-axial strains in n- and p-MOS inversion layers on (100), (110), and (111) Si," in *IEDM Tech. Dig.*, 2004, pp. 225-228.
7. B. Mereu, C. Rossel, E. P. Gusev, and M. Yang, "The role of Si orientation and temperature on the carrier mobility in metal oxide semiconductor field-effect transistors with ultrathin HfO$_2$ gate dielectrics," *J. Appl. Phys.*, **100**, 014504 (2006).
8. H. S. Momose, T. Ohguro, K. Kojima, S. Nakamura, and Y. Toyoshima, "1.5-nm gate oxide CMOS on (110) surface-oriented Si substrate," *IEEE Trans. Electron Devices*, **50**, 1001 (2003).
9. C. D. Sheraw, M. Yang, D. M. Fried, G. Costrini, T. Kanarsky, W.-H. Lee, V. Chan, M. V. Fischetti, J. Holt, L. Black, M. Naeem, S. Panda, L. Economikos, J. Groschopf, A. Kapur, Y. Li, R. T. Mo, A. Bonnoit, D. Degraw, S. Luning, D. Chidambarrao, X. Wang, A. Bryant, D. Brown, C.-Y. Sung, P. Agnello, M. Ieong, S.-F. Huang, X. Chen, M. Khare, "Dual stress liner enhancement in hybrid orientation technology," in *Digest of Technical Papers – Symp. on VLSI Technol.*, 2005, pp. 12-13.
10. Q. Ouyang, M. Yang, J. Holt, S. Panda, H. Chen, H. Utomo, M. Fischetti, N. Rovedo, J. Li, N. Klymko, H. Wildman, T. Kanarsky, G. Costrini, D. M. Fried, A. Bryant, J. A. Ott, I. M. Ieong, and C.-Y. Sung, "Investigation of CMOS devices with embedded SiGe source/drain on hybrid orientation substrates," in *Digest of Technical Papers – Symp. on VLSI Technol.*, 2005, pp. 28-29.
11. M. Yang, V. W. C. Chan, K. K. Chan, L. Shi, D. M. Fried, J. H. Stathis, A. I. Chou, E. Gusev, J. A. Ott, L. E. Burns, M. V. Fischetti, M. Ieong, "Hybrid-orientation technology (HOT): Opportunities and challenges," *IEEE Trans. Electron Device*, **53**, 965 (2006).
12. J.-H. Ahn, H.-S. Kim, K. J. Lee, Z. Zhu, E. Menard, R. G. Nuzzo, and J. A. Rogers, "High-speed mechanically flexible single-crystal silicon thin-film transistors on plastic substrate," *IEEE Electron Device Lett.*, **27**, 460 (2006).
13. G. M. Cohen, P. M. Mooney, V. K. Paruchuri, and H. J. Hovel, "Dislocation-free strained silicon-on-silicon by in-place bonding," *Appl. Phys. Lett.*, **86**, 251902 (2005).
14. H.-C. Yuan, Z. Ma, M. M. Roberts, D. E. Savage, and M. G. Lagally, "High-speed strained-single-crystal-silicon thin-film transistors on flexible polymers," *J. Appl. Phys.*, **100**, 013708 (2006).
15. D. K. Schroder, *Semiconductor Material and Device Characterization*, p. 549, John Wiley & Sons, Inc., New York (1998).

ECS Transactions, 6 (4) 145-150 (2007)
10.1149/1.2728853, ©The Electrochemical Society

Instability of Threshold Voltage of Flexible Single-Crystal Si TFTs

H. Pang[a], H.-C. Yuan[a], M. G. Lagally[b], G. K. Celler[c], and Z. Ma[a,*]

[a] Department of Electrical and Computer Engineering,
[b] Department of Material Science and Engineering,
University of Wisconsin-Madison, Madison, WI 53706, USA
[c] Soitec USA, 2 Centennial Dr., Peabody, MA 01960, USA
* mazq@engr.wisc.edu

N-type, top gate, single-crystal Si thin-film transistors (TFTs) using evaporated silicon monoxide (SiO) as the gate dielectric layer are fabricated on flexible plastic substrate. The threshold voltage (V_T) instability of the device is studied by applying high gate voltages. V_T shift to higher values under positive gate bias stress is observed. The logarithmic dependence of V_T shift on time indicates that electrons are injected and trapped in the SiO layer from Si channel. The subthreshold swing degradation is also found after the stress, suggesting the generation of interface states at Si/SiO surface. However, under negative gate stress, V_T firstly decreases and then increases after longer stress time. This is believed to be the combined effects of both holes injection from Si substrate and electrons emission from metal gate, for which the latter become dominant under a long-term stress.

Introduction

Flexible microelectronics has gained much attention in the past decades owing to its potential capability in portable and wireless applications. The amorphous silicon (α-Si) and poly-crystal silicon have long been used for the applications of flexible displays and radio-frequency identification (RFID) tags, which mainly concerns about the low cost and large area fabrication [1]-[4]. Nowadays, high quality, single-crystal silicon with high mobility can be integrated onto polymer substrate [5]-[8], providing great prospects for realizing flexible integrated circuits (ICs). High performance top-gated TFTs made on plastic substrate employing single-crystal Si membranes which are released from silicon-on-insulator (SOI) have been recently reported with record cut-off frequency f_T and maximum oscillation frequency f_{max} of 1.9 and 3.1 GHz, respectively [9]. The gate dielectric layer employed in these devices [9] is amorphous silicon monoxide (SiO), which has been used in metal-insulator-metal capacitors in microwave circuits [10]. The most significant advantage of SiO in the fabrication of TFTs on plastic substrate is its availability of low temperature deposition. However, previous studies [11], [12] showed that evaporated SiO contains high density of traps and defects, which brings about the reliability concerns to the TFTs. In this paper, the threshold voltage (V_T) instability of the single-crystal Si TFTs using SiO as the gate dielectric is studied by applying both high positive and negative gate voltage stresses.

Device Fabrication

The fabrication of the TFTs used in this study was detailed previously [9]. In brief, 200nm Si template layer on (001) SOI substrate was patterned into 30μm wide strips, followed by phosphorous ion implantation to create low-resistance source and drain regions. Buried oxide (BOX) layer was selectively removed by concentrated (49%) HF. The Si strips were then transferred onto 175μm-thick flexible PET plastic substrate with SU8 spun on top serving as an adhesive intermediate layer. A top-gate structure, consisting of 200nm amorphous SiO, 40nm Ti and 300nm Au, was formed after plasma dry etching that defines the active regions of the TFTs. The SiO layer was evaporated with e-beam at a pressure below 10^{-6} torr at room temperature. The utilization of SiO as the gate dielectric ensures that the entire process temperature is below 120°C, which is compatible with the PET substrate. Finally, metal pads with 40nm Ti and 500nm Au were formed on top of the Si strips as source and drain electrodes. Figures 1(a) and 1(b) show the optic-microscopic image and the schematic cross-section of a finished TFT, respectively.

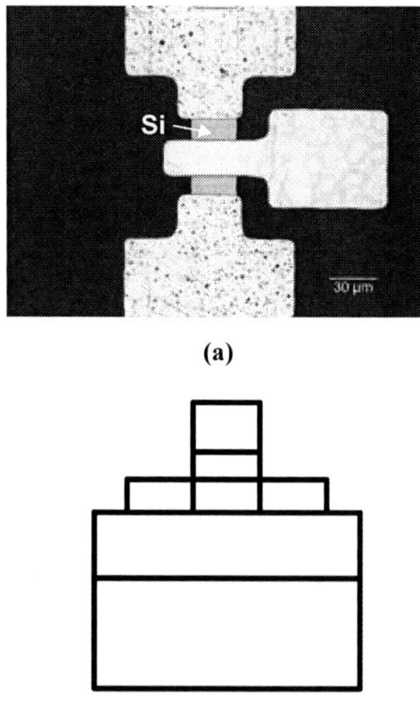

(a)

(b)

Figure 1. (a) Optic-microscopic image and (b) Schematic illustration of the cross section of the top-gate single-crystal Si TFT on plastic substrate.

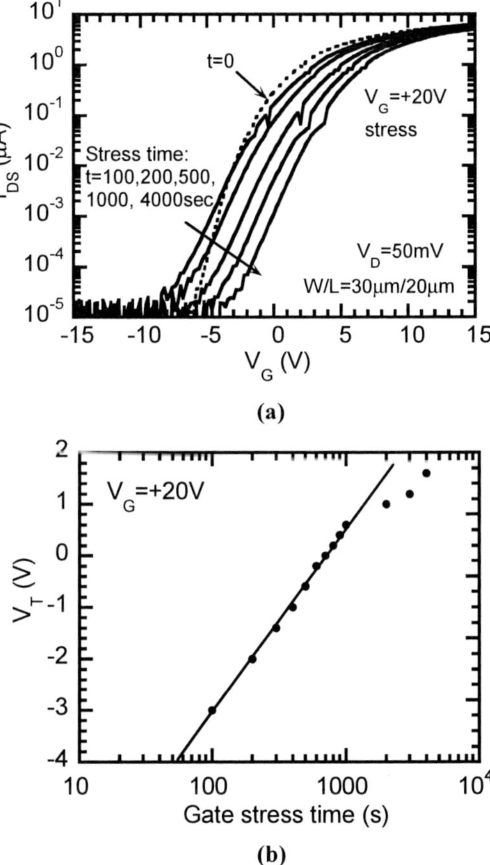

Figure 2. (a) Transfer characteristics of a Si TFT with W/L=30µm/20µm before (dotted line) and after (solid lines) +20V gate stress. The stress time is up to 4000sec. V_D= 50mV. (b) Time dependence of V_T shift.

Results and Discussions

Positive Gate Stress

To study the V_T instability, the device was stressed under both positive and negative high gate voltages, while keeping both the source and drain electrodes grounded. We define the V_T to be the gate voltage (V_G) at which the drain current I_D is 10nA during the sweeping of V_G from the off-state to the on-state at the drain voltage of V_D=50mV. Figure 2(a) shows the transfer characteristics of a TFT device with W/L=30µm/20µm in the logarithmic scale, before and after +20V gate voltage stress up to 4000s. A clear rightward shift of V_T is observed, which indicates the charge trapping in SiO layer, and the maximum value of the shift (ΔV_T) is 4.2V. The slight sub-threshold swing degradation after stress is speculated to be caused by the generation of interface states along the Si/SiO surface due to electron tunneling. It is interesting to notice that the V_T shift does not follow the logarithmic trend after longer stress time [13], as can be seen from the last three data points shown in Fig. 2 (b). Since the hole emission from metal to dielectric is not expected to occur under room temperature due to the large barrier height and a very low gate current of only 10^{-11}A order is monitored here during the stress, we can exclude the possibility of gate leakage. We believe that the deviation of V_T shift from the logarithmic trend is related to the accumulation of electrons in SiO. Further study of this phenomenon is underway.

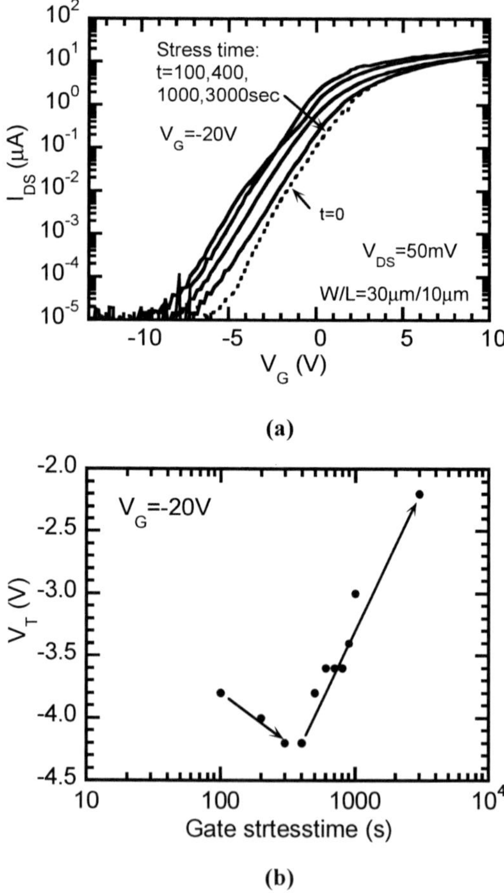

Figure 3. (a) Transfer characteristics of Si TFT with W/L=30μm/10μm plotted on log scale before and after -20V gate stress. The stress time is up to 3000sec. V_D= 50mV. (b) Time dependence of V_T shift.

<u>Negative Gate Stress</u>

Negative gate stress was also applied on the device with $W/L=30\mu m/10\mu m$, however, V_T was found to shift firstly toward lower values and then higher values under -20V V_G stress up to 3000s, as shown in Fig. 3 (a) and (b). It is speculated that the entire V_T shift curve suggests two mechanisms. Under the negative stress, both the tunneling of holes from Si channel to SiO and the Schottky emission of electrons from metal gate [14]-[16] exist. Initially, the hole injection plays the dominant role until the saturation occurs. Then electron emission take the dominance and significantly surmounts the hole injection, which results in the eventual ascend of V_T. This implies that the SiO may contain more gate stress induced electron traps than hole traps.

Conclusion

N-type, top gate, single-crystal silicon thin-film transistors were fabricated on the flexible plastic substrate, using the SiO as the gate dielectric layer. The V_T instability was investigated by stressing the device under high gate voltages. The enhancement of V_T was observed under positive gate stress, and the logarithmic time-dependence of V_T shift indicated the electron being injected and trapped in the SiO. It was also noticed that the generation of interface states along Si/SiO surface caused the subthreshold swing degradation during the stress. However, under the negative gate stress, V_T initially shifted to a lower value and then increases. This was believed to be the consequence of the collaboration of both holes injection from the Si channel and the electrons emission from metal gate, where, the latter mechanism eventually dominated in the longer term stress.

Acknowledgments

This work was partially supported by NSF under grant DMR0520527 and AFOSR. The Program Officer at AFOSR is Dr. Gernot Pomrenke.

References

1. P. F. Baude, D. A. Ender, M. A. Haase, T. W. Keeley, D. V. Muyres, and S. D. Theiss, *Appl. Phys. Lett.*, **82**, 3964, Jun 2, 2003.
2. Y. Chen, J. Au, P. Kazlas, A. Ritenour, H. Gates, and M. McCreary, *Nature*, **423**, 136, 2003.
3. T. K. Chuang, A. J. Roudbari, M. Trocolli, Y. L. Chang, G. Reed, M. Hatalis, J. Spirko, K. Klier, S. Preis, R. Pearson, H. Najafov, I. Biaggio, T. Afentakis, A. Voutsas, E. Forsythe, J. Shi, and S. Blomquist, *Proc. SPIE*, **5801**, 234, 2005.
4. C.-S. Yang, L. L. Smith, C. B. Arthur, and G. N. Parsons, *J. Vac. Sci. Tech. B.*, **18**, 683, 2000.
5. A. Tilke, M. Rotter, R. H. Blick, H. Lorenz, and J. P. Kotthaus, *Appl. Phys. Lett.*, **77**, 558, 2000.
6. E. Menard, K. J. Lee, D.-Y. Khang, R. G. Nuzzo, and J. A. Rogers, *Appl. Phys. Lett.*, **84**, 5398, 2004.
7. E. Menard, R. G. Nuzzo, and J. A. Rogers, *Appl. Phys. Lett.*, **86**, 093507, 2005.
8. Y. Sun, S. Kim, I. Adesida, and J. A. Rogers, *Appl. Phys. Lett.*, **87**, 083501, 2005.
9. H.-C. Yuan and Z. Ma, *Appl. Phys. Lett.*, **89**, 212105, 2006.

10. Z. Ma, S. Mohammadi, L.-H. Lu, P. Bhattacharya, L. P. B. Katehi, S. A. Alterovitz, and G. E. Ponchak, *IEEE Microw. Wirel. Compon. Lett.*, **11**, 287, 2001.
11. G. A. St John and S. W. Chaikin, *IEEE Trans. on Parts, Materials and Packaging*, **PMP-2**, 1/2, 1966.
12. F. Irrera, and F. Russo, *IEEE Trans. Electron Devices*, **46**, 2315, 1999.
13. R.H. Walden, *J. Appl. Phys.*, **43**.1178 (1972).
14. T. E. Hartman, J. C. Blair, and R. Bauer, *J. Appl. Phys.*, **37**.2468, 1966.
15. M. Rosmeulen, E. Sleeckx, K. De Meyer, *IEDM*, pp.189-192, 2002.
16. J. G. Simmons, *Physical Review*, Vol. 155, No. 3, pp. 657-660, Mar, 1967.

ECS Transactions, 6 (4) 151-157 (2007)
10.1149/1.2728854, ©The Electrochemical Society

Heat Generation and Transport in SOI and GOI Devices

Eric Pop

Micro and Nanotechnology Laboratory, Dept. of Electrical and Computer Engineering
University of Illinois Urbana-Champaign, Urbana IL 61801, USA

This paper surveys recent progress in our understanding of heat generation and transport in nanoscale transistors. Monte Carlo simulations show that under quasi-ballistic transport conditions, most Joule heat is generated in the device drain. Measurements and modeling find the device thermal resistance scales inversely with device size, reaching well over 100 K/mW for sub-100 nm transistors. This trend is partly also driven by decreased thermal conductivity of ultra-thin films, as well as the increased role of boundary thermal resistance. Somewhat surprisingly, the analysis shows that well-behaved GOI transistors ought to outperform SOI from a *thermal* viewpoint as well, highlighting the importance of electro-thermal co-design at device length scales approaching 10 nm.

Introduction

Modern CMOS device designs favor confined geometry transistors, such as Silicon-on-Insulator (SOI), FinFET, tri- or surround-gate, and ultimately nanowire (or nanotube) designs (1, 2). This trend is driven by the need for better electrostatic control of the device active region, and the need to lower parasitic capacitance. Another trend is to incorporate higher mobility materials (like germanium) into the active device channel.

The drive towards such device geometries has had several negative implications from a thermal point of view. The most commonly used electrical insulator (SiO_2) is also a very good thermal insulator (100x less thermally conducting than silicon). This has led to well known observations of severe steady-state self-heating in SOI transistors (3-6). Germanium itself has a bulk thermal conductivity only 40% that of silicon, while the thermal conductivity in very thin layers or nanowires is reduced even further, for both materials (1, 7, 8). Thin active layers and narrow device bodies confine the heat generation and dissipation regions, leading to higher local power densities, while a larger surface-to-volume ratio increases the role of interface thermal resistance between materials, impeding heat flow.

From a scaling point of view, it is also interesting to note that device operating voltages have been scaling linearly (or sometimes not at all) with the device dimensions (9), whereas the device volume and surface area scale cubically and quadratically with the linear dimension. As mentioned above, this leads to higher power densities *per device*, whose effects are compounded by the higher thermal resistances involved. This may lead, at the very least, to devices whose steady-state (I-V) characteristics suffer from significant self-heating and cannot be used to understand their dynamic operation. Or, at worst, to devices whose reliability, performance and leakage is compromised both in dynamic (digital) and steady-state operation, and whose thermal drawbacks may negate most advantages obtained from the advanced electro-static design.

Heat Generation in Nanoscale Devices

Heat generation takes an interesting shape in nanoscale devices, under non-equilibrium transport conditions. Monte Carlo (MC) analysis shows that due to quasi-ballistic transport in the active device region (channel < 50 nm), most Joule heat is generated in the drain region and near the contact (1, 10). This asymmetric heat generation is in contrast with lumped analyses of this problem which assumed uniform heat generation in the channel (5), or drift-diffusion (DD) based finite element modeling which places the heat generation region in sync with the local electric field (11).

The standard drift-diffusion approach calculates the heat generation as a dot product of the local current density and electric field ($H = \boldsymbol{J}\cdot\boldsymbol{E}$), whereas the Monte Carlo method implicitly accounts for all phonon generation and absorption events as electrons drift and scatter inside the device (10, 12). The former tends to overestimate the peak heat generation rate and generally predicts a narrower heat generation region, which follows the shape of the electric field. Figure 1 illustrates the heat generation profile computed along the 20 nm channel of a quasi-ballistic silicon device. The differences between the drift-diffusion and Monte Carlo computations are clearly evident at such short length scales (10). The Monte Carlo method yields a broader heat generation domain extending inside the device drain, and chiefly limited by the electron-phonon scattering rate there.

<u>Non-Equilibrium Heat Generation</u>

The Monte Carlo approach also shows that heat generation in silicon is not evenly divided among phonon modes, but that, rather, the acoustic phonon modes receive approximately one third and optical phonons the remaining two thirds of the Joule power. More specifically, the longitudinal optical (LO) *g*-type phonon receives approximately 60% of the total energy dissipation (10). The optical phonons have group velocities of 1000 m/s or less, and are thus much slower than the (zone center) acoustic phonons typically responsible for heat transport in silicon (group velocity 5000-9000 m/s). This non-equilibrium phonon generation has energy transfer bottleneck implications (10, 13). In other words, a significant non-equilibrium OP population may build up, particularly for

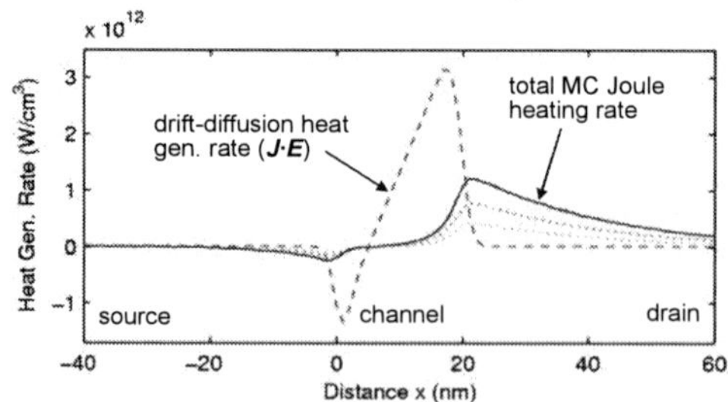

Figure 1: Heat generation in a n+/n/n+ quasi-ballistic silicon device with channel length L = 20 nm. The source and drain are doped to 10^{20} cm^{-3}, the applied voltage is 0.6 V. Unlike the classical (drift-diffusion) result, the Monte Carlo (MC) simulation shows that heat is dissipated far into the device drain. The dotted lines represent the optical phonon (upper) and acoustic phonon (lower) heat generation profiles from the MC result.

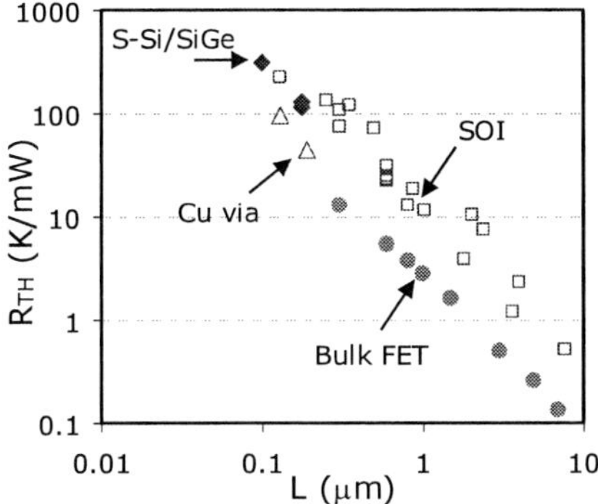

Figure 2. Thermal resistance data reported in the literature over nearly two decades of research across many bulk silicon FET (5, 14) and SOI device technologies (3-5, 15-18). The thermal resistances of single Cu vias and strained Si (on SiGe buffer) transistors (19, 20) are included for comparison. The x-axis represents the FET gate lengths, and the via diameters. The raw numbers have been normalized to a $W/L = 4$ device aspect ratio.

the g-type longitudinal optical (LO) mode. The generation rates for the other phonon modes are either smaller or their density of states (DOS) is larger (the DOS is proportional to the square of the phonon wave vector, which is largest at the edge of the Brillouin zone) and non-equilibrium effects are less significant. Assuming a 10 ps phonon lifetime, the occupation number of the g-type LO phonon would exceed $N_{LO} > 0.1$ and become comparable to unity for power densities greater than 10^{12} W/cm^3 (10). Such power densities are attainable in the drain of 20 nm (or shorter) channel length devices at operating voltages from the current ITRS guidelines (Fig. 1). Non-equilibrium phonon populations will increase electron scattering in the drain, leading (at the very least) to a magnification of the drain series resistance, and decreased device reliability.

Heat Transport in Nanoscale Devices

<u>Device Thermal Resistance</u>

Heat transport from a *lumped* electronic device can be quantified by measuring its thermal resistance (R_{th}) to the environment. This approach yields an average temperature rise of the lumped element as $\Delta T = PR_{th}$ where P is the dissipated power. In practice, the thermal resistance has been measured through noise thermometry (15), gate electrode electrical resistance thermometry (5, 14), pulsed I-V measurements (16, 19) or an AC conductance method (3, 4, 17, 18). Figure 2 presents a summary of this experimental data produced in the literature over the past sixteen years, covering a wide range of device dimensions and technologies. A clear trend emerges, showing that device thermal resistance increases as a power law of the (reduction in) device dimensions, and is reaching values well above 100 W/mK for sub-100 nm gate length devices. The simplest model illustrating the inverse proportionality of the device thermal resistance with its dimensions is $R_{th} = 1/(2kD)$ for the thermal spreading resistance of a heated disk (diameter D) on a semi-infinite plane with thermal conductivity k (21). This can be extended to the case of a

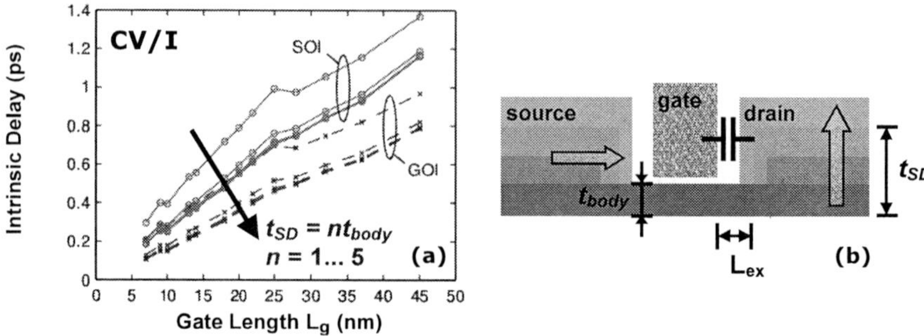

Figure 3. Electro-thermally self-consistent modeling of "well-behaved" SOI and GOI devices near the limits of scaling (26). A larger (e.g. elevated) source and drain design will alleviate heat dissipation problems, but eventually lead to an increase in parasitic capacitance which reduces the intrinsic speed gain. The heat dissipation was assumed to be entirely in the drain, as suggested by Monte Carlo simulations.

rectangular heat source (width W, length L) by replacing $D \approx (LW)^{1/2}$ and including additional 3-D heat spreading shape factors (14, 22). Several online tools are also available for quick, web-based spreading resistance calculations for various shapes and substrates (23). Many other models of varying sophistication have been published, all of which incorporate various inverse length and width dependencies (5, 14, 24-26). Naturally, the choice of such a model in practice depends on its complexity, and on the specific geometry of the device (for instance FinFET vs. bulk FET, to choose two fairly disparate cases).

<u>Devices with Thermally Insulating Substrates</u>

As expected, the thermal resistance of transistors built on thermally insulating substrates, like SOI and strained-Si on SiGe, can be particularly high, as shown in Fig. 2. For these devices, steady-state *I-V* measurements suffer from significant self-heating, and proper characterization must be done with the pulsed (16) or AC conductance methods (3). The trends in Fig. 2 suggest such characterization may need to be employed for near-10 nm *bulk* silicon FETs as well, and most certainly for surround-gate or FinFET type devices, although specific data are currently lacking.

Germanium-on-insulator (GOI) devices with high-κ dielectrics are very promising from an electrical point of view, but may also exhibit self-heating effects because the thermal conductivity of germanium is 60% lower than that of silicon. This would add to the thermal challenge already introduced by the buried insulator. Since data on well-behaved GOI devices is not yet available, this is an area where theoretical investigations are key at the moment. Such preliminary estimates (Fig. 3) suggest the high thermal resistance disadvantages of GOI may be mitigated by lower power dissipation (same current obtained at lower voltage), and also in part by a weaker temperature dependence of the germanium mobility (26). In other words, well-behaved ultra-thin body GOI devices are expected to maintain a performance advantage over similar SOI devices (27), assuming an adequate fabrication technique could be perfected.

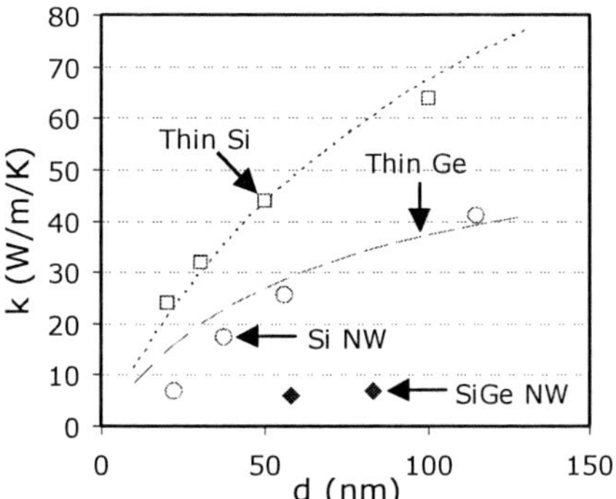

Figure 4. Thermal conductivity data along thin SOI layers, Si and SiGe superlattice nanowires (7, 8). The x-axis is the layer thickness or wire diameter, respectively. The dashed lines are simple models for Si and Ge (26). A significant decrease from the bulk thermal conductivity of Si (~140 W/m/K) and Ge (~60 W/m/K) is noted.

Reduced Thermal Conductivity of Thin Films and Nanowires

In addition to the thermally insulating substrate and boundary thermal resistance (1), thin-body or nanowire (NW) devices also suffer from reduced thermal conductivity owed to increased phonon-surface scattering in the active device region. Recently available data on such thin films and nanowires at room temperature is summarized in Fig. 4. Such data does not yet exist for thin Ge layers, thus a theoretical estimate (dash-dot line) is provided instead. Nevertheless, the most notable feature is the significant decrease in thermal conductivity (up to an order of magnitude) from bulk values for both Si and Ge. However, the decrease in Ge thin film thermal conductivity is expected to be proportionally less severe owing to the shorter phonon mean free path of this material (26). The lowest thermal conductivity is found along Si/Ge superlattice nanowires, where phonon conduction is particularly suppressed by scattering off the additional interfaces.

Transient Heat Conduction

The thermal resistance models mentioned above are sufficient for evaluating the *steady-state* behavior of semiconductor devices, i.e. relevant during *I-V* characterization, analog operation, or to estimate the temperature rise owed to device leakage. However, an understanding of *transient* heat conduction is necessary for short duration pulsed operation, such as during digital switching or electro-static discharge (ESD) events.

To first order, the temperature rise of a pulse-heated volume can be obtained from the energy of the heating pulse (E) and the heat capacity of the volume being heated (C): $E = Pt \approx C\Delta T$ where t is the pulse duration and P is its power. During digital operation, the duration of an inverter switching event is approximately $t \approx$ 50-100 ps, which is significantly faster than the thermal time constant of modern devices, $\tau \approx$ 50-100 ns (28). This is the "adiabatic limit," where the device is essentially thermally decoupled from its environment. In other words, while the device is ON there is little "smearing" of the heated volume outside the area where the actual heating takes place, i.e. the drain of the transistor. Any additional temperature spreading extends approximately $d \approx (\alpha t)^{1/2}$ outside the immediately heated volume (29), where α is the heat diffusion coefficient. This distance is of the order 10 nm into silicon or germanium, and 1 nm into immediately adjacent SiO_2 layers (like the top passivation layer, or the buried oxide below SOI). Typical estimates of the temperature rise during digital switching have shown this dynamic value does not exceed a few degrees (e.g., 5 K) for sub-micron device technologies (28, 30). However, the exact value for devices in the 10 nm range is unknown, and will be highly dependent on the ultimate choice of device geometry and materials.

For longer time scales, comparable to or larger than the device thermal time constants, several models have been proposed (22, 25, 31, 32). These bridge the time range from the adiabatic limit to the steady-state operation of a device, and are typically based on a Green's functions solution of the heat diffusion equation. This approach is faster and offers more physical insight than solutions based on finite-element (FE) solvers. The disadvantage of such methods vs. the FE approach is their applicability to only a limited range of geometries, like heated sphere, infinite cylinder or rectangular parallelepiped, not taking into account the full geometry and diverse materials making up a modern semiconductor device.

Conclusions

This paper summarizes recent advances in our understanding of heat generation and transport in sub-100 nm transistors. Particular attention is given to non-equilibrium effects, of importance in small thin-film silicon- (or germanium-) on-insulator (SOI/GOI) transistors. Heat generation predominantly occurs in the drain of such quasi-ballistic devices. The lumped thermal resistance of nanoscale transistors can easily surpass 100 K/mW for SOI, FinFET, as well as (smaller) bulk transistors. The thermal conductivity of thin films and nanowires is reduced by up to an order of magnitude from the bulk values in silicon and germanium, partly contributing to the increased thermal resistance of small devices. The electro-thermal co-design of nanoscale transistors is seen as an increasingly important area of research.

Acknowledgments

I acknowledge valuable discussions with Sanjiv Sinha and Kenneth Goodson. Support for some of this work was provided by the Semiconductor Research Corporation (SRC).

References

1. E. Pop, S. Sinha and K. E. Goodson, *Proc. IEEE*, **94**, 1587 (2006).
2. H.-S. P. Wong, *IBM J. Res. Dev.*, **46**, 133 (2002).
3. W. Jin, W. Liu, S. K. H. Fung, P. C. H. Chan and C. Hu, *IEEE Trans. Electron Devices*, **48**, 730 (2001).
4. B. Tenbroek, M. S. L. Lee, W. Redman-White, R. J. T. Bunyan and M. J. Uren, *IEEE Trans. Electron Devices*, **43**, 2240 (1996).
5. L. T. Su, J. E. Chung, A. D. A., K. E. Goodson and M. I. Flik, *IEEE Trans. Electron Devices*, **41**, 69 (1994).
6. K. E. Goodson and M. I. Flik, *IEEE Trans. Components, Hybrids, Manufacturing Technol.*, **15**, 715 (1992).
7. D. Li, Y. Wu, P. Kim, L. Shi, P. Yang and A. Majumdar, *Appl. Phys. Lett.*, **83**, 2934 (2003).

8. W. Liu and M. Asheghi, *J. Appl. Phys.*, **98**, 123523 (2005).

9. International Technology Roadmap for Semiconductors (ITRS), in *http://public.itrs.net*.

10. E. Pop, J. Rowlette, R. W. Dutton and K. E. Goodson, in *Intl. Conf. on Simulation of Semiconductor Processes and Devices (SISPAD)*, p. 307, Tokyo, Japan (2005).

11. G. K. Wachutka, *IEEE Transactions on Electron Devices*, **9**, 1141 (1990).

12. T. Sadi, R. Kelsall and N. Pilgrim, *IEEE Trans. Electron Devices*, **53**, 1768 (2006).

13. M. Artaki and P. J. Price, *J. Appl. Phys.*, **65**, 1317 (1989).

14. P. G. Mautry and J. Trager, in *Intl. Conf. on Microelectronic Test Struct.*, p. 221 (1990).

15. R. J. T. Bunyan, M. J. Uren, J. C. Alderman and W. Eccleston, *IEEE Electron Device Letters*, **13**, 279 (1992).

16. K. A. Jenkins and J. Y.-C. Sun, *IEEE Electron Device Letters*, **16**, 145 (1995).

17. T.-Y. Lee and R. M. Fox, in *IEEE Intl. SOI Conference*, p. 78 (1995).

18. M. Reyboz, R. Daviot, O. Rozeau, P. Martin and M. Paccaud, in *IEEE Intl. SOI Conference*, p. 159 (2004).

19. K. A. Jenkins and K. Rim, *IEEE Electron Device Letters*, **23**, 360 (2002).

20. G. Nicholas, T. J. Grasby, E. H. C. Parker, T. E. Whall and T. Skotnicki, *IEEE Electron Device Letters*, **26**, 684 (2005).

21. M. M. Yovanovich, J. R. Culham and P. Teertstra, *IEEE Trans. Components, Packaging and Manuf. Tech.*, **21**, 168 (1998).

22. R. C. Joy and E. S. Schlig, *IEEE Trans. Electron Devices*, **17**, 386 (1970).

23. Micro Heat Transfer Lab (U. Waterloo), in *http://www.mhtl.uwaterloo.ca/RScalculators.html*.

24. A. M. Darwish, A. J. Bayba and H. A. Hung, *IEEE Trans. Microwave Theory and Tech.*, **53**, 306 (2005).

25. N. Rinaldi, *IEEE Trans. Electron Devices*, **48**, 2796 (2001).

26. E. Pop, C. O. Chui, S. Sinha, K. E. Goodson and R. W. Dutton, in *IEEE Intl. Electron Devices Mtg. (IEDM)*, p. 411, San Francisco, CA (2004).

27. X. An, R. Huang, X. Zhang and W. Yangyuan, *Semicond. Sci. Technol.*, **20**, 1034 (2005).

28. K. A. Jenkins and R. L. Franch, in *IEEE Intl. SOI Conference*, p. 161 (2003).

29. K. Banerjee, A. Amarasekera, N. Cheung and C. Hu, *IEEE Electron Device Letters*, **18**, 405 (1997).

30. B. Tenbroek, M. S. L. Lee, W. Redman-White, C. F. Edwards, R. J. T. Bunyan and M. J. Uren, in *Proc. IEEE Intl. SOI Conf.*, p. 156 (1997).

31. V. M. Dwyer, A. J. Franklin and D. S. Campbell, *Solid-State Electronics*, **33**, 553 (1990).

32. Y. J. Min, A. L. Palisoc and C. C. Lee, *IEEE Trans. Components, Hybrids, Manufacturing Technol.*, **13**, 980 (1990).

ECS Transactions, 6 (4) 159-164 (2007)
10.1149/1.2728855, ©The Electrochemical Society

Electronic and Thermal Properties of Silicon Nanowires

E. B. Ramayya[a], D. Vasileska[b], S. M. Goodnick[b], and I. Knezevic[a]

[a] Department of Electrical and Computer Engineering, University of Wisconsin, Madison, Wisconsin 53706, USA
[b] Department of Electrical Engineering, Arizona State University, Tempe, Arizona 85281, USA

Electron mobility and thermal conductivity of silicon nanowires (SiNWs) of different cross sections were calculated by including scattering of electrons due to confined acoustic phonons, optical phonons, and surface roughness. A reduction in the acoustic phonon group velocity due to the spatial confinement in SiNWs is found to result in lower electron mobility and thermal conductivity compared to the bulk phonon approximation. Among the square SiNWs considered, a SiNW of cross section 6 nm x 6 nm was found to have the highest electron mobility due to the interplay of volume inversion and subband modulation.

1. Introduction

Among the emerging devices for future technology nodes, silicon nanowires (SiNWs) on SOI substrates have attracted a great deal of attention among researchers, due to their versatility to function as logic devices (1), interconnects (2), and biological sensors (3). SiNWs on silicon-on-insulator (SOI) substrate are expected to retain the inherent advantages of SOI-based devices, such as lower parasitic capacitance, lower power consumption, lower delay, and higher integration capability than conventional bulk MOSFETs. These advantages clearly address the two most important challenges facing the future semiconductor device industry: reducing power consumption and increasing device performance.

The importance of accounting for the modification of acoustic phonon spectrum due to spatial confinement when calculating thermal properties of SiNWs has been established both theoretically and experimentally (4-7). The lattice thermal conductivity of a 10 nm x 10 nm SiNW was found to decrease by a factor of five with respect to its bulk value (7). This decrease is primarily because of a decrease in the acoustic phonon group velocity in SiNWs. However, the importance of phonon confinement for electronic properties has not yet been accounted for in the literature (8).

In this paper, we present a calculation of the electron mobility and lattice thermal conductivity of SiNWs of square cross sections, which accounts for scattering due to confined acoustic phonons, optical phonons, and roughness of the Si/SiO$_2$ interface. The results were obtained using a Poisson-Schrödinger-Monte Carlo solver developed previously (8), with the *xyz* algorithm of Nishiguchi *et al.* (9) implemented to compute the confined phonon spectra. The lattice thermal conductivity was calculated by solving the phonon Boltzmann transport equation under relaxation time approximation as described in Ref. (7), with a modification of the group velocity calculation. The paper is organized as follows: in section 2, we discuss the importance of considering acoustic phonon confinement when calculating the mobility in SiNWs, and then calculate the electron mobility of wires with different cross sections to determine which wire gives the maximum mobility. In section 3, we present preliminary data for the thermal conductivity of SiNWs by properly accounting for the decrease in acoustic phonon group velocity due to the spatial confinement. Concluding remarks are given in section 4.

2. Electron Mobility and Confined Acoustic Phonons

In ultra-thin and ultra-narrow structures, such as the one originally proposed by Majima *et al.* (10) and considered in this work, the acoustic phonon spectrum is modified due to a mismatch of the sound velocities and dielectric constants between the active layer and the surrounding material (11). This modification of the acoustic phonon spectrum becomes more pronounced as the dimensions of the active layer approach the phonon-phase coherence length scales. Since the dimensions of the SiNWs considered in this work are of the order of the phonon mean free path, it is imperative to account for the modification in the acoustic phonon spectrum when calculating the electron mobility.

Calculation of Confined Acoustic Phonon Dispersion

Using the adiabatic bond charge model (microscopic calculation), Hepplestone *et al.* (12) have shown the validity of the elastic continuum model (macroscopic calculation) for wire dimensions greater than 2.5 nm. Hence, in this work we have used the macroscopic approach to get the modified phonon spectrum. Most of the previous studies of acoustic phonon confinement in nanowires have used approximate hybrid modes proposed by Morse (13) to calculate the dispersion spectrum. The basic assumption in the Morse formalism is that the width of the wire should be at least twice the thickness of the wire. But, Nishiguchi *et al.* (9) calculated the dispersion spectrum using the so-called *xyz* algorithm and found that the Morse formalism is valid only for the lowest phonon subband. Hence, in this work we have used the approach of Nishiguchi *et al.* (9) to calculate the acoustic phonon dispersion, although it is computationally intensive (14).

Importance of Acoustic Phonon Confinement

For an 8 nm x 8 nm SiNW, the confined acoustic phonon scattering rate is roughly three times the acoustic scattering rate calculated using bulk phonons. The increase in the acoustic phonon scattering is primarily due to a decrease in the group velocity (slope of the dispersion curve) of the confined acoustic phonons (14). Confined acoustic phonons spend relatively more time in the vicinity of the electrons than bulk phonons, due to the reduction in their group velocity; hence, electron-phonon scattering rate increases. Fig. 1 shows the electron mobility of an 8 nm x 8 nm SiNW, calculated by properly taking into account the change in the dispersion due to the acoustic phonon confinement. The mobility calculated with confined acoustic phonons is almost 50 % lower than that calculated with bulk acoustic phonons. Such a drastic reduction in mobility occurs not only due to the increase in the scattering rates stemming from confinement, but also because confined acoustic phonons must be treated as an inelastic scattering mechanism [in contrast to the scattering from bulk acoustic phonons, whose low energies enable the elastic approximation (8)]. From these results, we clearly see that confined acoustic phonons need to be properly included in the study of charge transport in SiNWs.

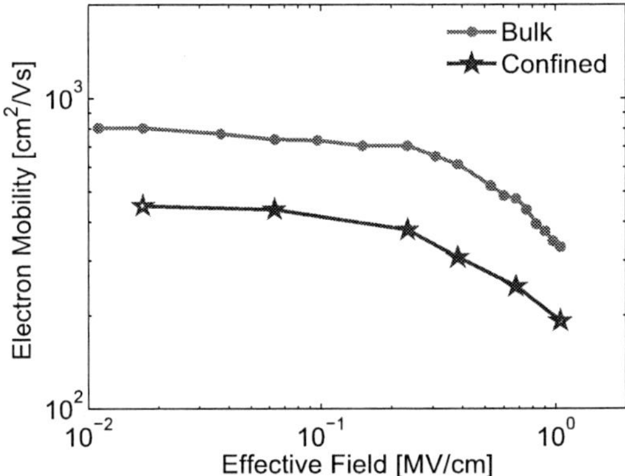

Figure 1. Variation of the field-dependent mobility for an 8 nm x 8 nm SiNW assuming bulk acoustic phonons (red curve) and confined acoustic phonons (blue curve).

<u>Electron Mobility Variation with Transverse Dimensions</u>

In our recent work (8), we found that phonon scattering and surface roughness scattering (SRS) had opposite effects on the electron mobility in SiNWs. With increasing spatial confinement, the SRS was found to decrease due to the onset of *volume inversion*, whereas phonon scattering was found to increase with increasing confinement. Since the total mobility in a SiNW is determined by the interplay of the SRS and phonon scattering, we expect the variation of the mobility with cross section to have a peak for a particular cross section. In order to find the cross section of the SiNW which gives the maximum electron mobility, we varied the cross-section of the wire from 8 nm x 8 nm to 3 nm x 3 nm. The variation of electron mobility with varying nanowire cross section at low (1.4 $\times 10^{-2}$ MV/cm), moderate (2.4 $\times 10^{-1}$ MV/cm) and high (1.04 MV/cm) transverse (gate) fields is plotted in Fig. 2. When the width and thickness are varied from 8 nm to 3 nm, irrespective of the transverse field, the mobility increases initially, attains a maximum at around 6 nm and then starts to decrease.

Three factors contribute to the increase in mobility when going from 8 nm to 6 nm: a) suppression of intersubband scattering due to a reduction in the number of occupied subbands with decreasing wire cross-section; b) suppression of intervalley phonon scattering due to depopulation of the 2-fold degenerate valleys that have a heavier effective mass; c) suppression of the SRS due to an increase in the average distance of carriers from the interfaces (onset of volume inversion). These three effects dominate over the increase in the intrasubband phonon scattering due to the increase in the electron-phonon overlap. As the thickness and width are reduced below 6 nm, the intrasubband, intersubband and intervalley scattering follow the same trend. However, in this region, although the electrons are distributed throughout the silicon channel due to complete volume inversion, their average distance from the interfaces decreases, and consequently the SRS starts to increase. In this region, the degradation in mobility caused by the increase in SRS and intrasubband scattering dominates over the enhancement in mobility due to the suppression of intervalley and intersubband scattering, so the overall mobility starts to decrease.

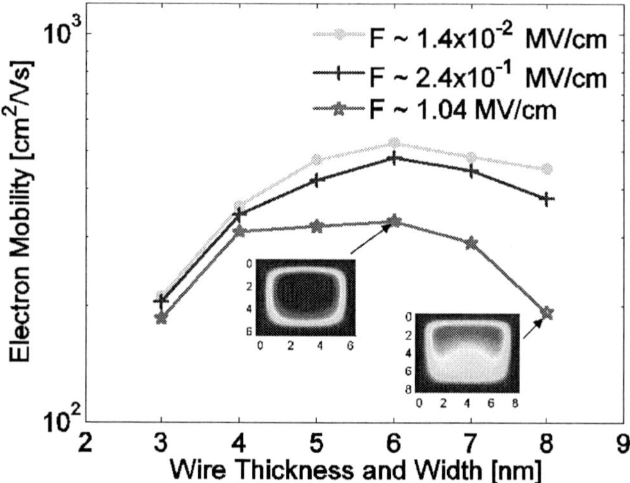

Figure 2. Electron mobility in square SiNWs as a function of the wire characteristic dimensions at a low (green curve), moderate (blue curve), and high (red curve) transverse effective field. Insets: high-field carrier density across the wire for the 6x6 nm^2 and 8x8 nm^2 wires.

3. Thermal Conductivity in SiNWs

As acoustic phonons are the principal carriers of heat out of the device, the reduction in their group velocity due to spatial confinement will directly degrade the lattice thermal conductivity of SiNWs. A reduction in thermal conductivity in SiNWs has been shown previously by several research groups (4-7). In this work, we calculated the thermal conductivity in SiNWs by following the procedure outlined in Ref. (7), which basically uses the relaxation time approximation to solve the phonon Boltzmann transport equation, with an improvement in the calculation of the group velocity. Debye model is used to calculate the lattice thermal conductivity and the effect of spatial confinement is accounted for by introducing the reduced acoustic phonon group velocity in the calculations. However, when calculating the acoustic phonon group velocity, we have accounted for the non-uniform energy gap between neighboring phononic bands, whereas previous studies (4,6,7) have used an average energy gap between phononic bands. This modification is very important, because the energy gap between different phononic bands strongly depends on the wavevector of the phonon. Fig. 3 shows our preliminary data on room-temperature thermal conductivity in SiNWs. In this calculation, we have used the specular scattering fraction of $\varepsilon = 0.5$. As the wire cross section decreases, the acoustic phonon group velocity decreases, which in turn results in the thermal conductivity decrease observed in Fig.3.

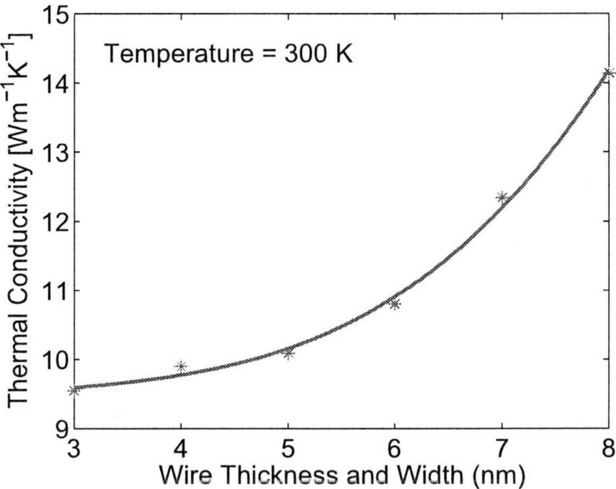

Figure 3. Variation of thermal conductivity with SiNW cross section, calculated at 300 K.

4. Conclusion

Electron mobility and thermal conductivity of square silicon nanowires (SiNWs), determined by the scattering of electrons with confined acoustic phonons, optical phonons, and surface roughness, were calculated using a Poisson-Schrödinger-Monte Carlo simulator. A reduction in the acoustic phonon group velocity due to the spatial confinement results in lower electron mobility and thermal conductivity of the wires compared to unconfined (bulk) silicon. Due to the interplay of volume inversion and subband modulation, among the square SiNWs with dimensions of order a few nanometers, the nanowire of cross section 6 nm x 6 nm was found to have the highest electron mobility

Acknowledgments

This work is supported by the Wisconsin Alumni Research Foundation (WARF) and the National Science Foundation through the University of Wisconsin MRSEC.

References

1. Y. Cui and C. Lieber, *Science,* **291**, 851853 (2001).
2. U. Landman, R. N. Barnett, A. G. Scherbakov, and P. Avouris, *Phys. Rev. Lett.,* **85**, 1958 (2000).
3. M.-W. Shao, H. Yao, M.-L. Zhang, N.-B. Wong, Y.-Y. Shan, and S.-T. Lee, *Appl. Phys. Lett.,* **87**, 183106 (2005).
4. J. Zou and A. Balandin, *J. App. Phys.,* **89**, 2932 (2001).
5. D. Lacroix, K. Joulain, D. Terris, and D. Lemonnier, *App. Phys. Lett.,* **89**, 103104 (2006).

6. Y. Chen, D. Li, J. R. Lukes, and A. Majumdar, *J. Heat Transfer,* **127**, 1129 (2005).
7. X. Lü and J. Chu, *J. App. Phys.,* **100**, 014305 (2006).
8. E. B. Ramayya, D. Vasileska, S. M. Goodnick, and I. Knezevic, *IEEE Trans. Nanotech.,* **6**, 113 (2007).
9. N. Nishiguchi, Y. Ando, and M. N. Wybourne, *J. Phys.: Condens. Matter,* **9**, 5751 (1997).
10. H. Majima, H. Ishikuro, and T. Hiramoto, *IEEE Electron Dev. Lett.,* **21**, 396 (2000).
11. E. P. Pokatilov, D. L. Nika, and A. A. Balandin, *Superlattices and Microstructures,* **33**, 155 (2003).
12. S. P. Hepplestone and G. P. Srivastava, *Phys. Status Solidi C,* **1**, 2617 (2004).
13. R. W. Morse, *J. Acoust. Soc. Am.,* **22**, 219 (1950).
14. E. B. Ramayya, D. Vasileska, S. M. Goodnick, and I. Knezevic, *submitted for publication.*

ECS Transactions, 6 (4) 165-170 (2007)
10.1149/1.2728856, ©The Electrochemical Society

Analysis of PD SOI pMOSFET Device Performance Enhancement due to Direct-Tunneling Current in the Partial n+ Poly Gate

G. Guegan[a], J. Pretet[b], R. Gwoziecki[a], O. Gonnard[b], G. Gouget[b], P. Touret[a], C. Raynaud[a] and S. Deleonibus[a]

[a] CEA LETI-MINATEC, 17 rue des Martyrs, 38054 Grenoble Cedex 9, France
[b] STMicroelectronics, 850 rue Jean Monnet, 38926 Crolles, France

In this paper, we investigate and analyze the impact of the body potential on the performance enhancement of BC pMOSFET when the body contact is floating. Many geometries and layouts of PD SOI pMOSFET with 1.6 nm physical gate oxide have been compared. The body potential is mainly controlled by direct tunneling mechanism due to ECB current from the partial n+ poly-gate into the conduction band of the n type silicon substrate. Narrow device and H gate layout provide the highest Ion gain.

Introduction

The advantages of SOI technology, high speed, low power dissipation, latch-up and soft-error immunities, co integration of digital and analog/RF circuits (1) (2), arise from the presence of a buried oxide in silicon. However, partially depleted (PD) SOI MOSFETs exhibit specific characteristics associated with either floating body (kink effect, transient bipolar effect, history-dependent delay) or self-heating effects. Scaling MOSFET devices to very-deep submicron dimensions has resulted in an aggressive shrinking of the gate oxide. This gate oxide scaling is responsible for a new floating body effect due to gate tunneling current. This gate-induced floating-body effect (GIFBE) which has been observed in both partially depleted SOI MOSFETs (3) and fully depleted transistors (4), leads to a second peak in transconductance. The predominant components which flow between the gate and the channel, have been investigated extensively (5). Several papers have reported a specific floating body effect induced by direct-tunneling current from the partial n+ poly gate of pMOSFET (6) (7), but only one device geometry was studied. The purpose of this work is to deeply study the impact of both layout and geometry of pMOSFET devices on performance enhancement when the body contact is floating.

Device Structure

PD-SOI MOSFETs were fabricated on conventional 300 mm Unibond® wafers with a 1.6 nm physical gate oxide, a 70 nm silicon film and a 145 nm buried oxide. The power supply voltage is one volt. These SOI MOSFETs were designed with various layouts: 3-terminal for no body tied devices, or body-contacted (BC) devices with either one body tied and T-gate design or two bodies tied with H-gate layout (Figure 1). The lateral poly-gate of BC pMOSFETs is partially covered by n+ implant to form the extra body terminal. Therefore, the body contact of pMOSFET which is covered by n+ implant, is defined by the stack n+ poly-gate, thin gate oxide and n type silicon film. The channel lengths and widths which are in the range 100 nm – 10 μm, allow to investigate the dimensional aspects of these new floating body effects due to various ratios between n+ poly-gate area and p+ poly-gate area.

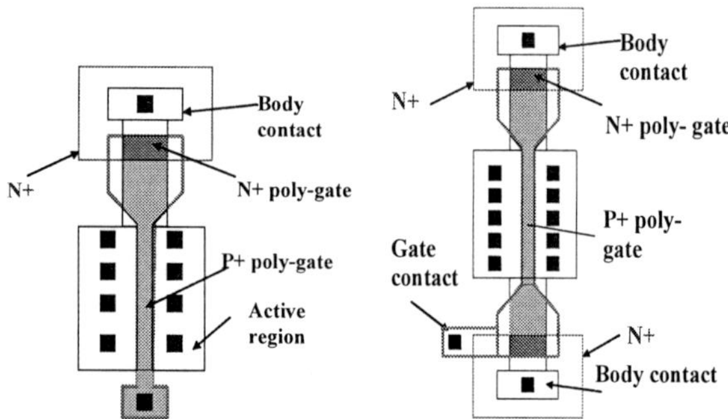

Figure 1. Layouts of T and H gate PD SOI pMOSFETs.

Experimental Result

Figures 2 and 3 show I-Vg characteristics for respectively pMOSFET and nMOSFET with T-gate structure used for the body contact. The body current characteristics of nMOSFET and pMOSFET are different: nMOSFET Ib is controlled by impact ionization mechanism while pMOSFET body current doesn't depend on drain bias. In the case of pMOSFET, the global behaviors of gate and body currents are similar because they are induced by direct-tunneling mechanisms through the thin gate oxide. According to various geometry device measurements, the body current of pMOSFET which is drain voltage insensitive and highly gate voltage dependent, is proportional to the partial n+ poly-gate area (Figure 4). This body current is induced by the conduction-band electron tunneling (ECB) from the partial n+ poly-gate into the n type silicon substrate through the thin gate oxide. The main direct tunneling currents and the energy-band diagram in strong inversion are shown in figures 5 and 6: EVB and HVB are respectively the electron tunneling from p+ poly-gate into silicon and the hole tunneling from the valence band of the n type silicon substrate. HVB and EVB dominate the p+ gate leakage of the inverted PMOS device because of the negligible concentration of electrons in the conduction band of the p+ polysilicon gate. In spite of the smaller area of the n+ polysilicon as compared to the p+ polysilicon, the ECB tunneling from the partial n+ polysilicon gate is the dominant mechanism since the barrier height is only 3.1 eV and the density of electrons in the conduction band of the n+ polysilicon is high. Consequently, this ECB tunneling current induces a large amount of electron charges into the neutral region. For nMOSFET, impact of gate tunneling currents from the partial p+ poly-gate on floating-body potential and body current under one volt power supply voltage is negligible due to the high HVB tunneling barrier of the p+ polysilicon gate: ϕ_b = 4.5 eV (8). Electron tunneling from either the conduction-band (ECB) or the valence-band (EVB) are the other direct tunneling components of the large area n+ polysilicon gate in the nMOSFET. For the nMOSFET biased to inversion, ECB is the largest component since the barrier height (3.1eV) is small compared to EVB barrier (4.2 eV). However, EVB tunneling currents can impact the floating body effect because they result in hole injection in the body (9). On the contrary of the ECB tunneling current associated with n+ polysilicon of the pMOSFET, the impact of these nMOSFET tunneling components on the neutral region is limited under one volt in normal operation.

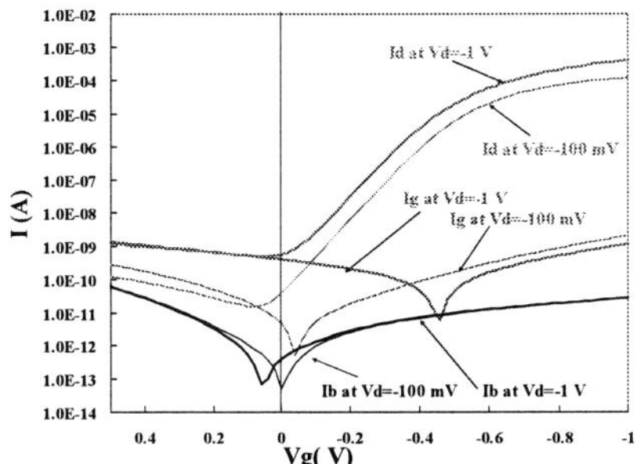

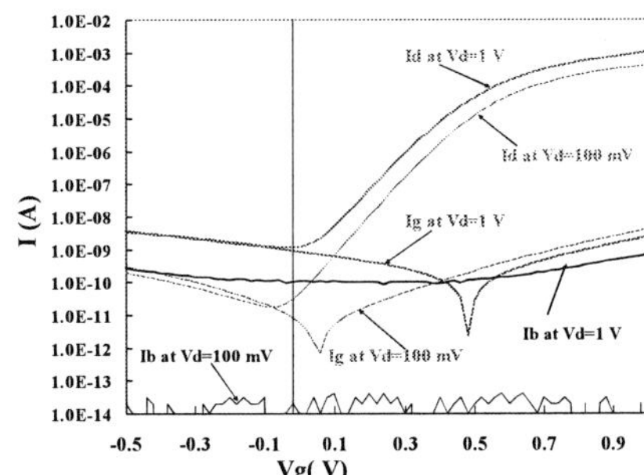

Figure 2. Id-Vg, Ib-Vg and Ig-Vg at Vds=-100 mV and -1 V characteristics for pMOSFET (W/L=2.5µm/0.11 µm).

Figure 3. Id-Vg, Ib-Vg and Ig-Vg at Vds=100 mV and 1 V characteristics for nMOSFET (W/L=2.5µm/0.11 µm).

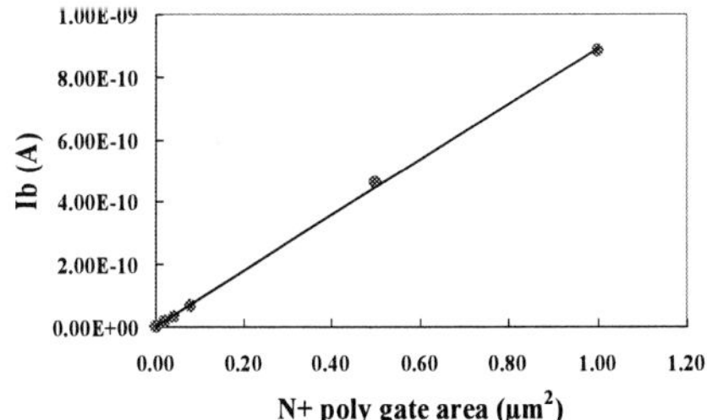

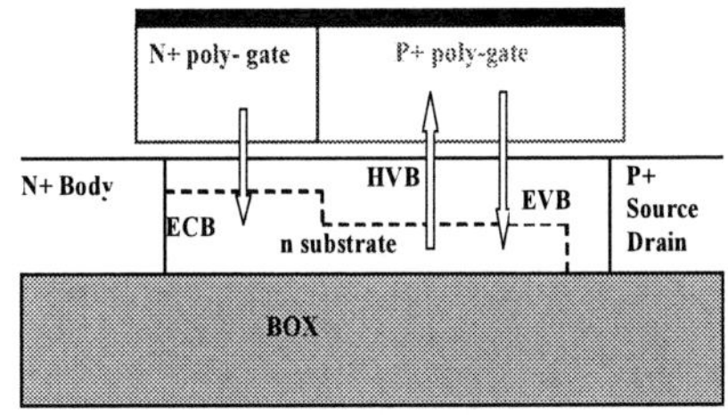

Figure 4. Variation of pMOSFET body current magnitude with partial n+ poly-gate area.

Figure 5. Dominant direct tunneling mechanisms in PD SOI BC pMOSFET under inversion.

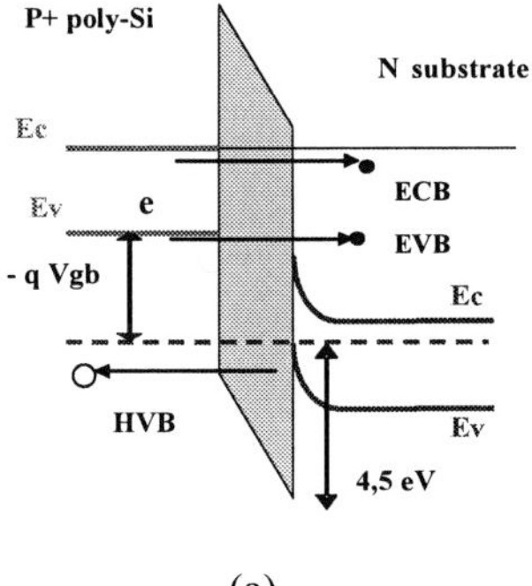

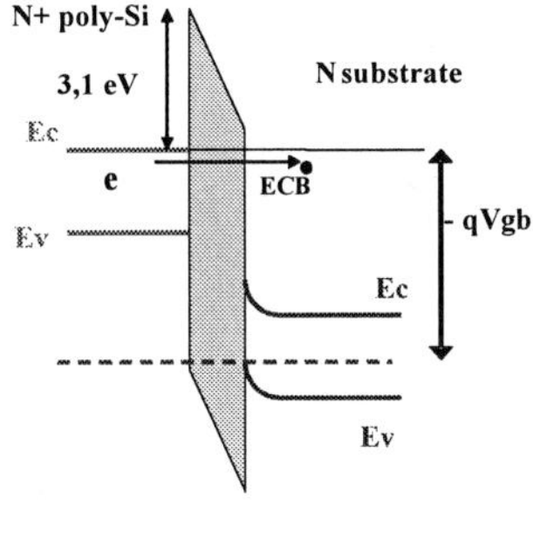

(a)   (b)

Figure 6. Energy bands and components of direct gate tunneling currents in the PD SOI pMOSFET biased in inversion region with (a) p+ polysilicon (b) n+ polysilicon.

The body potential characteristics of nMOSFET and pMOSFET, which are investigated by forcing to zero the body current and measuring Vbody, are different. As shown in figure 7, nMOSFET Vbody which is gate and drain voltage dependent, is mainly controlled by an impact-ionization mechanism. In contrast to nMOSFET, the body potentials of pMOSFET are drain voltage insensitive on a wide range of drain voltages (Figures 8 and 9). From -0.8 volt, the body potential of short gate length devices increases with drain voltage due to an impact-ionization mechanism at the front surface near the drain (Figure 9). On the other hand, this body potential increases with narrower channels due to the voltage drop under the p+ poly-gate when the device is wide (Figure 10).

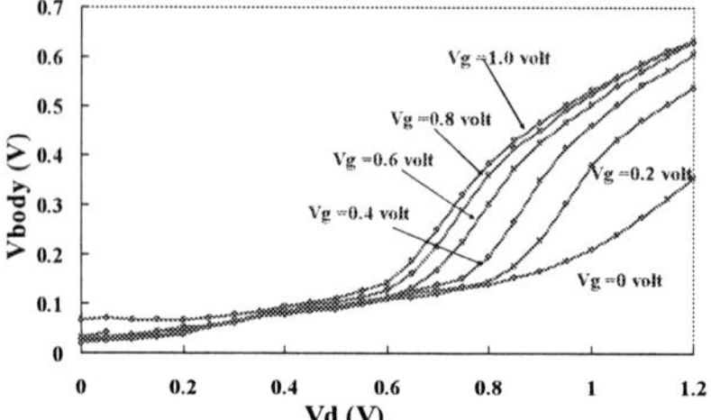

Figure 7. Body potential as a function of drain voltage for BC SOI nMOSFET (W/L=5μm/0.13 μm).

Figure 8. Body potential as a function of drain voltage for long and wide BC SOI pMOSFET (W/L=5μm/10 μm).

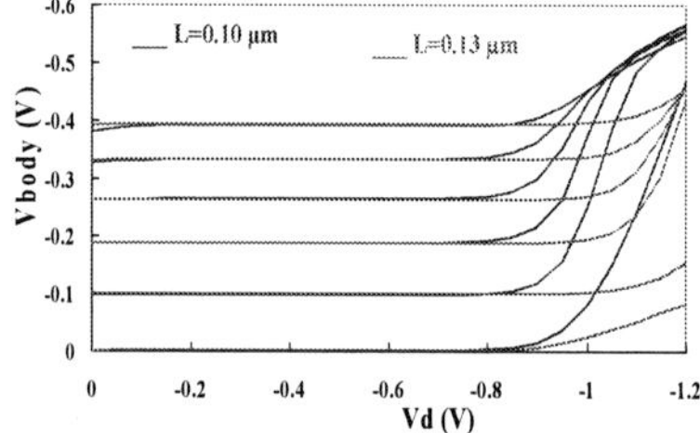

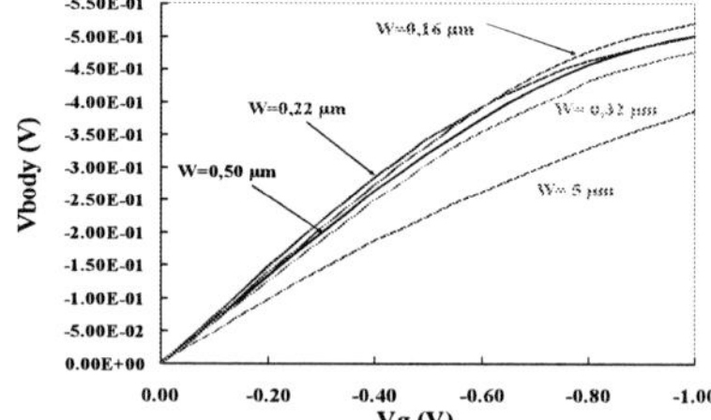

Figure 9. Body potential as a function of drain voltage for short and wide BC SOI pMOSFET (W/L=5μm/0.10μm and 5μm/0.13μm).

Figure 10. Body potential as a function of gate voltage at – 500 mV drain voltage for various channel width BC SOI pMOSFET (L=0.13μm).

Based on these body potential measurements, the drive capabilities of T-gate and H-gate with either body tied or with floating body configuration were compared in strong inversion (Figure 11). The floating body potential rises explain the 24% and 30% gain of Ion with respectively T and H gate design for floating body configuration. Furthermore, this increase of floating-body potential leads to higher drain current than a 3-terminal device without any 'kink' effect (Figure 12).

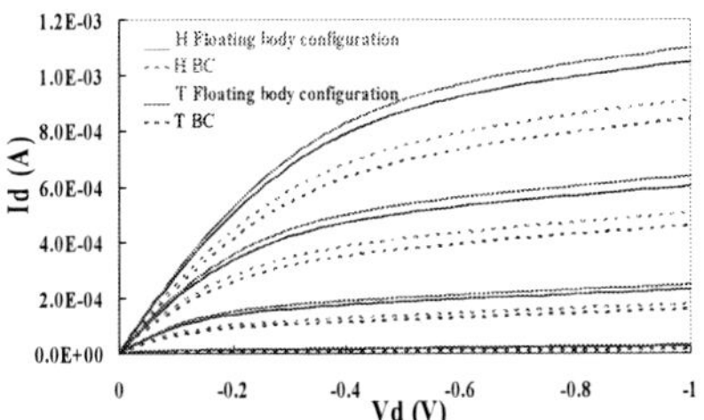
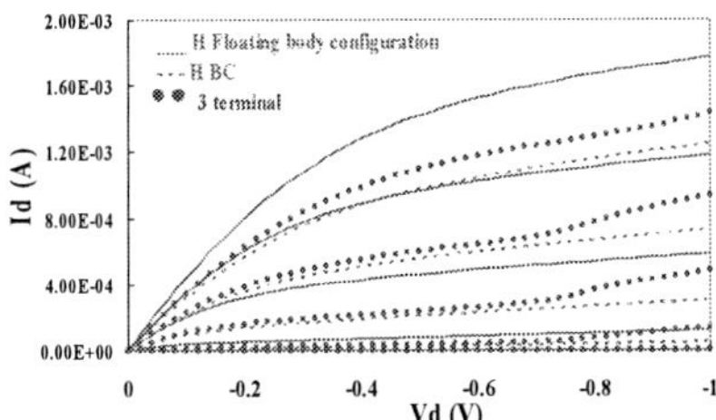

Figure 11. Output characteristics of H-gate, T-gate with body tied and FB configuration at Vgs=0, -0.2, -0.4, -0.6, -0.8 and -1 V (W/L=5µm/0.13 µm).

Figure 12. Output characteristic comparison of H-gate with either body tied or floating-body configuration and 3 terminal pMOSFET (W/L=5µm/0.11µm).

On the other hand, Figure 13 shows a lower subthreshold slope for both low and high drain voltages due to a lower depletion capacitance variation with floating body configuration in weak inversion. Furthermore, the comparison with 3 terminal device characteristics shows a one decade off-state leakage current reduction due to low floating body effect at high drain voltage. A second peak of transconductance due to GIFBE is observed with the 3 terminal device due to both direct tunneling currents between the channel and the p+ polysilicon gate and a sudden change of the body potential (Figure 14). In contrast to this 3 terminal device, the body voltage of BC pMOSFET with floating body configuration slowly increases with gate voltage (figure 10). Consequently, a higher transconductance with lower threshold voltage and no second peak is observed. Due to the high body potential, the drive capability increase has been proved on various device geometries if the body contact is floating. The Ion gain increases with narrower channels and H gate design. For instance, we have measured on devices with a 0.11 µm gate length, 24% and 39% gains of Ion for respectively 5 µm and 500 nm channel widths. Statistical measurements have confirmed on the one hand the drive capability enhancement of BC pMOSFET with floating body configuration and on the other hand, higher Ion gain with H gate design (Figure 15).

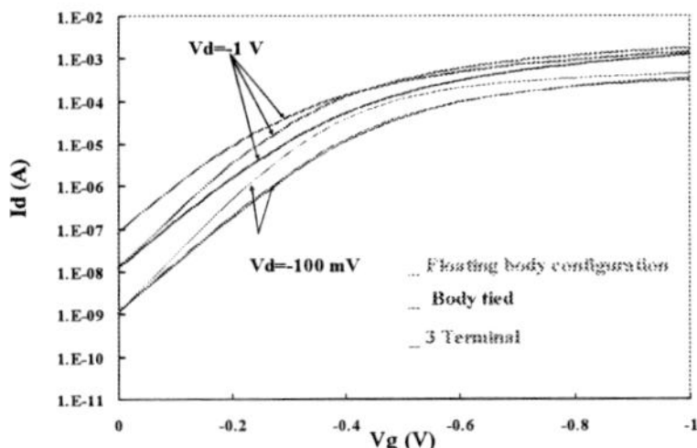
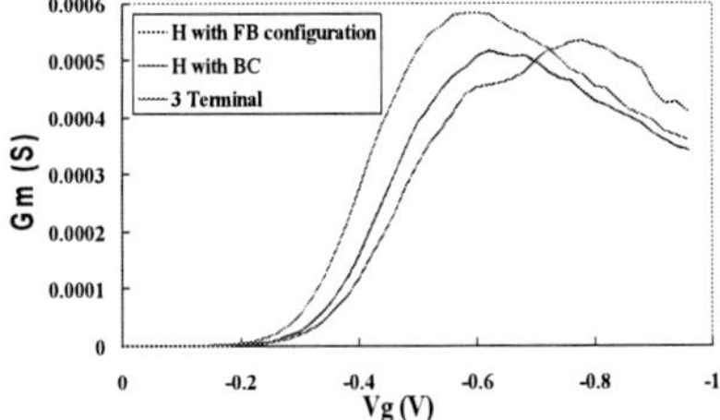

Figure 13. Id-Vg comparison of H-gate with either body tied or floating-body configuration and 3 terminal pMOSFET (W/L=5µm/0.11µm).

Figure 14. Gm(Vg) comparison of H-gate with either body tied or floating-body configuration and 3 terminal pMOSFET at Vds=-100 mV (W/L=5µm/0.13µm).

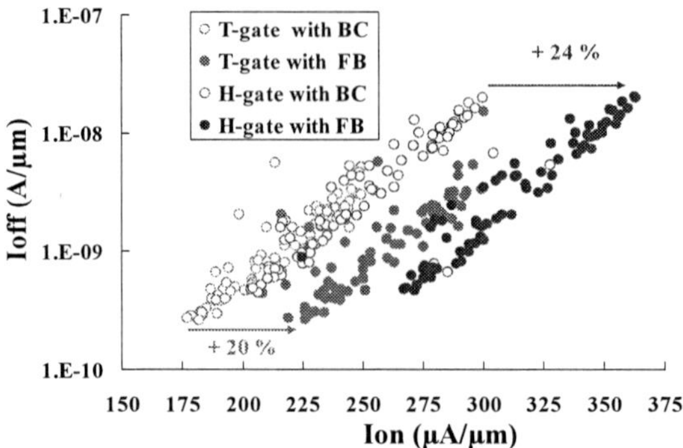

Figure 15.   Ion-Ioff comparison of H-gate (W/L=5µm/0.11 µm) and T-gate (W/L=2.5µm/0.11 µm) SOI pMOSFET with body tied and FB configuration.

Conclusion

We have reported a detailed analysis of the body potential impact on the performance enhancement of BC pMOSFET when the body contact is floating. Narrow device and H gate provide the highest gain. The drive capability increase with lower subthreshold slope has been explained with the ECB tunneling current from the partial n+ poly-gate into the conduction band of the n type silicon substrate through the thin gate oxide.

Acknowledgments

The authors would like to thank ST Microelectronics for device processing.

References

1. C. Raynaud et al, Proceedings of the 207th Electrochemical Society conference, volume 2005-03, p. 331 (2005).
2. J.-O. Plouchart et al, IEEE Transactions on Electron Devices, volume 52, no 7, p. 1370, (2005).
3. J. Pretet et al, Proceedings of the 32nd European Solid-State Device Research Conference, p. 515, (2002).
4. M. Cassé et al, Solid-State Electronics, volume 48, p. 1243, (2004).
5. W.-C Lee and C. Hu, IEEE Transactions on Electron Devices, volume 48, no 7, p. 1366, (2001).
6. S.-S. Chen, S. Huang-Lu and T. H Tang, IEEE Transactions on Electron Devices, volume 51, no 5, p. 708, (2004).
7. S. -S. Chen, S. Huang-Lu and T. H Tang, IEEE Electron Device Letters, volume 25, no 4, p. 214, (2004).
8. W. -C. Lee and C. Hu, IEEE Transactions on Electron Devices, volume 28, no 7, p. 1366, (2001).
9. J. –W. Yang, J. G. Fossum, G. O. Workman and C.-L. Huang, Solid-State Electronics, volume 48, p.259, (2004).

SESSION 4

POSTER PAPERS

ECS Transactions, 6 (4) 173-177 (2007)
10.1149/1.2728857, ©The Electrochemical Society

Non-Contact Photoelectric Method For Thin SOI Characterization

E. Tsidilkovski and K. Steeples

QC Solutions, Billerica, Massachusetts 01821, USA

A non-contact photoelectric measurement method for accurate and rapid determination of silicon film thickness in thin SOI has been developed. A simple model, which explains the measurements, is proposed. An excellent correlation of the measurement results to the data obtained with a standard optical technique has been demonstrated.

Introduction

SOI silicon film thickness is a critical parameter of the ultra-thin fully depleted SOI structures, since its fluctuations directly affect the transistor threshold voltage. Currently there are two common techniques used to measure film thickness: spectral ellipsometry and spectral reflectometry. These methods are sensitive to the native oxide variations, surface roughness and require additional information or reference (reflectometry) for accurate analysis. Electrical characterization of SOI, e.g. CV method, also faces serious challenges since it measures total capacitance of the SOI film plus substrate, and is essentially destructive.

The small signal ac surface photo voltage (SPV) technique traditionally used for bulk silicon and epi wafers [1] can be applied, as well, to film characterization [2]. However, until now the use of SPV-based methods was limited to thick films/SOI only [3]. The criteria for this condition can be stated as follows: the film thickness has to be greater than the sum of the two depletion layers associated with the surface and the front BOX interface. In this case (partially depleted SOI) the SOI measurement with the SPV technique is essentially similar to the bulk wafer measurement, which allows determination of material parameters, such as doping concentration and carrier lifetime.

We extend the applicability of the SPV based methods to thin SOI with the film thickness under 0.1um that is significantly smaller than the maximum depletion layer width W_d.

Method

The frequency-dependent small signal SPV measurement is well understood for the bulk single crystal semiconductor with the uniform doping. Charge on the wafer surface induces a space charge of the width W_d. Incident light with the energy higher than the band gap generates electron hole pairs in the space charge and neutral bulk regions, depending on the light wavelength. Depending on the modulation frequency, the measured photo-voltage is correlated either to the surface potential or to the photo-carrier lifetime.

Generation of a photo voltage in SOI film only, requires that the light energy should be high enough to allow full light absorption within the silicon film. We use high frequency modulation of light that allows 1) exclusion of surface/interface states lifetime effects and 2) correlation of the measured photo voltage to the film space charge capacitance. From the standard analysis of a small signal high frequency SPV in bulk uniformly doped silicon [1] we obtain $V_{SPV} \sim \dfrac{e\Phi}{\varepsilon\omega} W_d$. Here V_{SPV} is a surface photo voltage, e is the elementary charge, Φ is a light flux, ε is a silicon permittivity, ω is a light modulation frequency.

The specifics of SPV in a thin film/ SOI wafer is that even a short wavelength direct band-gap light that is fully absorbed in a silicon film may generate photo voltages in both the surface (V_{SPV}) and the BOX/silicon interface (V_{IPV}). When both space charge regions – surface and interface – are in depletion, the signs of the respective photo-voltages are opposite. Then a measured photo voltage V_M (Figure1) can be presented as a difference of the two photo voltages $V_M{=}V_{SPV}{-}V_{IPV}$ that yields: $V_M \sim \dfrac{e\Phi}{\varepsilon\omega}[W_{d1} - W_{d2}\exp(-\alpha W_{d1})]$,

where α is a light absorption coefficient, W_{d1} and W_{d2} are the depletion depths associated with the surface and the interface.

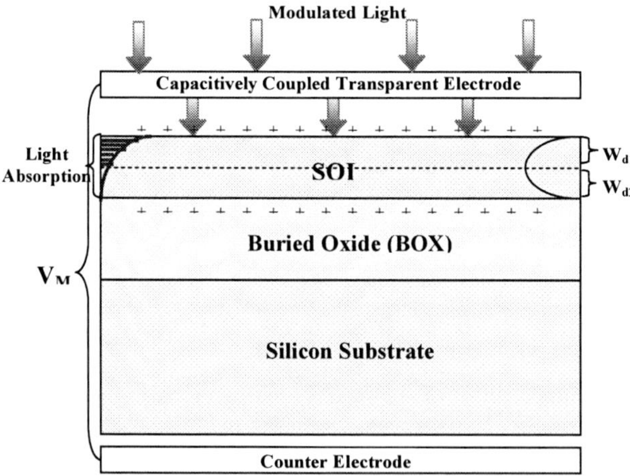

Figure 1. Thin SOI measurement with small signal ac-SPV method

In the thin SOI the top silicon film is fully or nearly fully depleted. This means that due to the absence of the majority carriers in the film the depletion layers are fixed. In ideal case of fully depleted SOI ($W_{d1}/W_{d2}{=}$const), a measured SPV signal yields the depletion layer width, and a silicon film thickness can be calculated through calibration to the reference value.

In reality, the minimal amount of remaining majority carriers still allows changing the ratio of W_{d1}/W_{d2} when a surface or interface potential is changed. Surface potential depends on a sample resistivity and surface condition (density of surface states, environment, etc.). Surface voltage unlike interface potential can be adjusted independently by various methods, such as chemical treatment, external bias or corona charge. To maintain a true non-contact approach we use a corona charge for a surface voltage modulation. By applying corona charge of the appropriate polarity a surface depletion layer can be modified from the maximum value (determined by the BOX charge) to the minimum, close to zero, value (flat band condition). In addition, charging a wafer with corona to a known reference state, such as inversion or flat band/accumulation, allows elimination of a measurement dependence on the surface charge variations.

In the first case, variations in the BOX charge may dominate the SPV signal, allowing calculation of the uniformity distribution of the BOX charge from the analysis of the SPV wafer map. When surface depletion layer is minimized, the SPV measurement yields the interface depletion width $V_M \sim -\dfrac{e\Phi}{\varepsilon\omega}W_{d2}$, independent of the BOX charge variations. Since the silicon film thickness $d=W_{d1}+W_{d2}$, and $W_{d1}\sim0$ in this condition, one can accurately determine the SOI film thickness by measuring SPV.

Results

A set of 200mm bonded SOI wafers[*] with similar BOX thickness ($\sim0.15\mu m$) and varying silicon film thickness was measured on QCS 7000 series Surface Charge Profiler. A thin ($\sim15A$) oxide layer was grown on the SOI wafer surface by a photo-oxidation process that allowed subsequent charging of the wafers with ionizing corona. The full wafer maps (about 2000 points per wafer) were taken and analyzed.

It is shown that for the tested wafers with the film thickness $\sim$ 50-100nm, only measurement with a short wavelength light ($<0.4um$) can be correlated to the SOI film properties. Charging the surface of SOI with corona changed a measured photo-signal dramatically (5-10 times depending on the sign of the charge). We can explain this result assuming that even the rather thin SOI layers may not be fully depleted but rather a small fraction of mobile charge carriers remain in the film. The almost immediate saturation of the response to charging indicates that these SOI films are *nearly* fully depleted.

The p-type doped SOI wafers when charged with positive corona prior to the measurement show no correlation to SOI film thickness, evidently due to the overwhelming variation of the BOX charges. On the other hand, charging the same wafers with negative corona results in a perfect correlation of a measured depletion layer thickness to the SOI thickness measured by spectral reflectometry technique (Figure 2). The effect is due to the flow of negative ions bringing the surface of the p-type wafer to accumulation/flat band condition, collapsing the surface depletion layer and, hence, changing the direction of the minority carrier current. This result corroborates the assumption we made earlier in the paper that the surface potential can be reduced by applying corona charge to the almost flat band condition, so that the measured depletion width practically equals to the SOI film thickness.

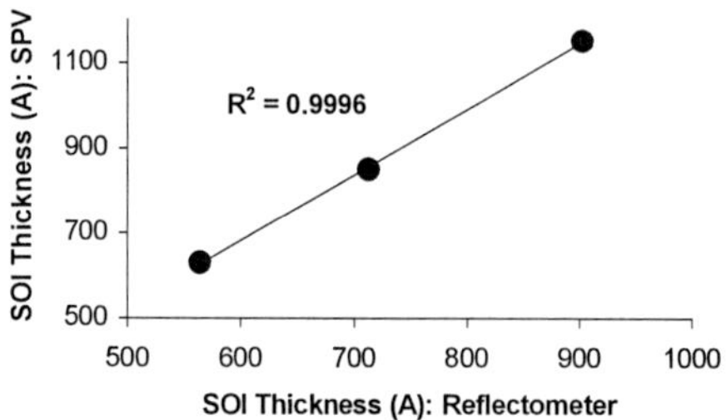

Figure 2. SOI Thickness Measurement: Correlation to spectral reflectometry. Three SOI wafers with different film thickness measured after negative corona charging

Figure 3 shows high-resolution thickness uniformity maps of the bonded SOI wafers. An adjustable resolution of the instrument allows detection of various non-uniformities without compromising the throughput.

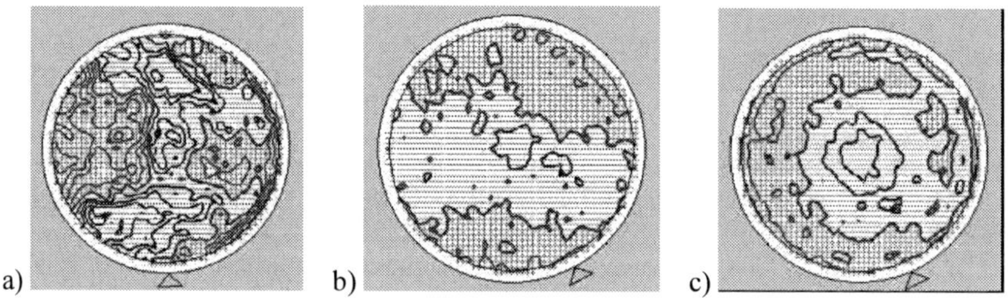

Figure 3. 200mm SOI wafers measured by SPV technique. The bold green line shows a mean value, areas with pluses and minuses represent the above and below mean values. The contour interval of the maps is 2%. The mean values and standard deviation of the maps are a)630A and 4.3%, b)880A and 1.5%, c)1170A and 1.8%, respectively.

Summary

The developed method of SOI characterization allows accurate determination of thin SOI layer thickness across the wafer. Incorporation of the corona charging technique makes the method essentially insensitive to the surface roughness/charge variations. Measurement data obtained with the method correlate perfectly with spectral reflectometry results.

Acknowledgments

*) Authors would like to thank SOITEC, France for providing wafers

References

1. R. S. Nakmanson, *Solid State Electronics* **18,** 617 (1975).
2. M. Leibovitch, et al *J. Appl. Phys.* **79,** 8549 (1996).
3. L. Lukasiak, P. Roman, A. Jakubowski, J. Ruzyllo, *Solid-State Electron.* **45,** 95 (2001).

ECS Transactions, 6 (4) 179-184 (2007)
10.1149/1.2728858, ©The Electrochemical Society

Implant Metrology for Bonded SOI Wafers Using a Surface Photo-Voltage Technique

Adam Bertuch[a], Wesley Smith[a], Ken Steeples[a], Robert Standley[b], Anca Stefanescu[b], and Ron Johnson[c]

[a] QC Solutions Inc., Billerica, Massachusetts 01821, USA
[b] MEMC Electronic Materials Inc., St Peters, Missouri 63376, USA
[c] Innovion, Gresham, Oregon 97230, USA

Small signal Surface Photo-voltage (SPV) measurement techniques have been applied to monitor ion implants typical of those used for layer-transfer SOI processes. This SPV wafer mapping technique was investigated for sensitivity to dose, implant uniformity, and repeatability for hydrogen and helium implants into (100) silicon wafers through a 1450Å surface oxide.

Introduction

With continued CMOS scaling, problems such as active and passive power dissipation, short channel effects and SRAM single event upsets are becoming increasingly intractible. Engineered substrates, in particular SOI wafers, provide an important avenue to managing these problems and enabling further scaling. This is driving the large-scale manufacturing of state-of-the-art, large diameter SOI wafers having very thin top silicon layers with very tightly controlled layer thickness and uniformity. The two dominant methods for producing such SOI wafers, SIMOX and bonded, ion implant-assisted layer transfer, both rely on ion-implantation technology (1). This places increased emphasis on monitoring and process control methods for ion implantation of non-dopant species. This paper presents preliminary results of an implant monitor technology applicable to donor wafers prior to the bond-and-layer-transfer process.

The donor wafers in bonded, layer-transfer SOI manufacture are generally subjected to implantation of H^+ (2) or co-implantation of H^+ with He^+ (3) to define a cleave plane. Prior to ion implantation, a thermal oxidation of the donor wafer is usually performed. Subsequent to implantation, the donor wafer is bonded to the handle wafer and layer transfer is effected. Proper dose, uniformity, and damage level are critical parameters for proper transfer of the layer from the donor wafer.

Historically, SIMS, Hydrogen Forward Scattering (HFS) or Nuclear Reaction Analysis (NRA) have been used to estimate H^+ dose. These techniques give depth profile information about the implant, but have rather low dose resolution, and are inherently single point measurements. They are also destructive tests, and require costly and complex measurement apparatus. Presently, non-destructive, in-line, whole-wafer mapping of layer-transfer implants is not widely used. In some cases, other metrology techniques are performed after layer transfer, or at the End of Line (EOL) on completed SOI wafers to indirectly infer and monitor the implant performance.

Surface Photo-voltage (SPV) measurement of as-implanted wafers is an established technique to produce high-resolution wafer maps of critical dopant implant parameters, without the need to thermally activate the implanted dopant (4). The small signal SPV technique has proven to be highly sensitive to such critical implant parameters as species, dose, and energy (5,6); sensitivity being defined as is typical, percentage change in metrology device signal divided by percentage change in input variable. Such normalized dose and energy sensitivities have been observed for non-activated dopants with values ranging from 1 to 4. Here, we extend this SPV technique to non-destructive, in-line monitoring of ion implants characteristic of those used for layer-transfer bonded SOI processing.

Experimental - SPV Measurement

A QC Solutions ICT300 SPV system was used for the measurements in this study. This technique is a low intensity, high modulation frequency, ac-SPV measurement. The light source and sensing probe are capacitively coupled to the wafer, approximately 100µm above the wafer surface. The wafer measurements are non-contact and non-destructive allowing product wafers to be measured, as is currently done in epi wafer applications of the system. A blue light photo source was used for the measurements, and all SPV measurement values are reported as dynamic charge (C/m^3) from the ICT300 system. This above-bandgap, modulated light is strongly absorbed, creating electron hole pairs in the near-surface, Si space charge region. Drift and diffusion of these photo-carriers in the space charge region modulates the surface potential, giving rise to the dynamic charge, Qd. At the modulation frequency used, the induced photo-carriers have sufficient time to recombine, creating a dynamic charge (Qd) that depends on the depletion width and the recombination time in the near surface region. The implanted ions introduce damage that acts as recombination centers, reducing photo-carrier lifetime and thus affecting the SPV-measured dynamic charge (4,5).

The wafer samples tested were ~10 ohm·cm, p-type, (100) crystalline substrates with 1450Å thermal oxide. The wafers were implanted with either hydrogen or helium through the oxide, at doses typical for thin SOI layer transfer processes (2,3). The implanted wafers were measured with the 1450Å thermal oxide intact, representative of a donor wafer immediately after ion implantation. Samples were prepared in both 200mm and 300mm wafer sizes.

Results and Discussion

<u>Wafer Map Uniformity</u>

In Figure 1 300mm medium resolution wafer maps (>4400 points/wafer) are presented for wafers implanted with 42 keV He^+, shown to the left, and with 24 keV H^+, shown on the right. Implanted doses of 1 - $1.5x10^{16}/cm^2$ were used, typical of the co-implantation method (3). Both wafers were measured through a 1450Å oxide on as-implanted donor wafers and show similar map signatures for the He^+ and H^+ implants. These wafer features were observed on all wafers measured from multiple implant lots from this particular implanter, indicating the map is characteristic of the implanter. The dynamic charge statistics for each wafer map are presented in Figure 1; average map value (X), range (R), and standard deviation (S). The dynamic charge value for the He^+ implant is >1 order of magnitude larger which may be attributed in part to the larger damage component and deeper implant profile for this implant. Map uniformity and wafer features are quite similar between the two implanted species that were investigated. In this study helium and hydrogen were investigated individually, and co-implanted wafers were not evaluated.

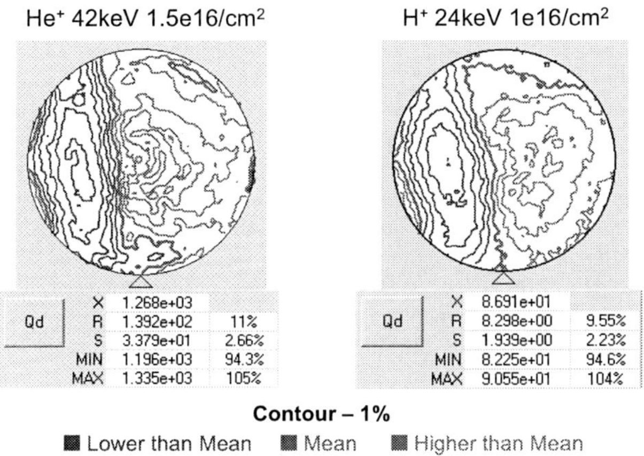

Figure 1. 300mm wafer Map of He$^+$ 42 keV, 1.5E16/cm^2 and H$^+$ 24 keV, 1E16/cm^2 with 1450Å surface oxide (contour interval = 1%).

The SPV measurement sensitivity to implantation can be understood with reference to Figure 2. The photo source illumination is absorbed within approximately 2 microns of silicon below the oxide, creating photo-carriers in the Si near surface region. For wafers doped with ~10^{15}/cm^3 boron, the depletion region extends ~8000Å below the oxide. At 24keV H$^+$ implant energy, the implanted hydrogen depth of ~ 3000Å from the surface produces an extensive damage profile lying within this Si depletion region as well as in the surface oxide. This Si damage directly affects the measured dynamic charge in the near-surface region by creating recombination centers that reduce minority carrier lifetimes, and by creating charged defects that affect the space-charge density, and hence minority carrier drift during illumination. For the 42 keV He$^+$ implant, the implant depth is slightly deeper at ~3500Å, but the damage density is considerably larger due to the fourfold increase in the ion mass. A higher damage concentration profile from the He+ implant gives and increase in the measured dynamic charge signal compared to the H+ measurement.

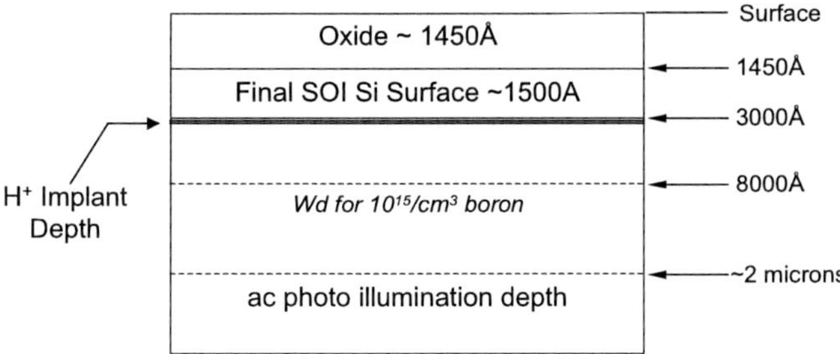

Figure 2. Depth Profile of SOI donor wafer after typical H$^+$ implant.

Since the SPV technique relies upon light transmission through the SiO_2-Si interface, the thickness and the transmission factor of the oxide need to be accounted for. The ICT300 software incorporates a recipe-specific light transmission algorithm, which corrects for variation in oxide thickness. The impact on the measurement of the implant damage in the oxide and oxide thickness variations within-wafer were investigated.

Wafers with a high H^+ implant dose, 6.6×10^{16}/cm^2, typical of mono-implant layer transfer (2) were prepared. SPV measurements were initially performed with the thermal oxide intact. After initial measurement, the wafer oxide was removed using a 5% HF:DI strip for 5 minutes, followed by a DI rinse and spin dry. The wafer surface after wet processing was hydrogen terminated. Direct comparison of an implanted wafer with the 1450Å oxide in place and after an HF strip is presented in Figure 3. By removing the oxide, the effect of the surface oxide damage on the SPV measurement can be eliminated and the response from the implant/substrate alone evaluated. The comparison before and after HF strip is presented on a single 200mm wafer. The implant map characteristic is similar before and after HF strip with similar range (R) and std deviation (S) for the wafer map summaries. The map average for the HF stripped wafer is shifted to a larger Qd value, which is attributed to altered light transmission into the Si through the SiO_2. For the data presented the light transmission algorithm was not applied to the measurement in order to directly evaluate the magnitude and wafer map statistics. The impact of the oxide on light transmission through the SiO_2 for the oxide thickness and light wavelength in the present case is fairly small, <10%.

It is important to note that the measured SPV signal is generated from the implanted Si region, and does not show a dependence on the surface oxide or damage within the SiO_2 region, allowing accurate measurement of the implant without removing the 1450 Å oxide.

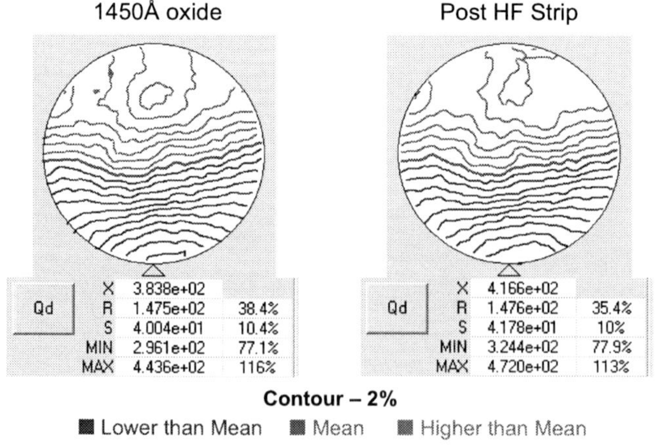

Figure 3. 200mm SPV wafer maps, H^+ 24keV, 6.6E16/cm^2, with and without a 1450Å oxide.

Dose Sensitivity

Figure 4 shows dose sensitivity for He$^+$ 42keV and H$^+$ 25keV implants into oxidized donor wafers. Implant dose was varied around the reported nominal dose values as reported with a normalized value along the x-axis. The dose ranges investigated are ~1.5x10^{16}/cm^2 for the He$^+$ implants, typical of the co-implantation method, and ~6x10^{16}/cm^2 for H$^+$ implants, typical of the mono-implantation method. Average Qd values for full wafer maps were used in the evaluation and were measured through the surface oxide on the as-implanted donor wafers. The measured sensitivity to implanted dose are shown as a normalized Qd sensitivity of ~1.2 for the two cases investigated.

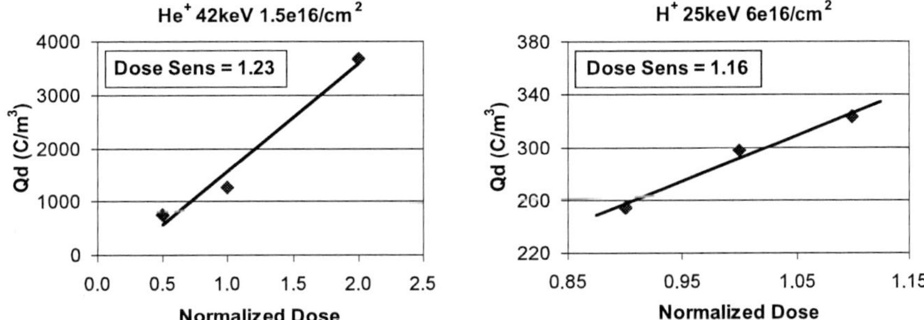

Figure 4. Qd response as a function of normalized dose for He$^+$ 42 keV 1.5E16/cm^2 and H$^+$ 25 keV 6E16/cm^2 with 1450Å surface oxide.

Repeatability

Measurement repeatability was evaluated for the H$^+$ implant with the 1450Å thermal oxide intact. Measurements were collected hourly for the test period of 5 days. The SPV measurements were full wafer maps using the standard wavelength light for the ICT300. Repeatability of the SPV measured technique shows a standard deviation (1-sigma) repeatability of 0.21% for the five day test period. This level of stability within the measurement system, would suggest that a dose variation of 1% can be resolvable using the investigated SPV monitor system with a 2 sigma tolerance level.

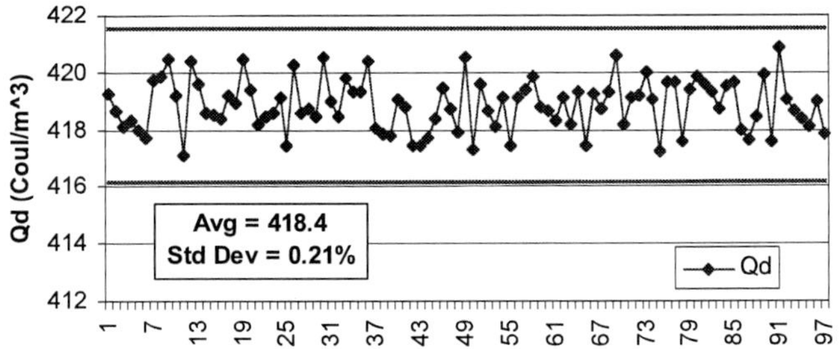

Figure 5. 5 Day Repeatability H$^+$ 24keV, 6.6E16/cm^2

Summary

The small signal SPV measurement technique presented is a capable high-resolution full wafer map technique for monitoring SOI implants, as implanted through a thick surface oxide. Standard hydrogen and helium layer transfer implants were evaluated for sensitivity to dose with a normalized sensitivity of approximately 1.2 for the investigated cases. A demonstrated repeatability of much less than 0.5% 1-sigma was observed for a multi-day period on hydrogen implanted wafers. This non-destructive technique produces high-resolution wafer maps of as-implanted, oxidized donor wafers. The H^+ and He^+ implants can be monitored non-destructively on standard, oxidized donor wafers without altering the SOI manufacturing process flow, allowing in-line SPC control of this critical step within the SOI layer transfer manufacturing process.

References

1. G.K. Celler and Sorin Cristoloveanu, *J. Appl. Phys.* **93**, 4955 (2003)
2. M. Bruel, *Electron. Lett.* **37**, 1201 (1995)
3. A. Agarwal, T.E. Haynes, V.C. Venezia, O.W. Holland and D.J. Eaglesham, *Appl. Phys. Lett.* **72**, 1086 (1998)
4. Tsidilkovski, E., Crocker, K., Steeples, K. Ion Implant Process Monitoring with a Dynamic Surface Photo-charge Technique. *Advanced Semiconductor Manufacturing Conference*, (2004).
5. K. Steeples, E. Tsidilkovski, Photoelectric Measurment Method for Implanted Silicon: A Phenomenological Approach, CP866, *Ion Implantation Technology*, 558-561 (2006).
6. R.S. Nakmanson, *Solid State Electronics* **18**, 617-626 (1975)

ECS Transactions, 6 (4) 185-190 (2007)
10.1149/1.2728859, ©The Electrochemical Society

Low Temperature Properties of Multi-Independent-Gate FET (MIGFET) Operating in Single-Gate and Double-Gate Modes

K-I Na [a, b], S. Eminente [b, c], S. Cristoloveanu [b], L. Mathew [d], A. Vandooren [d], Y-H Bae [e], and J-H Lee [a]

[a] School of Electrical Engineering and Computer science, Kyungpook National University, 1370, Sankyuk-dong, Buk-gu, Daegu 702-701, Korea
[b] IMEP, Minatec INP Grenoble, B.P. 257, 38016 Grenoble Cedex, France,
[c] ARCES Center, Bologna Italy
[d] Freescale Semiconductor Inc., Austin, USA and Crolles, France.
[e] Division of Electronics Eng., Uiduk University, Gandong, Gyeongju, 780-713, Korea

We investigated the temperature dependence of the electrical properties of the MIGFETs, such as the drain current, the transconductance, the threshold voltage, the subthreshold slope and the field effect mobility. Devices with a gate length of 30 nm, a gate width of 30 nm and a silicon fin height of 100 nm were characterized at a temperature range between 77 K and room temperature. The MIGFETs were operated and compared in double-gate (DG) and single-gate (SG) modes in order to reveal the coupling effects between the various gates. The subthreshold slope in SG mode improves with decreasing temperature, but remains far from the ideal value measured in DG mode. The variations of the transconductance and mobility as a function of temperature are explained by strong coupling and series resistance effects. In DG mode, the mobility is highly improved due to volume inversion and reduced surface roughness scattering.

Introduction

The double-gate (DG) FinFET is considered as a promising structure for scaling CMOS down to the decananometer size because of its performance and superior control of short-channel effect [1,2]. In DG FinFETs the two gates are normally tied together. However, more recently, reports of MIGFET (multiple independent-gate FET) fabrication have been presented [3-5]. The MIGFET is a FinFET where the two lateral gates are independent and can be controlled separately. Therefore, this device can be operated in DG mode (interconnected lateral gates) and SG mode (separated gates). The MIGFETs are very attractive thanks to their flexibility which can be used for reconfigurable integrated circuits. They offer: dynamic threshold voltage (V_{th}) control, transconductance modulation, fine threshold voltage tuning by appropriately biasing the opposite gate, simultaneous processing of two different signals applied to the lateral gates, innovative memory schemes, etc. In MIGFETs, 3-D coupling effects occur making the transistor operation rather complex. Low temperature measurements can clarify the device operation and transport properties.

In this paper, we report investigations of the temperature dependence of the MIGFET properties, such as drain current, transconductance, threshold voltage and subthreshold slope. We explore and discuss the main features related to SG mode operation at low temperature, and compare them to DG operation at room temperature.

Device Fabrication and Measurement

SOI wafers were used to fabricate the MIGFET devices. The wafers were patterned to make fins of about 100 nm height with two independent gate regions on the two lateral sides of the fin. The separation between these regions was achieved through a top spacer. The gate stack was composed of a thin oxide layer (t_{ox} = 2.4 nm) and polysilicon gate contacts. Test devices featuring gate length of 30 nm and fin width of 30 nm were probed. All devices were n-channel with undoped body. The buried oxide (BOX) thickness was $t_{box} \approx$ 100 nm. The detailed fabrication process of MIGFET devices were described elsewhere [3].

Figure 1 shows the simplified structure and defines the position of the gate electrodes. In DG mode V_{G1} = V_{G2}, whereas in SG mode V_{G1} was the controlling gate with variable bias applied on V_{G2} and substrate (V_{sub}). All characteristics remained identical after interexchanging the roles of V_{G1} and V_{G2}. The drain voltage was maintained in the linear region ($V_D \approx$ 100 mV).

The performance and the basic properties of MIGFETs were explored from room temperature down to 77 K. A cryogenic system, where the devices are in the dark with a continuous flow of liquid nitrogen cryostat, was used for the measurements.

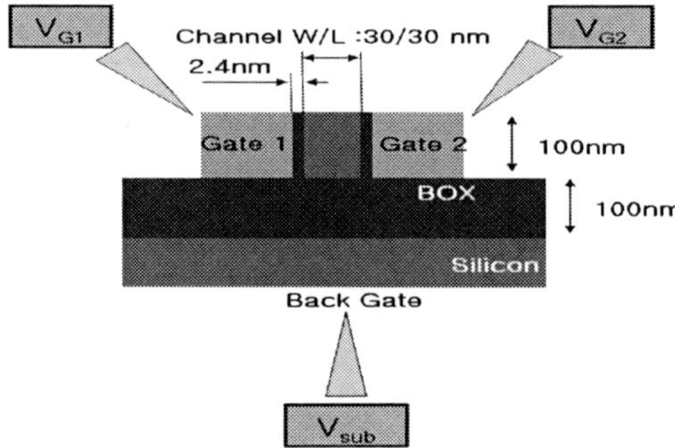

Fig. 1. Simplified MIGFET structure defining the position of the gate electrodes.

Characteristics of Device

From systematic measurements, we determined the threshold voltage, the subthreshold slope, the transconductance and the mobility as a function of gate voltage for various temperatures. Figure 2 shows the variation of threshold voltage V_{th1} with opposite gate voltage V_{G2}, in SG mode at 300 K; the substrate voltage was also varied from -10 V (accumulation wise) to +10 V (inversion wise). For decreasing gate voltage V_{G2}, threshold voltage increases since the opposite channel tends to change from inversion to accumulation. This coupling effect is similar to the one observed in regular planar SOI MOSFETs when V_{th1} is monitored as a function of the substrate bias and yields [6]: $\Delta V_{th1}/\Delta V_{sub} \approx t_{ox}/t_{box}$. In MIGFETs, the coupling coefficient is very strong $\Delta V_{th1}/\Delta V_{G2} \approx$ 1, simply because the two lateral oxides are identical. Note also in Figure 2 that the formation of an accumulation layer facing the inversion layer is difficult, as explained in [7].

An interesting aspect is that the threshold voltage value and coupling coefficient $\Delta V_{th1}/\Delta V_{G2}$ do not change with the substrate bias (Fig. 2). The explanation is two-fold:

(i) The lateral coupling between V_{G1} and V_{G2} is much stronger than the vertical coupling between $V_{G1,2}$ and V_{sub}.

(ii) The surface potential at the fin-BOX interface is primarily defined by the fringing field between the lateral gates [8], rather than by the substrate bias (as in planar SOI MOSFETs).

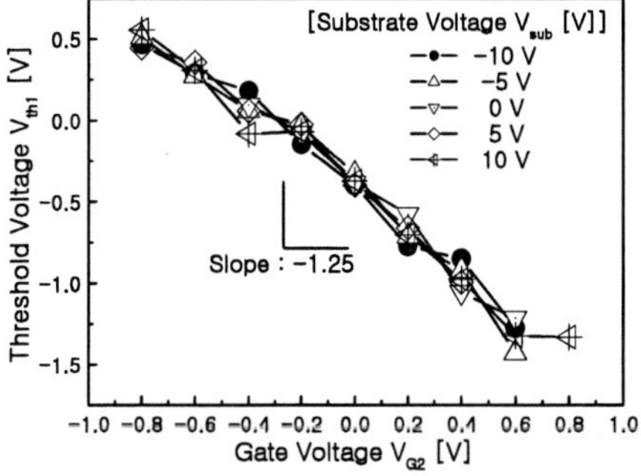

Fig. 2. Threshold voltage in SG mode versus opposite gate voltage for various substrate bias, at room temperature.

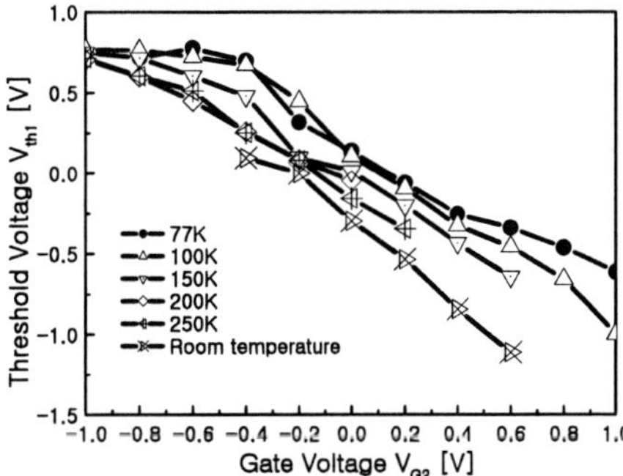

Fig. 3. Threshold voltage in SG mode as a function of opposite gate voltage and temperature with grounded back gate.

The threshold voltage in SG mode as a function of opposite gate voltage and temperature, with grounded back gate, is shown in Figure 3. The value of threshold voltage at $V_{G2} = -1$ V tends to saturate and a plateau is observed at low temperature, revealing the gradual formation of the accumulation layer. The variation of threshold voltage with temperature is accentuated for positive V_{G2} bias. However, the slope is more or less temperature independent because the coupling coefficient is governed by the oxide thickness.

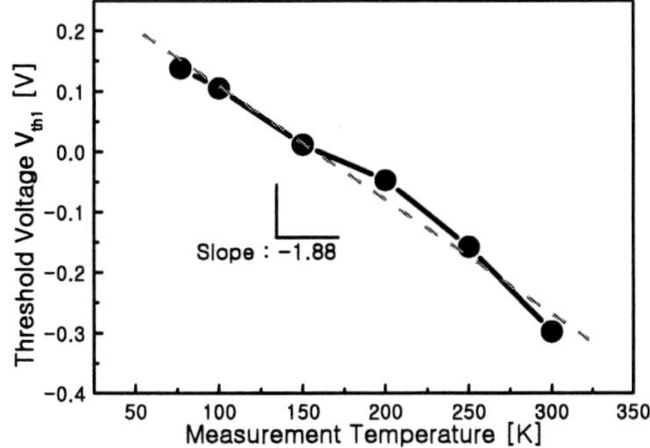

Fig. 4. Threshold voltage as a function of temperature.
The opposite gate and substrate were grounded.

The increase in threshold voltage at low temperature, measured by keeping the opposite gate and the substrate grounded, is illustrated in Figure 4. The slope is nearly constant, confirming that the MIGFET operates in full depletion regime in the whole temperature range. The variation of the threshold voltage with temperature can be expressed as [9-11]

$$\frac{dV_{th}}{dT} = \frac{d\phi_F}{dT}\left[\alpha\left(\frac{q\varepsilon_{Si}N_a}{\phi_F C_{ox}^2}\right)^{1/2} + 2 + \frac{qD_{it}}{C_{ox}}\right]$$

where $\alpha = 1$ for a partially depleted film and $\alpha = 0$ for a fully depleted film, ϕ_F is the Fermi potential, N_a the Si doping concentration, C_{ox} the gate oxide capacitance and D_{it} the interface state density. Indeed, the threshold voltage (V_{th}) temperature coefficient of about -1.88 mV/K is experimentally observed for our device, in fair agreement with the theoretical value calculated from the equation above: -1.84 mV/K [12].

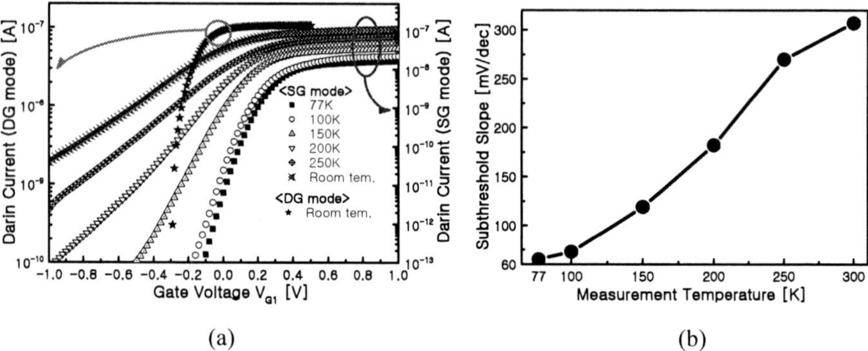

(a)   (b)

Fig. 5. (a) Subthreshold drain current versus gate voltage in a MIGFET: SG mode with $V_{G2} = 0$ at various temperatures, and DG mode at 300 K. The substrate was grounded. (b) Summarized subthreshold slope with varying temperature in SG mode.

Figure 5 (a) shows the semi-logarithmic drain current as a function of gate voltage with varying the temperature, in SG mode. The drain current decrease, observed when the temperature is lowered down to 100 K, is dominated by the threshold voltage increase. The subthreshold slope in SG mode improves from 307 mV/decade at 300 K to 65 mV/decade at 77 K, as illustrated in Figure 5 (b). The variation is nearly linear following the usual kT/q decay. The subthreshold slope is modest in SG mode because of the lateral coupling effect between the two gates which feature identical oxide thickness. This mode of operation is therefore not suitable for circuit applications requiring fast switching from OFF-state to ON-state. By contrast, in DG mode, the swing is greatly improved and reaches the ideal theoretical value (60 mV/dec at 300 K). It would be interesting to evaluate the contribution of volume inversion effect in this ideal swing value. Volume inversion means that the channel forms in the fin volume, far from Si/SiO$_2$ interfaces, and implies a reduced impact from interface traps [13].

Figure 6 shows the transconductance curves in SG and DG modes. The striking message is the clear transconductance gain (300%) at room temperature, achieved by changing from SG mode to DG mode. This gain exceeds the 200% enhancement that can be expected from the parallel combination of two independent channels. Similar results, reported earlier by Ernst et al [14], were explained by the volume inversion effect. Indeed, in DG transistors, the impact of surface roughness scattering is reduced and the effective field cancels at mid-distance between the gates. The electron mobility can therefore be higher than that in SG mode.

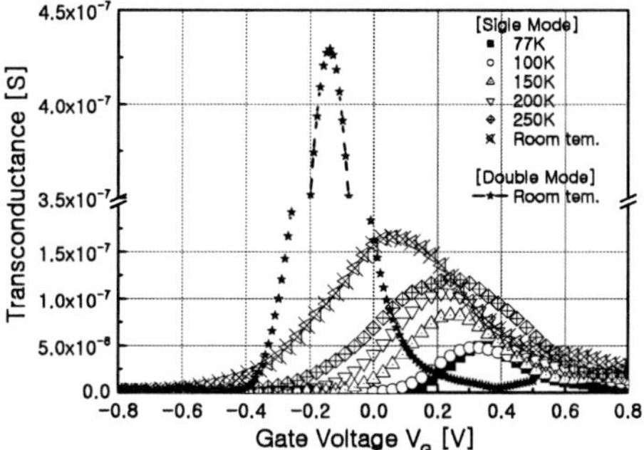

Fig. 6. Transconductance versus gate voltage for various temperatures in SG mode ($V_{G2} = V_{sub} = 0$) and in DG mode at room temperature.

A less predictable feature observed in Figure 6 is the transconductance degradation with temperature in SG mode. It is well known that the carrier mobility increases at low temperature as a consequence of reduced phonon scattering from 300 K to 77 K. If so, why would the transconductance show an opposite trend? We suspect that several issues may be important:

- The measurement conducted with constant V_{G2} bias does not maintain a constant field and surface potential at the opposite interface. Since threshold voltage (V_{th}) increases at low temperature, the surface potential for $V_{G2} = 0$ tends to reach the flat-band condition, which is known to be a cause of transconductance degradation [6].

- The series resistance can increase significantly at low temperature due to incomplete dopant activation. This scenario needs to be further investigated by probing long channel MIGFETs.

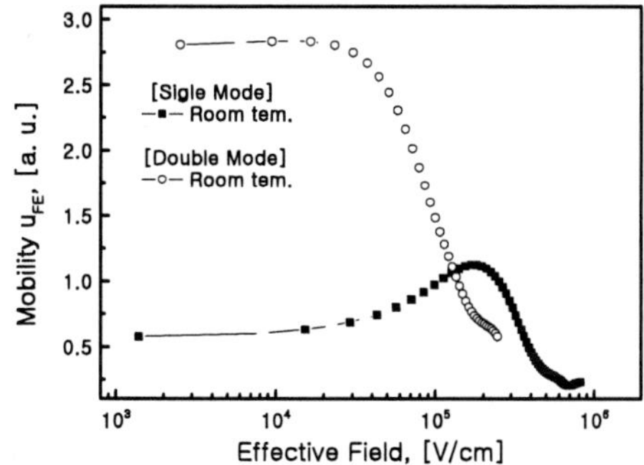

Fig. 7. Field-effect mobility versus effective field at room temperature in SG and DG modes.

Figure 7 compares the field-effect mobility as a function of effective field at room temperature. The mobility was deduced from the transconductance whereas the effective field was computed as in reference [15]. The difference in mobility between SG and DG modes is very large, confirming the advantage of DG operation. Nevertheless, one can question the validity of using the same extraction procedure for SG and DG modes.

Conclusion

The electrical properties of MIGFETs, operated in either SG or DG mode, have been investigated between room and liquid nitrogen temperature. A number of results are in agreement with the existing models developed for planar fully depleted SOI MOSFETs. At low temperature, the threshold voltage increases and the subthreshold slope improves. The coupling between the lateral gates is temperature independent and very strong, able to offset the substrate bias effect (*i.e.*, vertical coupling). DG mode offers a considerable improvement in transconductance and subthreshold slope. The transconductance behavior is very complex and difficult to explain by simple mobility considerations. Measurements on long devices, free from series resistance effects, would be useful for clarifying the transconductance trends.

Acknowledgments

This research was supported by Brain Korea 21 (BK21) of the Ministry of Education and the Korea Science and Engineering Foundation(KOSEF) through the National Research Lab. Program funded by the Ministry of Science and Technology (No. M10600000273-06J0000-27310)

References

1. W. Zhang et al., *IEEE Trans. Electron Devices*, vol. 52, no.10, pp. 2198–2206, (2005)
2. G. Pei et al., *IEEE Trans. Electron Devices*, vol. 50, no.10, pp. 2135–2143, (2003)
3. L. Mathew et al., in Proc. *IEEE Int. SOI Conf.*, pp. 187–189 (2004)
4. Y. X. Liu *et al.*, in *IEDM Tech Dig.*, pp. 986–988, (2003)
5. D. M. Fried *et al.*, *IEEE Electron Device Lett.*, vol. 25, no. 4, pp. 199–201, (2004)
6. S. Cristoloveanu, S.S. Li, *Electrical Characterization of Silicon On Insulator Materials and Devices*, Kluwer Academic Publishers, New York, 1995
7. S. Eminente et al., *Solid-State Electronics*, in press (2007)
8. S. Cristoloveanu et al., *Int. J. of High Speed Electronics and Systems*, 16(1), p. 9–30 (2006)
9. F. Balestra, et al., *Device and Circuit Cryogenic Operation for Low Temperature Electronics*, p.45-46, KLUWER ACADEMIC PUBLISHERS, Boston (2001)
10. G. Groeseneken et al., *IEEE Electron Device Lett.*, vol. 11, no. 8, pp. 329–331, (1990)
11. T. Elewa et al., *IEEE Trans. Electron Devices*, vol. 37, no.4, pp. 1007–1019, (1990)
12. J-P. Colinge, *SILICON-ON-INSULATOR TECHNOLOGY: Materials to VLSI, 3rd Edition*, p.262, Kluwer Academic Publishers, Boston/Dordrecht/London (2004)
13. F. Balestra et al., *IEEE Electron Device Lett.*, vol. 8, no. 9, pp. 410–412, (1987)
14. T. Ernst et al., *IEEE Trans. Electron Devices*, 50(3), p. 830–838 (2003)
15. Y. Taur et al., *Fundamentals of Modern VLSI Devices*, p.133, Cambridge University Press, New York (2002)

ECS Transactions, 6 (4) 191-196 (2007)
10.1149/1.2728860, ©The Electrochemical Society

Analysis of Trench Gate Power LDMOS Transistors in Thin SOI Technology

I. Cortés, P. Fernández-Martínez, D. Flores, S Hidalgo and J. Rebollo

Centro Nacional de Microelectrónica (CNM-CSIC),
Campus UAB, 08193 Bellaterra, Barcelona, Spain.
Tel. + 3493 594 7700, Fax. +3493 580 1496, e-mail: ignasi.cortes@cnm.es

The on-state performance of conventional thin SOI power LDMOS transistors can be further improved by implementing a wide trench gate partially filled with polysilicon, leading to asymmetric oxide characteristics. The on-state resistance of the analyzed TGLDMOS structure is lower than that of the LDMOS counterpart, but the structure design has to be optimized to minimize the added contributions to the parasitic capacitances. For this purpose, a modified TGLDMOS is proposed to reduce the gate-drain capacitance and to increase the cut-off frequency. The trench is modeled and simulated in this work with special emphasis on the gate-drain capacitance. An extensive simulation study has corroborated the expected electrical performance improvements of both TGLDMOS and modified TGLDMOS power transistors for 80 V switching and amplifying applications.

Introduction

LDMOS transistors are widely used as switches in Power ICs and RF power amplifies. One of the main constrains in switching applications is the reduction of the specific on-resistance (R_{on-sp}) while a high transconductance (g_m) over a wide range of gate voltages is required for RF amplifiers. In addition, power SOI technologies provide clear advantages over Bulk ones in terms of dielectric isolation and parasitics reduction at the cost of degraded thermal performances. Several techniques have been proposed to minimize R_{on-sp} in LDMOS devices. For instance, the resistance of the drift region necessary to provide the required breakdown voltage (V_{br}) has to be minimized. The superjunction concept (1) applied to LDMOS structures allows an increase of the drift doping level at a given V_{br} value. A second approach is the reduction of the total drift region length by placing an oxide trench in the LDMOS drift region (2). Finally, some authors have proposed the use of a vertical trench gate instead of the conventional lateral one (3, 4) leading to a further reduction of the LDMOS cell length.

This paper is addressed to the analysis of trench gate LDMOS (TGLDMOS) transistors to show the improvement of the electrical characteristics in comparison with the conventional LDMOS architecture in Thin SOI substrates. The R_{on-sp}/V_{br} trade-off as well as thermal resistance and capacitance behavior are analyzed by means of 2D numerical simulations to explore the suitability of the TGLDMOS structure for 80V applications.

The TGLDMOS structure

The cross sections of a conventional LDMOS and the TGLDMOS structure in thin SOI substrates are shown in Fig. 1 (a) and (b), respectively. The main features of the analyzed

TGLDMOS are its asymmetric oxide in the trench gate ($T_{ox2} > T_{ox1}$), and its oxide thickness between the polysilicon gate edge and the drift region ($T_{trench} - L_{poly}$) to avoid as much as possible the creation of an accumulation region, thus increasing the V_{br} value. Besides, the Silicon path from drain to source created along the trench sidewalls allows the reduction of the TGLDMOS cell length (L_{cell}) without degrading the V_{br} value.

Table I shows some of the geometrical parameters of both structures. Note that the drift region length in the LDMOS structure (L_{LDD}) is equal to the Silicon path along the trench sidewalls ($L_{Trench} + T_{Trench}$) in the TGLDMOS case. Both structures have been extensively analyzed by means of 2-D numerical simulations performed with Sentaurus (5) in order to compare their static and dynamic electrical characteristics.

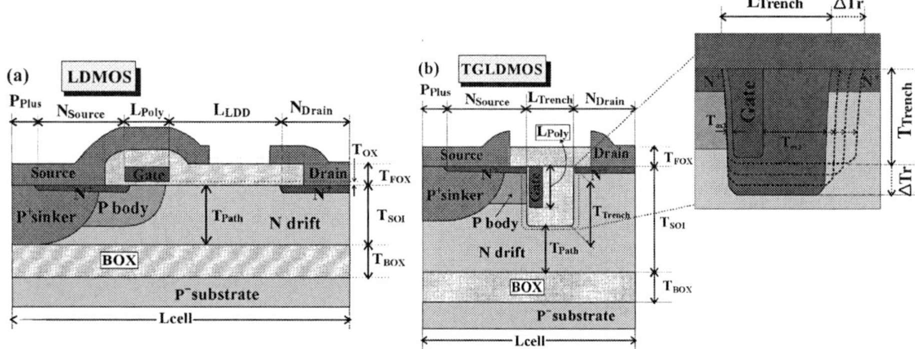

Figure 1. Cross-section of Thin-SOI LDMOS (a) and TGLDMOS (b) transistors.

TABLE I. Geometrical and Technological Parameters.

Transistor	T_{OX}	T_{FOX}	T_{BOX}	T_{Trench}	L_{Poly}	L_{LDD}	L_{Trench}
LDMOS	40 nm	0.7 μm	1 μm	--------	1 μm	3 μm	--------
TGLDMOS	40 nm	0.7 μm	1 μm	1.5 μm	1 μm	--------	1.5 μm

TABLE II. Simulated Electrical Performances.

Electrical Parameter	LDMOS (L_{cell}=9 μm)		TGLDMOS (L_{cell}=6 μm)	
	$N_{dr}=5\times10^{15}$ cm^{-3}	$N_{dr}=1\times10^{16}$ cm^{-3}	$N_{dr}=5\times10^{15}$ cm^{-3}	$N_{dr}=1\times10^{16}$ cm^{-3}
V_{br} (V)	94.5	93.3	97.5	97.4
R_{on-sp} ($\Omega\times$cm^2)	2.38×10^{-3}	2.42×10^{-3}	1.8×10^{-3}	1.42×10^{-3}
V_{br}^{2}/ R_{on-sp}	3.75×10^{6}	3.40×10^{6}	5.30×10^{6}	6.70×10^{6}

Static electrical characteristics

The main attention has been put on the analysis of V_{br} and R_{on-sp}. Fig. 2 shows the V_{br} evolution versus T_{Path} for two drift doping levels (N_{dr}) in LDMOS and TGLDMOS structures. Similar breakdown conditions are achieved in both cases with the same N_{dr} value. The maximum V_{br}, which corresponds to the optimal RESURF, is obtained at a lower T_{Path} value in TGLDMOS structures, this value decreasing when increasing N_{dr}. The optimal T_{Path} values giving the maximum V_{br} in the LDMOS case are 1.6 and 3 μm for N_{dr} = 5×10^{15} and 1×10^{16} cm^{-3}, respectively. In the equivalent TGLDMOS, these optimal values are 1.6 and 0.9 μm at identical N_{dr} values. Moreover, in the TGLDMOS structure, V_{br} sharply decreases once the V_{br} peak value is reached, this behavior being more remarkable with N_{dr} = 1×16 cm^{-3}. This is due to the increase of the N charge beneath the P^+ as T_{Path} increases. The depletion region results from the combination action of P^+ sinker, P body diffusion and the BOX/Substrate layer, and the T_{Path} increase accounts for a large area to be depleted, thus resulting in a V_{br} degradation. The simulated thermal resistance ($R_{th} = T_{max}/V_d \cdot I_d$) values at $V_g = 4$ V and $V_d = 28$ V are also represented in Fig. 2 as a function of T_{Path}. The TGLDMOS shows higher R_{th} values than the LDMOS counterpart due to the L_{cell} reduction. However, the simulated R_{on-sp} values indicated in Table II show that the TGLDMOS structure could improve the V_{br}^2/R_{on-sp} figure of merit by a factor of 2.

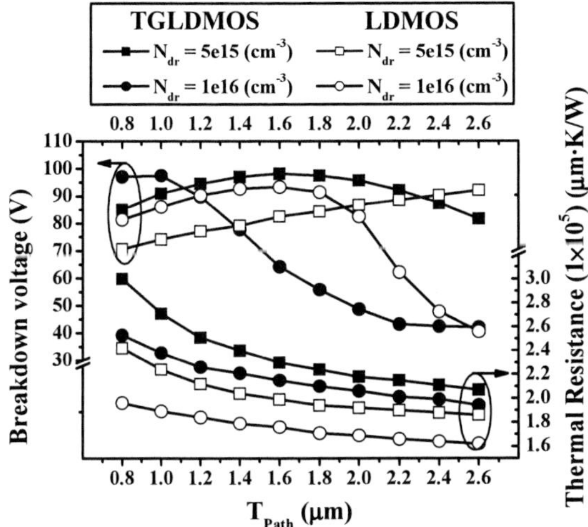

Figure 2. Simulated V_{br} and R_{th} versus T_{Path} in LDMOS and TGLDMOS transistors.

TGLDMOS dynamic characteristics

Power LDMOS transistors suitable for RF applications must exhibit a high transconductance (g_m) at low gate voltages and g_m should be as flat as possible over a wide range of gate voltages. This is crucial when the LDMOS is used to implement RF power amplifiers since a broad cut-off frequency (f_T) plateau is beneficial for low noise high gain and large signal operation (6). Hence, a linear transfer characteristic (I_d vs V_g) and a high maximum drain current (I_{dmax}) referred to as the compression current is desirable. Furthermore, the RF performance is enhanced when the gate-drain (C_{gd}) and gate-source (C_{gs}) capacitances are reduced as much as possible. In a conventional SOI LDMOS structure, the feedback capacitance C_{gd} is defined as the gate/drift overlap capacitance (C_{ov}) in series with the gate/drift junction capacitance (C_j). The C_{ov} value basically depends on the extension of the N drift under the polysilicon gate and the gate oxide thickness (T_{ox}). The C_j value depends on the N_{dr} doping concentration and the P body/N drift junction depth (7). There is another contribution to the C_{gd} due to the interaction between gate-drain metallic interconnections (C_{gdm}). In the case of an LDMOS structure with a source field plate (SFP), this capacitance does not contribute to the total C_{gd} value since it is like a Faraday shield which avoids the coupling between gate and drain terminals. Unfortunately, the overlap between the polysilicon gate and the source metal due to the SFP adds a capacitance contribution (C_{gsm}) to the total C_{gs} value.

As far as TGLDMOS structure is concerned, the asymmetric gate trench gives rise to two different oxide thicknesses, T_{ox1} and T_{ox2} as illustrated in Fig. 1. Then, the total C_{gd} value in the TGLDMOS is modified by a new component, C_{ov2}, which mainly depends on T_{ox2}. On the other hand, the C_{gs} component is only affected by the gate oxide component (C_{ox}). In order to alleviate the further increase of the C_{gd} value in the TGLDMOS structure while maintaining the same drift path length between drain and body, L_{Trench} and T_{Trench} have been increased and decreased Δ_{Tr}, respectively, as illustrated in Fig. 1. This trench modification accounts for a thicker T_{ox2}, and the subsequent decrease of C_{ov2} component. The simulation results show that in spite of the C_{ov2} decrease as Δ_{Tr} increases, there is also an increase of C_{ov1} (trench gate/drift overlap capacitance) that could compromise the C_{gd} evolution. Concretely, in the case of $\Delta_{Tr} = 0.4$ µm, the increased overlap between polysilicon gate and N drift region strongly affects the C_{gd} value at low V_d values. The evolution of V_{br} and $R_{on\text{-}sp}$ versus Δ_{Tr} is plotted on Fig. 3 for two different N_{dr} values. As it can be seen, $R_{on\text{-}sp}$ continuously decreases with the Δ_{Tr} increase due to the higher accumulation region in the trench corner vicinity close to the polysilicon layer. V_{br} exhibits a slight decrease for Δ_{Tr} up to 0.3 µm and a pronounced degradation at higher Δ_{Tr} values due to the enhanced high electric field peaks at trench corner.

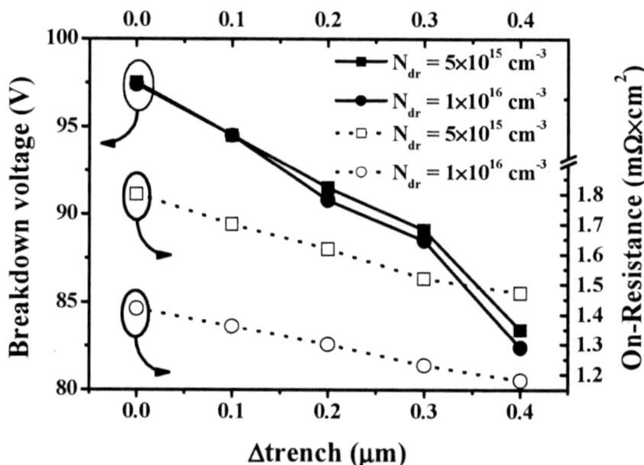

Figure 3. V_{br} and $R_{on\text{-}sp}$ evolutions versus Δ_{Tr} in TGLDMOS transistors.

According to the above mentioned results, the TGLDMOS structure should be modified in order to be used as an amplifier by lowering R_{th}, rising and widening g_m and decreasing C_{gd}. The T_{Path} increase at a given N_{dr} value leads to an I_{dmax} increase, thus improving g_m. However, as already stated, this solution degrades the V_{br} value. A strategy to avoid that is the definition of a deep P^+ sinker through Boron high energy multi-implantation to further deplete the N drift area. As far as a C_{gd} decrease is concerned, the drain diffusion should be separated a certain distance Δ_L from the trench gate, placing a SFP over it. These modifications can be seen in Fig. 4 where the cross-section of the modified TGLDMOS and a detail of all the resultant parasitic capacitance components are illustrated.

The SFP contributes to the N drift depletion between the trench and the N^+ drain diffusion, thus decreasing the total C_{gd} value. To explain the C_{gd} variation, an additional capacitance (C_{dep}) has been placed in series with C_{ov2}. On the other hand, the SFP adds and additional capacitance component (C_{gsm}) to the total C_{gs}. The evolution of C_{gd} and C_{gs} versus V_{ds} for LDMOS, TGLDMOS and modified TGLDMOS structures with Δ_L=1.5 µm is represented in Fig.5. As it can be seen, the C(V) evolution shows a better reduction of C_{gd} in the case of the LDMOS structure, although the modified TGLDMOS shows a better behavior than the TGLDMOS. The SFP position close to the gate edge and the lower drift area in the LDMOS structure accounts for a quicker depletion process in the P body/N drift junction, thus further decreasing C_j and C_{gd} with respect to the TGLDMOS structure. Concerning C_{gs}, the lower interaction between the source metal and the polysilicon gate in the TGLDMOS with respect to the LDMOS leads to a C_{gsm} reduction, and accounts for the smaller C_{gs} smaller in the TGLDMOS.

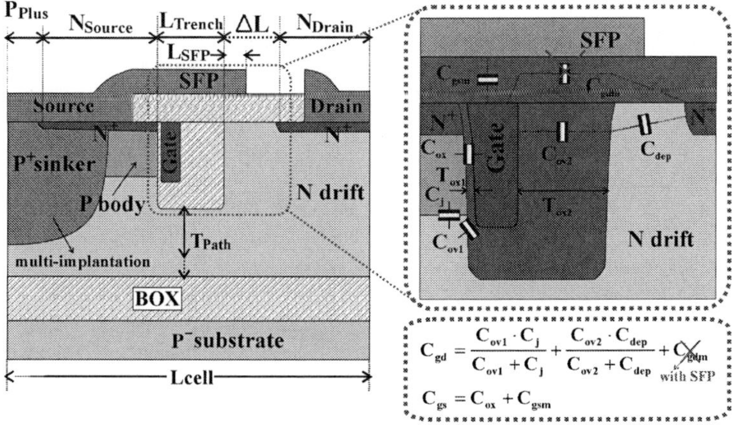

Figure 4. Cross-section of the modified TGLDMOS structure.

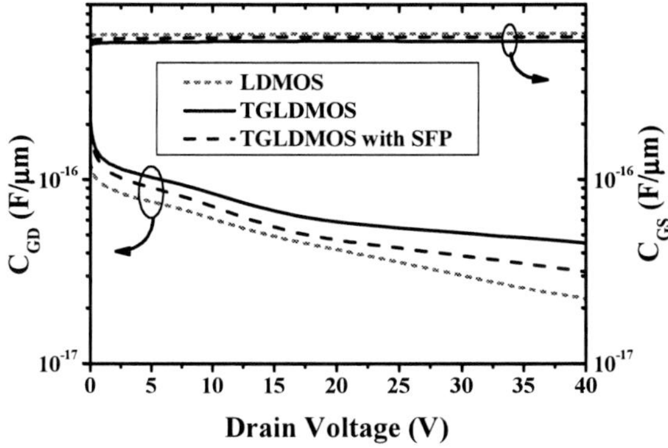

Figure 5. Simulated C_{gd} and C_{gs} evolutions versus V_d for the three analyzed LDMOS structures. TGLDMOS: T_{Path} = 0.9 µm, Δ_L = 1.5 µm and L_{SFP} = 0.5 µm.

According to the f_T simulation results, the modified TGLDMOS structure can operate as an amplifier by increasing T_{Path}, although T_{Path} values higher 1.6 µm lead to V_{br} degradation. Table III summarizes the simulated results of V_{br}, $R_{on\text{-}sp}$ and $f_{T\,max}$ obtained in both LDMOS and modified TGLDMOS structures with $T_{Path} = 1.6$ µm.

TABLE III. Simulated Electrical Performances.

Electrical Parameter	LDMOS (L_{cell}=9 µm) N_{dr}=1×10^{16} cm^{-3}	TGLDMOS (L_{cell}=7.5 µm) N_{dr}=1×10^{16} cm^{-3}
V_{br} (V)	93.3	81.3
$R_{on\text{-}sp}$ (Ω×cm^2)	2.42×10^{-3}	1.34×10^{-3}
f_{Tmax} (GHz)	6.2 at V_G=2.94 V	6.8 at V_G=3.26 V

Conclusions

This paper analyzes by means of numerical simulations the electrical performance of the trench gate LDMOS (TGLDMOS) addressed to 80 V applications, and provides a comparison with the conventional LDMOS structure in Thin SOI substrates. The TGLDMOS is characterized by the use of a trench for the channel formation, which leads to a device cell length reduction. Simulation results show that the TGLDMOS structure improves the $V_{br}^2/R_{on\text{-}sp}$ figure of merit by a factor of 2 compared to the LDMOS. Hence, this structure is an interesting device for switching applications in SOI based Power ICs due to the $R_{on\text{-}sp}$ reduction. However, thermal and capacitive behaviors limit the suitability of this structure in RF power amplifiers. In this sense, a modified TGLDMOS is also proposed to overcome these drawbacks. The main features of the modified TGLDMOS are the definition of a deep P^+ sinker through Boron high energy multi-implantation, the spacing of the trench gate from the N^+ drain diffusion, and the source field plate over the trench gate and part of the N drift region. Simulation results evidence the C_{gd} reduction in comparison with the TGLDMOS. Improvements on $R_{on\text{-}sp}$ and $f_{T\,max}$ of the modified TGLDMOS structures prove the suitability of this structure for RF power amplifiers applications

Acknowledgments

This work is supported by the Ministerio de Educación y Ciencia, Spain, under grant TEC2005-07511.

References

1. T. Fujihira, *Jpn. J. Appl. Phys.*, **36**, 6254 (1993).
2. J. M. Park, S. Wagner, T. Grasser and S. Selberherr, *Solid State Electronics*, **48**, 1007 (2004).
3. P.H. Wilson, S. Sapp and N. Thornton, US Patent 7033891 (2006).
4. Seung-Chul Lee, Jae-Keun Oh, Min-Koo Han and Yearn-Ik Choi, *Physica Scripta*, **T101**, 58 (2002).
5. TCAD TOOL Suite, Synopsys (2006).
6. E. Khon, A. Lepore, H. Lee and M. levy, in *Proc. IEEE/Cornell Conf. on Adv. Concepts in High Speed Semicond. Dev. and Circuits*, 91 (1989).
7. S.M. Xu, P.D. Foo, J.Q. Liu F.J. Lin and C.H. Ren, in *Proc. IEEE Int. Electron Devices Meeting (IEDM)*, 201 (1999).

ECS Transactions, 6 (4) 197-203 (2007)
10.1149/1.2728861, ©The Electrochemical Society

Assessment of the CBR quantum transport simulator on experimentally fabricated nano-FinFET

H. Khan, D. Mamaluy and D. Vasileska

Department of Electrical Engineering, Arizona State University, Arizona 85287, USA

We have utilized fully self-consistent quantum mechanical simulator based on Contact Block Reduction (CBR) method, to simulate experimentally fabricated 10nm FinFET device. A series of simulations have been performed with varying S/D extension length and doping profile to match the experimental data. The simulation results have been found to be in good agreement with the experimental data in the subthreshold regime. Small signal analysis has been performed to extract device capacitances and to compare the intrinsic propagation delay to that of experimental device.

INTRODUCTION

The accurate modeling of nano-scale devices, especially with gate length approaching 10 nm, necessitates using of fully quantum mechanical approach. As device dimensions are becoming smaller than the mean free path and the dephasing length, ballistic transport starts playing the main role in device behavior. Devices that exhibit quasi-ballistic transport include DG-MOSFETs, modulation doped 2DEGs, resonant tunneling devices, etc., and most recently, quasi-planar FinFETs. In this work we utilize an efficient method based on Green's function approach to calculate self-consistently transport properties in nanoscale FinFET devices and to match the simulated results to the corresponding experimental findings. This method is termed as the Contact Block Reduction (CBR) method [1], [2].

The CBR method allows one to calculate the ballistic transport properties of a two- or three-dimensional device that may have any shape, potential profile, and any number of leads. In this method, quantities like the transmission function and the charge density of the open system can be obtained from the eigenstates of a corresponding closed system $H^0 |\alpha\rangle = \varepsilon_\alpha |\alpha\rangle$ and the solution of a very small linear algebraic system for every energy step E . Importantly, it has been shown that when using generalized von Neumann boundary conditions [2], only a very small percentage of the total number of eigenstates is necessary to accurately construct the open system solution. The self-consistent solution of the ballistic or quasi-ballistic transport properties of an open device requires repeated solution of the Schrödinger and Poisson equations. To improve the convergence of this highly non-linear set of coupled equations we utilize the *predictor-corrector* approach [3], [4], which has been adopted for the open systems [5]. The flow chart for our in-house self-consistent simulator showing major steps is presented in Figure 1. After the initial guess for the potential and the *initial* number of device eigenstates, the CBR loop is started. For each CBR-Poisson iteration the following tasks are performed: 1) transverse lead modes are calculated; 2) eigen-problem is solved for closed-system with von Neumann boundary conditions at the contacts; 3) open-system solution is constructed. The simulator has been developed to incorporate the automatic determination of the required number of device eigenstates and lead modes for each iteration to yield desired accuracy ε. The accuracy ε also determines the upper error norm for the functional F; if $\|F\| < \varepsilon$ then the solution is considered to be converged and the next bias point can be processed, otherwise the predictor-corrector approach is invoked to determine correction $\Delta\varphi$ to the potential. With updated potential φ CBR routine is called again and the loop continues until convergence is achieved. Note that the CBR module is called for each non-equivalent Si valley to obtain the LDOS and transmission function for each valley; then the total charge density, currents, etc. are calculated as the corresponding sums.

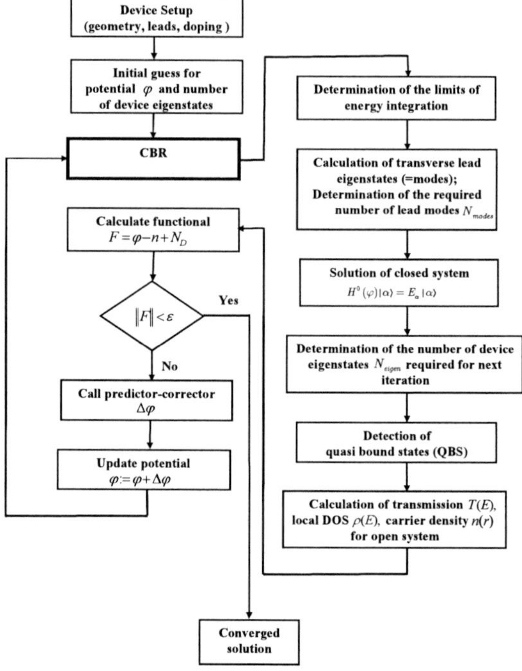

Figure 1. Flow chart showing major steps in CBR algorithm.

The goal of this work is to simulate an experimentally fabricated 10 nm FinFET device with fin width of 12 nm [6]. The corresponding transport simulation area for this device is about 50 nm (device length) by 25 nm (device width). While the CBR method for quantum transport simulation can be used with any multi-band Hamiltonians, including the tight-binding and $k \cdot p$ [1], in this work, we choose to adopt the effective mass model and finite difference discretization to be able to simulate this relatively 'large' FinFET device within a reasonable frame of time. We have modified our CBR simulator in such a way that semiconductor devices on wafers of *arbitrary crystallographic orientation* can be simulated. This was necessary to match the experimental data [6] for a FinFET device fabricated on (110) wafer plane.

A known peculiarity of ultra-scaled nano-transistors is that source, drain and gate regions are usually heavily doped, therefore it is important to include quantum-mechanical effects of exchange and correlation. In this work this is done via the local density approximation (LDA). The phenomenological scattering on the phonons using the relaxation time approximation has been taken into account. Since this phonon scattering model rely on phenomenological parameters, in the work we present results that include into account this phenomenological scattering on phonons as well as purely ballistic ones (that do not depend on such parameters).

SIMULATION RESULTS

FinFET device description

Figure 1 (left panel) depicts the geometry of the FinFET device being investigated in this study. The fin is usually made thin enough when viewed from above so that both gates simultaneously control the entire fully depleted channel film. Usually the top surface of the fin is covered by a thicker oxide than the oxide of the side gates (front and back); therefore channels form only along the vertical surfaces of the fin. The fin thickness (t_{si}) is considered as the most important process parameter as it controls the carrier mobility as well as threshold voltage [7]. Thickness of *Si* fin is usually required to be smaller than the gate length, for efficient suppression of short channel effects. In this work, however, we are simulating an experimentally fabricated 10 nm FinFET structure [6] that did not satisfy this criterion and had the geometry shown in the right panel of Figure 2, with the gate length of 10 nm and fin thickness of 12 nm.

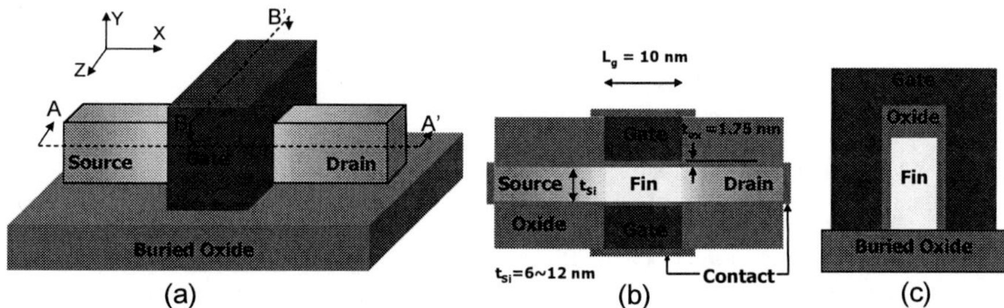

Figure 2. (a) – 3D schematic view, (b) – top view along A-A' cross-section (not to scale), and (c) –side view along B-B' cross-section.

In the experiment the gate electrode consisted of n^+ polysilicon, and the side gate-oxide thickness was 1.75 nm. In our simulations we are using the same device parameters. The effect of top gate on transport is assumed to be negligible due to much thicker gate oxide compared to side gate oxide.

Comparison with experimental results

A comparison of the simulated transfer characteristics at $V_D = 0.1V$ with experimental data from Ref. [6] is presented in Figure 3. The simulation results closely match the experimental data over the subthreshold regime. However, above threshold, as the gate voltage increases, the deviation between simulated and experimental data increases. Our subsequent simulation shows that while the introduction of phenomenological scattering with phonons (that has been done here using the relaxation time approximation, with the scattering parameter $\eta = \hbar/\tau_r = 2 \times 10^{-3}$ eV) somewhat reduces the on-state current values (Figure 3), it can not however, fully explain the large difference between the experimental and simulated current values for the high gate voltages. Therefore, it is reasonable to assume that other scattering mechanisms, that could affect the on-state current only, are responsible for the discrepancy.

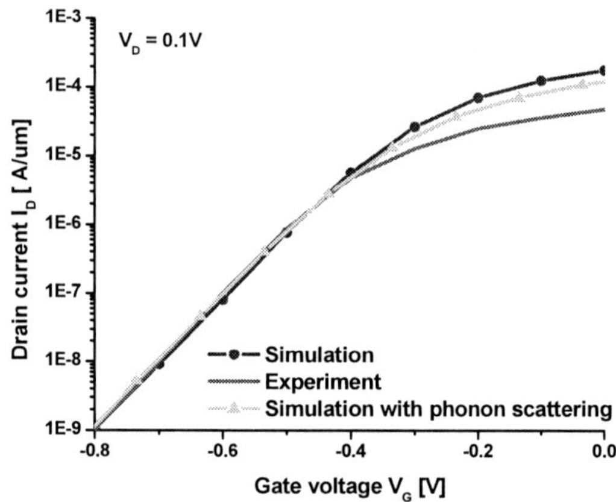

Figure 3. Comparison of simulated transfer characteristics to the experimental one at $V_D = 0.1\text{V}$.

It is well-known that in nanoscale devices, the presence of an unintentional dopant in the channel is highly probable [8]. Depending on its position and applied bias this unintentional dopant can significantly alter the device behavior, particularly when the channel is very lightly doped. An unintentional dopant sitting at a random location within the channel introduces a localized barrier which impedes the carrier propagation. The impact is significantly larger for an unintentional dopant sitting at the beginning of the fin near the source end compared to other probable positions [9]. When the drain bias is low the lateral electric field is small. For gate voltage much below the threshold voltage the intrinsic barrier is already high enough (as shown in Figure 4) so that the effects of the localized barrier introduced by the unintentional dopant are negligible. Therefore, over the subthreshold regime, we see a very good correspondence between simulation and experiment. For higher gate voltages the intrinsic barrier is reduced significantly (Figure 4) and the localized barrier due to unintentional dopant is expected to play a dominant role in determining the drain current. Consequently for $V_D = 0.1\text{V}$ and high gate voltages (i.e. in the on-state) the deviation between simulated drain current and experimental value increases.

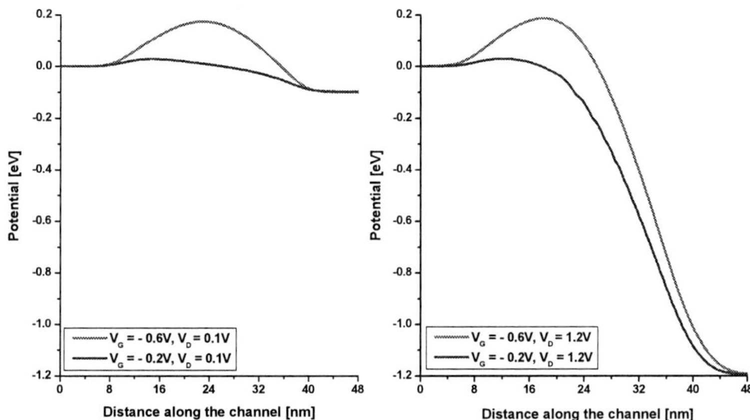

Figure 4. Potential profile along the channel in subthreshold and on-state for a drain voltage of 0.1 V (left panel) and 1.2 V (right panel)

For higher values of drain voltage carriers are rapidly accelerated due to high lateral electric field and can overcome the localized barrier introduced by the unintentional dopant. Therefore, the effect of the unintentional dopant on carrier propagation at high drain and gate bias is not as strong as for the low drain voltages. In the presence of an unintentional dopant in real device, one can expect that above threshold, higher drain bias would yield smaller reduction in drain current (relative to the corresponding experimental value) than the one with low drain bias. These considerations are supported by the simulation results obtained for $V_D = 1.2\text{V}$ and presented in Figure 5. One can see that, in this case the simulation results (that did not take into account unintentional doping) in the on-state are closer to the experimental data, when compared to those obtained for $V_D = 0.1\text{V}$. However, a full 3D simulation that takes into account both unintentional and discrete doping will be necessary to make a positive conclusion on the nature of the current deviation in the on-state.

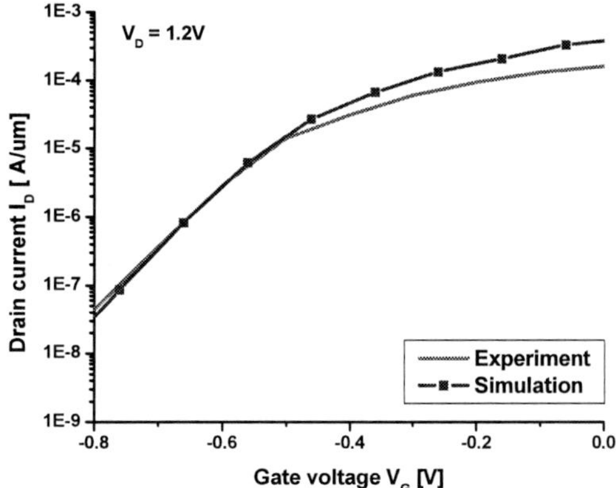

Figure 5. Comparison of simulated transfer characteristics to the experimental one at $V_D = 1.2\text{V}$.

<u>Capacitances and intrinsic propagation delay</u>

To estimate the intrinsic delay of the FinFET we calculated the channel capacitance in the linear regime of operation ($V_{DS} = 0\,\text{V}$). In this case the potential profile along the channel is symmetric with respect to the middle (x=24 nm) of the channel. The corresponding electron density has two local maximums in the middle of the channel (x=24 nm) with z=14 nm and z=22 nm as shown in Figure 6. That reflects the fact that the FinFET device with fin width of 12 nm has two distinct channels [5].

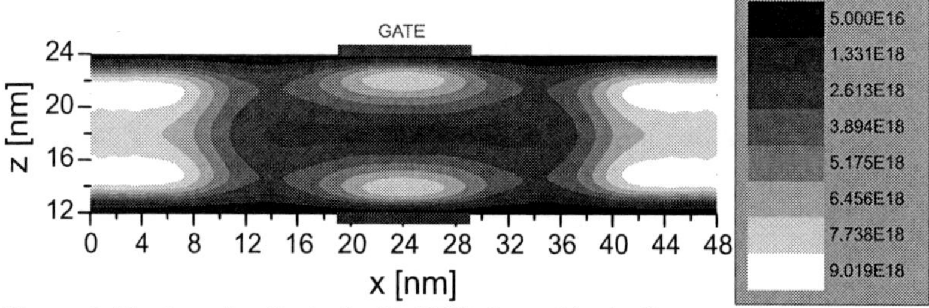

Figure 6. Electron density in the FinFET channel in the linear regime of operation.

The capacitance distribution, defined as $C(r) = \partial Q(r)/\partial V_G$, is shown in Figure 7. We see that the peaks in the capacitance are closer to the interface than the peaks of electron density: the corresponding positions are z=13 nm and z=23 nm. The fringe capacitance can be defined as the change of charge (with the small change of gate voltage) outside the region between the gates (x<19 nm and x>29 nm). Importantly, even in linear regime of operation, the relative contribution of the fringe capacitance is significant: the contribution from source (and drain) side is around 20% of the total capacitance value.

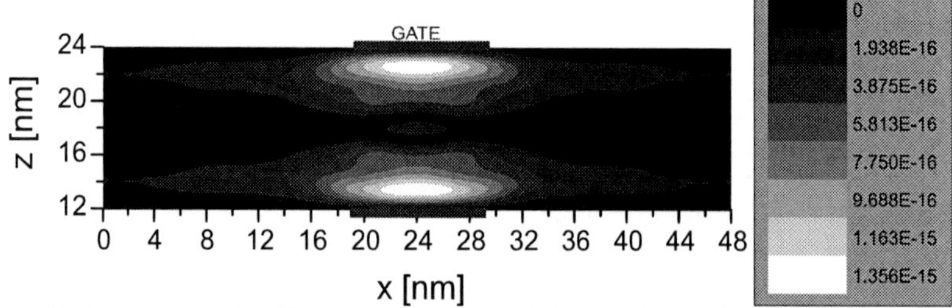

Figure 7. Capacitance distribution in the FinFET channel in the linear regime of operation.

The total channel capacitance has been found to be $C_{linear} = 2.6\,\text{pF}/\text{cm}$. Finally, the intrinsic delay, defined as $\tau = C_{linear}V_{DD}/I_{ON}$, can be calculated from our simulation results using $I_{ON} = 15.5\,\text{A}/\text{cm}$ (for $V_{DD} = 1.2\text{ V}$). The resulting value is 0.2 ps. The experimental value of τ is 0.34 ps [6]. This difference between the experiment and simulation is due to over-estimation of the ON-current mentioned above.

CONCLUSION

In this work, we utilize the self-consistent CBR method has been used for investigating ballistic and quasi-ballistic transport in nano-scale FinFET devices. To the best of our knowledge, this is a first study presented in the literature that makes a comparison between experimental data and theoretical fully quantum-mechanical Green's function calculations in FinFET devices. The agreement between the two is very good in the subthreshold regime, and there are significant deviations in the on-state, which in our opinion, can be explained by the presence of random unintentional dopants in the experimentally fabricated FinFET. Similarly, the intrinsic delay value obtain from our simulations is lower than the experimental one, due to the overestimation of the ON-current. Our simulations show quite significant influence of the fringe capacitance, which may be as high as 40% of the total value of capacitance, even in the linear regime of operation.

We conclude that to obtain better correspondence with the experiment, a fully 3D, self-consistent quantum transport simulator that takes into account discrete doping, random impurities in the channel is necessary. In addition, the inclusion of the parasitic source-drain resistance is desirable for not very highly doped ($\sim 10^{19}\,\mathrm{cm}^{-3}$) devices.

References

1. D. Mamaluy, D. Vasileska, M. Sabathil, T. Zibold, P. Vogl, "Contact block reduction method for ballistic transport and carrier densities of open nanostructures", Phys. Rev. B 71, pp. 245321-1-245321-14 (2005).
2. D. Mamaluy, M. Sabathil, P. Vogl, "Efficient method for the calculation of ballistic quantum transport", J. Appl. Phys. 93, pp. 4628-4633 (2003).
3. A. Trellakis, A. T. Galick, A. Pacelli, U. Ravaioli, "Iteration scheme for the solution of the two-dimensional Schrödinger-Poisson equations in quantum structures", J. Appl. Phys. 81, pp.7880-7884 (1997).
4. R. Lake, G. Klimeck, R. C. Bowen, D. Jovanovic, D. Blanks, M. Swaminathan, "Quantum Transport with Band-Structure and Schottky Contacts", phys. stat. sol. (b) 204, pp. 354-357 (1997).
5. H. Khan, D. Mamaluy, D. Vasileska, "Quantum Transport Simulation of Experimentally Fabricated Nano-FinFET", Tran. El. Dev. (accepted for publication in April's issue, 2007).
6. Bin Yu, Leland Chang, Shibly Ahmed, Haihong Wang, Scott Bell, Chih-Yuh Yang, Cyrus Tabery, Chau Ho, Qi Xiang, Tsu-Jae King, Jeffrey Bokor, Chenming Hu, Ming-Ren Lin, David Kyser, "FinFET Scaling to 10nm Gate Length", IEDM Tech. Digest (IEEE, Piscataway, NJ, 2002), pp. 251-254.
7. Leland Chang, Yang-kyu Choi, Daewon Ha, Pushkar Ranade, Shiying Xiong, Jeffrey Bokor, Chenming Hu, Tsu-Jae King, "Extremely Scaled Silicon Nano-CMOS Devices", Proc. IEEE 91, pp. 1860-1873 (2003).
8. T. Mizuno, J.Okamura, A. Toriumi, "Experimental Study of Threshold Voltage Fluctuation Due to Statistical Variation of Channel Dopant Number in MOSFET's", IEEE Trans El. Dev. 41, pp. 2216-2221 (1994).
9. H. R. Khan, D. Vasileska and S. S. Ahmed, "Modeling of FinFET: 3D MC Simulation Using FMM and Unintentional Doping Effects on Device Operation", J. Comp. El. 3, pp. 337–340 (2004).

ECS Transactions, 6 (4) 205-209 (2007)
10.1149/1.2728862, ©The Electrochemical Society

Simple Analytical Model to Study the ZTC Bias Point in FinFETs

M. Bellodi[b,*], L. M. Camillo[a], J. A. Martino[a], E. Simoen[c], C. Claeys[c,d]

[a] Department of Electricity, LSI/PSI/USP, University of São Paulo, São Paulo, Brazil
[b] Department of Electricity, Centro Universitário da FEI, S.B.C, São Paulo, Brazil
[c] IMEC, Kapeldreef 75, B-3001 Leuven, Belgium
[d] E.E. Dept., KU Leuven, Leuven, Belgium
*bellodi@fei.edu.br

In this work we present a simple analytical model to study the Zero Temperature Coefficient (ZTC) bias point in FinFETs operating from room temperature up to 573 K. Three-dimensional simulations are carried out and compared with experimental results to qualify the results.

Introduction

SOI CMOS technology has been used in high temperature applications due to low leakage currents, a steeper subthreshold slope and suppression of the latch-up phenomenon compared to bulk devices (1).

The Zero Temperature Coefficient (ZTC) bias point is very important for the stability of the circuit operation over a wide temperature range. Some researchers have studied the ZTC point in bulk MOSFETs (2), Partially Depleted (PD) (3) and Fully Depleted (FD) (4) SOI devices. It is expected that the mutual compensation of the mobility degradation (c) and the threshold voltage (Vth) temperature dependence may result in a single ZTC bias point i.e.,a V_{ZTC} value with its associated drain current I_{ZTC}. However, sometimes this mutual compensation is not observed, which affects the stability of the ZTC.

The goal of this work is to propose a simple analytical model to study for the first time the ZTC bias point in FinFETs using the same approach of reference (4) in linear and saturation regions. Experimental and three-dimensional simulations are used to validate the model proposed.

FinFET Characteristics

The n-type triple gate FinFETs are fabricated starting from an SOI wafer with buried oxide thickness t_{BOX} = 145 nm, the top silicon film thickness H_{FIN} is 60 nm, the Fin width W_{FIN} = 120 nm and the silicon film doping is N_a=1.0x10^{15} cm^{-3}. The gate oxide consists of 1 nm thermal oxide and 2 nm HfO$_2$. A 5 nm thick TiN layer is then deposited and a polysilicon cap completes the gate stack. Nickel silicidation is used for contacting the device electrodes (5).

Figure 1 shows a schematic of the FinFET in a correspondent three-dimensional scheme and its main dimensions.

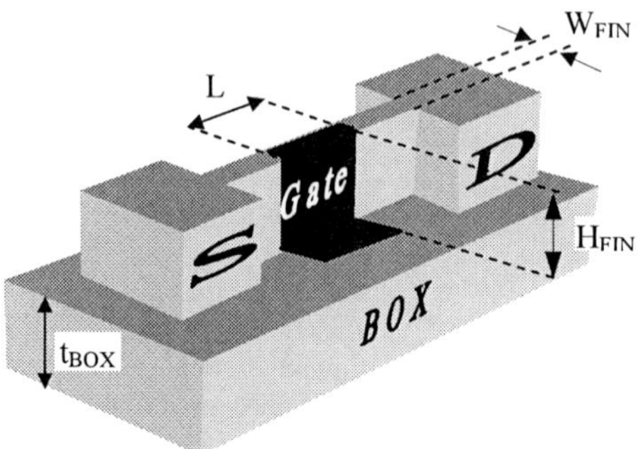

Figure 1. Three dimensional scheme of the FinFET.

Description of the ZTC Analytical Model

By definition, the Zero Temperature Coefficient point represents the gate voltage (V_{ZTC}), which insures that the drain current (I_{ZTC}) remains constant with temperature variations. This behavior is illustrated in figure 2 for a FinFET operating in the linear region. It is worthwhile to mention that similar results are observed when the device is operating in the saturation region, however, changing the drain current and V_{ZTC} values due to the high values of the drain bias in that regime.

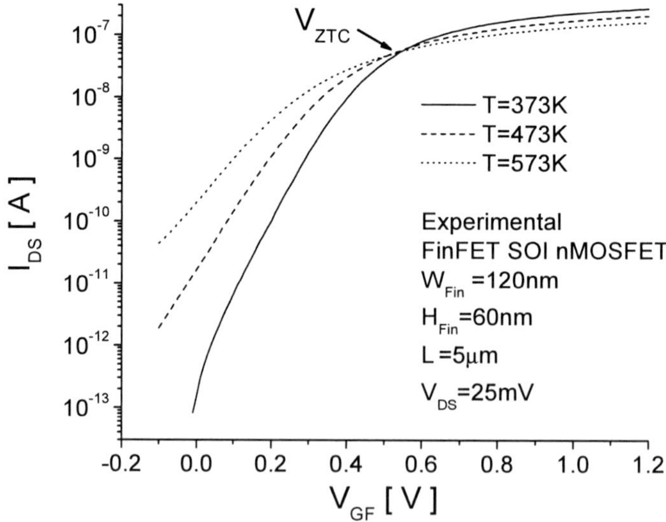

Figure 2. Experimental curves for a FinFETs operating at high temperatures.

To study the ZTC stability, V_{ZTC} is obtained by the cross point between the drain current I_{DS} versus the gate voltage V_{GF} curves, where $V_{ZTC\ 1,2}$ means between temperatures T_1 and T_2 as shown in figure 3.

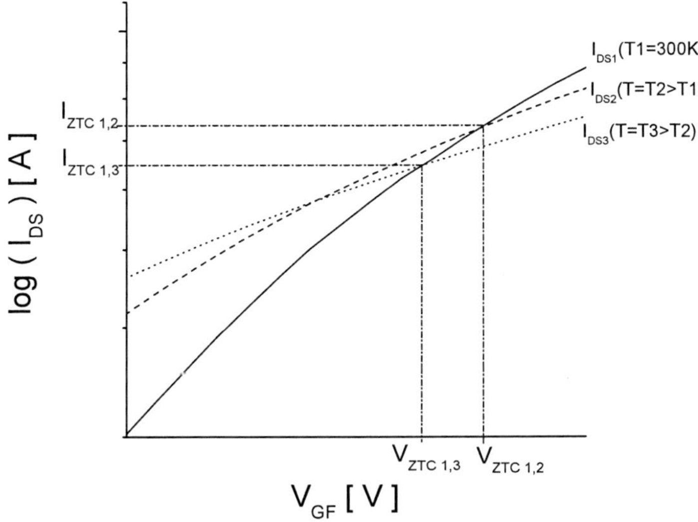

Figure 3. The ZTC point between temperatures T1, T2 and T3.

Therefore, $V_{ZTC\ 1,2}$ can be calculated as shown in equations [1] and [2] for the linear and the saturation region, respectively.

$$V_{ZTC\ 1,2\,(LIN)} = V_{th1} + \frac{\left(V_{th_1} - V_{th_2}\right)\left(gm_2 / gm_1\right)}{1 - \left(gm_2 / gm_1\right)} + \frac{V_{DS}}{2} \qquad [1]$$

$$V_{ZTC\ 1,2\,(SAT)} = V_{th_1} + \frac{\left(V_{th_1} - V_{th_2}\right)\cdot\left(gm_2 / gm_1\right)}{1 - \left(gm_2 / gm_1\right)} + {}$$
$$+ \sqrt{\left[V_{th_1} + \frac{\left(V_{th_1} - V_{th_2}\right)\cdot\left(gm_2 / gm_1\right)}{1 - \left(gm_2 / gm_1\right)}\right]^2 - \frac{V_{th_1}^2 - V_{th_2}^2\left(gm_2 / gm_1\right)}{1 - \left(gm_2 / gm_1\right)}} \qquad [2]$$

where: V_{th1} and V_{th2} are the threshold voltage at temperatures T_1 and T_2 respectively, gm_1 and gm_2 are the transconductances at the respective temperature and V_{DS} is the drain voltage applied to the device. Also, the transconductance ratio $gm_2/gm_1 \cong (\ T1/T2\ )^c$.

Analyzing the V_{ZTC} in function of increasing T_2 for T_1 as a reference, (Vth_1-Vth_2) increases and the transconductance ratio (gm_2 / gm_1) decreases, which shows the two competing terms to maintain dV_{ZTC} / dT near zero (ZTC stable).

Results and Discussion

Figure 4 shows the $V_{ZTC\ 1,2}$ values obtained experimentally and by the proposed model (as described in figure 3), in function of T_2 (T_1 =300K) for a FinFET operating in the linear and the saturation regime. The best c-value to fit the experimental data is 1.61. A reasonably close agreement with experimental results is observed and the maximum error found is 11% in the linear and 4% in the saturation region.

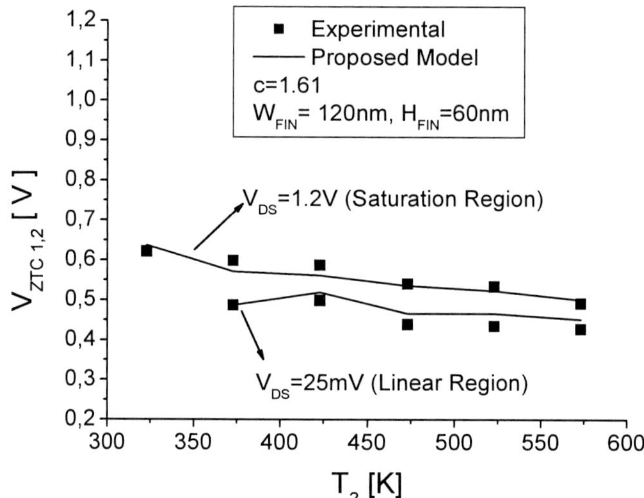

Figure 4. Experimental and modeled V_{ZTC} results for FinFETs.

The stability of the ZTC point in the FinFETs studied in this work is examined and compared to the planar FD SOI devices of ref (4). The results show that $|\ dV_{ZTC} / dT\ |$ is ≈ 0.1 mV/K in FD SOI (4) and approximately 0.5 mV/K in FinFET in the worst case (saturation region). The FinFETs show a 5 times less ZTC stability than in planar FD SOI. These results show that in FinFETs the mutual compensation between the mobility degradation and the threshold voltage temperature dependency is not well balanced.

Recently, another study reported in the literature pointed out that the $|\ dV_{ZTC} / dT\ |$ in FinFETs is around 0.6 mV/K (6), which is lower than the value observed in undoped FD SOI ($\sim$ 1.2 mV/K).

Three-dimensional numerical simulations using ATLAS (7) were done and the agreement between the simulation curves and the model applied to the simulation data is very good and confirms the tendency observed experimentally. The maximum error found is around 4% in the linear and 3% in the saturation region for the temperature range studied. These results are reported in figure 5.

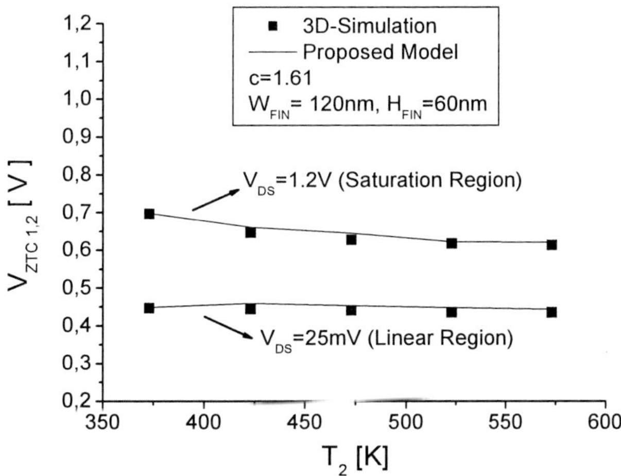

Figure 5. 3D-simulated and modeled V_{ZTC} results for FinFETs.

Conclusions

A simple analytical model to evaluate the Zero Temperature Coefficient (ZTC) bias point in n-type triple-gate FinFETs operating from room temperature up to 573 K has been reported. The results point out that the FinFET show a 5 times lower ZTC stability when compared to the Fully-Depleted SOI nMOS operating at the same conditions (bias and temperature).

The study performed with the proposed model shows a good agreement between experimental and simulations results in linear and saturation region, whereby the maximum error found is around 11%.

References

1. D. S. Jeon and D. E. Burk, *IEEE Trans. Electron Devices*, **38**, 2101 (1991).
2. Z. D. Prijic et al., *Micr. Reliab.*, **32**, 6, 769 (1992).
3. A. A. Osman et al., *IEEE Trans. Electron Devices*, **42**, 9,1709 (1995).
4. L. M. Camillo, J. A. Martino, E. Simoen, C. Claeys, *Micr. Journal*, **37**, 952 (2006).
5. N. Collaert et al., *Symp. VLSI Tech.*, 108 (2005).
6. V. Kilchytska et al., in *EUROSOI 2007- Conference Proceedings/2007*, p. 30, Leuven (2007).
7. *ATLAS Device Simulator*, v.5.10.0.R, Silvaco Int. (2005).

ECS Transactions, 6 (4) 211-216 (2007)
10.1149/1.2728863, ©The Electrochemical Society

Temperature Influences on FinFETs with Undoped Body

M. A. Pavanello[1,2], J. A. Martino[2], E. Simoen[3], R. Rooyackers[3],
N. Collaert[3] and C. Claeys[3,4]

[1] Centro Universitário da FEI,
Av. Humberto de Alencar Castelo Branco, 3972
09850-901, São Bernardo do Campo, Brazil
pavanello@fei.edu.br

[2] Laboratório de Sistemas Integráveis, Universidade de São Paulo
Av. Prof. Luciano Gualberto, trav. 3 n. 158, 05508-900, Sao Paulo, Brazil

[3] IMEC, Kapeldreef 75, B-3001 Leuven, Belgium

[4] E.E. Dept., KU Leuven, Kasteelpark Arenberg 10, B-3001 Leuven, Belgium

This work presents a study, based on DC measurements, of the temperature influence on the performance of nMOS triple-gate FinFETs with high-κ dielectrics, TiN gate material and an undoped body. FinFETs show smaller threshold voltage variations with temperature than planar fully-depleted SOI MOSFETs. The subthreshold slope reduced with the temperature and approached the ideal value at lower temperatures. In the temperature range under study the mobility increases linearly as the temperature is reduced and the dominating mobility degradation factor is phonon scattering. The DIBL has been evaluated and no temperature dependence has been found. Finally, the series resistance has been also extracted and demonstrates a reduction as the temperature is reduced due to the mobility improvement.

Introduction

FinFETs are well established as a promising technological solution for achieving high-performance sub-100 nm Si MOSFETs due to their excellent scalability as well as easy accommodation in standard Silicon-On-Insulator (SOI) processes (1). Figure 1 presents a cross- sectional view of the FinFET indicating the channel length (L), the fin height (H_{Fin}) and the fin width (W_{Fin}). Depending on the thickness of the top gate oxide this device operates as a vertical double-gate or as a triple-gate one. In the latter, the current conduction on the top contributes to the overall device performance.

In order to improve the gate control in extremely scaled devices, generally narrow fins are used in parallel configuration to enhance the drain current drive.

Beyond the performance enhancements provided by the FinFETs, the temperature reduction can boost these improvements due to increased carrier mobility and velocity saturation, reduced inverse subthreshold slope, finally resulting in better switching characteristics than at room temperature.

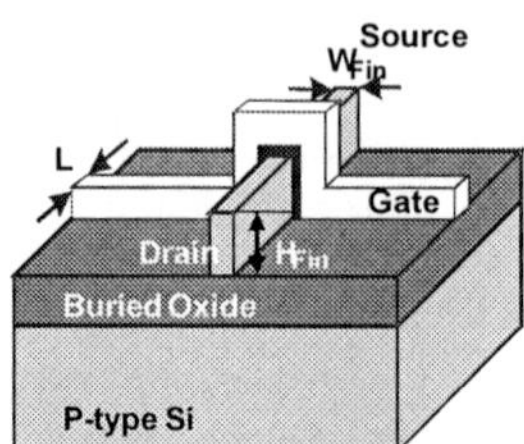

Figure 1 - Schematic representation of a FinFET.

This work studies the operation of undoped body FinFETs with HfO_2 gate dielectric and TiN gate material in the temperature range of 150 K up to 380 K.

Device Characteristics

The triple-gate n-type FinFETs were fabricated starting from SOI wafers with 145 nm buried oxide thickness according to the process described in ref. (2). The top Si layer thickness is decreased down to 60 nm, which is the fin height (H_{Fin}). The fin width (W_{Fin}) is 20 nm. After the silicon film definition a 1 nm thick interfacial thermal oxide is grown followed by the deposition of 2 nm HfO_2. Subsequently a 5 nm thick TiN film is deposited and a 100 nm polysilicon capping layer completes the gate stack. No channel doping or halo implantation is applied during the processing. Nickel silicidation is used for the device electrodes. The final structure consists of 30 fins (Nfins) in parallel configuration.

The digital characteristics of this technology have been presented in ref. (2) whereas its potential for achieving high performance analog circuits is reported in ref. (3).

Measurements and Discussion

The samples were cooled down to 150 K using the Variable Temperature Micro Probe System model K20 from MMR Technologies and the experimental curves were obtained with a Keythley 4200 semiconductor parameter analyzer using a medium integration time. All the measurements were performed starting from the lowest temperature.

Figure 2 presents the subthreshold slope (S) and threshold voltage (V_T) for FinFETs with channel length (L) of 150 nm, 90 nm and 60 nm extracted from the measured drain current (I_{DS}) *versus* gate voltage (V_{GF}) with a drain bias (V_{DS}) of 50 mV. In all the measurements the gate current remains negligible at any temperature. For the V_T extraction the double derivative technique was applied.

The threshold voltage linearly decreases with the temperature rise, independent of the channel length under study, at a rate of -0.58 mV/K. For a planar single-gate fully-depleted SOI MOSFET the threshold voltage variation with temperature is described as $\Delta V_T/\Delta T = \Delta \Phi_F/\Delta T$, where Φ_F is the Fermi potential, independent of silicon film thickness. The variation of $|\Delta V_T/\Delta T|$ for FinFETs is smaller than the Fermi potential reduction with temperature increase, in the order of -0.96 mV/K. This indicates that the V_T of FinFETs is clearly more thermally stable or less temperature-dependent than for single-gate devices. A degradation of about 20 mV is observed for the threshold voltage of 60 nm long transistors in comparison to L=150 nm at the same temperature.

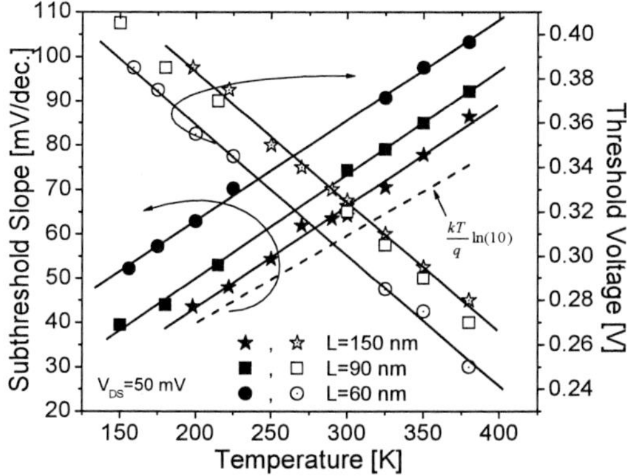

Figure 2 – Extracted Subthreshold Slope (S) and Threshold Voltage (V_T) for L=150 nm, 90 nm and 60 nm FinFETs as a function of the temperature. Closed symbols refer to S and open symbols to V_T.

In order to get some insight in the reasons for the smaller $\Delta V_T/\Delta T$ in FinFETs, the analytical model proposed in ref. (4) for double gate MOSFET has been used. The top conduction of triple-gate FinFETs has been negleted due to the narrow W_{Fin} under study. According to this model, the V_T is described as a function of the surface potential (ϕ_S^*):

$$V_T = \phi_S^* + V_{FB} + \frac{qN_A W_{Fin}}{2C_{ox}}\sqrt{1 + \frac{2kTC_{ox}}{q^2 N_A W_{Fin}}} \quad [1]$$

$$\phi_S^* = 2\Phi_F + \frac{kT}{q}\ln\left(\frac{C_{ox}}{4C_{Si}}\frac{1}{1 - \exp\left(-\frac{q^2 N_A W_{Fin}}{8C_{Si}kT}\right)}\right) \quad [2]$$

where V_{FB} is the flatband voltage, q is the electron charge, k is the Boltzmann constant, T is the absolute temperature, C_{ox} is the gate oxide capacitance per unit of area, C_{Si} is the silicon film capacitance per unit of area and N_A is the silicon film doping concentration.

For the analysis the temperature influences on n_i, Φ_F and N_A were also implemented. The V_{FB} term in Eqn. [1] increases with the temperature at a rate of ~1.1mV/K, whereas ϕ_S^* decreases so that due to their competition an overall V_T increase is obtained for reduced temperatures as the contribution of the third term of Eqn. [1] is almost temperature-independent and negligible. The temperature influences both terms but the fin width influences only ϕ_S^*. According with Eqn. [2], the wider the fin the larger $|\Delta\phi_S^*/\Delta T|$. Thus, for narrow FinFETs, the reduction on the surface potential variation with the temperature is the responsible for the smaller variation of V_T with the temperature.

Similar V_T variations with temperature for narrow FinFETs have been reported in ref. (5) for temperatures ranging from 300 K up to 550 K.

Regarding the subthreshold slope, the results of figure 2 show that for L=150 nm FinFETs the measured values are slightly higher than the ideal one (dashed line indicated in figure2). The temperature reduction tends to reduce the difference between the theoretical limit and the experimental results for any L, which can be associated to the better coupling at lower temperatures improving the subthreshold slope. As the channel length is reduced an S degradation is observed at any temperature. The FinFET with L=60 nm presents an S degradation of about 20 mV/dec with respect to L=150 nm at the same temperature.

This small degradation of V_T of L=60 nm FinFET compared to the strong degradation of S suggests that coupling provided by the fin width is effective on alleviating the V_T reduction with L but is not that effective for the S control.

Figure 3 presents the I_{DS} x V_{GF} curves measured with V_{DS} of 0.1 V and 1.1 V for the FinFETs with L=90 nm and L=60 nm at 200 K and 380 K.

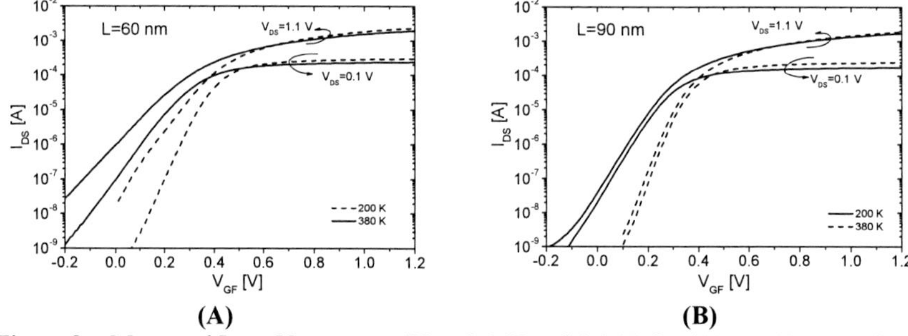

Figure 3 – Measured I_{DS} x V_{GF} curves (V_{DS}=0.1 V and 1.1 V) for (A) L= 60 nm and (B) 90 nm FinFETs at 200 K and 380 K.

Using the data of figure 3 one can study the temperature influence on the Drain Induced Barrier Lowering (DIBL) as demonstrated in Table 1. The DIBL has been extracted considering a constant current level of I_{DS}=100*(W/L) [nA].

Table 1 - Calculated Drain Induced Barrier Lowering as a function of the temperature for L=90 nm and L=60 nm FinFETs

Temperature [K]	DIBL [mV/V]	
	L=90 nm	L=60 nm
200	18	125
380	20	120

Independently of the channel length the DIBL is practically temperature insensitive in the studied range of temperatures. On the other hand, there is a clear degradation on the performance of L=60 nm FinFETs.

Figure 4 presents the maximum transconductance (g_m) extracted in linear region (V_{DS}=50 mV) for the FinFETs with L=90 nm and L=150 nm.

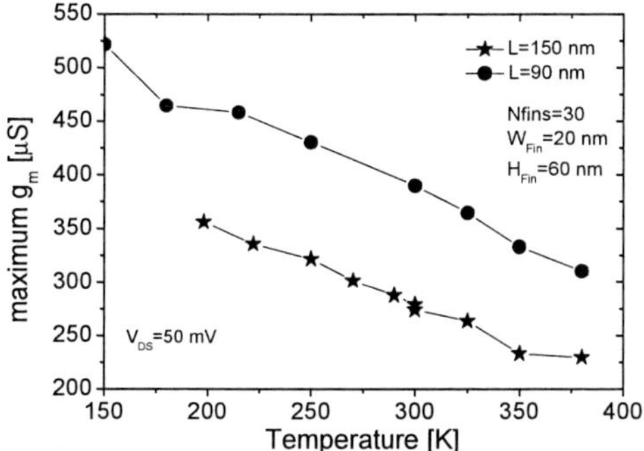

Figure 4 - Maximum transconductance (g_m) extracted in linear region (V_{DS}=50 mV) for the FinFETs with L=90 nm and L=150 nm

The maximum g_m increases almost linearly with the temperature for both channel lengths at a rate of $|dg_m/dT|$=0.74 µS/K for L=150 nm and $|dg_m/dT|$=0.78 µS/K for L=90 nm. With the data of figure 4 the effective mobility (μ_N) variation with temperature for the long-channel FinFET is presented in figure 5. For this extraction the effective channel width width has been estimated as W=(2*H_{Fin}+W_{Fin})*Nfins.

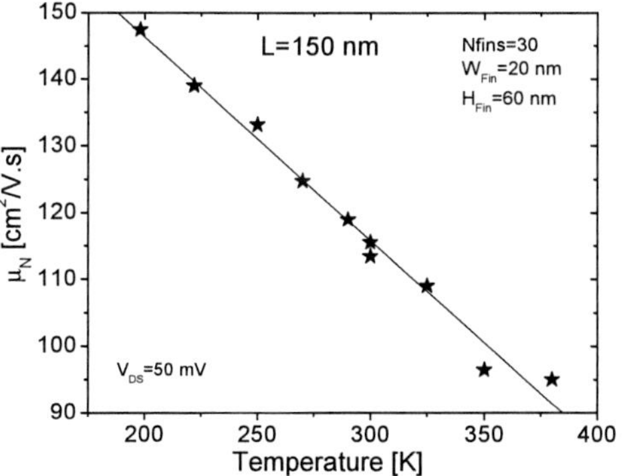

Figure 5 – Extracted electron mobility (V_{DS}=50 mV) as a function of the temperature for L=150 nm FinFET.

In the measured temperature range the mobility increase is almost linear with the temperature reduction, similar as for the transconductance. This typical straight degradation of μ_N with temperature rise indicates that the dominating mobility degradation factor still being the phonon scattering mechanism in the studied temperature range (6) and not the Coulombic scattering due to the interface and oxide charges. Due to narrow fin width the dominating current conduction occurs in the <110> plane rather than in the typical <100> for planar devices. This is the reason for the relatively low values of μ_N obtained (7).

One of the problems associated with the use of narrow FinFETs is the increase of the parasitic series resistance (R_S). The extracted R_S as a function of temperature is plotted in figure 6. For the R_S extraction the method proposed in ref. (8) has been adopted.

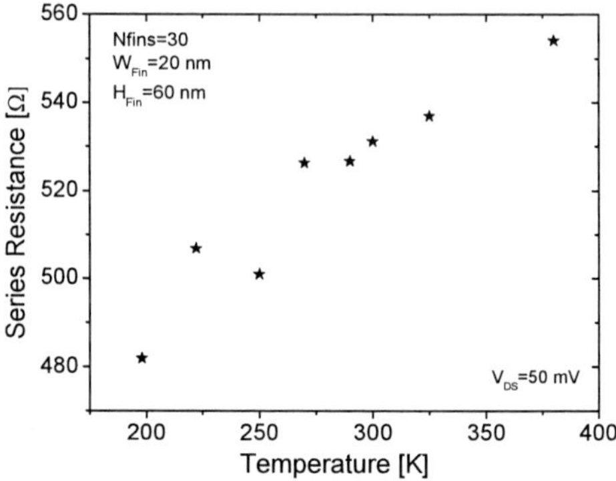

Figure 6 – Extracted series resistance (V_{DS}=50 mV) as a function of the temperature.

The temperature reduction improves R_S as the carrier mobility in the extensions is increased and no carrier freeze-out has been observed for the studied temperatures.

Conclusions

The influence of temperature variations on some parameters of nMOS triple-gate FinFETs with high-κ dielectrics, TiN gate material and undoped body has been presented. The FinFET threshold voltage is more thermally stable than for planar fully-depleted SOI MOSFETs. As the temperature decreases the subthreshold slope reduces and approaches the theoretical limit indicating improvement of the channel control. A linear transconductance and mobility increase have been found in the studied temperature range indicating that the dominating mobility degradation factor is phonon scattering. The DIBL has been evaluated and no temperature dependence has been observed. The series resistance decreases as the temperature is reduced due to the mobility improvement.

Acknowledgements

M. A. Pavanello and J. A. Martino acknowledge to CNPq for the financial support.

References

1. J. Park and J.-P. Colinge, *IEEE Trans. on Electron Devices*, 49, 2222 (2002).
2. N. Collaert *et al.*, in *Symposium on VLSI Technology Digest of Technical Papers*, 108 (2005).
3. D. Lederer *et al.*, *Solid-State Electronics*, 49, 1488 (2005).
4. P. Francis *et al.*, in *IEEE Trans. on Electron Devices*, 41, 715 (1994).
5. V. Kilchytska *et al.*, in *EUROSOI 2007 Conference Proceedings*, 30 (2007).
6. J.-P. Colinge *et al.*, *IEEE Electron Device Lett.*, 27, 120 (2006).
7. T. Rudenko *et al.*, *Microelectronic Engineering* 80, 386 (2005).
8. A. Dixit *et al.*, *IEEE Trans. on Electron Devices*, 52, 1132 (2005).

ECS Transactions, 6 (4) 217-222 (2007)
10.1149/1.2728864, ©The Electrochemical Society

Application of Double Gate Graded-Channel SOI in MOSFET-C Balanced Structures

R. T. Doria[a], M. A. Pavanello[a,b], A. Cerdeira[c], J. P. Raskin[d] and D. Flandre[e]

[a]Centro Universitário da FEI
Av. Humberto de Alencar Castelo Branco, 3972
09850-901 – São Bernardo do Campo, Brazil
*e-mail: rtdoria@fei.edu.br

[b]Laboratório de Sistemas Integráveis – Universidade de São Paulo
Av. Prof. Luciano Gualberto, trav.3 n.158, 05508-900, São Paulo, Brazil

[c]Sección de Electrónica del Estado Sólido (SEES), CINVESTAV
Av. IPN No. 2508, Apto. Postal 14-740, 07300 DF, México

[d]Laboratoire d'Hyperfréquences
[e]Laboratoire de Microélectronique, Université Catholique de Louvain
Place du Levant 3, Maxwell Building, B-1348 Louvain-la-Neuve, Belgium

This work studies the linearity of conventional and Graded-Channel (GC) Gate-All-Around (GAA) devices when applied in 2-MOS and 4-MOS balanced structures operating as tunable resistors. The study has been performed through device characterization and two-dimensional process and device simulations. Total harmonic distortion (THD) and third order harmonic distortion (HD3) have been evaluated. When taking into account similar on-resistance, the use of the GC GAA transistors in both 2-MOS and 4-MOS structures improves the linearity. The use of GC GAA devices in 2-MOS balanced structures allows a reduction of the gate overdrive voltage of 22.5% without degrading THD and HD3. On the other hand, the use of GC GAA devices in 4-MOS structures leads to an improvement in both HD3 and THD by 7 dB for devices with similar channel length at the same gate voltage overdrive.

Introduction

The Gate-All-Around (GAA) nMOSFETs feature a channel region fully surrounded by gate oxide and metal electrode (Figure 1). These devices behave as double gate transistors since the contribution of the vertical channels of the silicon film to the current flow is negligible. Consequently, the devices behavior is controlled by the gates in the top and bottom sides of the silicon film. Different studies have reported advantages of the GAA technology in relation to conventional single gate devices such as volume inversion, improved transconductance, nearly ideal subthreshold slope, higher gain and better control of channel charges (1).

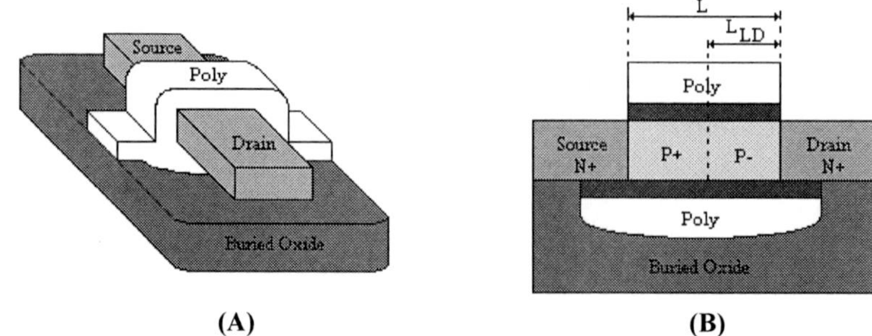

(A)   **(B)**

Figure 1 – (A) Completed Gate-All-Around device and (B) Cross-section of a Graded-Channel GAA transistor

The Graded-Channel (GC) SOI nMOSFET is an asymmetric channel device, which presents a region with length L_{LD} that preserves the natural wafer doping near the drain, while the rest of the channel receives the threshold voltage ion implantation (2). In this device the effective channel length can be approximated to $L_{eff} = L\text{-}L_{LD}$, where L is the mask channel length. GC GAA devices have demonstrated remarkable improvements for analog operation as amplifiers, strongly reducing the output conductance and leading to an improvement of the open-loop voltage gain from 67 dB in conventional GAA (uniformly doped channel) up to 90 dB (3).

For purely analog circuits such as continuous-time filters required for any analog-to-digital conversion, MOSFET-C filters implemented by MOSFETs operating in the triode region as tunable circuits appear as a good alternative (4 - 5). The source/drain regions are used as resistor terminals and the gate voltage is tuned for a certain on-resistance (R_{ON}). Generally long-channel transistors are used to obtain high R_{ON}. However, the obtained current output is strongly nonlinear. Thus, the 2-MOS and 4-MOS balanced structures presented in Figure 2 have been proposed to improve the circuit harmonic distortion (HD) as they suppress even order harmonics such as the second order distortion (HD2), which usually dominates in single transistor circuits (6). As a result, the nonlinearities are given by the third order harmonic distortion (HD3), which is equal to the total harmonic distortion (THD) in this case.

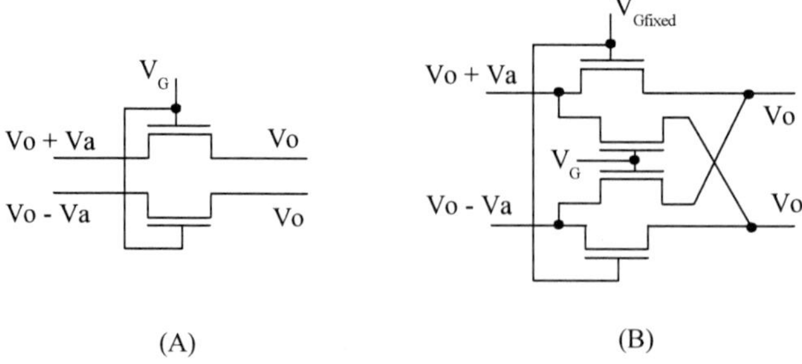

(A)   (B)

Figure 2 - Resistive structures used in MOSFET-C filters: (A) 2-MOS structure and (B) 4-MOS structure.

This paper presents the application of double-gate GC devices in MOSFET-C balanced structures and discusses some design aspects involved in GC GAA use. The third order harmonic distortion and the on-resistance are variables of huge interest and have been adopted as figures of merit. The study is based in process/device and circuits simulations calibrated from experimental results.

Devices Characteristics and Measurements

The studied GC GAA devices were fabricated according to the process described in ref. (7). Devices with 3 parallel fingers were fabricated each one with L = W = 3 µm (W is the channel width) and different L_{LD}/L ratios. The studied devices present silicon film thickness of 80 nm, gate oxide of 30 nm, buried oxide of 390 nm. The total harmonic distortion (THD) has been determined for the devices biased at $V_{DS} = 0$ V and varying the sinusoidal amplitude (Va) as presented in Figure 3 for a fixed gate voltage overdrive ($V_{GT} = V_{GS} - V_{TH}$, V_{GS} being the gate voltage and V_{TH} the threshold voltage) of 2 V. The distortion properties were extracted using the Integral Function Method (IFM) (8 - 9) since this technique allows nonlinearity characterization only from DC transfer characteristics, without the need of AC characterization as in e.g. Fourier based methods.

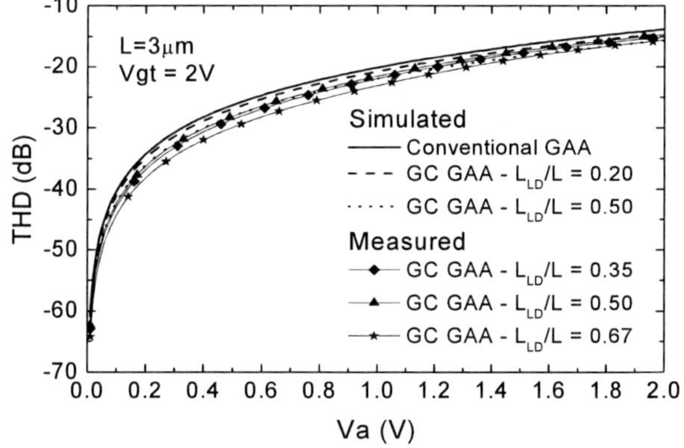

Figure 3 - Measured and simulated curves of THD as a function of the input signal (Va) for the GAA and GC GAA devices.

Process and devices simulations of transistors with similar dimensions were performed using the Athena (10) and the Atlas (11) tools, respectively. Default simulator coefficients were used. Analytical models for the mobility degradation due to vertical and lateral electric field, doping-dependent carrier lifetime, bandgap narrowing and impact ionization were included in the simulation files. Then, the results for THD were determined and have been also presented in Figure 3. In this case total harmonic distortion is practically equal the second order harmonic distortion (HD2). As demonstrated in this figure, the use of GC GAA provides a slight improvement in the THD, which is larger as the L_{LD}/L ratio rises. Also Figure 3 validates the simulations performed due to the good agreement between THD of experimental and simulated data.

Balanced Structures Evaluation

After validating the process/device simulations at the device level by comparison with experimental results, the circuit module of Atlas simulator has been used for the 2-MOS and 4-MOS circuit configurations.

As previously mentioned, R_{ON} and HD3 were taken as figures of merit for the balanced structures analysis. Nevertheless, for most common applications R_{ON} must be adjusted for a target value (12). So that, we analyzed the structures for a fixed R_{ON} of 37 kΩ ±5% which represents a $V_{GT} = 2$ V in conventional GAA for the 2-MOS structure and represents a $V_{GTfixed} = 2.8$ V and $V_{GT} = 1$ V for the 4-MOS. R_{ON} has been determined at $V_{DS} = 0.1$ V for all the studied structures. For both 2-MOS and 4-MOS structures we simulated conventional and GC GAA devices with several L_{LD}/L and $L = 10$ µm. These devices were simulated operating at zero bias voltage (Vo = $V_{DS} = 0$ V) with an input signal amplitude Va varying from 0 to 0.5V.

In order to eliminate HD2 and get only HD3, balanced circuits with MOSFETs are used with two transistors, 2-MOS, or four transistors, 4-MOS. In these circuits transistors are operating as quasi-linear resistors. Distortion was determined applying the IFM to the I_{DS}-V_{DS} characteristics for the bias voltages previously described.

The distortion analysis was performed for a fixed R_{ON}, which in the case of the 2-MOS structure required a reduction of V_{GT} in GC GAA for the target R_{ON} when increasing the L_{LD}/L ratio. This happened due to the shorter L_{eff} provided by the GC architecture since the lightly doped region absorbs part of the applied V_{DS} (13). As a result the 2-MOS structure allows a reduction of V_{GT} that can reach 22.5% for devices with $L_{LD}/L = 0.6$ as shown in Table I.

Table I: Values of V_{GT} obtained for several devices at a fixed R_{ON} in 2-MOS structures			
	Mask L_{LD}/L	R_{ON}(kΩ)	V_{GT}(V)
	0.1	36	2.00
	0.2	37	1.90
GC GAA	0.3	36	1.85
	0.4	37	1.75
	0.5	37	1.65
	0.6	37	1.55
Conventional GAA	$L = 7$ µm	37	1.40
	$L = 10$ µm	36	2.00

This reduction, however, has insignificant influence on HD3 as can be seen in Figure 4 where are plotted the HD3 curves as function of Va. Reducing the channel length of conventional GAAs to $L = 7$ µm in the 2-MOS structure, where a similar reduction on V_{GT} is required for reaching the target R_{ON}, one can see a degradation of the linearity in conventional GAA that can reach 5dB, i.e. almost a factor of 2.

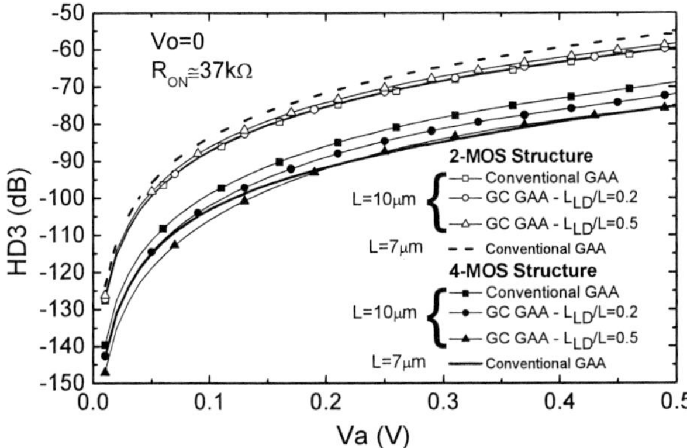

Figure 4 - Curves of HD3 *versus* Va for 2-MOS and 4-MOS balanced structures.

HD3 has also been extracted for 4-MOS structures at a fixed R_{ON} of 37 kΩ ±5%. Nevertheless, for this structure the target R_{ON} could be reached without reducing V_{GT} for any studied device, as presented in Table II. As can be seen in Figure 4, the results obtained for the referred circuit show an improvement of the linearity provided by the GC transistors. As the L_{LD}/L ratio increases, a reduction that achieves 7 dB for L_{LD}/L = 0.5 can be observed in HD3. When diminishing the conventional device channel length to 7 μm, a reduction on HD3 is perceived. However, the proposed R_{ON} can only be attained for this structure through an increase of 50% in V_{GT} as shown in Table II. Besides that, this improvement on HD3 is not greater than the one obtained for the GC structure with L_{LD}/L = 0.5. Therefore, the GC architecture allows for the use of longer devices at lower gate voltages, which is of interest for low voltage circuits.

Table II: Values of R_{ON} and V_{GT} obtained for several devices at a fixed R_{ON} in 4-MOS structures. ($V_{GTfixed}$ = 2.8 V)			
	Mask L_{LD}/L	R_{ON}(kΩ)	V_{GT}(V)
GC GAA	0.2	38	1.0
	0.5	37	1.0
Conventional GAA	L = 7 μm	37	1.5
	L = 10 μm	38	1.0

According to the HD3-Va curves shown in the Figure 4 we can note that for any studied device the 4-MOS structure presented lower distortion levels overcoming the 2-MOS response. Comparing the results from both balanced structures, an improvement of at least 10 dB can be seen in any 4-MOS structure.

Conclusions

This work performed an evaluation of the application of double gate Graded-Channel devices in MOSFET-C balanced structures with 2-MOS and 4-MOS devices using the third order harmonic distortion (HD3) and on-resistance (R_{ON}) as figures of merit. HD3 has been determined for both 2-MOS and 4-MOS structures for a targeted R_{ON}. The results obtained for the 2-MOS structure show a dependence of R_{ON} on V_{GT}

when varying L_{LD}/L. Thus, in order to fix R_{ON} a reduction of V_{GT} is required for devices with larger L_{LD}/L ratio. For the transistor with $L_{LD}/L = 0.6$, the 2-MOS allows for a reduction of V_{GT} of 22.5% without degrading HD3 characteristics, which is of huge interest for low voltage applications. Besides that, in a conventional GAA with $L = 7\ \mu m$, where a similar reduction on V_{GT} is required for reaching the target R_{ON}, a degradation of 5 dB in HD3 can be observed. On the other hand, the 4-MOS structure analysis show no dependence between R_{ON} and V_{GT} when varying L_{LD}/L. However, an improvement in HD3 is provided by the GC architecture. According to the results obtained, for structures composed by devices with greater L_{LD}/L ratios a reduction in HD3 of 7 dB can be observed. Although a reduction in HD3 is perceived when diminishing L of the conventional device to $7\ \mu m$ in the 4-MOS structures analysis, the improvement obtained is similar to the one obtained for the longer GC device with $L_{LD}/L = 0.5$ with the penalty of a 50% increase in V_{GT} consequently one note that the GC architecture allows the use of longer devices keeping the same HD3 characteristics at lower voltages. Furthermore, an additional improvement of at least 10 dB is obtained over the 2-MOS structure.

Acknowledgments

Rodrigo T. Doria and Marcelo A. Pavanello acknowledge the Brazilian research-funding agencies CAPES and CNPq for the financial support.

References

1. A. Kranti, T. M. Chung, D. Flandre and J.-P. Raskin, *Solid-State Electron.*, **48**, 947 (2004).
2. M. A. Pavanello, J. A. Martino and D. Flandre, *Solid-State Electron.*, **44**, 917 (2000).
3. M. A. Pavanello, J. A. Martino, J.-P. Raskin and D. Flandre, *Solid-State Electron.*, **49**, 1569 (2005).
4. G. Groenewold, B. Monna and B. Nauta *et al.*, "Micro-power analog filter design," in *Analog Circuit Design: Low-Power Low-Voltage, Integrated Filters and Smart Power*, R. van de Plassche *et al.*, Eds. Norwell, MA: Kluwer,. 1995.
5. M. Banu and Y. Tsividis, *IEEE J. Solid-State Circuits*, **SC-18**, 644 (1983).
6. L. Vancaillie, V. Kilchytska, J. Alvarado, A. Cerdeira and D. Flandre, *IEEE Transactions on Electron Devices*, **53**, 263 (2006).
7. A. Vandooren, J. P. Colinge and D. Flandre, *IEEE Transactions on Nuclear Science*, **46**, 1242 (1999).
8. A. Cerdeira, M. A. Alemán, M. Estrada, D. Flandre, B. Parvais, G. Picún, in *18th Symposium on Microelectronics Technology and Devices*, p. 131 (2003).
9. A. Cerdeira, M. A. Alemán, M. Estrada, D. Flandre, *Solid-State Electronics*, **48**, 2225 (2004).
10. Athena Users' Manual, Edition 10, 2004.
11. Atlas Users Manual, 2004.
12. A. Cerdeira, M. A. Alemán, M. A. Pavanello, J. A. Martino, L. Vancaillie and D. Flandre, *IEEE Transactions on Electron Devices*, **52**, 967 (2005).
13. M. de Souza, M. A. Pavanello, B. Iñiguez and D. Flandre, *Solid-State Electronics*, **49**, 1683 (2005).

SESSION 5

MATERIAL CHARACTERIZATION

ECS Transactions, 6 (4) 225-233 (2007)
10.1149/1.2728865, ©The Electrochemical Society

SOI Metrology and Characterization in Modern Wafer Production.

O.Kononchuk, F.Brunier, and M. Kennard

SOITEC, Parc Technologique des Fontaines, 38190 Bernin, France.

Overview of characterization techniques used for process control and process development in modern SOI wafer manufacturing is presented. Examples of electrical characterization using Psi-MOSFET, SOI layer structural defect characterization, stress measurements in sSOI wafers are given. Advantages and limitations of the techniques are discussed.

Introduction

Silicon on insulator (SOI) wafers have become today one of standard substrate choices for high-end CMOS applications. Among numerous SOI manufacturing technologies, the Smart Cut technology was the first, which allowed mass production of SOI wafers. The quality of SOI wafers is approaching the quality of high end bulk Si products. The SOI manufacturing process requires monitoring of dozens of parameters, both on-line during wafer processing and off-line on the finished product. Most of the parameters can be controlled using standard silicon processing metrology. Examples include geometrical parameters like flatness, nanotopology, edge shape, or crystal quality like doping, orientation, metal contamination etc. There are also techniques, which are widely used in silcon wafer manufacturing, but require adjustment to specific SOI structure. One dramatic example is particle inspection on the front side of SOI wafers. Conventional laser particle inspection tools, like SP1, use visible light (488nm), which leads to interference of reflected light from the top Si surface and Si/BOX interface. As a result, LPD size threshold of SP1 on modern SOI wafers is limited to 0.15um. Introduction of new inspection tools based on UV light (e.g. KLA-Tencor SP2 tool uses laser at 355nm wavelength) essentially solved this problem for SOI layers down to at least 20nm thickness and allowed a significant decrease of the detection threshold [1], which now meets the requirements of ITRS 45nm device generation.

In this paper we will focus on the characterization techniques developed specifically for SOI wafer characterization. We will give some examples of the techniques, which are used in SOITEC characterization laboratory for SOI process monitoring and process development.

Metrology for Off-Line Process Monitoring.

A number of SOI product parameters can be assessed only using destructive metrology techniques. Among those, which are widely used in SOI wafer manufacturing, are the techniques for measuring electrical quality of BOX and its interfaces and for inspection of crystalline defects in top Si layer.

Wafer Characterization by Psi-MOSFET Technique.

Electrical quality of BOX/Si interface of the finished SOI wafer greatly depends on BOX formation method as well as SOI manufacturing steps. Characterization of this parameter with a simple, fast turnaround method is vital for statistical process control. One of the techniques, which are suited for these purposes, is electrical characterization of the interface using pseudo-MOSFET (Psi-MOSFET) method. This technique has been described in detail (see for example [2,3]). It is based on measurements of characteristics of an upside-down MOSFET structure created between two point contacts at the top of the Si layer with the SOI substrate acting as a gate. Electron and hole mobility, interface trap density and fixed oxide charge can be extracted from traditional drain current and transconductance ($I_D(V_G)$ and $g_m(V_G)$) curves. Extracted mobility values should not be directly compared with the carrier mobility measured in conventional PMOS or NMOS structures, but rather serve as a process control parameter.

An example of sensitivity of the extracted parameter to BOX quality is illustrated in the Fig.1. It is known that bonding of silicon wafers with relatively thin oxides produces interface defects during bonding anneal due to water/hydrogen trapped at the bonding interface, if no special process optimization is done [4]. A set of the SOI wafers with different BOX thickness was prepared to vary bonding interface quality. LPD measurements by SP1 on these wafers, shown in the figure 1, confirm presence of micro voids in the SOI stack for the BOX with thickness below 500A. The results of Psi-MOSFET measurements correlate very well with SP1 findings, demonstrating that not only micro bubbles are generated at the interface, but also interface trap density is degraded, which causes apparent decrease of extracted electron mobility. Thus, Psi-MOSFET measurements can provide additional complimentary information for SOI process development.

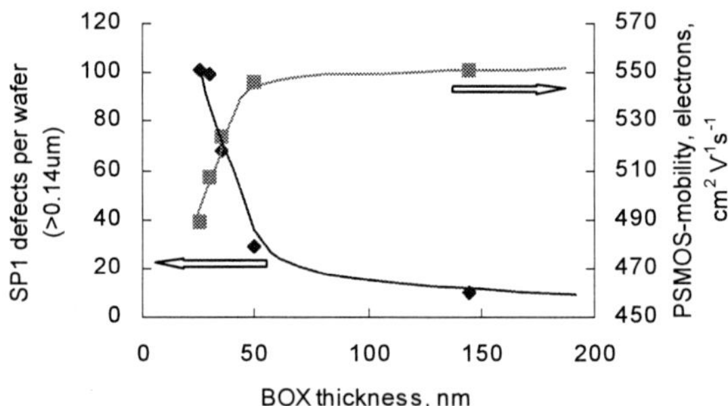

Figure1.   Dependence of Psi-MOSFET extracted electron mobility and SOI wafer defectivity on BOX layer thickness.

We would like to emphasize that extracted mobility values depend not only on interface/BOX quality, but also on measurement conditions [5], surface preparation [6] and parameters of the SOI structure. Fig.2. shows dependence of extracted mobility on top silicon layer thickness [7], demonstrating difficulty of measuring mobility in thin SOI structures. Although few variations of pseudo-MOSFET technique have been proposed, such as Hg – probe measurements [3] or ring – FET structures [8], still questions have been raised concerning the application of these techniques to ultra-thin SOI layers [9].

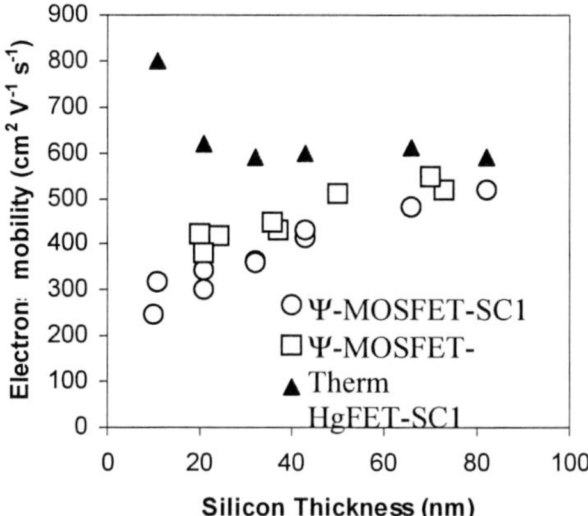

Figure 2. Extracted Psi-MOSFET electron mobility measured on the wafers with different SOI layer thickness. SOI layer thickness was varied by two different thinning techniques [7].

Recent studies showed that top surface charge can strongly influence the extracted parameter values. This impact is increasing with the film thickness reduction. Special surface passivation is needed to improve the surface control and a new parameter extraction model is currently under development to adapt this technique to thin films [10]. Nevertheless, in the present state the metrology can be used only for comparative studies or statistical process control.

Preferential Chemical Etching of Structural Defects.

Another key parameter, traditionally monitored in SOI wafer production, is structural defectivity of top Si layers. For the delineation of crystalline defects in Si bulk and final SOI wafers, strained Silicon epitaxial layers and final sSOI products, Secco based etching solutions have been historically used [11]. This etching solution can be diluted to adapt etching rate to thin SOI layers. It has high selectivity for etching of extended defects such as dislocations and, especially stacking faults. It results in small and narrow etch pits, which are usually highlighted by second etching step, selective to underlying layer, e.g. HF in case of SOI. This two-step etching results in well-developed etch pits which can be clearly seen in an optical microscope. The etch removal is uniform and a smooth surface is obtained after etching.

However, the Secco etching solution is based on toxic hexavalent chromium (Cr^{6+}), which is going to be restricted in the future for environmental protection reasons. The development of chromium-free etching solutions for surface defect decoration on different materials and stacks is a current challenge for volume SOI manufacturing. Recently, the development of Cr-free solution based on CP-4 modifications showed promising results to replace the Secco for SOI process monitoring [12].

Latest improvements in sSOI technology allowed production of high quality sSOI layers with defect densities in the $10^5 cm^{-2}$ range [13] and donor Si/SiGe epitaxial wafers with the density of threading dislocations even below $10^4 cm^{-2}$. Such defect densities are well below detection limit of conventional TEM techniques. Metrology based on selective etching with ability to provide the information about structure of the defects similar to TEM is required. One of the promising techniques is HCl gas phase etching in epi reactor [14,15]. Figure 3 shows an example of partially relaxed sSOI wafers etched by standard Secco (a) and HCl (b) techniques. Partially relaxed samples were chosen to increase defect density to be able to correlate etching results with TEM defect analysis.

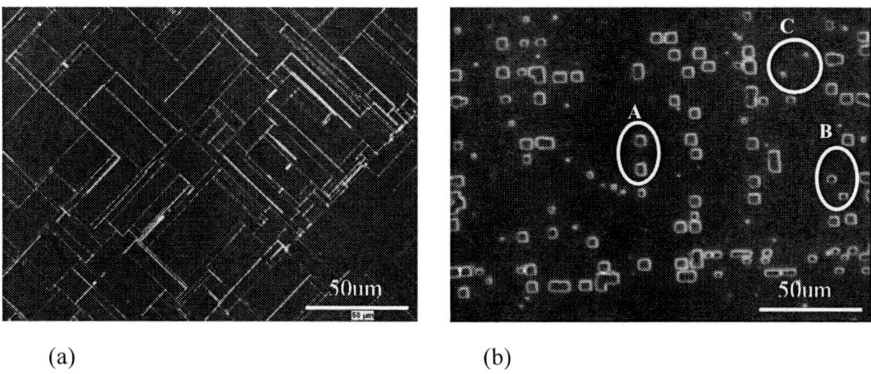

(a) (b)

Figure 3. Optical microscope images of Secco (a) and HCl (b) etched partially relaxed sSOI wafer with stress equivalent to $Si/Si_{0.8}Ge_{0.2}$ mismatch. (Images are rotated at 45° with respect of each other).

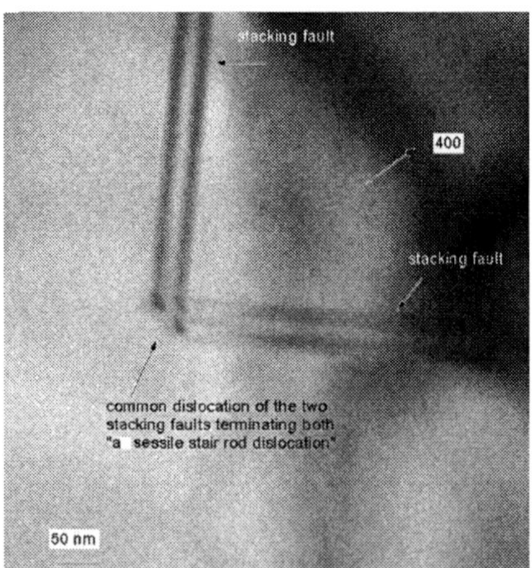

Figure 4. Bright field TEM image of sSOI layer, showing an example of formation of stair rod dislocation structure. SSOI wafer is from the same set shown in figure 3.

It is known, that Si layers under high tensile stress can relax by generation of the perfect and 90° dislocations loops, depending on the stress and thickness of the layers [16]. It results in threading dislocation segments of 60° a/2<101> and 90° a/6<211>

types, as well as stacking faults produced by passage of Shockley dislocations. When threading dislocation density becomes high, Shockley partial dislocations can react with each other producing very stable Lomer-Cotrell dislocation with Burger's vector a/6<110>, which is illustrated in Fig.4, where an example of such reaction is shown. Comparative analysis of the images of Fig 3(a) and (b) shows that, while Secco etching reveals mostly stacking faults, HCl etching produces square, crystallographically oriented etch pits of different size (marked A,B,C on the image), but SF contrast is less pronounced.

Figure 5 shows etch pit size distribution for the sample etched by HCl. Clearly, three different size populations are seen, corresponding to different defect types. The group with larger sizes corresponds to Lomer-Cotrell dislocations, as such pits are located at the intersection of the stacking faults. Middle group corresponds to Shockley partials. We can speculate that the group with the smallest size of the etch pits is related to 60° perfect dislocations. Thus, HCl technique allows to gain additional information on defect structure compared to Secco etching. This becomes possible due to the differences between Secco and HCl etching mechanisms. While Secco etching has 3 times higher etching rate of the material very close to the defect core [17] and produces very narrow etch pits along the dislocation or SF plane, HCl etching has very high anisotropy, near 100:1, of lateral versus perpendicular etching rates resulting in large and shallow pits as illustrated in Figure 6. In the case of two step Secco etching, the size of the etch pits visible in the optical microscope is defined by second HF etching step. It leads to loss of the information about defect structure. HCl etching does not require second etching step because it does not etch buried oxide and has high lateral selectivity, thus preserving information on selectivity of the etching rate to different types of the defects. Detailed analysis of the etching behaviour will be published elsewhere [18].

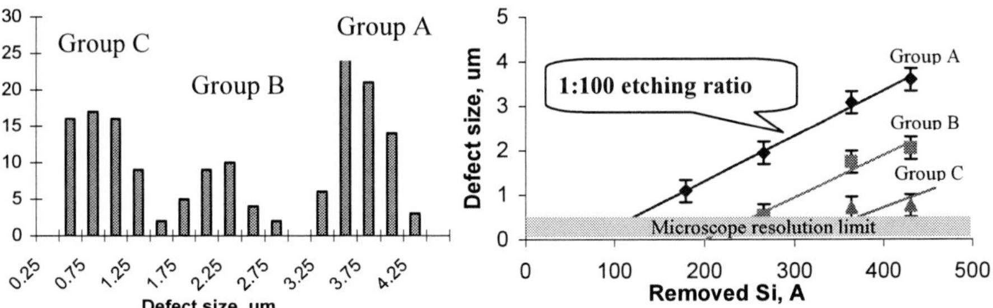

Figure 5. Size distribution histogram of the etch pits after HCl etching measured in optical microscope.

Figure 6. Evolution of size of etch pits for different types of the defects during HCl etch.

One should mention the limitations of the etching techniques. Both techniques have practical limit of threading dislocation density and stacking faults. For dislocations one can measure densities in the range $10^2 - 10^7 cm^{-2}$. Linear SF density can be measured accurately only in the case of very low densities $< 10^3 cm^{-1}$. In the case of large strain relaxation of highly stressed Si layers, single stacking faults may not be resolved by chemical etching because the layers relax by generation of SF from surface nucleation sites producing SF separated only by few atomic planes [19]. One should also take precautions to prevent additional relaxation of the layers during HCl etching taking into account relatively high temperature (800°C) of the treatment.

<u>Silicon Crystal Grown-in Defect Assessment using HF Inspection of SOI Layers.</u>

Interesting application of well known HF etching technique became possible due to very high quality of modern SOI processes. Traditional HF characterization consists of etching an SOI wafer in HF solution followed by wafer inspection using LPD tools. The concentration of HF solution and etching time have to be adopted to the SOI and BOX layer thickness and, in general, fall in the range of 10-30% and 20-60min, respectively. This technique can detect microscopic protrusions in SOI layer either induced by SOI process or coming from crystal quality of the top Si material. Historically, one of the results of using HF characterization was the optimization of top Si material to minimize COP density. Current SOI wafers typically have HF defect densities below 0.05 defects per cm^2 with typical values in the range of $0.01cm^{-2}$. We propose to use this technique to study density and size distribution of oxygen precipitates in top Si wafers. Indeed, if oxygen precipitates are embedded in top Si layer with the size larger than the thickness of the layer, they can be detected by HF inspection technique. Applying consequent steps of HF delineation and layer thinning, it is possible to measure oxygen precipitate size distribution. For controlled layer thinning, chemical etching by SC1 solution can be used as an example. Temperature and composition of the solution should be optimized to reduce the difference in etching rates of Si and SiO_2.

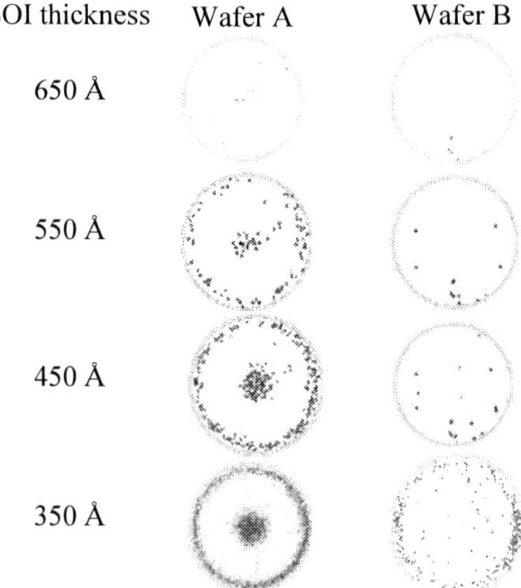

Figure 7. LPD maps of SOI wafers after subsequent HF etching and SC1 thinning steps. Si wafers used for the top SOI layer were subjected to oxygen precipitation treatment before SOI process.

Figure 7 shows examples of LPD maps of the same 300mm wafers after series of SOI layer thinning by SC1 treatment, each followed by HF defect revelation. Prior to SOI process, Si wafers used for the top Si layer transfer were subject to oxygen precipitation heat treatment. The results demonstrate that this technique provides information on defect size distribution as well as their radial distribution. The advantage of this technique is that the defects as small as 10nm in size can be measured with densities as low as 10^4 cm^{-3}, which is beyond the detection limit of traditional techniques, such as Laser Scattering Tomography (LST) [20].

<u>Stress measurements in sSOI wafers.</u>

One of the key parameters defining functionality of sSOI wafers is stress in the top Si layer. Control of stress in final sSOI product is a mandatory part of manufacturing process. Few appropriate techniques are available for the metrology of stress in thin top Si layer in sSOI wafers. Lattice parameter measurement with traditional high resolution X-ray diffraction can be difficult in very thin 10-20nm layers. However, grazing incidence HRXRD was shown to be very promising technique to measure in-plane lattice parameter in sSOI layers [21]. Another technique, photoreflectance spectroscopy, allows direct measurements of critical point transitions in semiconductor band structure. This technique was successfully applied to Si/SiGe heterostructures and showed ability to measure Ge concentration in SiGe layers as thin as 50A with accuracy better than 1% [22,23].

Raman spectroscopy has been extensively used for strain measurements in thin Si layers. It can be employed to control Ge concentration and strain in top Si layer in Si/SiGe structures simultaneously, measuring frequency of Si-Si and Si-Ge vibrational modes [24]. Utilizing UV light (325nm) as an excitation source allows to extract Si-Si signal only from the top Si layer, which improves accuracy in case of thin 10-20nm sSOI layers. One of the advantages of Raman spectroscopy is the ability to obtain stress distribution on the micron length scale using focused excitation beam (micro Raman mode) [25]. Figure 8 shows stress distribution in 40nm sSOI layer on the micron scale deducted from micro Raman measurements. Stress nonunniformity modulated along <110> directions is clearly seen.

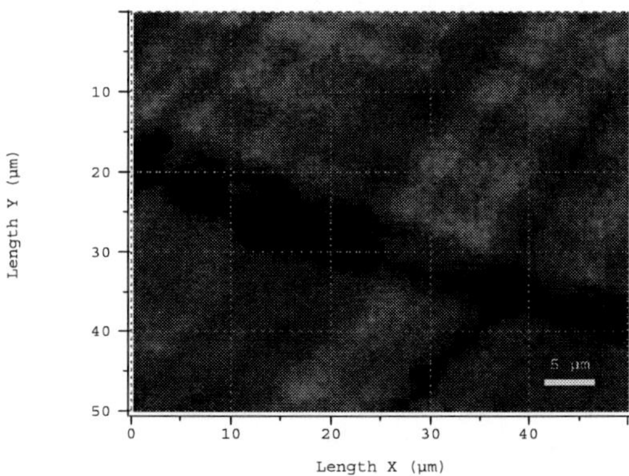

Figure 8. Micro Raman image of stress distribution in 40nm top Si layer of sSOI [25].

This stress modulation has been reported earlier in SGOI structures [26], where it has been attributed to lateral fluctuations in Ge concentration. Figure 9 shows that the origin of this stress nonuniformity cannot be attributed to variation of Ge concentration.

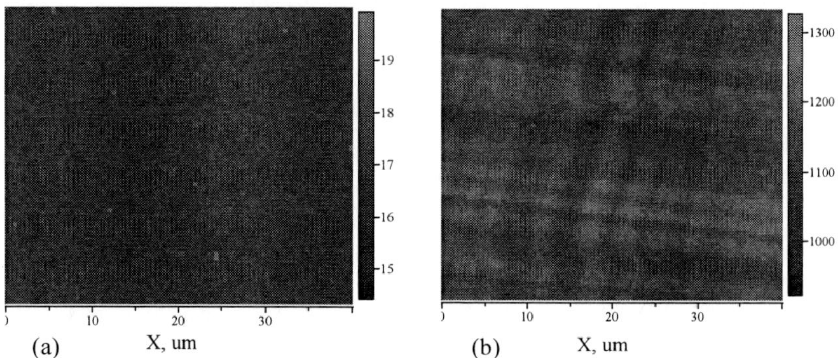

Figure 9. Ge concentration (%) map (a) and map of the stress (Mpa) in top Si layer (b) measured in Si/SiGe structure used as a donor wafer for sSOI.

It results from the stress from dense misfit dislocation network in SiGe graded layer below SiGe buffer layer with constant Ge content. It is interesting, that this stress is transferred into sSOI layer and survives full layer transfer process. It should be noted that such variations of the stress should not affect functionality of sSOI wafers, as dependence of electron mobility in Si on stress significantly decreases in the range of stress equivalent to $Si/Si_{0.8}Ge_{0.2}$ mismatch.

Acknowledgments

The authors would like to thank F.Allibert for electrical characterization results, A.Abbadie for the help with Secco and HCL etching, P.Reynaud for the experiments with HF/SC1 etching and Y.-M. Le Vaillant for micro Raman measurements.

References

1. C.Moulin, D.Delprat and C.Maleville, *2005 IEEE International SOI Conference*, 146 (2005).
2. S Cristoloveanu et al, *IEEE T.E.D.* **47**, 1018 (2000).
3. H.J. Hovel, *Sol. Stat. Elect.* **47** 1311 (2003).
4. K.Mitani, V.Lehman, R.Stengl, D.Feijoo, U Goesele, and H.Z.Massoud, *Jpn. J. Appl. Phys.*, **30**, 615 (1991).
5. S.Williams, S.Cristoloveanu, and G.Campisi, *Materials Sci. Eng.*, **B12**, 191 (1992).
6. D. Munteanu et al, *El. & Sol. St. Lett.* **2, 242** (1999).
7. F. Allibert, N.Bresson, K.Bellatreche, C.Maunand-Tussot, and S.Cristoloveanu, *2005 IEEE International SOI Conference*, 59 (2005).
8. H.J.Hovel, T.E.McKoy, US patent 2005 009 2985.
9. F. Allibert et al, *2002 IEEE International SOI Conference*, (2002).
10. G. Hamaide, F. Allibert, H.J. Hovel and S. Cristoloveanu, to be published.
11. L.F.Giles, A.Nejim, P.L.F.Hemment, *Vacuum*, **43**, 297 (1992).

12. J.Mahlis, A.Abbadie and B.Kolbesen, *these proceedings*.
13. I. Cayrefourcq, A.Boussagol and G.Celler, *ECS Trans.* **3**, (7) 399 (2006).
14. Y. Bogumilowicz, J.M. Hartmann, R. Truche, Y. Campidelli, G. Rolland and T. Billon, *Semicond. Sci. Techn.* **20**, 127 (2005).
15. S. Kreuzer, F. Bensch, R. Merkel, G. Vogg, *Mat. Sci. In Semicond. Proc.* **8**, 143 (2005).
16. E.P.Kvam, R.Hull, *J.Appl.Phys.*, **73**, 7407 (1993).
17. S.W. Bedell, H. Hovel, A. Domenicucci, K. Fogel, A. Reznicek and D.K. Sadana, *Electrochem. Soc. Proc.* **2005-03**, 345 (2005).
18. A.Abbadie, S.W.Bedell, J.M.Hartmann, D.K. Sadana, F. Brunier, C. Figuet and I. Cayrefourcq, submitted to *J. Electrochem. Soc.*
19. W.Wegscheider and H.Cerva, *J. Vac. Sci. Technol. B*, **11** 1056 (1993).
20. K.Moriya, K.Hirai, K.Kashima and S.Takasu, *J. Appl. Phys.* **66**, 5267 (1989).
21. D. Kozemura, K.Yamasaki, S.Tanaka, Y.Kakemura, T.Yoshida and A.Ogura, *2006 IEEE International SOI Conference Proceedings*, 53 (2006).
22. D.J.Hall and R.T.Carline, *Appl.Surf.Sci.* **125**, 1 (1998).
23. C.Chen, P.V.Kelly, Z.Liu, W.Huang, W.Dou and P.H.Tsien, *Metals and Materials Int.*, **10**, 489 (2004).
24. J.C.Tsang, P.M.Mooney, F.Dacol and J.O.Chu, *J. Appl. Phys.*, **75**, 8098 (1994).
25. A.Tiberj, V.Paillard, C.Aulnette, N.Daval, K.Bourdelle, M.Moreau, M.Kennard and I.Cayrefourcq, Mat. Res. Soc. Symp. Proc. Vol. 809 (2004).
26. K.Kutsukake, N.Usami, T.Ujihara, K.Fujiwara, G.Sazaki and K.Nakajima, *Appl. Phys. Lett.*, **85**, 1335 (2004).

ECS Transactions, 6 (4) 235-244 (2007)
10.1149/1.2728866, ©The Electrochemical Society

Investigation of Apertureless NSOM for Measurement of Stress in Strained Silicon

Colin McDonough, Jacob Atesang, Yunfei Wang, and Robert E. Geer

College of Nanoscale Science & Engineering, University at Albany, SUNY
Albany, NY 12203

Strain in blanket and patterned silicon-on-insulator (SOI) structures have been investigated via apertureless near-field scanning Raman spectroscopy, specifically, to investigate the efficacy of so-called tip-enhanced Raman scattering (TERS) for Si strain characterization and metrology for integrated circuit (IC) devices. The current study compares TERS generation using Ag-coated W tips in a 45° incident beam, fixed-tip geometry and Ag-coated SiO_2 capillary tips in a normal incident, scanning-tip geometry. The latter demonstrates superior performance and is used in a differential scheme to investigate strain profiles in patterned SOI structures. Specifically, a blanket device layer with an engineered strain of 0.075%, is patterned to form arrays of isolated mesa structures (2 μm diameter). These mesa structures exhibit a stress relaxation of 117 MPa over a region extending ~ 200 nm from the island edge.

Introduction

Illumination of a nanometer scale metallic tip in the immediate vicinity of a Raman-active surface exhibits enhancement of Raman scattered light similar to so-called surface-enhanced Raman scattering. Such effects have given rise to the development of apertureless near-field scanning optical microscopy (a-NSOM) for which a metallic or metal-coated nanoprobe is utilized to generate local (< 50nm) scattering enhancement at a surface (1-5). This approach has been investigated as a route to provide high spatial resolution profiling of stress in strained-Si device structures. Stress metrology in Si-based device structures has become an increasing critical issue for integrated circuit (IC) manufacturers which utilize strained Si channels in metal-oxide semiconductor field-effect transistors (MOSFETs) (6). We have carried out investigations of tip-enhanced Raman scattering (TERS) to determine the efficacy of this approach. For these measurements Veeco Aurora-3 and Nanonics MV2000 NSOMs have been integrated with a Renishaw Raman spectrometer. In place of conventional coated optical fibers metallized tips have been used as apertureless probes. Probes are maintained within 2-5 nm of the surface via a resonant vibration feedback loop. TERS imaging has been applied to blanket and patterned SOI test structures to investigate spatial and spectroscopic resolution. Results are discussed in detail below. While preliminary, these studies demonstrate the attractive potential of TERS-based approaches for strain metrology in Si-based device structures.

Experimental –NSOM System Schematics

The two a-NSOM systems employed in this study differ, primarily, in two aspects. Firstly, the Aurora system employs a 'shear-force' approach for a fixed-tip feedback in which the metallized tip is mounted on one leg of a vertically-oriented tuning fork that is mounted rigidly in the system (Fig. 1), i.e. the tip position is fixed with respect to the far-field illumination beam with the sample position being controlled by a conventional piezoelectric tube scanner. In contrast, the metallized tip in the MV2000 system is mounted to a horizontally oriented tuning fork which utilizes a 'normal-force' feedback approach and is controlled by a separate scanner so that it may be positioned at various points with respect to the sample and far-field illumination beam. Sample positioning within the MV2000 likewise employs piezoelectric positioners. Secondly, the Aurora system geometry employs a 45° angle of incidence between the far-field beam and the sample normal. The MV2000 operates in normal incidence mode.

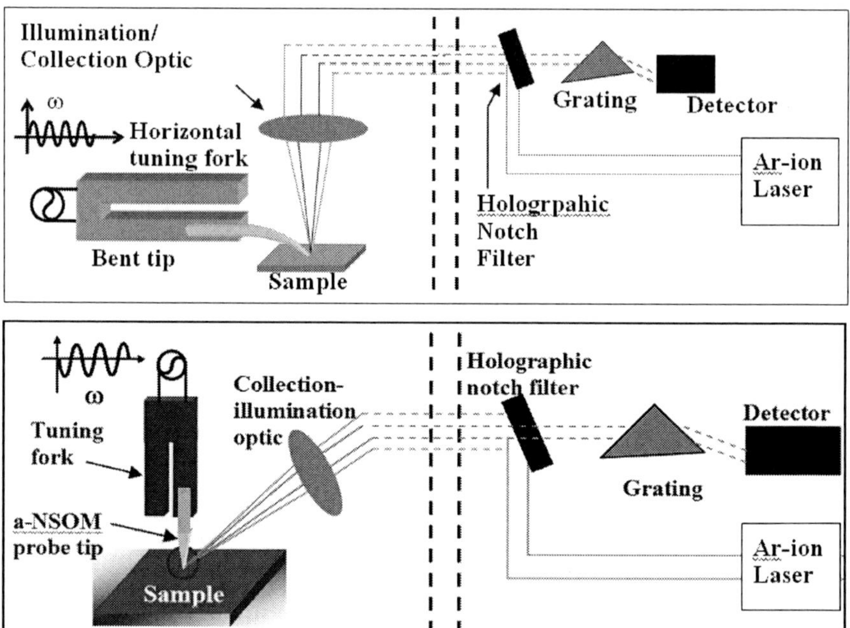

Figure 1. (Upper panels) Component schematic of the Nanonics MV2000 NSOM system integrated with a Renishaw optical spectrometer. The holographic notch filter assembly, in addition to rejecting Rayleigh-scattered light also acts as a high-rejection beam splitter for laser illumination of the tip. (Lower panels) Component schematic of the Aurora 3 NSOM system integrated with a Renishaw optical spectrometer.

The Renishaw spectrometer system (RM-100) integrated with the MV2000 and the Aurora systems is equipped with a Leica DM/LM optical microscope. A long-working distance 50X objective was used for our experiments in both systems. This lens focuses the incident laser (Ar-ion, 514 nm) down to spot sizes of 2 μm and 3 μm diameters, respectively for the MV2000 and Aurora systems. The collected Raman signal is detected by a charge coupled device (CCD) cooled by a Peltier cooling system. The spectrometer uses a 3000 lines/cm grating for spectra acquisition. A notch filter is used for rejecting the unwanted Rayleigh scattering in addition to serving as an efficient beamsplitter.

SOI Sample: Characterization and Test Structure Fabrication

A schematic of the SOI stack (supplied by SOITEC, Inc.) utilized for this work is shown in the left panel of Fig. 2. Prior to investigation with apertureless NSOM-based Raman the SOI sample was characterized using x-ray photoelectron spectroscopy (XPS), x-ray diffraction (XRD), and microRaman (tip retracted). The XPS elemental analysis (not shown) confirms the absence of significant contamination or other compositional variations that may effect the Raman spectra. XRD measurements were carried out on blanket SOI samples to independently characterize the stress state. A diffuse diffraction spot at h = -0.012, l = 4.023 (indexed to the bulk Si (004) peak) in the x-ray diffraction pattern represented the (004) peak corresponding to the strained Si layer (7). The displacement in the h-l plane results from the inherent strain of the SOI top layer and the slight misorientation inherent in the bonding process of the strained layer to the Si handle wafer. The position of this peak indicates a change in the lattice spacing, $\Delta c/c$, of 0.0057 of the device layer. Assuming biaxial strain and using the accepted elastic constants for Si, the in-plane strain is estimated at 0.746%. This matches well with the SOITEC specifications which list an in-plane strain of 0.75% for this sample.

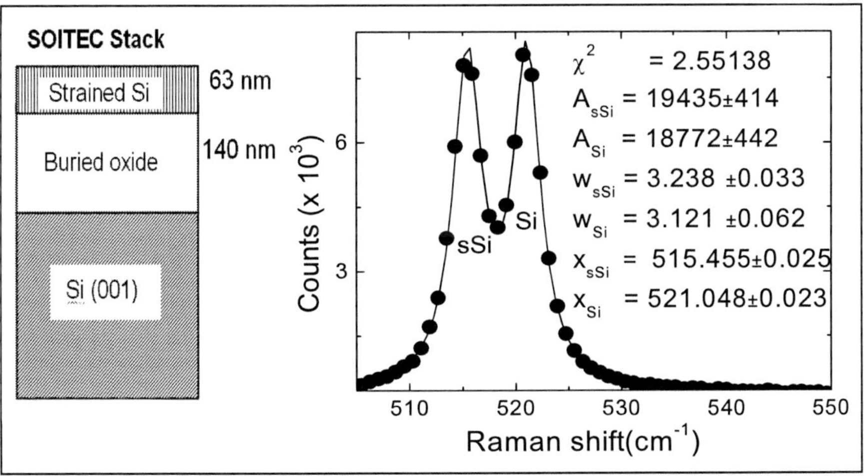

Figure 2. (Left) Schematic profile of the SOI stack used. (Right) Raman far-field or 'micro' Raman spectra of blanket SOI sample. The data was fit to a double Lorentzian. The fit parameters are listed in the plot. A, w, and x refer to the Lorentzian amplitude, width, and center position, respectively. The subscripts 'sSi' and 'Si' denote the strained and bulk Si peaks, respectively.

The SOITEC SOI sample was evaluated using the Veeco Aurora NSOM in micro-Raman mode for which the tip was retracted but the sample maintained in the focal plane of the illumination objective. Figure 2 shows a typical Raman spectrum from the blanket SOI wafer. As expected two clear peaks are resolved, one corresponding to the strained

Si device (top) layer, the second corresponding to the bulk. The Raman shift of 5.6 cm^{-1} corresponds to a biaxial strain $(\sigma_{xx}+\sigma_{yy})$ of 2.55 GPa using the models described by De Wolf and coworkers and references therein (8-10). This yields an in-plane strain of 0.0071 ± 0.0005 and an out of plane strain of 0.0056 ± 0.0004, in reasonable agreement with the XRD data.

Measurements of TERS: 45° Incidence Geometry

To investigate the presence of TERS from SOI samples a series of experiments was undertaken to investigate Ag-coated W tips in the Veeco Aurora NSOM (45° incident beam geometry). Electropolished W tips (tip radii 20-80 nm) were sputter-coated with Ag (30s) (5). Due to the irregular morphology of the etched W tips it is nontrivial to characterize an average sputtered thickness of Ag. Also, it is not experimentally feasible to determine the exact point on a tip which provides enhancement and, therefore, difficult to determine a specific, optimal coating thickness.

Tips were attached to a vertically mounted tuning fork and evaluated for TERS using the SOI sample. Raman spectra were collected with the tip in feedback and with the tip retracted. To avoid slight changes in Raman intensity associated with misalignment of the optical focal plane the Veeco Aurora NSOM stage position was fixed at the focal plane of the NSOM objective lens via direct control of the vertical piezo scanner of the sample. This ensures that the position of the SOI sample with respect to the NSOM focal plane is fixed during tip-evaluation.

Figure 3 illustrates SOI Raman peaks acquired with the Aurora tip in feedback (left panel of Fig. 3.) and out of feedback (right panel of Fig. 3). A measurable increase in signal is present for the data acquired in feedback compared with the data acquired with the tip retracted. This increase (after background subtraction) represents a raw enhancement of approximately 12%. Although modest in absolute terms, this increase in measured intensity corresponds to an enhancement factor of approximately 10^{3} assuming an effective tip radius of 50 nm. Although the complex morphology of the tips makes an accurate estimation of the effective tip-radius difficult the enhancement factor is in the range typically reported in the literature (1-5).

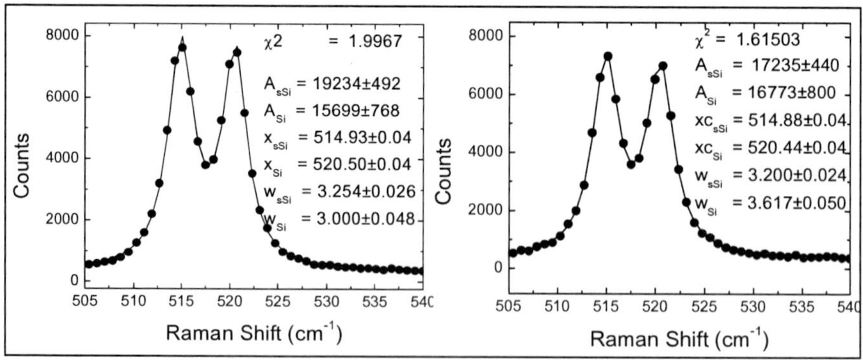

Figure 3. Raman spectra from SOI sample with tip in feedback (left) and tip out of feedback (right). Each data set was fit using a double Lorentz peak model. Fitting parameters are shown in the plot.

Figure 4 shows the strained Si Raman peak height both with the tip in and out of feedback as a function of time. Each signal is exceedingly stable. This stability is critical in that it underscores the substantial (although relatively modest) change in intensity with and without the Ag-coated W tip in feedback. Likewise, the Raman peak position is extremely stable over the measurement time. Figure 5 plots the peak positions of the strained Si and Si Raman peaks as a function of time (with a Ag-coated W tip in feedback). The peak positions are extremely stable. The variation (1 sigma) over the measurement time is approximately 0.019 cm^{-1}.

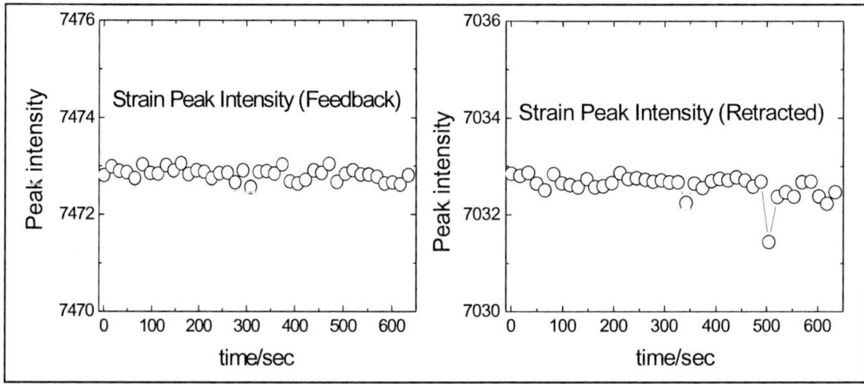

Figure 4. Peak intensity for the strained Si peak from the SOI sample with a Ag-coated tip in feedback (left) and out of feedback (right) as a function of time. Both signals exhibit excellent stability (< 0.007% variation) over an 11 minute observation time.

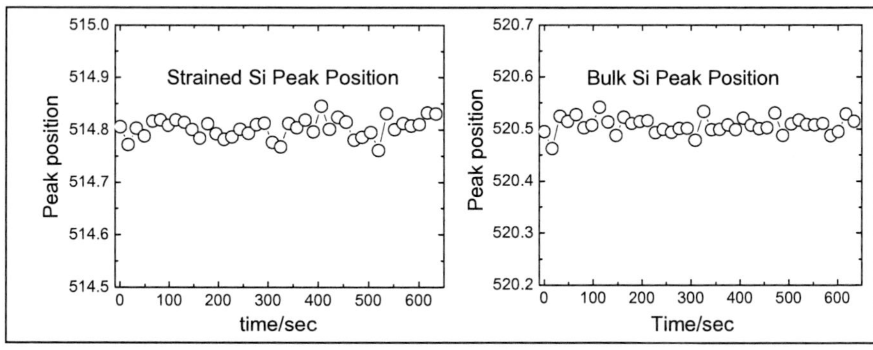

Figure 5. Strained (left) and bulk (right) peak positions from a blanket SOI sample with a Ag-coated tip in feedback as a function of time. Both signals exhibit excellent stability (< 0.019 cm^{-1} error) over an 11 minute observation time.

Counter intuitively, both strained Si and bulk Si Raman peaks as measured with the Ag-coated W tip exhibit enhancement. This is expected for the strained (surface) Si peak, but not for the bulk. In fact, for the data shown in Fig. 3 the bulk Si Raman peak exhibits a raw TERS enhancement of 6%, half that for the strained Si Raman peak. (As shown in more detail below this behavior was not observed for other optical configurations). However, for all the SOI Raman measurements undertaken with the Veeco Aurora NSOM bulk Si Raman enhancement accompanied strained Si Raman enhancement. This observation mirrors recent reports by other research groups in this field (11). It is likely that these apparent inconsistencies arise from optical interference effects associated with the SOI structure itself. The presence of the tip and effects associated with optical shadowing may exacerbate this effect. This effect is not seen for the normal incident TERS data described below.

Measurements of TERS: Normal Incidence Geometry

As a complement to the work described in the preceding section, measurements were undertaken with Ag-coated bent glass fiber tips on a Nanonics MV2000 NSOM. Unpatterned sections of the SOI stack investigated with the Aurora NSOM system were evaluated using the MV2000 system. This data is shown in Fig. 6. Note the magnitude of the surface peak relative to the bulk has decreased significantly compared to data acquired from the Aurora system. This is a function of the optical illumination geometry. Preliminary examination has attributed this feature, in part, to an interferometric effect. A quantitative description of this effect is under development. Also note that the relative TERS enhancement documented in Fig. 6 ($\sim$ 20%) is substantially larger compared to the data of Fig. 3. And more notably, the bulk peak actually sees a 13% reduction in intensity compared to data acquired with the tip retracted. This reduction is due to the shadowing of the sample by the metallized tip. The shadowing effect has been confirmed by displacing the tip away from the center of the beam while in feedback.

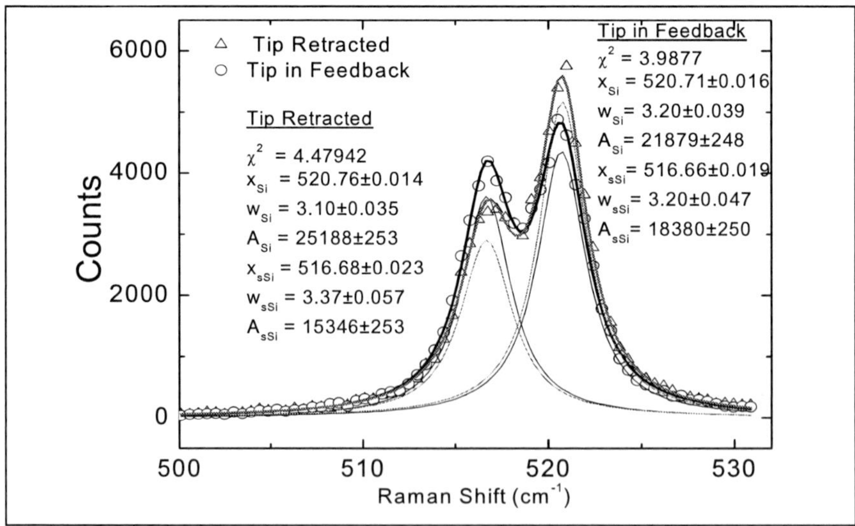

Figure 6. Raman spectra in feedback and retract mode from the SOI sample on a MV2000 NSOM. Enhancement of the strained Si Raman peak relative to the bulk is clearly evident.

From the data presented above we conclude that TERS provides a measurable enhancement for SOI blanket samples for both the off-axis and normal incidence illumination geometries. Moreover the data show that a stable enhancement can be achieved over spectra acquisition times suitable for high-resolution spatial and spectral imaging. Ostensibly, the normal axis geometry provides the most promising results from the point of view of SOI Raman peak enhancement. This not unexpected since the tip-scanning feature of the MV2000 permits optimal placement of the tip in the optical illumination beam. The Aurora system does not permit this and, as a consequence, there is no methodology to effectively optimize tip position/orientation for that system. Based on the results described above, the normal incidence geometry will be utilized to investigate patterned SOI test structures discussion in the following section.

TERS investigation of patterned SOI test structure.

The SOI test structures used to evaluate normal-incidence TERS scanning consist of isolated 'mesas' of strained Si patterned on the buried oxide. A conventional 'via' mask was combined with a contrast inversion photolithography process to pattern a thin (50 nm) layer of Cr islands across the SOI wafer. A TMAH wet etch step resulted in the removal of unprotected Si. The buried oxide layer constituted a highly selective etch stop. The Cr was removed via a wet process which resulted in a field of strained-Si 'mesas' on a buried oxide substrate. A SEM image of a patterned strained-Si mesa is shown in Fig. 7. The octagonal shape results from the anisotropic TMAH wet etch used to fabricate the test structure. Each mesa is approximately 2 microns in diameter. The thickness of the SOI device layer is approximately 63 nm and the thickness of the buried oxide layer is approximately 140 nm.

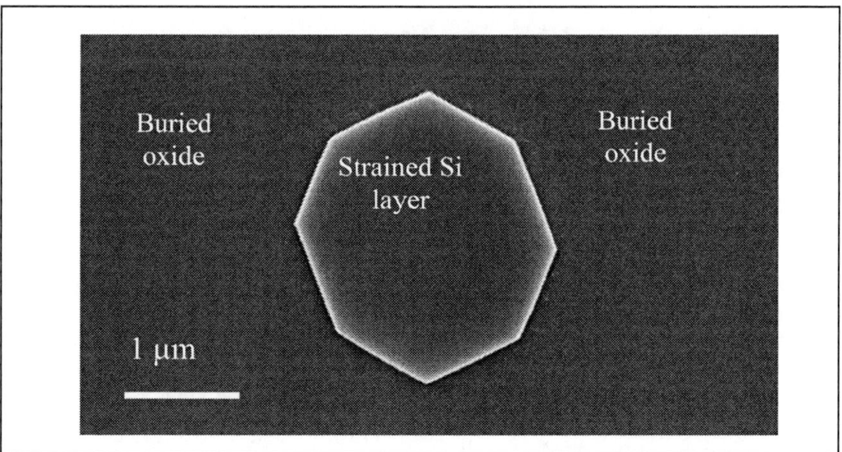

Figure 7. SEM micrograph of a patterned SOI test structure consisting of a circular (octatgonal) 'mesa' of strained Si (approximately 63 nm in thickness). The surrounding material consists solely of the buried oxide layer.

The SOI test structures have been used to evaluate normal incidence TERS using the MV2000 NSOM. To separate the tip-specific component of the Raman-shifted light a differential approach has been employed to isolate the near-field generated Raman signal from the far-field generated Raman signal. In this mode a series of spatially-resolved Raman spectra are acquired at a 2D array of points on the substrate with the tip in feedback. The same measurements are repeated with the tip retracted and the differential is taken to isolate the surface-specific TERS signal. Using this approach, topographic and spectral Raman imaging of the patterned SOI test structure was undertaken for comparative evaluation. Topographic imaging agreed in detail with the SEM image shown in Fig. 7. Spectral imaging results are shown in Fig. 8. Specifically, the intensity of the Raman peak associated with the strained-Si peak (see Fig. 6) is mapped. The gray-scale map and vertical height corresponds to linear intensity variation. Treating the edge of the mesa as a vertical boundary and the corresponding broadening due to a Gaussian resolution function a lower limit of the spatial resolution is estimated at 90-100 nm. Considering the tip used to obtain this image has a diameter of nearly 175 nm it is likely that the resolution can be reduced below 45 nm with sub-100nm diameter tips.

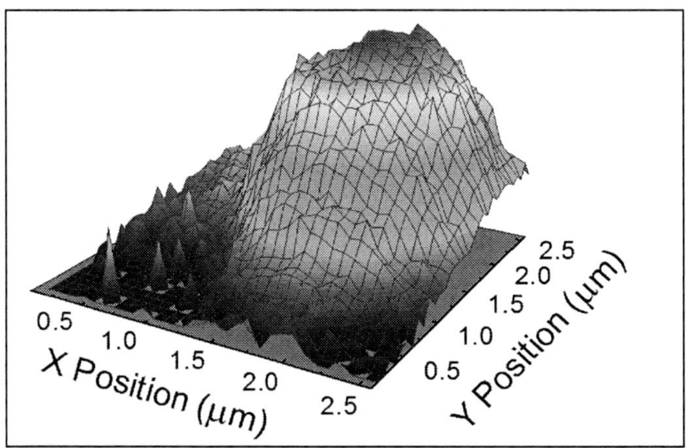

Figure 8. Intensity map of the peak intensity of the strained-Si Raman peak in the vicinity of the strained-Si mesa.

Although the maximum spatial gradient observed in the strained-Si peak height corresponds to the topographically-determined mesa position, a Raman 'halo' is present about the mesa, evident in Fig. 8. This halo consists of relatively low intensity strained-Si Raman signal extending away from the mesa over several hundred nanometers. This likely resulted from the conventional far-field Raman signal associated with the spot size of the illuminating laser. The presence of residual far-field Raman scattering in the differential signal implies modification of the far-field (or background) signals due to the presence of the tip. It is probable that shadowing effects, drift, and diffuse Raman signal generation from the glass capillary combine to produce the 'halo.' Current efforts have centered on increased efficiency for far-field Raman signal rejection using modified scan geometries and polarization selection.

The mesa test structure offered the opportunity to investigate edge-induced strain relaxation and the efficacy of measuring such relaxation via TERS. Shifts in the strained-

Si Raman peak position were investigated as a function of spatial position across the mesa along a line passing through the mesa center. Figure 9 illustrates this dependence by plotting the position of the strained-Si Raman peak as a function of spatial position across the mesa. The mesa edge occurs at approximately the 1.0 μm point in the scan. At this edge the strained-Si Raman peak position is 517.37±0.06 cm^{-1}. This compares to an average value of 516.84±0.06 cm^{-1} over the center region of the mesa. This corresponds to a reduction in the biaxial tensile stress of 117±18 MPa (along a single in-plane principal axis). This represents a 9% reduction in strain at the mesa edges. The profile in Fig. 9 is fit to a Gaussian-smeared step with a characteristic width of 228±38 nm for empirical purposes only. It is likely that the same halo-effect seen from the strained-Si Raman peak intensity maps is contributing to the Gaussian width of the profile in Fig. 9. Investigations into this are continuing as well as finite-element modeling of the stress relaxation in the mesas. Preliminary analysis implies small amounts of residual compressive stress in the underlying Si associated with the mesa relaxation. Detailed studies of this effect are continuing.

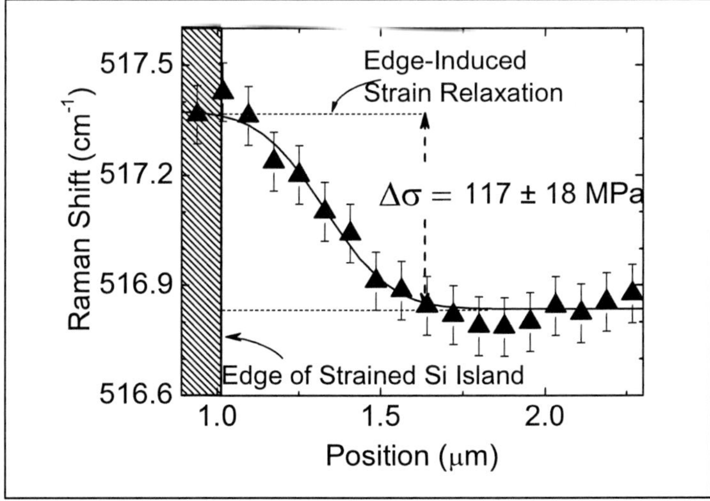

Figure 9. Spatial profile of the strained-Si Raman peak position across the center line of the mesa structure depicted in Fig. 7. A stress relaxation of approximately 117 MPa is evident. The solid line is an empirical fit to a Gaussian-smeared step and served to characterize the spatial variation of the profile and as a guide to the eye.

Although preliminary, these results suggest great promise in the use of a-NSOM Raman imaging as a tool for strain/stress characterization in Si devices. Although spatial resolution must still be improved the initial results from differential Raman mapping of the SOI test structures described above demonstrate the technique's potential.

Conclusion

Strain in blanket and patterned silicon-on-insulator (SOI) structures have been investigated via apertureless near-field scanning Raman spectroscopy. A normal-incident illumination geometry demonstrated promising results for TERS-based stress characterization in SOI-based mesa structures. TERS-based stress imaging in the SOI mesa structures revealed a stress relaxation of 117 MPa over a region extending ~ 200 nm from the island edge.

Acknowledgements

Support is gratefully acknowledged from the Semiconductor Research Corporation and DARPA through the SRC FRCP Interconnect Focus Center, New York Center for Advanced Interconnect Science and Technology and the New York Office for Science, Technology, and Academic Research. It is a pleasure to acknowledge SOI test wafers and technical discussions with G. Cellar, J. Rinderknecht, M. Hecker, L. Zhu and E. Zschech. It is also a pleasure to acknowledge technical NSOM assistance from H. Taha, test structure fabrication from N. Tokranova, B. Altemus, A. Gracias, and L. Clow, and XRD measurements from R. Matyi.

References

1. L. Novotny, E. J. Sanchez, and X. S. Xie, *Phys. Rev. Lett.*, **82**, 4014 (1999).
2. A. Hartschuh, E. Z. Sanchez, X. S. Xie, and L. Novotny, Phys. Rev. Lett., **90**, 9 (2003).
3. H. Watanabe, Y. Ishida, N. Hayazawa, Y. Inouye, and S, Kawata, *Phys. Rev. B,* **69**, 155418 (2004).
4. W. X. Sun, Z. X. Shen, *Ultramicroscopy,* **94,** 237 (2003).
5. Jacob Atesang, Robert Geer *Proceedings of SPIE,* Robert E. Geer, Norbert Meyendorf, George Y. Baaklini, Bernd Michel, Editors, Vol. 5766, p. 134, SPIE Press, Bellingham, WA (2005).
6. K. Rim, *Solid State Electronics,* **43**, 1133 (2003).
7. R. Matyi, unpublished.
8. I. De Wolf, *Semicond. Sci. Technol.* **11**, 139 (1996).
9. I. De Wolf, H. E. Maes, and S. K. Jones, *J. Appl. Phys.* **79**, 7148 (1996).
10. E. Anastassakis, A. Cantarero, and M. Cardona, *Phys. Rev. B* **41**, 7529 (1990).
11. A. Sokolov, unpublished.

ECS Transactions, 6 (4) 245-250 (2007)
10.1149/1.2728867, ©The Electrochemical Society

Effective Control of Strain in SOI by SiN Deposition

Daisuke Kosemura[a], Kosuke Yamasaki[a], Yasuto Kakemura[a], Tetsuya Yoshida[a]
Atsushi Ogura[a], Hidetsugu Uchida[b], Hideki Naruoka[b], and Masaki Yoshimaru[b]

[a] School of Science and Technology, Meiji University, 1-1-1 Higashimita, Tama-ku,
Kawasaki, Knagawa 214-8571, Japan
[b] Semiconductor Technology Academic Research Center, 3-17-2 Shinyokohama,
Kouhoku-ku, Yokohama, Kanagawa 220-0033, Japan

Strain introduction in SOI substrates by SiN capping film was
evaluated by UV-Raman spectroscopy. Induced strain became
larger with increasing SiN film thickness. SOI substrates originally
had tensile strain and the SiN cap shifted the strain toward the
compressive direction. The induced strain was larger in the thinner
SOI substrates with the same thickness of the SiN capping film.
We also evaluated a substrate after SiN patterning with a high
spatial resolution of 200-nm, and found strain enhancement at the
pattern edge.

Introduction

In state-of-the-art LSI (large-scale integrated circuit) technologies, strain management
in Si is recognized as one of the most important technologies to achieve high
performance operation. This is because carrier mobility can be enhanced by introducing
appropriate strain in the channel region of metal-oxide-semiconductor
field-effect-transistors (MOSFETs) [1]. Two techniques are proposed for strain
introduction. One is "local strain", in which strain is introduced only in a desired region
during LSI fabrication. A SiN capping film and embedded SiGe source/drain (S/D)
structures are proposed for the local strain technology. The other is "global strain", which
means using a wafer with strained-Si film as a starting material such as a strained-Si on
insulator (SSOI), SiGe on insulator (SGOI) and so on. It has been reported that a
tensile/compressive contact-etch-stop-layer (CESL) in short channel devices is more
effective than a strained-Si substrate for mobility enhancement [2, 3]. We believe the
important factors are the absolute value and the uniformity of strain at the CESL/Si
interface. The strain in Si has been characterized by Raman spectroscopy, X-ray
diffraction, electron diffraction, and other methods. However, it has been difficult to get a
detailed profile of the strain in the very thin layer at the CESL/Si interface. In this study,
we measured the strain by high-resolution UV-Raman spectroscopy with expecting the
strain field induced by CESL to be near the surface [4]. Furthermore, we investigated the
mechanism responsible for introducing strain by comparing the strain before and after
SiN patterning.

Experiment

Sample

SiN capping films with 0 (reference), 20, 40, 60, and 80-nm thick were deposited on
(001) SOI substrates by low-pressure chemical vapor deposition (LP-CVD). The SOI
film thicknesses were 30, 50, and 100-nm. A conventional (001) bulk Cz-Si substrate was

also evaluated for comparison. We also prepared a substrate with the patterned SiN capping film to simulate device structure more precisely.

<u>UV-Raman measurement</u>

The UV-Raman spectroscopy system used in this study utilized resonant effects for Si measurement by using a 364-nm Ar ion laser as an excitation source. The penetration depth of the UV laser is appropriate for evaluating thin channel layers less than 5-nm [5, 6]. Moreover, the resonant effect using a 364-nm excitation laser is very attractive, because a strong Raman signal can be obtained without significant heating of the sample [7]. We carefully selected the incident laser power so that the red shift due to the elevated temperature was completely suppressed by checking the Raman peak as a function of laser power, and consequently selected the incident laser power of approximately 0.5-mW. A strong signal due to the resonant effect made this verification process easy. We used Rayleigh scattering as a standard for the in-situ calibration of the wavenumber. Rayleigh scattering can be detected by separating the concave mirror (dual focusing mirror) in the spectrometer into two parts, and successfully obtained both signal, Raman and Rayleigh scattering, with the same charge-coupled-device (CCD) detector simultaneously with keeping high resolution. During the measurement any mechanical components including grating in the monochromator did not move at all. Finally, we obtained the resolution of wavenumber less than 0.1 cm^{-1}. Detailed explanations of the UV-Raman system were described elsewhere [8]. The experiments were performed at room temperature in backscattering geometry from the (001) Si substrate. The incoming light is polarized along a [110] direction. In some cases, the polarization dependence for the scattering light was examined. A 90× objective lens was used to focus the laser beam on the sample, resulting in a beam spot diameter of about 0.5-μm. A Lorentz function is fitted to each spectrum in order to determine the peak position of the Si Raman line. A quasi-line light source using Galvanometer was also used for the high spatial resolution measurement.

Results and Discussion

<u>SOI with SiN capping film</u>

The peak positions of Raman spectra were plotted in Fig. 1 (a) after converting the wave-number into biaxial strain ($\Delta d/d$) for bulk Cz-Si substrates with the SiN capping films (20, 40, 60 and 80-nm). In the case of biaxial strain in the x-y plane of (001), the strain components ε_{xx} and ε_{yy}, can be described as follows [9-11],

$$\Delta\omega = (1 / \omega_0) [pS_{12} / (S_{11} + S_{12}) + q] \, \varepsilon_{xx} \, (\varepsilon_{yy}) \qquad [1]$$

or

$$\Delta\omega \, (cm^{-1}) \sim -723 \times \varepsilon_{xx} \, (\varepsilon_{yy}) \qquad [2]$$

$$S_{11} = 7.68 \times 10^{-12} \, Pa^{-1},$$
$$S_{12} = -2.14 \times 10^{-12} \, Pa^{-1},$$
$$S_{44} = 12.7 \times 10^{-12} \, Pa^{-1},$$
$$p = -1.43\omega_0^2,$$
$$q = -1.89\omega_0^2.$$

Here $\Delta\omega$ is Raman frequency shift from un-strained Si, ω_0 is Raman frequency of un-strained Si, p and q are material constants, so-called phonon deformation potentials, and S_{ij} are the elastic compliance tensor elements of Si. In Fig. 1, the horizontal axis indicates the SiN capping film-thickness and the vertical one is the strain calculated by the Raman shift. Five points for each substrate were measured in 100-μm^2 area to reduce the error in the measurement which can be estimated to be less than 0.003 % after Lorentz curve fitting with in-situ calibration using Rayleih scattering as a standard. We utilized (001) bulk Cz-Si substrate as an unstrained Si reference. The strains induced by 20 and 40-nm SiN film were very small, however the strains with 60 and 80-nm SiN film were clearly observed. From the Raman peak shift toward higher wavenumber, we can conclude the strains were compressive and probably biaxial. Fig. 1 (b) also shows strain shift for SOI (30-nm) substrates with SiN capping layers (0, 20, 40, 60 and 80-nm). The Raman peak shifted toward high wave-numbers with increasing SiN film thickness from 0 to 80-nm, implying compressive strain was induced by the SiN layer deposition. It is apparent that the virgin SOI substrate already had a tensile strain and this strain shifted gradually toward the compressive direction by increasing the SiN film thickness from 0 to 80-nm. As can been seen in Fig. 1 (d), the dependence of SiN film-thickness was similarly confirmed for the substrates with 100-nm SOI film thickness. 50-nm-thick SOI wafers with 20, 40 or 60-nm SiN film (Fig. 1 (c)) seem to be exception probably due to fluctuations of Raman measurement within a resolution and/or sample fabrication. The original tensile strain was larger in thinner SOI.

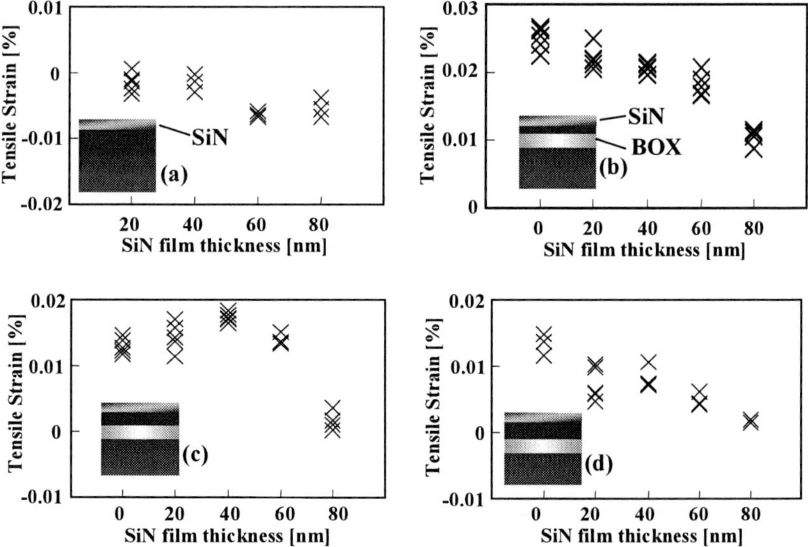

Figure 1.   Strain in SOI ((b) 30, (c) 50, (d) 100-nm) vs. SiN film thickness. (a) reference.

The effect of SOI film thickness on the strain introduction by the SiN layer deposition is summarized in Fig. 2. The strain shifts before and after the 80-nm-thick SiN layer deposition is plotted as a function of the SOI thicknesses of substrates. From Fig. 2 we can conclude the SiN cap is more effective for thinner SOI substrate in strain introduction. It is also very clear that the SiN capping film induced strain for SOI substrate more effectively than Bulk Si substrate.

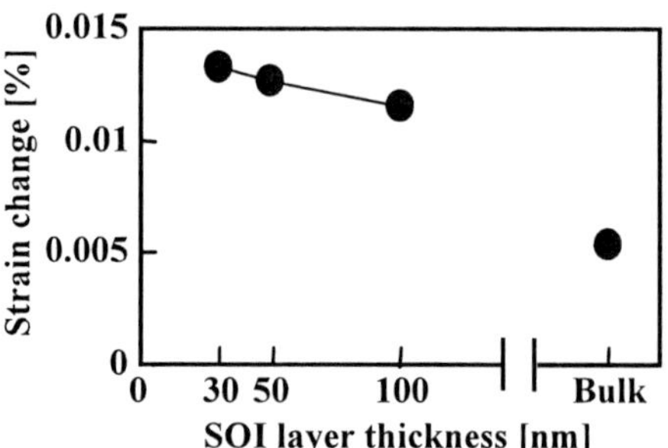

Figure 2.   Strain change by 80-nm SiN deposition on SOI with various film thicknesses.

<u>SiN patterning effect</u>

The measured strain induced by the SiN capping film was too small to improve device performance sufficiently. We assume this might be because there was no space to relieve strain in the same sample. We therefore evaluated strain line-profile in the Si substrate after SiN patterning into Line (SiN capping film)/Space (bare Si) along a [110] direction. We successfully obtained spatial resolution of 200-nm/pixel in Raman measurement using a quasi-line shape excitation light source and a CCD detector with reasonably short measurement time in spite of a larger beam spot diameter of approximately 500 nm [8]. Fig. 3 (left) shows the optical microscope image of the substrate after SiN patterning into Line (0.3-μm)/Space (0.6-μm) with a quasi-line shape laser illumination and Fig. 3 (right) shows the spectral image taken by the CCD detector attached on a monochromator. White and black contrast was clearly recognized, so the spatial resolution was high enough to evaluate strain profile in the substrate with Line (0.3-μm)/Space (0.6-μm) SiN patterning.

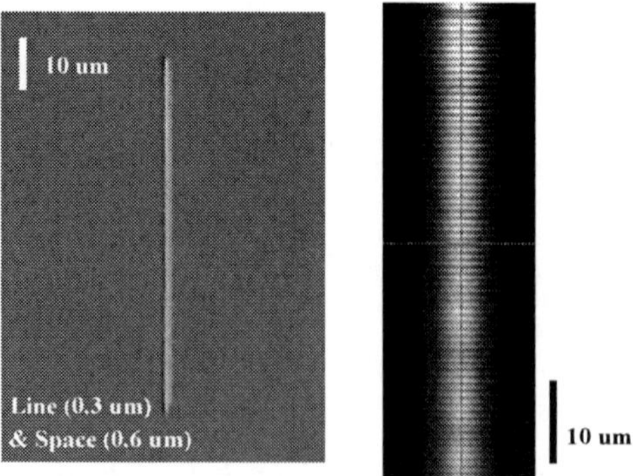

Figure 3.   Optical microscope (left) and spectral (right) image of the substrate after SiN patterning into Line (0.3-μm)/Space (0.6-μm) with illuminating a quasi-line shape laser.

Fig. 4 shows the dependence of the Raman frequency on the position across a 2-μm opening of SiN capping film, measured with 364-nm and 532-nm excitation light wavelengths. Using 364-nm excitation, $\Delta\omega$ (peak shift from un-strained Si Raman peak at 520 cm^{-1}) was clearly larger than that measured with 532-nm at the SiN pattern edge, indicating higher strain near the surface. This clearly proved UV-Raman measurement with excitation light of 364-nm wavelength is very effective to evaluate strain at the SiN/Si interface.

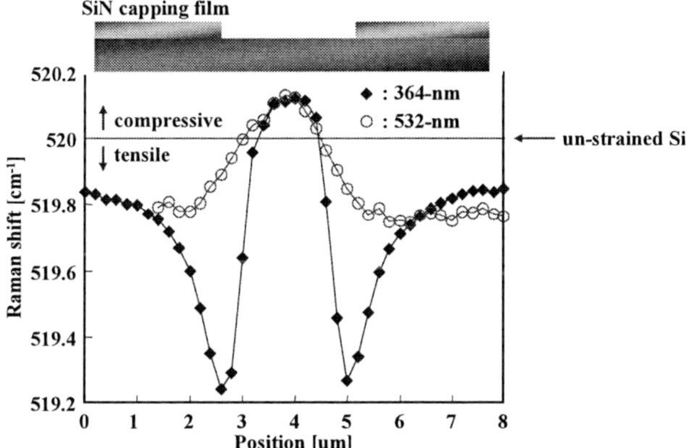

Figure 4.   Strain enhancement at SiN pattern edge measured by UV and visible laser excitation.

Fig. 5 shows the dependence of the SiN opening width for Raman shift. As shown in Fig. 5, large Raman shifts at the SiN pattern edge were observed. Moreover the Raman shift of opposite direction in the center of the opening was also appeared in Fig. 5 (b). This reverse strain in the opening might be a reaction of the pattern edge strain enhancement. We assume this reverse strain is an origin of mobility enhancement in a device using CESL.

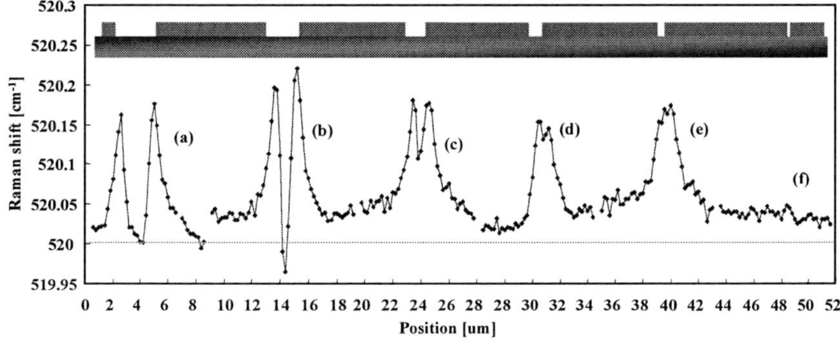

Figure 5.   Raman shift vs. opening width in the patterned SiN film.

Although it can be considered that the reverse strain should be enhanced by narrowing SiN opening width, it was disappeared in Fig. 5 (c), (d), and (e). Besides, significant increase of peak width (FWHM) was detected in Raman spectra. Two possibilities may be considered to explain these results. One is the presence of multiple Raman peaks overlapped due to the lack of spatial resolution. The reverse strain might be embedded in a main peak and disappeared. Another possibility is more complicated. At the pattern edge the strain-field should be asymmetry. This can cause an appearance of forbidden Raman peak originated in TO phonons, as well as allowed LO phonons [12, 13]. This can be a cause of the increase of FWHM or multiple peaks [14], and make a strain calculation more complicated [15]. We have confirmed, so far, the appearance of Raman peak under the forbidden polarization configuration.

The present study on patterning effect was examined only for bulk Cz-Si substrates. We are now preparing for the similar study on the SOI substrates in the near future.

Conclusion

SiN capping films were deposited on (001) SOI substrates by LP-CVD and the strain induced at the SiN/Si interface (5-nm) was evaluated by UV-Raman spectroscopy with a 364-nm Ar ion laser as an excitation source. The SiN capping film induced strain in SOI substrate more effectively than in Bulk Si substrate. Induced strain became larger with increasing SiN film thickness. SOI substrates originally had tensile strain and the SiN cap shifted the strain toward the compressive direction. The induced strain was larger in the thinner SOI substrates with the same thickness of the SiN capping film. We successfully obtained spatial resolution of 200-nm using a quasi-line shape excitation light source and a CCD detector with reasonably short measurement time. Large Raman shift was detected at the SiN pattern edge, the large strain was confined near the surface. We also observed reverse strain in the opening of SiN capping film.

References

1. S. Takagi *et al.*, F. Gibborns, J. Appl. Phys. **80**, 1567 (1996).
2. C. Gallon *et al.*, SSDM 2005 pp. 34–35.
3. F. Andrieu *et al.*, IEEE 2005.
4. K. F. Dombrowski, A. Fischer, B. Dietrich, I. De Wolf, H. Bender, S. Pochet, V. Simons, R. Rooyackers, G. Badenes, C. Stuer and J. Van Landuyt, IEEE 1999.
5. M. Holtz, W. M. Duncan, S. Zollner and R. Liu, J. Appl. Phys. **88**, 2523 (2000).
6. S. Nakashima, T. Yamamoto, A. Ogura, K. Uejima and T. Yamamoto, Appl. Phys. Lett. **84**, 2533 (2004).
7. V. Paillard *et al.*, Appl. Phys. Lett. **73**, 1718 (1998).
8. A. Ogura *et al.*, Jpn. J. Appl. Phys. **45**, 3007 (2006).
9. I. De Wolf *et al.*, Semicond. Sci. Technol. **11**, 139 (1996).
10. E. Anastassakis *et al.*, Solid State Commun. **8**, 133 (1970).
11. S. Nakashima *et al.*, J. Appl. Phys. **99**, 053512 (2006).
12. I. De Wolf, J. Vanhellemont, A. Romano-Rodriguez, H. Norstrom and H. E. Maes, J. Appl. Phys. **71** (2), 898 (1992).
13. I. De Wolf, H. E. Maes and Stephen K. Jones, J. Appl. Phys. **79** (9), 7148 (1996).
14. I. De Wolf, H. Norstrom and H. E. Maes, J. Appl. Phys. **74** (7), 4490 (1993).
15. F. Cerdeira, C. J. Buchenauer, Fred H. Pollak and Manuel Cardona, Phys. Rev. B **5**, 580 (1971).

ECS Transactions, 6 (4) 251-255 (2007)
10.1149/1.2728868, ©The Electrochemical Society

Structural properties of tensile-strained Si layers grown on $Si_{1-x}Ge_x$ virtual substrates (x = 0.2, 0.3, 0.4 and 0.5)

J.M. Hartmann, D. Rouchon, J.P. Barnes, [b]M. Mermoux and T. Billon

CEA-LETI, Minatec, 17, Rue des Martyrs 38054 Grenoble Cedex 9, France.
[b] LEPMI / ENSEEG / INPG, 1130 Rue de la Piscine, 38402 St Martin d'Hères, France.

We have studied the structural properties of tensile-strained Si (sSi) layers grown on polished $Si_{0.8}Ge_{0.2}$ and $Si_{0.7}Ge_{0.3}$ virtual substrates as a function of their thickness. The surface morphology, the tensile-stress, the linear density of defects and the threading dislocation density have been quantified for different sSi layer thickness. Results are compared to those previously obtained on sSi layers grown on top of polished $Si_{0.6}Ge_{0.4}$ and $Si_{0.5}Ge_{0.5}$ virtual substrates (ECS Trans. 3, No. 7, 319 (2006)).

We have recently reported the properties of tensile-strained Si (sSi) layers deposited on top of $Si_{0.6}Ge_{0.4}$ and $Si_{0.5}Ge_{0.5}$ Virtual Substrates (VS) (1). We expand here this study by focusing on the specifics of sSi layers grown on top of polished $Si_{0.8}Ge_{0.2}$ and $Si_{0.7}Ge_{0.3}$ VS. Indeed, the electron (and hole) mobilities in bi-axially strained Si are nearly twice those in bulk Si(001) for Ge contents (of the VS underneath) ≥ 20% (2) (≥ 35% (3)). Those sSi layers can be peeled off and bonded to oxidized Si substrates using the SmartCut[TM] process, yielding high quality eXtra-strained Silicon-On-Insulator (XsSOI) substrates (4). Both "conventional" sSOI and advanced XsSOI substrates provide (when used as templates of Fully Depleted SOI devices) superior long and short channel Metal Oxide Semiconductor Field Effect Transistor performance (5-7).

Strained Si layers were grown at 700°C using SiH_2Cl_2 at 20 Torr in a Reduced Pressure – Chemical Vapour Deposition (RP-CVD) reactor. We have performed Specular X-ray Reflectivity (SXR) measurements to determine the sSi layer thickness (in-between 10.2 and 36.7 nm). The measured sSi growth rate at 20 Torr, 2.30 nm min.$^{-1}$, does not depend upon the Ge content of the SiGe VS underneath.

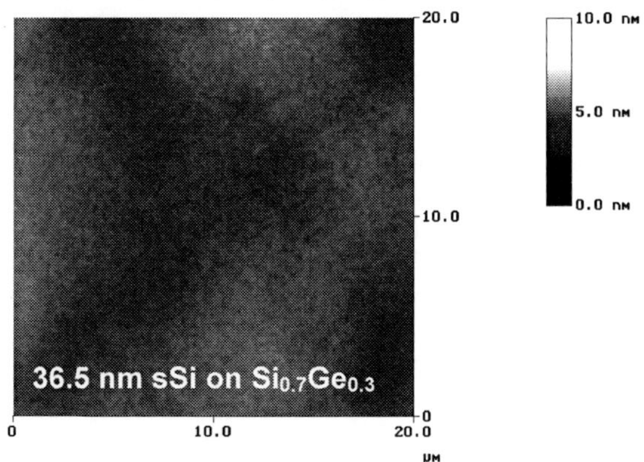

Figure 1 : Tapping force AFM image of the surface of a 36.5 nm sSi layer grown on top of a polished SiGe virtual substrate. Image sides are roughly aligned with the <100> directions.

Figure 1 shows an Atomic Force Microscopy image of the sSi surface. Some small spatial wavelength roughness (a few hundred nms) is clearly seen. No other features (such as lines running along the <110> directions) could be observed, even for the thickest sSi layers. This implies that the Stacking Fault linear density is low. The root mean square roughness (the Z ranges = Z max. – Z min.) associated to 20 μm x 20 μm images of the surface of those sSi layers increase slightly as the Ge content increases, from 0.35 nm (2.60 nm) for 20% up to 0.40 nm (2.84 nm) for 30%.

We have measured the stress level in the sSi layers with UV-Raman spectroscopy. The tensile strain is fairly stable as the sSi layer thickness increases (see **Figure 2**). It is equal to 1.32 GPa for sSi on $Si_{0.8}Ge_{0.2}$ VS and to 1.87 GPa for sSi on $Si_{0.7}Ge_{0.3}$ VS. These values are very close to the ones expected for pseudomorphic sSi layers. Indeed, the constant composition SiGe layers grown on top of the VS have a mean Ge concentration (a macroscopic degree of strain relaxation) equal to 20.8% (97.4%) and 29.0% (99.3%), as determined by X-Ray Diffraction (XRD).

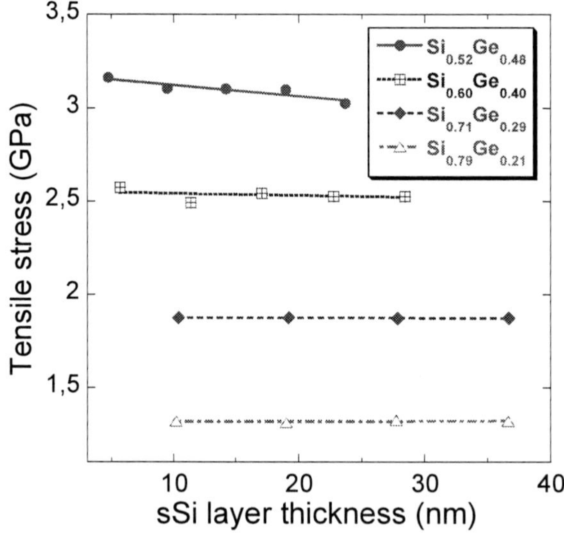

Figure 2 : Tensile stress in the sSi layers as a function of their thickness. The Ge content of the SiGe VS underneath is indicated in the insert for each set of data points.

An area map around the (004) order of the same sample whose surface is shown above can be found in **Figure 3**. The tensile stress associated to the thickest sSi layers has been extracted from such maps. The values obtained, 1.32 GPa and 1.95 GPa, are close to the UV-Raman ones.

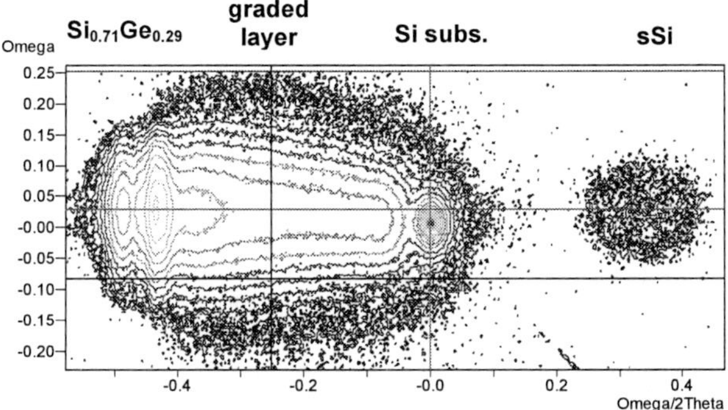

Figure 3 : Area map around the (004) diffraction order of a 36.5 nm thick sSi layer grown on top of a Si$_{0.71}$Ge$_{0.29}$ virtual substrate. The mosaïcity coming from the strain relaxation in the graded layer broadens along the omega direction the constant composition Si$_{0.71}$Ge$_{0.29}$ layer. Given that the tensile strain induces some lattice parameter shrinking along the (001) direction, the peak associated to the sSi layer is at higher angles than the one associated to the unstrained Si(001) substrate.

We have also studied the chemical abruptness of the interface between sSi and SiGe by Secondary Ions Mass Spectrometry. The slopes of the Ge decay profiles, 0.96 nm / dec. (sSi on Si$_{0.8}$Ge$_{0.2}$) and 0.97 nm / dec. (sSi on Si$_{0.7}$Ge$_{0.3}$), are in-between the slopes associated with sSi on Si$_{0.6}$Ge$_{0.4}$ (1.06 nm /dec.) and sSi on Si$_{0.5}$Ge$_{0.5}$ (0.73 nm / dec.). This might mean that the surface of the SiGe layers prior to the growth of the sSi layers was very slightly cross-hatched. Indeed, some very long wavelength undulations were present on the surface of our Si$_{0.6}$Ge$_{0.4}$ VS, whereas no such features were present on Si$_{0.5}$Ge$_{0.5}$ VS, explaining the differences in chemical abruptness (1.06 ⇔ 0.73 nm / dec.). We have otherwise imaged the thickest sSi layer on a polished Si$_{0.71}$Ge$_{0.29}$ virtual substrate in High Resolution Transmission Electron Microscopy. The sSi layer thickness found, ~ 37 nm, is in very good agreement with the SXR thickness, 36.5 nm. The atomic planes of the Si$_{0.71}$Ge$_{0.29}$ VS are prolonged in the sSi layer, yielding (at least locally) some high crystalline quality epitaxial stack. The interface in-between sSi and Si$_{0.71}$Ge$_{0.29}$, which is rather flat, seems chemically not that abrupt. This might be linked to some out-diffusion of Ge into Si.

Finally, we have performed Secco revelation of the dislocations presents in the sSi layers followed by optical microscopy and defect counting (see **Figure 4**). The Threading Dislocations Density (TDD) increases slightly as the sSi layer thickness increases, from 10^5 up to 2.5×10^5 cm^{-2}. It is however very close to the TDD associated to the starting SiGe VS : $1.05 \pm 0.36 \times 10^5$ cm^{-2}. Lines (which are thought to be Stacking Faults (SF) and Misfit Dislocations (MD) coming from the propagation of pre-existing threading dislocations) also appear for the thickest sSi layers. For a given sSi layer thickness, the LLD becomes larger as the Ge content of the SiGe VS increases from 21% up to 29% and finally 40%. sSi layers grown on top of Si$_{0.52}$Ge$_{0.48}$ VS are characterized by much smaller LLDs than the ones associated to sSi on Si$_{0.60}$Ge$_{0.40}$, however (0 up to 300 cm^{-1} ⇔ 0 up to 800 cm^{-1}). The only difference that might explain such a discrepancy is the surface roughness of the polished SiGe virtual substrates used as templates for the growth (very long wavelength, small amplitude surface cross-hatch on the Si$_{0.6}$Ge$_{0.4}$ VS ⇔ featureless surface for the Si$_{0.5}$Ge$_{0.5}$ VS).

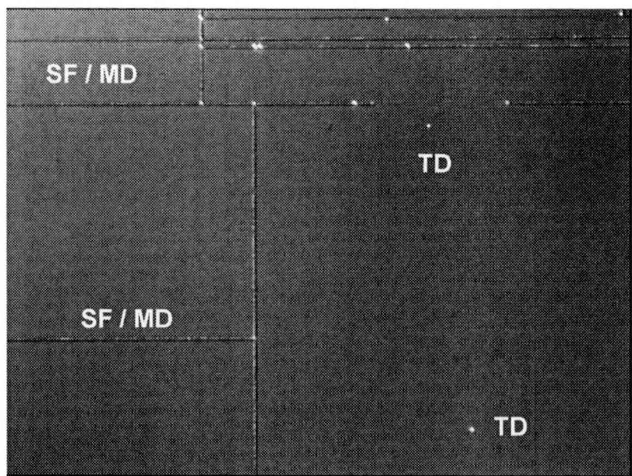

Figure 4 : *130 μm x 100 μm optical microscopy image of the surface of a 36.5 nm thick sSi layer grown on top of a $Si_{0.71}Ge_{0.29}$ virtual substrate after a Secco defect revelation.*

As for TDD, no brutal increase of the LDD (which would be a signature of some plastic strain relaxation) occurs as the sSi layer increases (see **Figure 5**). Were we to adopt as critical thickness the ones for which the LDD exceeds 50 cm^{-1}, we would find values around 12 nm and 10.5 nm for Ge contents equal to 21% and 29%, respectively.

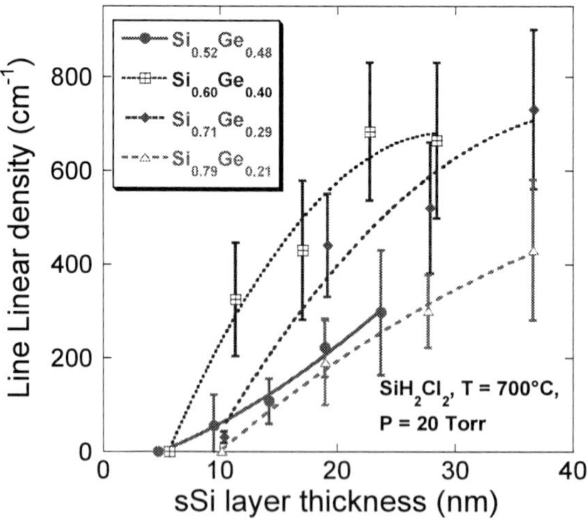

Figure 5 : *linear defects linear density (as revealed by Secco and counted by optical microscopy) in miscellaneous thickness sSi layers grown on top of polished SiGe virtual substrates. The Ge content of the SiGe VS underneath is indicated in the insert for each set of data points.*

Acknowledgements

This work, carried out in the NaNoTech and the Silicon Technology Platform Departments of CEA-LETI, was supported by the European Union IST Pullnano and Silonis Medea[+] Projects. L. Saidi, J.P. Colonna, F. Gonzatti, L. Masarotto, S. Rousseau, P. Roche, B. Assie and J.P. Mazzon are gratefully acknowledged for their help in operating and maintaining the Epi Centura tool.

References

1. J.M. Hartmann, A. Abbadie, D. Rouchon, Y. Guinche, P. Holliger, G. Rolland, M. Buisson, M. Mermoux, F. Brunier, C. Defranoux, F. Pierrel and T. Billon, *ECS Trans.* Vol. **3**, no. **7**, 309 (2006).
2. M.T. Currie, C.W. Leitz, T.A. Langdo, G. Taraschi, E.A. Fitzgerald and D.A. Antoniadis, *J. Vac. Sci. Technol. B* **19**, 2268 (2001).
3. C.W. Leitz, M.T. Currie, M.L. Lee, Z.-Y. Cheng, D.A. Antoniadis and E.A. Fitzgerald, *J. Appl. Phys.* **92**, 3745 (2002).
4. T. Akatsu, J.M. Hartmann, C. Aulnette, Y.-M. Le Vaillant, D. Rouchon, A. Abbadie, Y. Bogumilowicz, L. Portigliatti, C. Colnat, N. Boudou, F. Lallement, F. Triolet, Ch. Figuet, M. Martinez, P. Nguyen, C. Delattre, K. Tsyganenko, C. Berne, F. Allibert, Ch. Deguet, M. Kennard, E. Guiot, F. Metral and I. Cayrefourcq, *ECS Trans.* Vol. **3**, no. **6**, 107 (2006).
5. K. Rim, K. Chan, L. Shi, J. Ott, N. Klymko, F. Cardone, L. Tai, S. Koester, M. Cobb, D. Canaperi, B. To, E. Duch, I. Babich, R. Carruthers, P. Saunders, G. Walker, Y. Zhang, M. Steen and M. Ieong, Proceedings of the 2003 IEDM Conference, Washington (USA), p. 49.
6. F. Andrieu, C. Dupré, F. Rochette, O. Faynot, L. Tosti, C. Buj, E. Rouchouze, M. Cassé, B. Ghyselen, I. Cayrefourcq, L. Brévard, F. Allain, J.C. Barbé, J. Cluzel, A. Vandooren, S. Denorme, T. Ernst, C. Fenouillet-Béranger, C. Jahan, D. Lafond, H. Dansas, B. Previtali, J.P. Colonna, H. Grampeix, P. Gaud, C. Mazuré and S. Deleonibus, 2006 Symposium on VLSI Technology Digest of Technical Papers, p. 134.
7. F. Andrieu, O. Faynot, F. Rochette, J.-C. Barbé, C. Buj, Y. Bogumilowicz, F. Allain, V. Delaye, D. Lafond, F. Aussenac, S. Feruglio, J. Eymery, T. Akatsu, P. Maury, L. Brévard, L. Tosti, H. Dansas, E. Rouchouze, J.M. Hartmann, L. Vandroux, M. Cassé, C. Fenouillet-Béranger, F. Boeuf, F. Brunier, I. Cayrefourcq, C. Mazuré, G. Ghibaudo and S. Deleonibus, accepted for publication in the Proceedings of the VLSI 2007 Conference (Kyoto (Japan), June 2007).

ECS Transactions, 6 (4) 257-262 (2007)
10.1149/1.2728869, ©The Electrochemical Society

X-ray Absorption Spectroscopy of Strained-Si Nanomembranes

C. Euaruksakul, Z. Li, D. E. Savage, and M. G. Lagally

University of Wisconsin-Madison, Madison, Wisconsin 53706, USA

We use x-ray absorption (XAS) total electron yield measurements to investigate strain-induced splitting of the conduction band minimum in very thin Si layers and membranes. We use synchrotron generated X rays with variable energy to excite electrons from the 2p core level of strained Si to empty electronic states in the conduction band minima. The Auger electrons emitted in the relaxation process make the method extremely surface sensitive and reflect the position and density of states of the conduction band. We determine these parameters for elastically strained Si nanomembranes and strained-Si-on-insulator. We demonstrate significant lateral nonuniformity in the strain in strained-Si-on-insulator, while the strain is uniform on strain sharing nanomembranes. Low-energy electron microscopy measurements support this conclusion.

Introduction

A primary use of strain in semiconductor devices is to enhance electronic transport properties. Tensile and compressive strains modify the band structures of Si and Ge. The changes in the band structure in turn influence the mobility of charge carriers. The mobility of electrons in tensilely strained Si increases because the reduction of degeneracy of the conduction band minima reduces the scattering rate in electron transport, increasing carrier mean free time, τ. The average effective mass also decreases. The mobility, μ, is given by

$$\mu = q\tau/m^*, \qquad [1]$$

where q is electron charge and m* is electron effective mass. Figure 1 shows schematically the Si band structure, the 6-fold degenerate sub-bands at the conduction band minimum, a schematic depiction of the level splitting under tensile strain, and the behavior of electron and hole mobilities with increasing tensile strain in Si.

Strain is typically introduced via growth on a lattice with different lattice constant. In the most interesting situations the strained layer is quite thin, and methods to determine strain quantitatively in these very thin films are desirable. X-ray diffraction (XRD) and Raman spectroscopy (RS) are perhaps the most widely used. Peak shifts in x-ray diffraction can be interpreted directly in terms of lattice constant changes. Raman spectra give information on strain via shifts of inelastic photon scattering peaks. One relates the shift to the magnitude and type of strain. In both XRD and RS, measurements become difficult when the film becomes very thin. Electrical measurements in electronic devices can give the charge carrier mobility for even the thinnest films, but are indirect in that they cannot say what produces the mobility change.

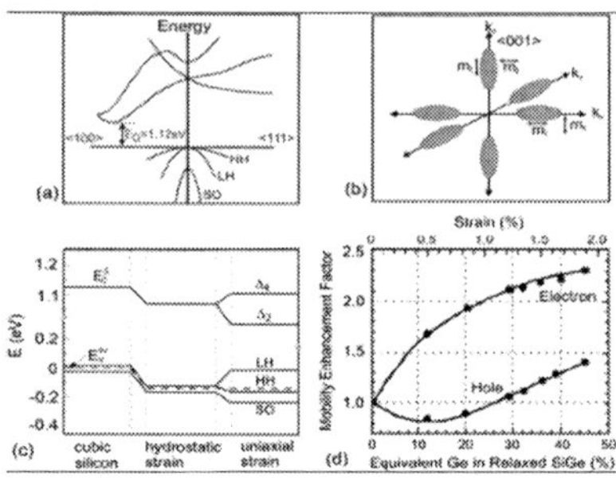

Figure 1. Si band structure and effect of strain. a) conduction and valence bands; b) degenerate conduction band valleys; c) level shift with hydrostatic and uniaxial strain; d) effect on mobilities (the abscissa is proportional to tensile strain in Si). Figure from (1) with permission. Data in panel c) from (2). Data in panel d) from (3).

X-ray absorption spectroscopy (XAS) can provide intermediate information, in that one can effectively observe the splitting of the conduction band degeneracies, the most important factor in electron mobility enhancement. By making a total electron yield measurement, one can create sensitivity to very thin films. The process is as follows: Synchrotron generated X rays with tunable energy, hv, excite electrons from shallow core levels (here the 2p level) of Si to empty conduction band states, as shown in figure 2. The excited atoms relax predominantly by an Auger process, emitting an electron with an energy characteristic of the levels involved. The spectrum of the Auger electrons with x-ray energy as the variable reflects the convolution of the densities of states of the 2p core level and the conduction band. If the density of states (DOS) of the 2p core level is comparatively confined (4) and its shape does not change with the strain, the spectrum provides an accurate picture of the density of states of the conduction band. The very shallow escape depth of Auger electrons at the energies involved in these XAS measurements (mean free path of the order of a few nm) enables measurement of energy level changes in very thin layers, so thin that quantum size effects are observable (5). The depth resolution in our XAS study is around 2-3 nm, thicknesses at which other techniques either fail or become extremely difficult.

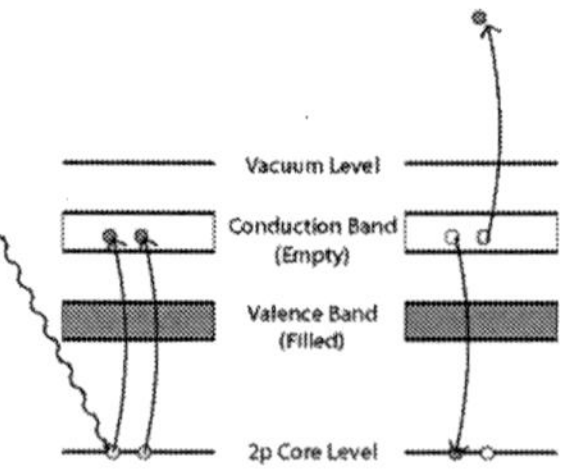

Figure 2. Auger process in XAS total electron yield measurement.

When tensile strain is applied to Si, the 6 conduction band minima near X-points, which are degenerate without strain, shift and split their degeneracy depending on the direction of the strain. The shift is due to the hydrostatic component of the strain while the degeneracy splitting is due to the uniaxial component and does not change the average conduction band energy. Available calculations disagree in the sign of the shift. Reference (6) reports a negative shift of the levels and the more recent Reference (7) reports a positive shift. In our experiment, biaxially tensile strain on Si (001) splits the conduction band into 2-fold and 4-fold degenerate valleys (Δ_2 and Δ_4, respectively) as already shown in figure 1. All calculations agree on the splitting. We analyze XAS total electron yield spectra of several strained Si samples with known nominal strain magnitudes to find the shift and the splitting of the conduction band, in order to establish the capability of this technique for quantitative strain measurements in very thin membranes.

Experiment

We are particularly interested in measuring strain in very thin films because we have developed methods to fabricate elastically strain sharing Si nanomembranes and transfer them to new hosts (1), (8), providing the opportunity for a range of novel applications (9), (10), in which strain will be very important. We prepare elastically strain-sharing membranes according to (8) (figure 3). First, we grow layers of Si/SiGe/Si by chemical vapor deposition in ultra-high vacuum (UHV-CVD) on a Si-on-insulator (SOI) substrate. The thickness of all layers is less than the kinetic critical thickness, so that no dislocations can form during growth. We pattern an array of 5 μm^2 square holes on the sample with photolithography and plasma etching and immerse the sample into 49% HF solution to release the Si/SiGe/Si structure from the initial substrate. During the etching/release process, the SiGe layer, which initially contains compressive strain, elastically relaxes by increasing its in-plane lattice constant, causing the sandwiching, initially unstrained, Si layers to become tensilely strained. The relaxation of SiGe continues until the total strain energy of the structure is at its minimum. The final strain in Si layers can be approximated by

$$\varepsilon_{Si} = \Delta\varepsilon * t_{si}/(t_{si}+t_{sige}),\qquad\qquad [2]$$

where $\Delta\varepsilon$ is the strain mismatch between Si and SiGe, t_{si} is the total thickness of Si layers, and t_{sige} is the thickness of the SiGe layer.

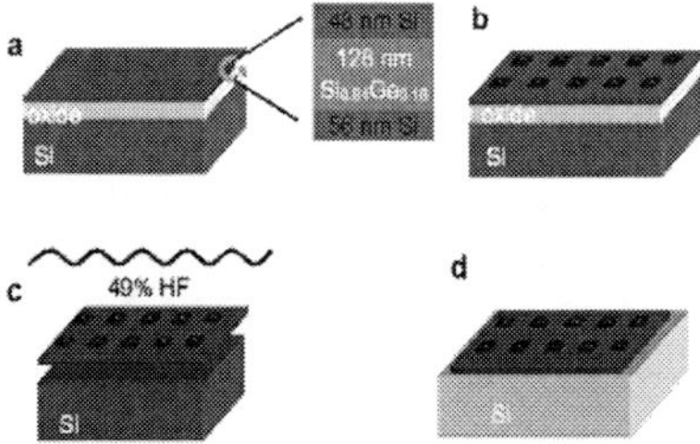

Figure 3. Fabrication process of an elastically strain sharing membrane. (a) Si/SiGe/Si structure is grown on a SOI substrate. (b) An array of holes is patterned on the structure as channels for HF etchant. (c) The patterned sample is immersed into HF solution to remove the buried SiO_2. (d) After SiO_2 is completely removed, the structure relaxes and is transferred onto a new host substrate, which can be anything.

We compare elastically strain sharing membranes with different strains to strained-Si-on-insulator (sSOI) provided by SOITEC. sSOI is fabricated by growing graded SiGe buffer layers on a bulk Si substrate, followed by a fully relaxed SiGe layer with constant Ge composition and finally tensilely strained Si. The tensilely strained Si is then bonded to a handling-Si substrate via an oxide layer. The initial substrate and buffer layer are removed by the smart-cut process or etched away by selective chemical etching and then the strained Si layer is smoothened by chemical mechanical planarization (CMP). We analyze the spectra from sSOI to compare the lateral uniformity of strain on sSOI with that on our fabricated membranes.

We clean the samples with Piranha solution ($H_2O_2+H_2SO_4$) following by diluted HF solution to remove the native oxide and terminate the surface with hydrogen. A sample is then loaded into an ultra-high vacuum (UHV) analysis chamber that connects to the Hermon beamline at the University of Wisconsin Synchrotron Radiation Center. The computer-controlled monochromator at the beamline provides soft X-rays with high energy resolution. We scan the X-ray energy from 99 to 103eV with a 10 meV step, covering the L_2 and L_3 absorption edge on Si. At each X ray energy step, we measure the total electron yield by a picoammeter that is electrically connected to the sample. The XAS electron yield is normalized with the electron yield from a gold mesh situated between the monochromator and the analysis chamber to compensate for changes of the intensity of the synchrotron-generated X rays over observation periods. In our scanning energy range, photoelectrons form a flat background in the spectra and do not disturb the data analysis from the Auger electron spectra.

Results and Analysis

Figure 4 shows normalized absorption spectra for strained Si (sSOI) compared to unstrained Si (regular SOI). Absorption becomes possible when X rays have enough energy to excite electrons from the $2p_{3/2}$ level to the bottom of the conduction band (L_3 transition). The transition in the unstrained SOI has a given position and shape; for the strained film, there will be transitions from the 2p core level to the lower Δ_2, and to the higher Δ_4 sub-bands, but with intensity scaled down to 1/3 and 2/3 for Δ_2 and Δ_4 sub-bands respectively. Because the splitting caused by the uniaxial component of strain does not change the average energy of the conduction band minimum, we know that the Δ_4 sub-band moves only half the distance of the Δ_2 sub-band. With the energy split (ΔE_Δ) and the average level shift (ΔE_{av}) as two variables, we reconstruct the spectrum by adding Δ_2 and Δ_4 peaks separated by ΔE_Δ and shifted by ΔE_{av} (11). We get excellent agreement, for standard samples, with predictions of sub-band splitting in (2), (6), (7).

Lateral strain inhomogeneity will cause broadening of the XAS spectra, averaged over the beam size. We find such broadening in the spectra from sSOI, but not from elastically strain sharing membranes (11). It is not surprising that the elastically strain relaxed Si nanomembranes show uniform strain laterally. The relaxation proceeds without the formation of dislocations (as long as each layer is grown thin enough), and the membrane can freely relax as it is released from the oxide. On the other hand, in the growth process used to make sSOI, described above, the ultimate strained Si layer grows on a series of dislocated SiGe layers. When the graded SiGe buffer layers relax to their final lattice constant they form dislocations. Dislocation nucleation is not uniform, and hence the SiGe layer can relax nonuniformly. This non-uniformity of the strain is transferred to the overgrown strained Si and again when the strained Si is bonded to the second substrate. If this nonuniform relaxation of the SiGe layers occurs, the resulting strained Si layer in sSOI will be laterally inhomogeneous in the strain.

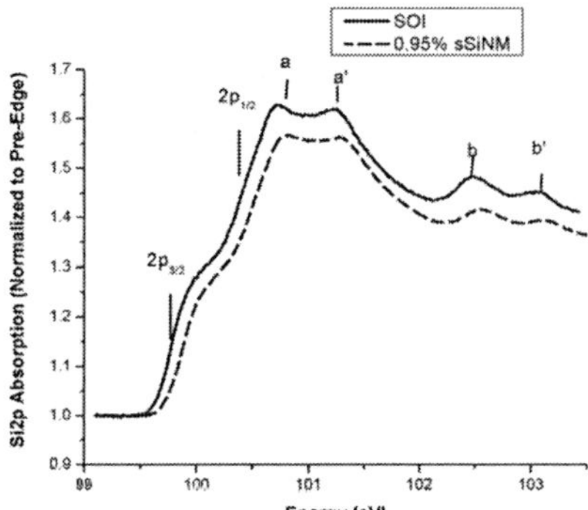

Figure 4. XAS total electron yield spectra from unstrained SOI and 0.95% strained sSOI. The signal comes only from the top several nm of the sample.

One can demonstrate structural differences in sSOI and elastically strained Si nanomembranes that relate to strain and its nonuniformity. Low-energy electron microscopy (LEEM) is sensitive to individual dislocations in a film because they leave a strain signature at the surface (12). Images of dislocations in different areas on a 50 nm thick sSOI sample with 0.8% tensile strain show many dislocation lines. That is expected, because the thickness of the strained Si is well beyond the critical thickness. The number of dislocations locally reflects the degree of local relaxation. Figure 5 shows LEEM images of two areas, demonstrating a large difference in dislocation density in different areas. One can thus expect that the strain will be nonuniform. The difference in dislocation density shows that strain relaxation in the Si layer of sSOI varies by +/- 0.035% from the mean (+/- 4% of the nominal strain) across a 100 μm x 100 μm area of surface. For elastically strain sharing membranes, strain applied to Si during the relaxation of SiGe layers is macroscopically uniform, as expected from a truly unconstrained structure. Very few dislocations are observed on these membranes (it is hard to find any).

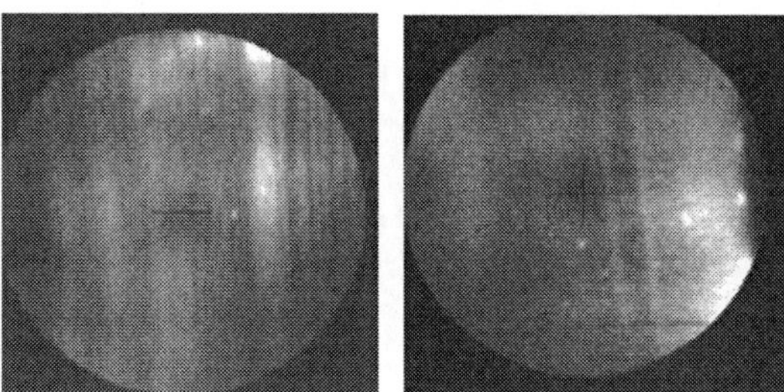

Figure 5. LEEM images of dislocation lines in different areas on a sSOI sample with 0.8% nominal tensile strain. The dislocations appear as dark and bright lines. The left image has many more dislocations than the right one. The field of view is 8 μm.

Summary

Strain influences the band alignment in Si and SiGe alloys, which in turn affects the carrier mobility. We have investigated, using XAS, the strain in elastically strain sharing Si nanomembranes and compared it to sSOI. We demonstrate that XAS total electron yield is an effective way to study the strain-induced splitting of the Si conduction band in very thin membranes and films. Experimental results agree with theoretical predictions of the level splitting with strain. Our results on level shifting support only the more recent prediction of (7). We demonstrate that the strain is laterally quite nonuniform in sSOI, a conclusion that is supported by LEEM images of the surfaces of sSOI samples.

Acknowledgments

This research was supported by NSF, DOE, and AFOSR. SOI and sSOI substrates were provide by G.K. Celler, SOITEC. C.S. Ritz fabricated elastically strain-sharing membranes for XAS measurements. We thank F. J. Himpsel, F. Zheng, and X. Liu for help with early XAS measurements and with data analysis.

References

1. S. A. Scott and M. G. Lagally, *J. Phys. D: Appl. Phys.*, 40, R75 (2007).
2. F. Schäffler, *Semicond. Sci. Technol.*, **12**, 1515 (1997).
3. K. Rim et al, *IEEE IEDM Tech. Dig.*, 3.1.1 (2003).
4. K. Hricovini, R. Günther, P. Thiry, A. Taleb-Ibrahimi, G. Indlekofer, J. E. Bonnet, P. Dumas, Y. Petroff, X. Blase, X. Zhu, S. G. Louie, Y. J. Chabal and P. A. Thiry, *Phys. Rev. Lett.*, **70**, 1992 (1993).
5. Y. K. Chang, H. H. Hsieh, W. F. Pong, M.-H. Tsai, F. Z. Chien, P. K. Tseng, L. C. Chen, T. Y. Wang, K. H. Chen, D. M. Bhusari, J. R. Yang and S. T. Lin, *Phys. Rev. Lett.*, **82**, 5377 (1999).
6. M. M. Rieger and P. Vogl, *Phys. Rev. B*, **48**, 14276 (1993).
7. S. Richard, F. Aniel, G. Fishman and N. Cavassilas, *J. Appl. Phys.*, **94**, 1795 (2003).
8. M. M. Roberts, L. J. Klein, D. E. Savage, K. A. Slinker, M. Friesen, G. K. Celler, M. A. Eriksson and M. G. Lagally, *Nature Materials*, **5**, 388 (2006).
9. H.-C. Yuan and Z. Ma, *Appl. Phys. Lett.*, **89**, 212105 (2006).
10. H.-C. Yuan, Z. Ma, M. M. Roberts, D. E. Savage and M. G. Lagally, *J. Appl. Phys.*, **100**, 013708 (2006).
11. Z. Li, C. Euaruksakul, F. Zheng, F. J. Himpsel, C. S. Ritz, D. E. Savage, X. Liu and M. G. Lagally, unpublished.
12. P. Sutter and M. G. Lagally, *Phys. Rev. Lett.*, **82**, 1490 (1998).

ECS Transactions, 6 (4) 263-269 (2007)
10.1149/1.2728870, ©The Electrochemical Society

Evaluation of different etching techniques in order to reveal dislocations in thick Ge layers

A. Abbadie[a], J.M. Hartmann[b], C. Deguet[b], L. Sanchez[b], F. Brunier[a] and F. Letertre[a]

[a] SOITEC, Parc Technologique des Fontaines, 38190 Bernin, France.
[b] CEA-LETI, Minatec, 17, Rue des Martyrs 38054 Grenoble Cedex 9, France.

We have used different defect etching techniques in order to assess the crystalline quality of Ge thick layers grown on top of Si(001) (used as templates for the formation of GeOI wafers). For the first time, 3 different methods were evaluated, namely a Chromium-free chemical solution, a chemical Secco solution and a gaseous HCl etch in an epitaxy reactor. We have obtained etch pits densities close to 10^7 cm^{-2}, independently of the amount of Ge removed and the defect revelation method used. The EPD obtained were also similar to the threading dislocations densities measured using plane view TEM, confirming the high efficiency of dislocations delineation by the preferential etching techniques studied. Similar values were obtained on final product (GeOI) after Secco etch. Finally, we checked the capability of the Cr-free solution in terms of defect delineation on bulk Ge and other Ge epitaxial layers by varying defect densities.

Two main paths are explored nowadays to continue improving the performances of Metal Oxide Semiconductor Field Effect Transistors (MOSFETs) : new architectures and new materials (1). Germanium (Ge), due to its better transport properties than Silicon (Si), is a promising candidate for sub-22nm nodes high performance MOSFETs (2). Ge presents also the property of being compatible with III-V materials (GaAS) which are key for high performance light detectors (photodiodes, solar cells) and emitters (lasers, LEDs). Germanium On Insulator (GeOI) substrates combine the advantages of bulk Ge substrates (intrinsically higher mobilities than in Si) and "On Insulator" substrates (such as a better electrostatic control, which is mandatory when down-scaling the MOSFET dimensions). For pure optoelectronic applications, GeOI is an attractive platform to integrate GaAs-based heterojunction bipolar transistor (HBTs) and may be a solution for space-based solar cells or optical devices with Ge (3). GeOI structures could then provide a solution for both electronic and optoelectronic functions on the same platform. Many papers recently reported promising performance in large and small dimensions transistors (mostly p-type) built either on bulk Ge substrates (4-5), on Ge thick layers epitaxially grown on top of Si(001) substrates (6) or on GeOI substrates (7-8). Those GeOI substrates were obtained using either the Ge condensation method (7) or the Smart Cut^{TM} layer transfer process (8-9), the donor wafers in the latter case being mainly thick Ge layers on Si(001), as in Ref. (6).

Ge substrates on Silicon are obtained using an epitaxial growth of thick Ge layers on Si(001) substrates. Due to the 4.2% lattice parameter mismatch between Ge and Si, the critical thickness for plastic relaxation is reached after only a few monolayers of Ge, above which misfit dislocations appear. The two techniques used to obtain such substrates with "reduced" TDD (Treading Dislocations Densities) are based on either Ge virtual substrates or Ge layers grown on Si substrates. The first technique consists in growing a $Si_{1-x}Ge_x$ graded buffer (with x varying from 0% up to 100%) on top of which a pure Ge layer is grown, leading to more than 10 μm thick stacks (10). The second technique requires the stacking of a high growth temperature Ge layer on top of a low growth temperature Ge layer, itself on Si (001), leading to a less than 3 μm thick epitaxial layer (11). The TDD obtained with the two techniques is expected to be between 1×10^6 and 1×10^7 cm^{-2}. Thick Ge layers grown directly on Si, due to their Si like mechanical behavior and their reduced cost (compared to Ge bulk substrates or Ge virtual substrates), are rather attractive as donor wafers for GeOI fabrication.

Given the impact of the crystalline quality on the electrical performances of MOSFET devices, it is of the utmost importance to have reliable defect revelation methods in order to quantify the defect density in thick Ge layers grown on Si(001) and in GeOI substrates. Such techniques have to be sensitive enough to track the process step improvements that would yield lower defect density layers. We will show in this paper that the development of efficient and well calibrated characterization methods are necessary for a proper evaluation of their crystalline quality.

We have used thick Ge films grown directly onto Si(001) substrates using an Epi Centura Reduced Pressure - Chemical Vapour Deposition system. A low growth temperature / high growth temperature strategy (400°C / 750°C) was adopted in order to obtain smooth Ge layers with lower defects densities. A 8 times {750°C, 10 minutes / 900°C, 10 minutes} thermal cycling under H_2 was used after the growth of the Ge layers with the goal of significantly reducing the threading dislocation density. Some cross-sectional Transmission Electron Microscopy images of a 2.8 μm thick annealed Ge layer are plotted figure 1.

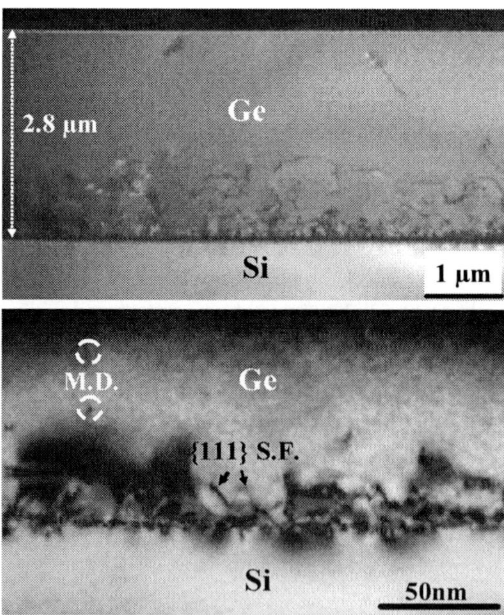

Figure 1: (top) Weak beam dark field (g = 004) TEM image of a 2.8 μm thick annealed Ge layer grown on top of a Si(001) substrate; (bottom) Dark field (g = 2-20) TEM image of the interface between Ge and Si;

{111} Stacking Faults as well as Misfit Dislocations running perpendicularly to the image plane can clearly be seen.

What can easily be seen is that the top 1.8 µm is relatively defect free. Most of the misfit dislocations are indeed confined in the bottom one micron thick part of the layer and especially in the first 20 nm of the low temperature Ge epilayer.

The surface of those thick Ge layers is rather smooth, especially when considering the large lattice mismatch in-between Ge and Si. The surface root mean square (rms) roughness and the Z range (= $Z_{max.} - Z_{min.}$) associated to annealed 2.5 µm thick Ge layers are indeed equal to 1.0 nm and 7.7 nm only.

We have investigated using chemical revelation the threading dislocations presents in the thick Ge layers. For the first time, we have demonstrated a real correlation between 3 techniques used for defects delineation : a Cr-free wet etching solution (12-14), a Secco wet-etching solution (15) and gaseous HCl etching (16). The Cr-free wet-etching solution consists in a dip in a HF / HNO_3 / CH_3COOH mixture followed by a short dip in H_2O_2 for defects delineation. Characterized by a relatively high etch rate (between 1900 and 2400 nm/min) and a good etching uniformity, the Cr-free solution has been optimized for thick Ge film analysis. Using an optical microscope in the Nomarski contrast mode and Atomic Force Microscopy (AFM), we have quantified the Etch Pit Density (EPD) and studied the etch pits shape as a function of the Ge etched thicknesses (from 0.09 µm up to 1.2µm). EPD values close to 10^7 cm^{-2} have been obtained irrespectively of the Ge etched thickness. A typical optical image ($130x100µm^2$) of the round-shape Ge defects after a 1.2µm Cr-free etch is shown figure 2.

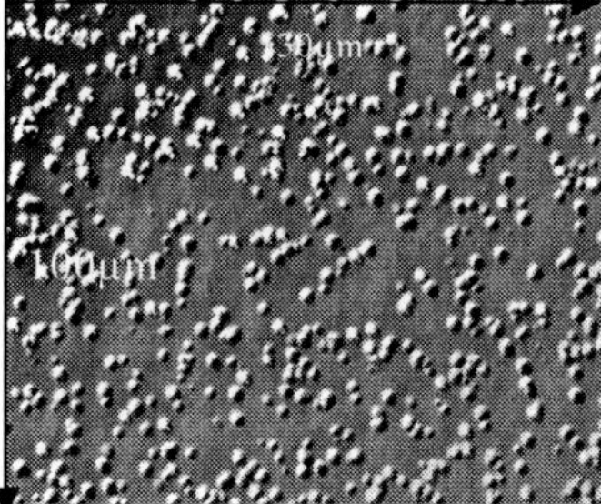

Figure 2: (130x100)µm² optical microscope image acquired in the Nomarski contrast mode of the surface of a thick Ge layer after a Cr-free etch corresponding to about 1.2µm of Ge etched.

Such a solution can be used as well on bulk Ge wafers. Preliminary EPD values around 10^4 cm^{-2} have been obtained. Such values will have to be kept in mind for the characterization of "bulk" GeOI wafers.

Each component of this Cr-free chemical mixture has a specific purpose: nitric acid is the oxidizing agent; hydrofluoric acid is necessary to dissolve the oxidation product; acetic acid is a diluant, needed for an homogeneous removal of the etched by-products. The oxidation mechanism, as already studied by Steinert et al (17) and recently Ackert et al (18), is based on several equilibria, implying the formation of oxidizing species, such as nitrogen oxide and nitrous acid. In addition, the dislocation etch pit formation is influenced by the local strain field and/or impurities which are embedded in dislocations ("decoration"). They generate (during the oxidation and dissolution of Ge) a surface potential between dislocation sites and the surrounding crystal. This surface potential leads to etch rates which are higher at defects sites. Such a mechanism, most likely reaction-rate limited as opposed to diffusion limited (19), results in circular etch pits.

Those EPD values ($\sim 10^7$ cm^{-2}) were compared with the Threading Dislocation Densities (TDDs) obtained thanks to plane view Transmission Electron Microscopy (figure 3). The mean TDD values found were in-between 1 and 2x10^7 cm^{-2}, confirming the high dislocation delineation efficiency of this Cr-free solution.

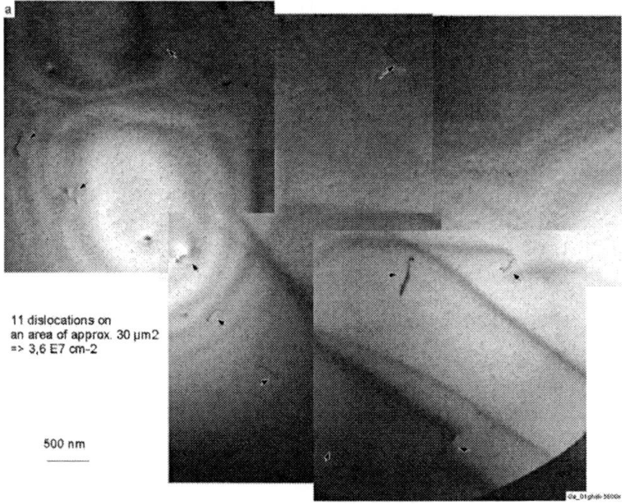

Figure 3: 200Kev Plane-View TEM images acquired in the bright field mode of a 2.5 µm Ge layer on Si(001), showing some dislocations in a area of approx.30µm² (corresponding to a TDD ~ 3.6x10⁷ cm⁻²).

We have investigated the etch pits on the same Ge/Si samples, using Secco etches after different dip times (from 2 up to 15 min.). The Secco Etch Pit Density is in good agreement with the one found after Cr-free etches (i.e. $\sim 10^7$ cm^{-2}). Similar defect shapes and sizes are also observed after both etches, confirming the reaction mechanism hypothesis detailed above for the Cr-free solution. The Ge surface after a 11min. etching time (corresponding to ~ 0.9µm of Ge etched) is represented on the 130x100µm^2 optical microscope image shown in figure 4.

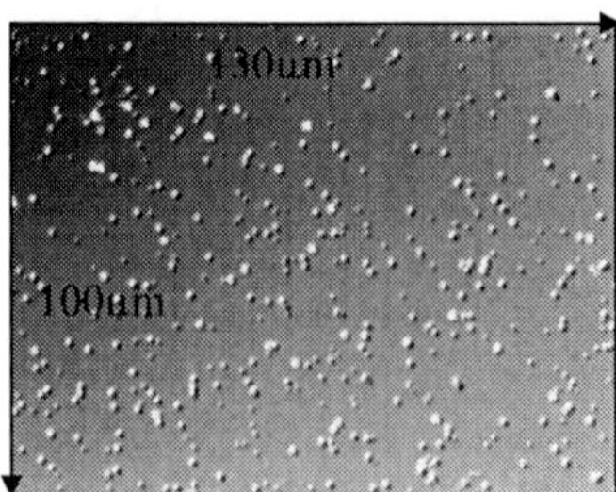

Figure 4: (130x100)µm² optical microscope image acquired in the Nomarski contrast mode of the surface of a thick Ge layer after the Secco etch of about 0.9µm of Ge.

Finally, others thick Ge layers were etched using gaseous HCl in a RP-CVD reactor in order to reveal defects. Two etched thickness, 93 and 280 nm, were tested. For each one, we have estimated the EPD using Atomic Force Microscopy (figure 5). Values are comparable to the ones obtained with the Cr-free and Secco wet-chemical etches (i.e. $\sim 10^7$ cm^{-2}).

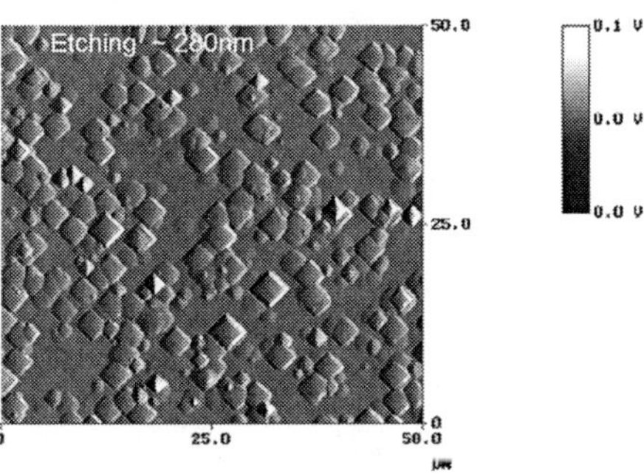

Figure 5: 50x50µm^2 AFM amplitude image of the surface of a Ge sample after some HCl etch.

HCl etch results in inverted pyramids with sides along the <100> directions and a low aspect ratio. A different mechanism from the last one is suggested, most likely dependant on the HCl (and Cl$^-$ ions) diffusion towards the Ge surface. A threading dislocation deforms the crystal locally and introduces additional strain. The etch pits delineating at the site of threading dislocations, the pits formation during the HCl etch would be mainly attributed to the inhomogeneous strain distribution next to the dislocation.

Figure 6 demonstrates the excellent correlation between all 3 techniques in terms of Etch Pit Density (and thus TDDs).

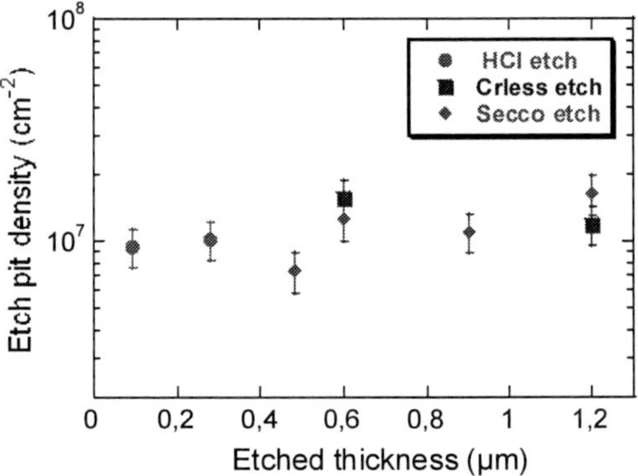

Figure 6: Etch Pit Density as a function of the Ge etched thickness, this after Cr-free or Secco wet etches or after a gaseous HCl etch.

In order to verify the suitability and efficiency of the Cr-free solution, we have quantified the EPD on 1.3 µm and 2.5 µm thick Ge layers grown on Si(001) without any subsequent thermal cycling under H$_2$. Rough surfaces with high defects densities were obtained whatever the techniques used (Cr-free or Secco solution), with TDDs equal to 7×10^7 (1.3 μm) and 4.7×10^7 cm^{-2} (2.5 μm), respectively (see figure 7). The EPD extracted as a function of Ge process characteristics (thickness and/or thermal cycling) are reported in figure 8.

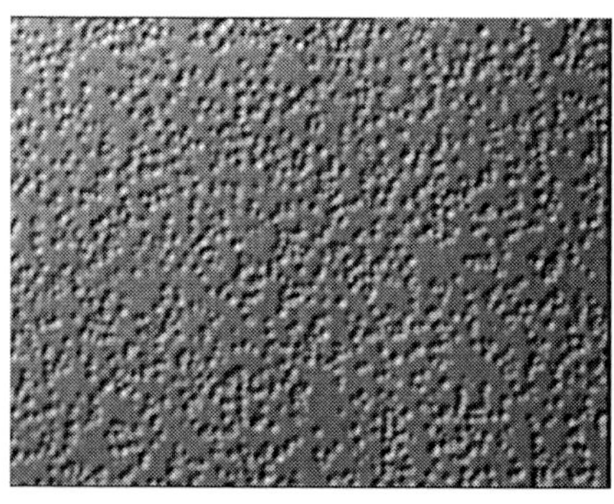

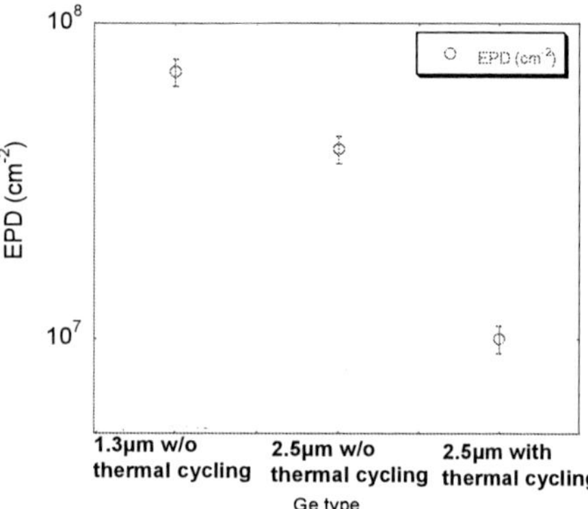

Figure 7: *(130x100)μm² optical microscope image of the surface of a 2.5μm thick Ge sample after a Cr-free etch.*

Figure 8: *EPD (cm⁻²) as a function of the Ge type (thickness and thermal cycling) after a Cr-free etch or a Secco etch.*

Finally, we have etched some GeOI wafers (fabricated thanks to the SmartCut^{TM} process with thick Ge layers on Si(001) as donor wafers) with a standard Secco solution. EPD close to the one of the starting material have been obtained, confirming the good quality of "on-insulators" structure obtained through the Smart Cut process and benefiting from continual improvements (20-21). Some additional etching results on GeOI wafers will be discussed in a forthcoming publication (22).

Acknowledgements

The authors would like to acknowledge all people from SOITEC and CEA-LETI who have contributed to this work. Special thanks to N. Cherkashin from CEMES and H. Cerva from Siemens for the TEM analysis.

References

1   The International Technology Roadmap for Semiconductors, Semiconductor Industry Association, http://www.itrs.net

2   S.W. Bedell, K. Fogel, A. Reznicek, J. Ott and D.K. Sadana, ECA Cancun, *The Electrochemical Society Proceedings Series*, Pennington (2006).

3   S.G. Thomas, E.S. Johnson, C. Tracy, P. Maniar, X. Li, B. Roof, Q. Hartmann and D.A. Ahmari, IEEE Electron Devices Letters, **26** 438 (2005).

4   W. P. Bai, N. Lu, J. Liu, A. Ramirez, D. L. Kwong, D. Wristers, A. Ritenour, L. Lee and D. Antoniadis, *VLSI Technology Digest of Technical papers,* p.121, (2003).

5   N. Wu, Q. Zhang, C. Zhu, D. S. H. Chan, A. Du, N. Balasubramanian, M. F. Li, A. Chin, J. K. O. Sin and D.-L. Kwong, *IEEE Electr. Dev. Lett.* **25**, 631 (2004).

6   P. Zimmerman, G. Nicholas, B. De Jaeger, B. Kaczer, A. Stesmans, L-Å Ragnarsson, D.P. Brunco, F.E. Leys, M. Caymax, G. Winderickx, K. Opsomer, M. Meuris and M.M. Heyns, *IEDM Digest of Technical papers*, p. 655 (2006).

7   T. Tezuka, S. Nakaharai, Y. Moriyama, N. Sugiyama, S. Takagi, *VLSI Technology Digest of Technical Papers*, pp. 198–199 (2004).

8   L. Clavelier, C. Le Royer, C. Tabone, J.M. Hartmann, C. Deguet, V. Loup, C. Ducruet, C. Vizioz, M. Pala, T. Billon, F. Letertre, C. Arvet, Y. Campidelli, V. Cosnier and Y. Morand, *Silicon Nanoelectronics Workshop Proceedings*, Kyoto, Japan, pp. 18-19 (2005).

9   C. Deguet, J. Dechamp, C. Morales, A.M. Charvet, L. Clavelier, V. Loup, J. M. Hartmann, N. Kernevez, Y. Campidelli, F. Allibert, C. Richtarch, T. Akatsu and F. Letertre, Vol. **809**, p. 153, *The Electrochemical Society Proceedings Series*, Pennington (2005).

10 Y. Bogumilowicz, J.M. Hartmann, C. Di Nardo, P. Holliger, A.-M. Papon, G. Rolland and T. Billon., *J. Cryst. Growth*, **290**, 523 (2006).

11 J.M. Hartmann, J.F. Damlencourt, Y. Bogumilowicz, P. Holliger, G. Rolland and T. Billon, *J. Cryst. Growth*, **274,** 90 (2005).

12 H.C. Luan, D.R. Lim, K.K. Lee, K.M. Chen, J.G. Sandland, K. Wada and L.C. Kirmeling, *Appl. Phys. Lett.* **75**, 2909 (1999).

13 Q. Li, Y.B. Jiang, H. Xu, S. Hersee and S.M. Han, *Appl. Phys. Lett.* **85**, 1928 (2004).

14 D.P. Malta, J.B. Posthill, R.J. Markunas and T.P. Humphreys, *Appl. Phys. Lett.* **60**, 844 (1992).

15 F. Secco d'Aragona, *J. Electrochem. Soc.* **119**, No. 7, 948 (1972).

16 Y. Bogumilowics, J.M. Hartmann, R. Truche, Y. Campidelli, G. Rolland and T. Billon, Semicond. Sci. Techn. **20**, 127 (2005).

17 M. Steinert, J. Ackert, A. Henaige and K. Wetzig, *J. Electrochem. Soc.* **152**, 12, C843 (2005).

18 J. Acker, M. Steinert, A. Henssge, A. Krause and K. Wetzig, *Meet. Abstract, Electrochem. Soc.* **501**, 742 (2006).

19 Rozgonyi, *Encyclopedia of Materials: Science and Technology*, p. 8524, Elsevier Science (2001).

20 C. Deguet et al, First Euro SOI Workshop 31 (2005).

21 F. Letertre, C. Deguet, C. Richtarch, B. Faure, J.M. Hartmann, F. Chieu, A. Beaumont, J. Dechamp, C. Morales, F. Allibert, P. Perreau, S. Pocas, S. Personnic, C. Lagahe-Blanchard, B. Ghyselen, Y.M. Le Vaillant, E. Jalaguier, N. Kernevez and C. Mazure, Mat. Res. Soc. Proc. Vol. **809**, The Material Research Society, B4.4.1 (2004).

22   Abbadie et al, to be published.

ECS Transactions, 6 (4) 271-277 (2007)
10.1149/1.2728871, ©The Electrochemical Society

A Chromium-free Defect Etching Solution for Application on SOI

J. Mähliß[a], A. Abbadie[b] and B. O. Kolbesen[a]

[a] Department of Inorganic and Analytical Chemistry, Johann Wolfgang Goethe-University, Frankfurt/Main, Germany
[b] Advanced Technology Department, SOITEC, Parc technologique des Fontaines, Bernin, France

A new defect etching solution is presented especially developed for application on silicon-on-insulator (SOI) substrates. This "*FS Cr-free SOI etching solution*" is free of toxic hexavalent chromium. Very efficient in revealing crystalline defects, it provides a low etch rate, which makes it a promising candidate for Secco replacement on SOI films. Excellent correlation of etch pit densities with Secco solution has been obtained. The mechanism and characteristics of this solution will also be discussed.

Introduction

New substrate materials like SOI (1) have been successfully introduced into the microelectronic industries. SOI offers many advantages (e.g. reduced junction capacitance, lower leakage current) compared to standard silicon substrates.

SOI structures consist of a single-crystalline Si film separated by a layer of SiO_2 from the bulk substrate. In this study we have used SOI wafers fabricated by the Smart Cut™ technology. It utilizes ion implantation as an atomic scalpel that cuts through the crystalline lattice and permits a transfer of a thin layer of material.

Defects that can be observed in SOI structures are of a conventional nature and originate from the CZ crystal growth process. The predominant grown-in defects are attributable to the aggregation of vacancies and self-interstitials (2, 3, 4). COPs (crystal originated particles), for instance, are octahedral voids formed by vacancy condensation.

Of course such crystalline defects in the active layer should be avoided as they may have a negative impact on the electrical properties of the device and reduce device yield.

For delineation of crystalline defects in SOI material and thereby for characterization of its quality the standard solution used is a diluted Secco (5, 6). This Secco etching solution works very well in combination with a subsequent HF step. At defect sites it produces well-developed etch pits which can be clearly seen under an optical microscope due to the underetching of the buried oxide (BOX). Furthermore its etch removal is uniform and a smooth surface is obtained after etching.

However, the Secco etching solution contains toxic hexavalent chromium (Cr^{6+}). As the usage of Cr^{6+} will be heavily restricted in the near future (already forbidden in Asia) a chromium-free etching solution, the *FS Cr-free SOI*, has been developed.

The *FS Cr-free SOI etching solution* fulfills all requirements: similar to the Secco etching solution, a good uniformity with well-developed etch pits at defect sites is obtained. The defect densities after *FS Cr-free SOI* etching correlate very well with those obtained after Secco etching on SOI with different defect densities. The etch rate is also within the range of that of the diluted Secco solution and is therefore attractive for etching on thin SOI films.

Experimental

The *FS Cr-free SOI etching solution* consists of 47.7 parts by volume of nitric acid (69 %), 5.9 parts by volume of hydrofluoric acid (49 %), 46.4 parts by volume of acetic acid (100 %) and bromine (one must add 0.5 ml bromine to 100 ml etching solution). The ratio of the components has been tested and adjusted accurately. Each component of the mixture has a specific purpose: nitric acid is an oxidizing agent for silicon, hydrofluoric acid is necessary to dissolve the silicon dioxide produced and acetic acid is a diluent which is needed for a uniform etch removal. Bromine is the key factor in this solution and is absolutely necessary for the etching of SOI as it initiates the oxidation reaction which is then carried on by nitric acid.

All experiments were carried out in 200 ml wide-mouthed reaction vessels made of polypropylene. With respect to the maximum volume of these vessels the volume of the etching solutions was calculated to ~ 150 ml. In addition the experiments were verified with volumes corresponding to those at production level. A diluted Secco etching solution was used as a reference. All etching solutions were mixed the day before they were used. Etching experiments were performed in a temperature-controlled water bath.

All SOI material used was made by the Smart Cut™ technology. Wafers were cut into squares of ~ 1.5 cm x 1.5 cm and cleaned with short blasts of nitrogen. For the etching experiments each sample was placed vertically in a Teflon™ sample-holder and dipped individually in the etching solution. The solution was not stirred during etching. The samples were then rinsed in ultra-clean Millipore water for two minutes, followed by a short dip in HF (50 %) to remove the BOX below the defect site. They were rinsed again in Millipore water for two minutes.

The thickness range of the SOI samples tested was between 50 and 100 nm. The etching times were calculated to give a remaining Si layer of ~ 30 nm. The thickness of the Si layer prior to and after etching was determined by ellipsometry.

Results and Discussion

General characteristics

The etch rate of the *FS Cr-free SOI etching solution* is 3.5 nm/s at RT on SOI. It decreases gradually with a decrease in temperature (Tab. I). As shown in the table, a change in the etch rate has no influence on the etch pit densities.

TABLE I. Etch rates and etch pit densities at different temperatures

Temperature [°C]	Etch rate [nm/s]	Etch pit density [cm^{-2}]
23	3.5	833 ± 394
13	2.1	878 ± 450
7	1.5	737 ± 355
3	1.1	855 ± 539

With diluted Secco etching solution (Cr^{6+} = 0.08 M) used as a reference (2.0 nm/s etch rate at 23 °C), an etch pit density of about 827 cm^{-2} has been found. This is in good agreement with the results shown in Tab. I and indicates that there is no dependence of the etch pit density on the etch rate within this range.

After etching a smooth homogeneous surface is produced and defects are revealed clearly (Fig. 1) like in the case of Secco etching (Fig. 2). The etch pits are roughly round in shape with a diameter of 4.5 to 5.5 μm (Figs. 3 and 5).

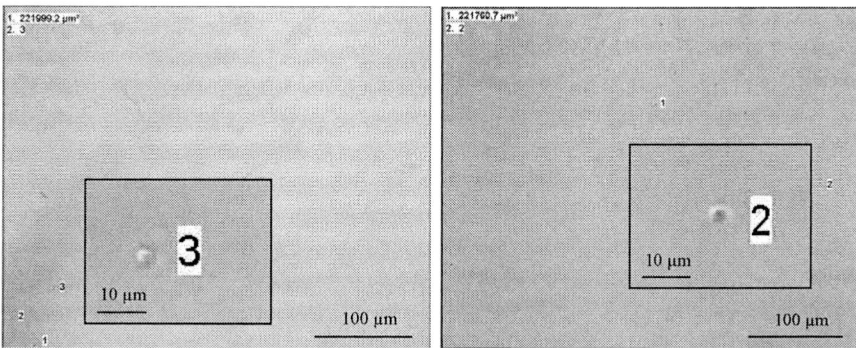

Figure 1. *FS Cr-free SOI.*   Figure 2. Secco.
Optical micrographs displaying etch pits after *FS Cr-free SOI* (Fig. 1) etching and Secco etching (Fig. 2) on SOI samples etched down to 30 nm enlarged in the inset. The etch pits in the center corresponding to the crystalline defects are surrounded by a halo produced by the underetching of the BOX.

Subsequent underetching of the BOX below the etch pit formed at a defect (producing a halo) by a short dip in a HF solution (50 %) is necessary to increase contrast for better observation of the etch pit under the optical microscope (Figs. 3 and 4). This requires that the pit reaches down to the BOX. Therefore the maximum allowed thickness of the remaining Si layer after etching is limited. No defects were made visible in SOI layers thicker than ~ 55 nm after etching indicating that the etch pits formed at the defects did not reach the BOX level. Hence, no BOX underetching takes place by the HF-dip preventing delineation of the defect in the optical microscope. For Si layers to be etched down to ~ 30 nm HF dipping times between 60 s and 90 s are necessary. The principle of the defect delineation process is the same for *FS Cr-free SOI* etching and Secco etching (Fig. 4).

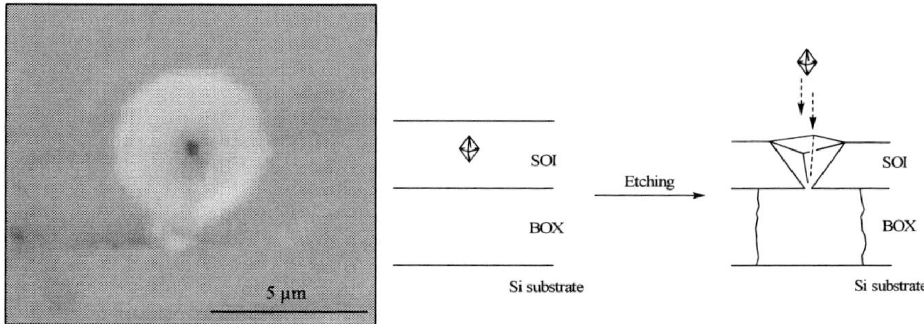

Figure 3. Optical micrograph of an etch pit and its halo on a SOI sample after *FS Cr-free SOI* etching and subsequent HF-dip.
Figure 4. Sketch of the defect delineation process: a) original location of defect (vacancy agglomerate); b) *FS Cr-free SOI etch* or Secco etch proceed preferentially at defect site reaching the BOX layer; in the subsequent HF-dip the BOX is underetched resulting in an improved defect visibility.

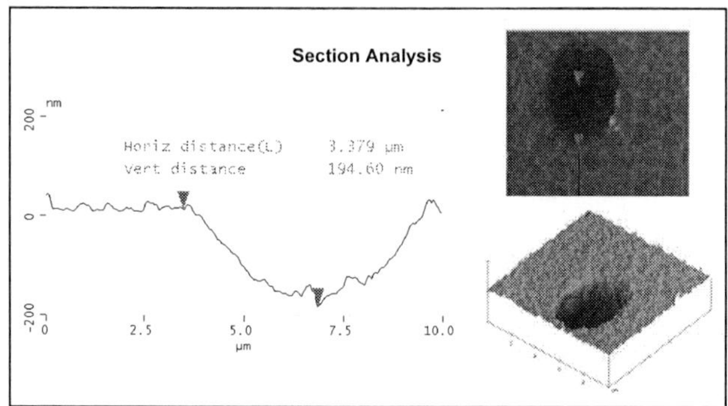

Figure 5. AFM line scan (10 µm x 10 µm) of an etch pit on a SOI sample etched down to ~ 30 nm using the *FS Cr-free SOI etching solution* (note depth in nm and size in µm). In the AFM image the halo surrounding the etch pit is not visible.

Figure 6 demonstrates the excellent correlation of the etch pit densities (EPDs) after a *FS Cr-free SOI* etching and a Secco etching used as a reference. Different SOI samples, with Si thicknesses between 50 nm and 100 nm, were tested. For a reliable correlation, SOI wafers with defect densities varying from 100 to 1000 defects/cm^2 were used.

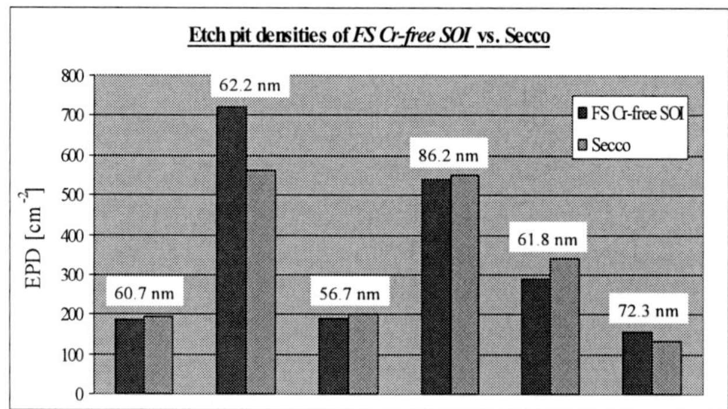

Figure 6. Correlation of etch pit densities (cm^{-2}) after a *FS Cr-free SOI* etching vs. a Secco etching.

Etching mechanism and the special role of bromine

The wet etching of silicon involves its chemical oxidation, which is brought about by bromine and nitric acid, followed by oxide layer dissolution with hydrofluoric acid.

In the *FS Cr-free SOI etching solution* the amount of hydrofluoric acid was successively reduced in order to slow down the dissolution of the oxidation product SiO_2 and therefore to decrease the etch rate. However, the absence of bromine combined with low HF concentrations resulted in no etching at all of SOI samples, whereas such mixtures etched bulk Si. This confirms that bromine is necessary in the reaction to initiate the oxidation mechanism on SOI.

While bromine is crucial to the process, Fig. 7 shows that the concentration of nitric acid strongly influences the etch rate of the solution as it contributes to the oxidation. Furthermore, nitric acid is indispensable to the *FS Cr-free SOI etching solution,* as crystalline defects are not revealed in its absence.

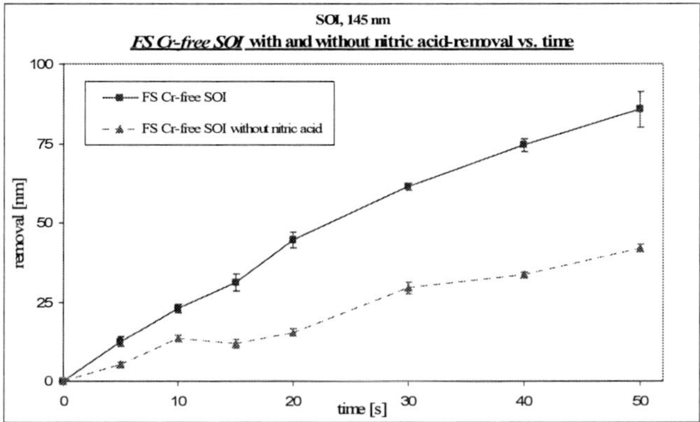

Figure 7. *FS Cr-free SOI* removal with and without nitric acid as a function of time.

As far as the mechanism of oxidation is concerned, S. Meltzer and D. Mandler (7) suggested a mechanism based on an interaction between silicon, bromine and hydrofluoric acid. In a mixture of HBr/HF they generated bromine electrochemically at the tip of an electrode very close to the Si surface. According to the authors after removal of the oxide layer by fluoride the exposed silicon reacts with bromine to form two Si-Br bonds (Fig. 8, step 1). The polarized Si-Br bond increases the chemical reactivity of the other Si back bonds which are, therefore, more easily attacked by HF (step 2). Finally, the activated Si-H bonds, which are also unstable in aqueous and fluoride media, will undergo nucleophilic attack evolving hydrogen (step 3).

Figure 8. Oxidation mechanism of silicon by bromine in the absence of nitric acid.

We studied (Fig. 9) the influence of the bromine content on the etch rate of the solution and on the uniformity on the surface etched. An increase in the bromine content increases the etch rate and vice versa. However, while an increase in bromine does not influence the uniformity of etching, when reduced to less than half of its normal concentration a significant damage to the surface has been observed.

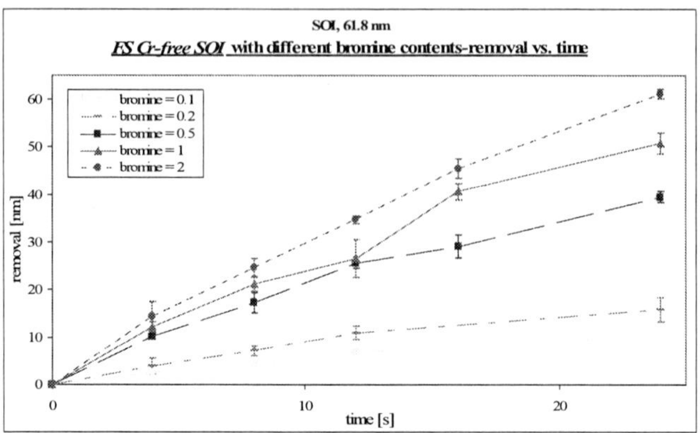

Figure 9. *FS Cr-free SOI etching solution* with different bromine contents.

<u>Replacement of the addition of elemental bromine</u>

Bromine is a strong oxidizing agent. It is, however, toxic to the environment and hazardous to health when inhaled. Moreover, precautions have to be taken in the storage and handling of bromine (storage at low temperatures, protection from light, difficulty of measurement of small volumes etc.). For these reasons, the addition of elemental bromine can be replaced by the addition of sodium salts of bromate and bromide to the etching solution. In acid solutions bromate and bromide generate bromine by comproportion reaction:

$$BrO_3^- + 5\,Br^- + 6\,H^+ \rightarrow 3\,Br_2 + 3\,H_2O. \tag{1}$$

Bromate and bromide are not as hazardous as bromine and much easier to store and handle.

Conclusions

The *FS Cr-free SOI etching solution* contains two strong oxidants (HNO_3, Br_2), each of which plays a decisive role in the silicon oxidation and crystalline defect delineation process. Bromine is necessary during the first step of etching to initiate the oxidation process which then is carried on by nitric acid. The ability of making crystalline defects visible is linked to the nitric acid.

The etch rate of the *FS Cr-free SOI etching solution* is 3.5 nm at 23 °C but it can easily be decreased by lowering the temperature. No relationship between etch rate and etch pit density has been observed. After a uniform etch removal, crystalline defects are made visible by a final HF-dip, which then enables their observation under an optical microscope. An excellent correlation of etch pit densities has been found after a *FS Cr-free SOI* etching and a Secco etching on SOI samples.

In conclusion, we have shown that the *FS Cr-free SOI etching solution* is a highly suitable alternative for the Secco etching solution on SOI with the additional advantage of being chromium-free.

Acknowledgments

This work was supported by SOITEC S.A., Parc technologique des Fontaines, Bernin. The authors would like to thank François Brunier and Oleg Kononchuk for useful discussions, Martin Lommel for the AFM images and Yvonne Filbrandt-Rozario for assistance with the text.

References

1. C. Malleville et al., IEEE SOI conf. (2001), 155.
2. S. Sama, M. Porrini, F. Fogale and M. Servidori, *J. Electrochem. Soc.* **148**, No. 9, G517-G523 (2001).
3. P. Papakonstantinou, K. Somasundram, X. Cao and W. A. Nevin, *J. Electrochem. Soc.* **148**, No. 2, G36-G42 (2001).
4. R. Falster and V. V. Voronkov, *Mater. Sci. Eng.*, B. **73** (2000) p.69.
5. F. Secco d'Aragona, *J. Electrochem. Soc.* **119**, No. 7, 948-951 (1972).
6. L. F. Giles, A. Nejim and P. F. L. Hemment, *Vacuum* **43**, 297 (1992).
7 S. Meltzer and D. Mandler, *J. Chem. Soc. Faraday Trans.*, **91**(6), 1019-1024 (1995).

ECS Transactions, 6 (4) 279-284 (2007)
10.1149/1.2728872, ©The Electrochemical Society

Mobility Characterization by Magnetoresistance in Ultra Thin SOI Ring-FETs

T.V. Chandrasekhar Rao[a,b], S. Cristoloveanu[c], J. Antoszewski[a], T. Nguyen[c], H. Hovel[d], P. Gentil[c], and L. Faraone[a]

[a] Microelectronics Research Group, School of Electrical, Electronic and Computer Engineering, The University of Western Australia, Crawley , WA 6009, Australia
[b] TPPED, Bhabha Atomic Research Centre, Bombay 400085, India
[c] IMEP, INPG Minatec, BP 257, 38016 Grenoble Cedex 1, France
[d] IBM, Yorktown Heights, USA

We report a novel use of magneto-resistance (MR) to estimate carrier mobility in ultra-thin virgin SOI wafers. Measurements were made in pseudo-MOSFET configuration with electrical contacts in Corbino geometry and substrate biasing. The use of high magnetic fields enabled accurate extraction of electron mobility as a function of gate voltage (effective field) and film thickness. The MR mobility values are large and correlate well with the field-effect mobility determined in pseudo-MOSFETs at zero magnetic field.

Introduction

Intrinsic carrier mobility has a critical bearing on advancing the frontiers of SOI technology. As thickness of silicon template layer is reduced to accommodate advanced CMOS circuits, mobility is limited by several additional factors. These include fluctuations in layer thickness, scattering by surface optical phonons etc. (1). Therefore, accurate evaluation of carrier mobility in ultra-thin SOI layers becomes even more imperative. Among the methods used to measure mobility, it has recently been shown that the geometric magneto-resistance (MR) is particularly suited for the evaluation of short-channel SOI MOSFETs (2).

For SOI process optimization, it is very important to measure the carrier mobility at the wafer level. The most successful method is the pseudo-MOSFET and its variants (Ring-FET and Hg-FET) (3-5). The aim of this paper is to demonstrate that these two methods, MR and pseudo-MOSFET, can be advantageously combined. Our strategy is to measure MR under high magnetic fields in Ring-FETs formed on SOI wafers. The Ring-FET electrical contacts have perfect Corbino configuration (Fig. 1), which is known to maximize the geometric MR (6) and eliminate edge effects in transport studies (7).

We present the results of mobility measurements on ultra-thin virgin SOI films using high-field MR technique in Ring-FETs with Corbino geometry. Measured MR shows quadratic dependence with magnetic induction (B) as expected, thus demonstrating the efficacy of this mobility extraction technique. Besides, this method is attractive because it is straightforward and does not need complex modeling or fitting parameters which would adversely affect the accuracy.

Experimental

MR is measured on SOI wafers with ultra thin Si films (T_{Si} = 10–27 nm) by pseudo-MOSFET (3). The wafers, fabricated by the SIMOX process, feature 90 nm thick buried oxide (BOX) and low doping. The SOI film thicknesses for the different samples were obtained by etching the starting wafers with TMAH (Tetramethyl Ammonium Hydroxide) and measuring the resulting thickness by spectroscopic ellipsometry. The Corbino geometry is achieved by evaporating circular ohmic contacts which serve as source and drain (Fig. 1) (4,5). The contacts consist of 100 nm Er capped with 200 nm Ag and are ohmic to electrons without the need for annealing. Positive substrate biasing induces an inversion layer of electrons at the film-BOX interface. Such a test device is named 'pseudo-MOSFET with Corbino ring configuration' or 'Ring-FET ' for short.

Magnetic field generated by an Oxford Instruments superconducting magnet (0–12 T) is applied perpendicular to the device and its electrical characteristics measured using a parameter analyzer (HP 4156A). MR is computed from the drain current-voltage '$I_D(V_G)$' characteristics measured under a range of magnetic fields and at low drain bias (ohmic region V_D = 0.1–0.5 V).

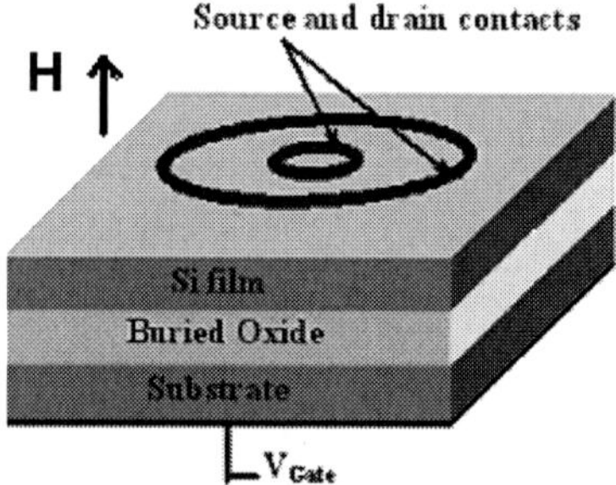

Figure 1. SOI Ring-FET showing contacts, magnetic field direction and biasing conditions.

Drain characteristics of one of the test devices in weak inversion region are shown in Figure 2. A high magnetic field does not modify the sub-threshold slope. Analysis of this slope together with the measured thicknesses indicates an interface state density DIT at the Si film / BOX interface of 3.5 - 5 x 10^{11} cm^{-2} eV^{-1} Figure 3 shows the same characteristics on linear scale (strong inversion) measured under a number of magnetic fields. We note the invariance of the threshold voltage and the expected quadratic dependence of the channel resistance, $R_B(B) = V_D/I_D$, with magnetic induction.

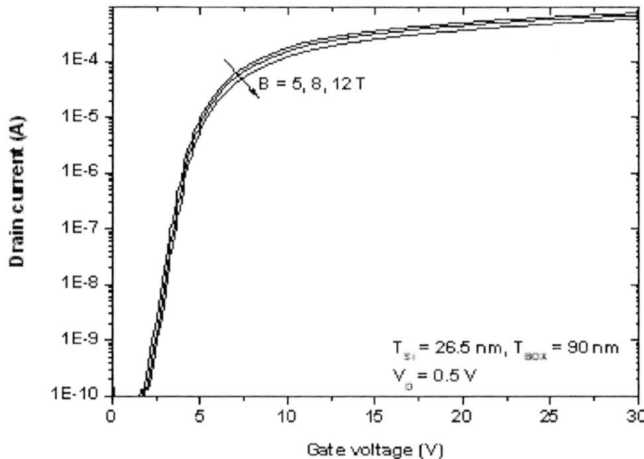

Figure 2. Typical I_D (V_G) curves for an SOI Ring-FET in the weak inversion region. SOI film is 26.5 nm thick and magnetic field was varied from 0 to 12 T.

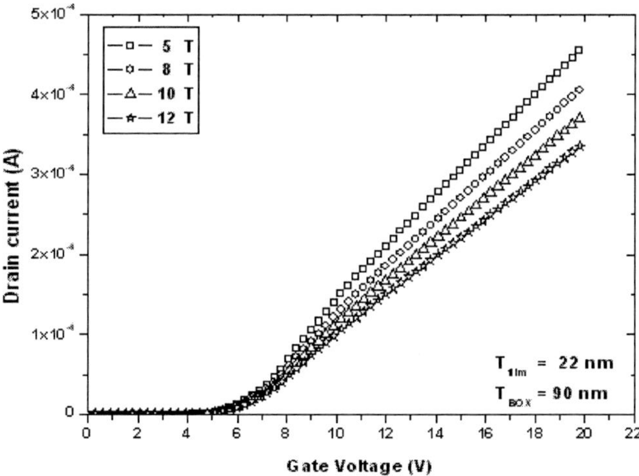

Figure 3. Drain current versus substrate voltage for a range of magnetic fields in Ring-FET fabricated on a 22 nm thick SOI film (V_D = 0.5 V).

<u>Analysis and Results</u>

In Corbino-like discs, the Hall field is totally suppressed and magneto-resistance R_B reaches a maximum value

$$R_B = R_0 (1 + \mu^2 B^2) \qquad [1]$$

where B is the magnetic induction with field applied normal to the wafer and μ is the MR carrier mobility. Channel MR ratio, $[R_B - R_0]/R_0$, is computed from drain current curves for each value of gate voltage V_G. Electron mobility is simply determined by plotting this ratio $(R_B/R_0 - 1)$ against B^2. Figure 4 shows that straight lines are indeed obtained, in agreement with Eq.(1), for a wide range of substrate bias V_G.

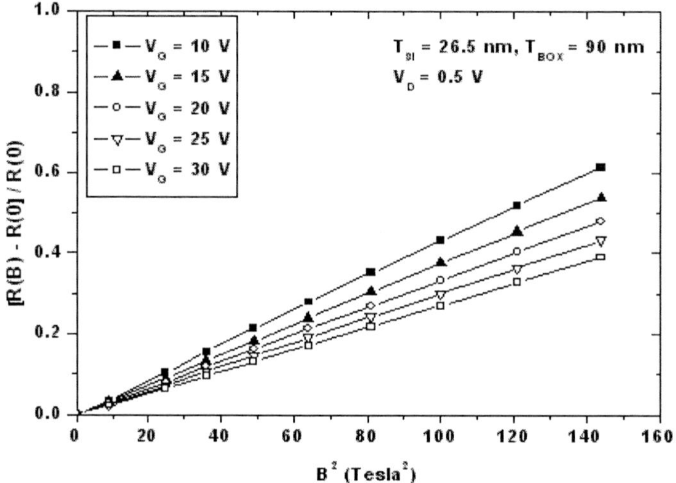

Figure 4. Magnetoresistance ratio versus magnetic field, for various gate voltages. SOI Ring-FET was fabricated on a 26.5 nm thick film ($V_D = 0.5$ V).

The slope of the straight line yields the carrier mobility. Note that this method is free from any geometrical parameters: channel length, width, aspect ratio, film and BOX thickness, drain and gate bias. The lone constraint is the requirement of high magnetic fields, as $B \sim 1/\mu$, in order to achieve a measurable MR ratio.

Figure 5 shows the results of MR measurements on two SOI devices of varying T_{Si}. Electron mobility value extracted from the slopes of these straight lines turns out to be $\sim$ 0.06 m^2/Vs, which is on expected lines for this SOI material. A weak mobility degradation is observed in the thinner film.

Besides MR and electron mobility, I_D (V_G) curves can be used to estimate several other useful parameters such as threshold voltage, transconductance, field-effect mobility, etc. Figure 6 shows a comparison between MR mobility and the effective mobility. The latter was determined from zero magnetic field transconductance measurements on the same Ring-FETs, using '$(\mu_{eff})^2 = \mu_0.\mu_{FE}$'. Transconductance yields field-effect mobility μ_{FE}, whereas pure mobility μ_0 is extracted from the Y function (3-5). In this case an appropriate aspect ratio of the Ring-FET is necessary as described earlier (4-5).

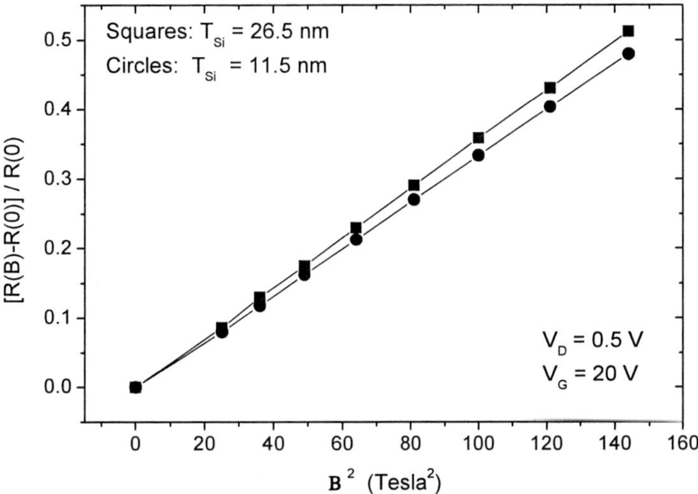

Figure 5. Normalized magnetoresistance plotted as a function of square of magnetic induction, in SOI films of different thicknesses. The slopes of straight lines yield electron mobility.

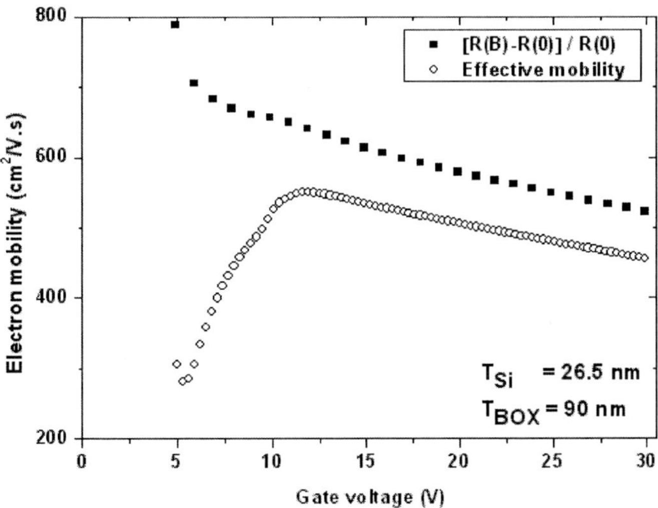

Figure 6. Magnetoresistance mobility and effective mobility as a function of gate voltage, in a 26.5 nm thick SOI film.

The low field mobility was also obtained from the slope of I_D / $G_M^{1/2}$ (3-5) in the same range of gate voltages. The mobilities obtained were 580 - 590 cm^2 / V.s, in excellent agreement with the results shown in Figure 6.

The curves of Figure 6 reflect an excellent correlation between μ_{eff} and μ_{MR} under strong inversion. The difference observed in weak inversion region is due to the inaccuracy of field-effect mobility extraction in this bias range. Note that the gate voltage can be converted into inversion charge or effective vertical field, which enables a further comparison of μ_{MR} with 'universal' mobility curves.

Conclusion

We have demonstrated a novel utility of both magnetoresistance and pseudo-MOSFET techniques to evaluate carrier mobility in ultra-thin SOI wafers. The addition of a magnetic field enriches the Ring-FET capabilities for SOI characterization. We have noted large values of electron mobility even in very thin SOI layers and nearly invariant in the range of film thicknesses investigated.

The MR measured as a function of gate voltage allows a direct comparison of MR mobility with the effective and field-effect mobilities measured on the same device. Such measurements conducted at low temperature would be useful to reveal and fully understand the basic scattering mechanisms controlling the carrier mobility behavior in ultra-thin SOI films.

References

1. D. Esseni and A. Abramo, *Semicond. Sci. Technol.*, **19**, S67 (2004).
2. W. Chaisantikulwat, M. Mouis, G. Ghibaudo, C. Gallon, C. Fenouillet-Beranger, D.K. Maude, T. Skotnicki, S. Cristoloveanu, *Solid-State Electronics*, **50(4)**, 637 (2006).
3. S. Cristoloveanu, S. Williams, *IEEE Electron Device Lett.*, **13(2)**, 102 (1992).
4. D. Munteanu, S. Cristoloveanu, H. Hovel, *Electro-chemical and Solid-State Lett.*, **2(5)**, 242 (1999).
5. H. Hovel, *Solid-State Electronics*, **47**, 1311 (2003).
6. L. Li, L. Fang, K.J. Liao, G. Z. Fu, F.F. Yang, G. B. Liu, R .J. Zhang, C. L. Cao, and W. M. Chen, *J. Cryst. Growth* **287**, 101 (2006).
7. Y. Paltiel, E. Zeldov, Y. Myasoedov, M. L. Rappaport, G. Jung, S. Bhattacharya, M. J. Higgins, Z. L. Xiao, E. Y. Andrei, P. L. Gammel, and D. J. Bishop, *Phys. Rev. Lett.*, **85**, 3712 (2000).

SESSION 6

TECHNOLOGY AND MATERIALS

ECS Transactions, 6 (4) 287-293 (2007)
10.1149/1.2728873, ©The Electrochemical Society

Strained Silicon Directly on Insulator (SSOI):
Biaxial, Uniaxial, or Hybrid Strain?

Aaron Thean, Ph.D.

Freescale Semiconductor Inc.
3501 Ed Bluestein Blvd., MD: K-10, Austin, TX 78721
USA.
Tel: 512-933-2816, Fax: 512-933-6962, Email: Aaron.Thean@freescale.com

This paper reviews the development of uniaxial and biaxial strain silicon technologies and how they may be combined through strain engineering of SSOI substrates. Through a novel method of selective biaxial-uniaxial strain hybridization, this work demonstrates a scalable enhanced SSOI CMOS technology. With this mixed-strain approach, nFET uniaxial strain can be amplified by the substrate strain platform while SSOI pFETs can be enhanced beyond conventional single unaxially-strained or biaxially-strained Si. This can be accomplished with minimal process complexity cost

Introduction

Facing unprecedented power-performance challenges at 32nm and beyond, the need to maintain leakage and performance at deep nanometer dimensions is rapidly forcing the classical transistor into a tight corner. To continue performance scaling, new materials and structures are making their way into the classical transistor. Embedding SiGe in the source/drain, overlayer stressors and stress-memory transfer are becoming mainstream methods to boost device mobility [1]. However, further scaling of the stressors will become a growing challenge as transistor gate-to-contact pitch and active dimensions are reduced to meet circuit density needs (Fig.1).

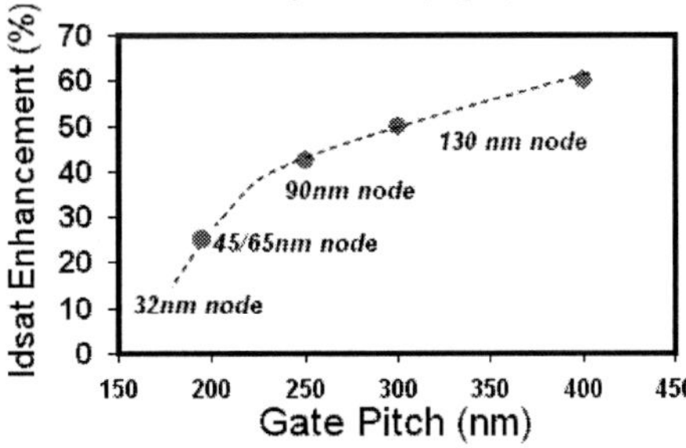

Fig.1 pMOS drive current improvement with compressive nitride stressor as a function of gate pitch.

Increasing overlayer film stress bring diminishing returns as inter-gate distances are shrunk (Fig. 2) [2],[3] limiting the overlayer-film-to-silicon contact and S/D volume for embedding SiGe. Removable spacers can delay the issue temporarily by maximizing the S/D contact region [4] (Fig. 3) but continuing scaling with this stressor approach may require relaxing the lay-out restrictions to achieve performance with penalties on density and complexity.

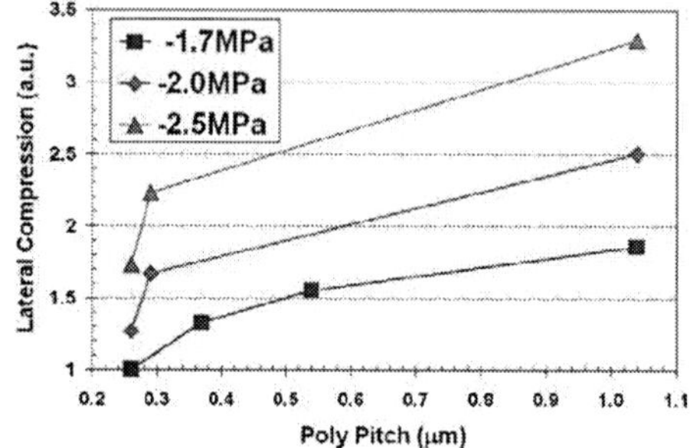

Fig. 2 Mechanical stress simulations for lateral compressive stress in the PMOS channel versus gate poly pitch as a function of nitride overlayer film stress (After ref. 2).

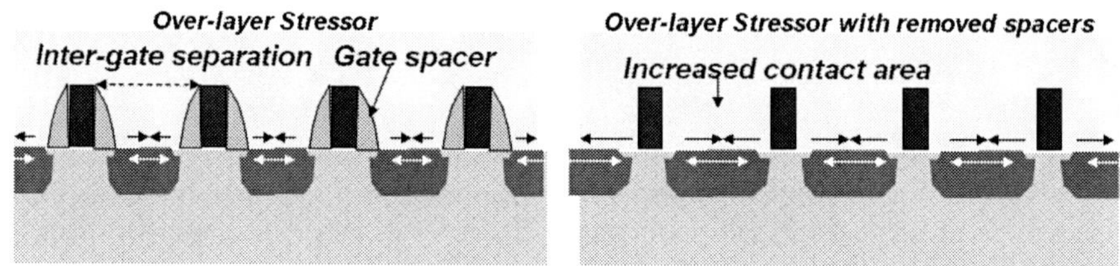

Fig. 3 Schematics of stacked transistors with off-set spacers (Left) and spacers removed intentionally to increase S/D contact region (Right).

Biaxial Versus Uniaxial Strained Silicon

Today, in-plane uniaxially-strained Si are achieved through different techniques of localizing biaxially-strained films. For example, compressive/tensile nitride overlayers and pseudomorphic SiGe on Si are intrinsically biaxially strained (Fig. 4). Through the interactions with the source/drain and gate/spacer topography, uniaxial stress are mechanically transferred into the channel region.

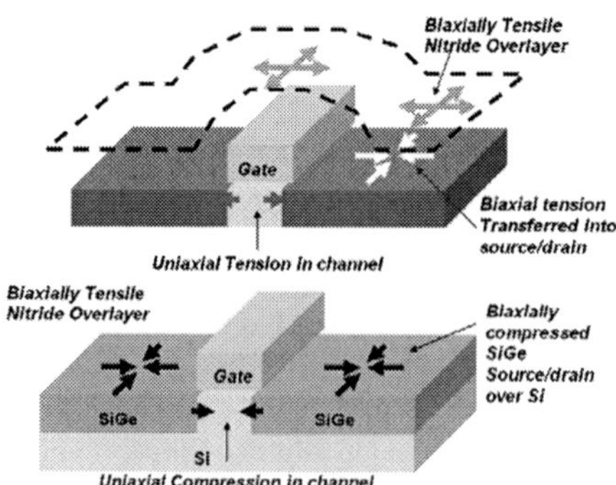

Fig. 4 Illustrations showing the localization of biaxially stressed films to achieve uniaxially strained Si MOSFET channels.

Due to fundamental limits on critical material thickness and mechanical integrity, aggressively scaled feature sizes will increasingly challenge effective localized transfer of the mechanical stress.

In contrasts, strained substrates provide the possibility of imposing high strain directly in the transistor channel at the wafer level. An essentially alternate channel approach, the challenge is to preserve the strain state throughout device processing. Although, strained substrates like SSOI are categorized as biaxial stressors, simulations and experiments have shown that the strain state of these materials can be dramatically changed through interactions with device geometry [5,6]. Three-dimensional mechanical stress simulations show that strong uniaxial strain can be derived from biaxially-strained SSOI structure by making the structure longer than wide (Fig. 5).

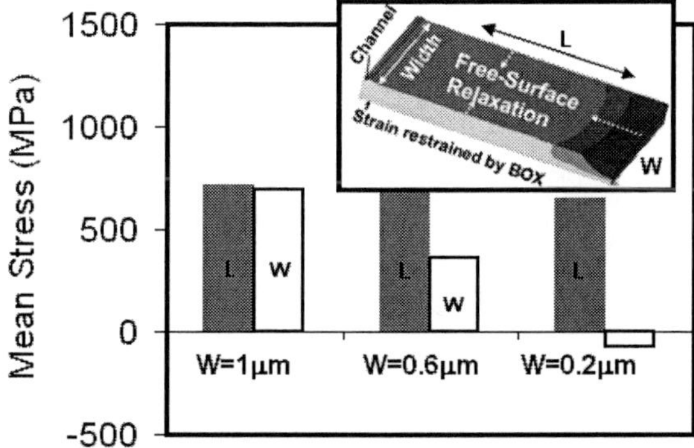

Fig. 5 Finite-element stress simulations showing the effects of SSOI free-surface relaxation along L & W as a function of dimension. Stress component along the width (W) can be reduced with moderate reduction of stress along L by shrinking the W dimension.

Application of such a techniques to achieve high uniaxial strain in devices like SSOI, SGOI (Silicon Germanium on Insulator) FinFETs and ultra-thin body devices have been recently demonstrated [6,7,8,9].

Being a strain platform, strained substrates like SSOI can work in tandem with conventional uniaxial stressors to attain higher channel stress (Fig.6).

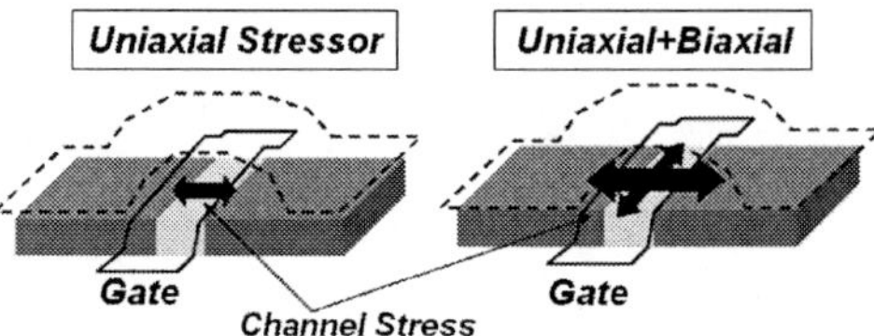

Fig.6 MOSFET illiustration showing the addition of strain due to strong combination of uniaxial and biaxial stressors

But, n- and pMOS desire different strain configuration and the fundamentally weak pMOS enhancement under pure biaxial tension (Figs. 7,8), on a homogeneously-strained substrate becomes an issue [2].

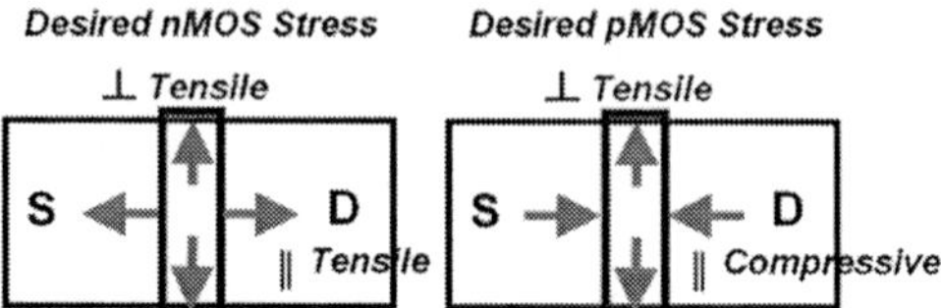

Fig.7 Desired CMOS stress configuration for enhanced device drive performance for Si(100)/<110> orientation

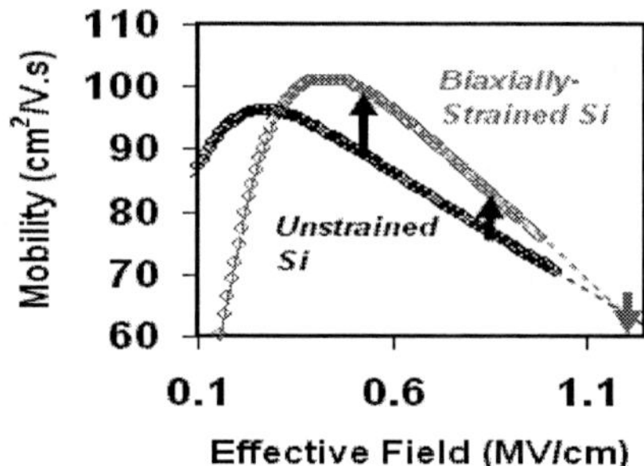

Fig.8 Hole mobility of biaxially-tensile pMOS relative to device with unstrained Si. The high sensitivity of biaxially-strained holes to effective field leads to degradation for scaled devices.

Hybrid Strained Silicon

The natural strain sensitivity of super-critically thick SSOI offers a solution to this problem. We show that substrate strain can be manipulated during device fabrication, enabling hybrid strain CMOS. It has been shown that through optimized ion implantation and associated thermal processes, the biaxial tension of a localized region of the SSOI substrate can be changed selectively (Fig. 9) with minimal impact on crystal quality.

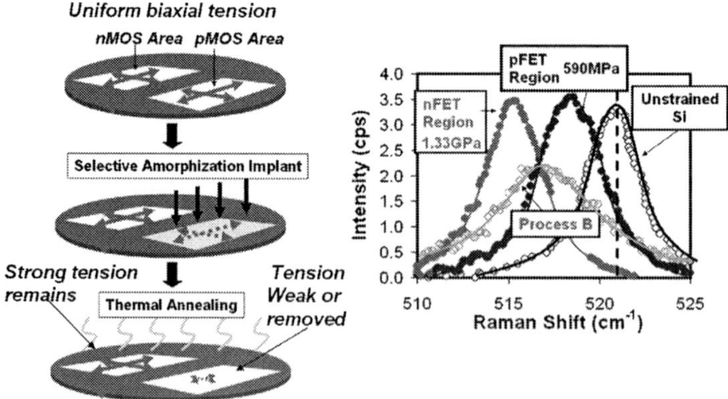

Fig.9 Left: Selective relaxation of local regions of SSOI wafer. Right: Raman spectra showing regions of different strain due to selective strain relaxation technique

With the ability to modulate the strain selectively on the SSOI wafer with implantation, it is now possible to change the strain state of devices to achieve the desired strain configuration through the combination with local strain techniques. This hybrid strain approach can be accomplished with minimal process complexity (Fig. 10) and provides a mean to extend the performance of partially-depleted SOI CMOS technology.

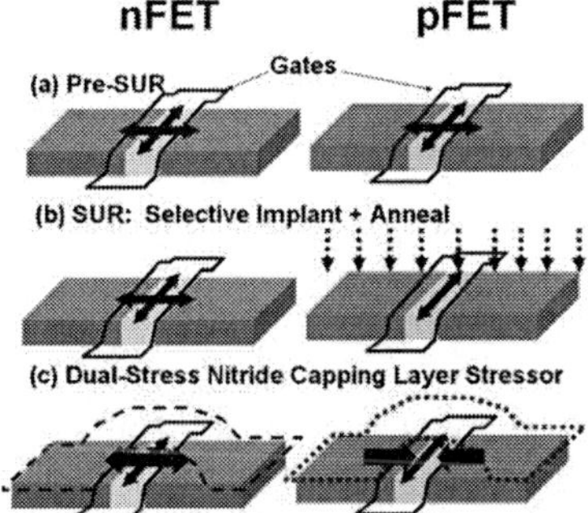

Fig.10 Schematics illustrating the process of Selective Uniaxial Relaxation & final CMOS strain configuration after dual-stress nitride capping layer

Performance enhancement through hybrid strain has been demonstrated (Fig. 11). Through careful strain engineering, popular pMOS uniaxial stressors (eg. compressive nitride liners and embedded SiGe source/drain) are fully compatible with the hybrid strain approach. Other groups have also shown that building enhanced short-channel pMOS devices on SSOI wafers is not an issue [10,11] once the intrinsic biaxial strain in the p-channel is uniaxially changed through processing.

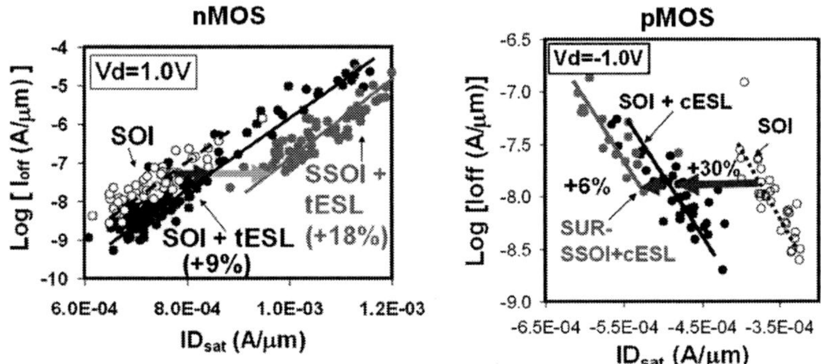

Fig.11 Left: Further nMOS enhancement due to addition of tensile nitride liner (tESL) with SSOI . Right: Further pMOS enhancement due to addition of compressive liner with SSOI through hybrid strain technique.

In fact, SSOI-eSiGe hybrid pMOS can be enhanced above the conventional eSiGe process by exploiting the beneficial strain components from the SSOI stress (Fig. 12)

[12]. Especially, when residual width tension of the pMOS can be maintained during the selective relaxation processes.

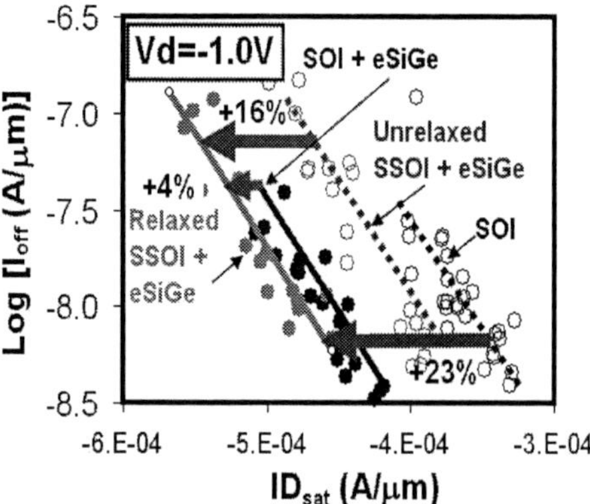

Fig.12 Further pMOS enhancement due to addition of embedded SiGe stressor with SSOI through hybrid strain technique.

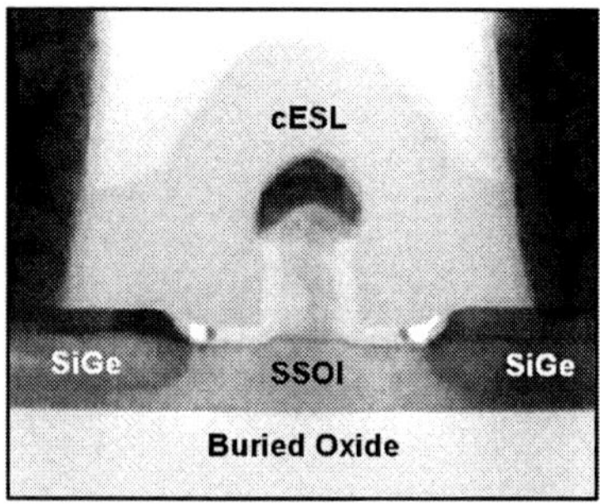

Fig.13 TEM cross-section image of eSiGe-SSOI hybrid strain pMOS transistor

Conclusion

Due to the unique strain stability of globally strained substrates like SSOI, a novel method of selective biaxial-uniaxial strain mixing can be implemented to achieve a scalable enhanced SSOI CMOS technology. With such a mixed-strain approach, nFET uniaxial strain can be amplified by the substrate strain platform while SSOI pFETs can be enhanced beyond conventional single uniaxially-strained or biaxially-strained Si, with minimal process complexity. This may be an effective way to overcome the diminishing gains from conventional local stressors in response to tighter device lay out to meet circuit density needs.

Acknowledgements

The author would like to acknowledge the Austin Novel Device Group and all the device and process support provided by the ASTS and ATMC staff, the management support from Bruce E. White, Jon Cheek, and Suresh Venkatesan. Special thanks go out to Bich-Yen Nguyen for the support and many fruitful discussions.

References

1. Thompson et. al, IEDM Tech Dig., p.61 (2002)
2.. P.Grudowski, Symp. VLSI Technology Tech Dig, p.76, (2006).
3. A. Oishi et. al., IEDM Tech. Dig., p.239 (2005).
4. X. Chen et. al., Symp. VLSI Technology Tech Dig, p.74, (2006).
5. A.V-Y Thean, IEDM Tech Dig. , p 515, (2005)
6. N. Collaert, Symp. VLSI Technology Tech Dig, p. 64, (2006)
7. F. Andrieu, Symp. VLSI Technology Tech Dig, p. 168 (2006)
8. T. Irisawa, et.al. IEDM Tech. Dig. p. 727 (2005).
9. T. Mizuno, et. al. IEDM Tech. Dig. p. 453 (2006).
10. A. Wei, et.al. ECS Transactions, 3(7) 719-725 (2006)
11. H. Yin, et. al. Symp. VLSI Technology Tech Dig., p. 94, (2006)
12. A. V-Y. Thean, et. al. VLSI Technology Tech. Dig., p. 164, (2006)

ECS Transactions, 6 (4) 295-307 (2007)
10.1149/1.2728874, ©The Electrochemical Society

Mixed orientation Si-Si interfaces by hydrophilic bonding and high temperature oxide dissolution: wafer fabrication technique and device applications

K.L. Saenger,[a] J.P. de Souza,[a] K. Fogel,[a] J.A. Ott,[a] A. Reznicek,[a] H. Yin,[b] C.Y. Sung,[a] and D.K. Sadana[a]

[a]IBM Research, T.J. Watson Research Center, P.O. Box 218, Yorktown Heights, NY 10598
[b]IBM Semiconductor Research and Development Center
Microelectronics Division, Hopewell Junction, NY 12533

In this paper we describe a "quasi-hydrophobic" bonding method in which ultrathin (<1-2 nm) oxide present on wafer surfaces during bonding is removed after bonding by a high-temperature oxide dissolution anneal to leave the desired direct Si-to-Si contact at the bonded interface. We will show that the direct-silicon-bonded (DSB) interfaces produced by this method are clean enough to allow implementation of a recently described amorphization/templated recrystallization technique for changing the orientation of selected DSB layer regions from their original orientation to the orientation of the underlying handle wafer, and then discuss some of the mechanisms and integration challenges associated with wafer and device fabrication by these techniques.

Introduction

Semiconductor device technology is increasingly relying on bonded Si wafer substrates to achieve desired performance targets. Most Si wafer bonding is hydrophilic, between wafer surfaces that are oxide-like. Hydrophilic bonding is clearly the technique of choice when an oxide is desired at the bonded interface, for example, when fabricating silicon-on-insulator (SOI) structures. However, certain fabrication schemes for bulk hybrid orientation substrates (1,2) require a direct Si-to-Si bond between Si surfaces having different surface orientations, with no oxide layer at the bonded interface.

Direct Si-to-Si bonding for direct-silicon-bonded (DSB) wafers is normally achieved with hydrophobic bonding (3), a technique with some difficulties. Hydrophobic (H-terminated) surfaces are more easily contaminated than hydrophilic ones, often making it necessary to perform the bonding in a vacuum environment. In addition, the widely used surface plasma treatments developed to allow room temperature bonding typically, though with some exceptions (4), introduce surface oxygen, making them incompatible with an oxide-free bonded interface. Bonding at higher temperatures can also present difficulties, since most cleaving processes for separating the bonded layer from the wafer to which it was originally attached are thermally activated and start occurring in the same temperature range as the bonding. Given these difficulties, it was thought that it might be preferable to fabricate DSB wafers with a "quasi-hydrophobic" bonding technique in which an ultrathin (1-2 nm) oxide would be present on one or both wafer surfaces *during* bonding (thereby allowing the bonding to be hydrophilic), but removed *after* the bonding (to leave the desired direct Si-to-Si contact at the bonded interface).

The stability of buried oxide layers in Si is a topic with relevance to both Si-based devices (5) (due to the need for tightly controlled thicknesses of thin insulating SiO_2 layers between electrically active Si regions) and silicon-on-insulator (SOI) substrate fabrication. SOI substrates fabricated by SIMOX (separation by implanted oxygen) are typically subjected to internal thermal oxidation (ITOX) anneals to thicken buried oxides formed by high dose oxygen ion implantation (6,7). In the context of Si wafer bonding, the thermal stability of thin (1 - 10 nm) interfacial oxides between bonded wafer surfaces has been examined as a function of Si wafer doping, Si wafer growth method (float-zone (FZ) or Czochralski (Cz)), and Si wafer surface orientation (7-10). It was concluded that dissolution of native oxides having a thickness >1 nm is not possible with annealing at temperatures in the range 1100 - 1200 °C, and that undesirable oxide islanding is typical, especially with Cz wafers. In contrast to these results, it was shown (4) that a few monolayers of interfacial oxide could be made to disappear in FZ wafers after annealing at 1150 °C. Unfortunately, FZ wafers are still very expensive, are relatively easily deformed during processing, and typically are used only in cases where high resistivity substrates are required. More to the point, it is expected that interfacial oxide layers will never be as thin as a few monolayers if the bonding is done in any environment other than high vacuum.

In this paper we describe a quasi-hydrophobic Si bonding method in which interfacial oxides remaining after bonding are removed by high temperature (1320 - 1325 °C) annealing. This is followed by results from a related study on the dissolution of oxide layers disposed between a Si substrate and a polycrystalline Si (poly-Si) overlayer. We will then discuss the mechanisms for oxide dissolution most likely to be operative for our process conditions. Finally, we will show that the DSB interfaces produced by this bonding method are clean enough to allow implementation of an amorphization/templated recrystallization (ATR) technique (1,2) for changing the orientation of selected DSB layer regions from their original orientation to the orientation of the underlying handle wafer, and briefly review how the ATR technique may be used with hybrid orientation DSB wafers for device applications.

Quasi-hydrophobic Bonding Method

The steps of the quasi-hydrophobic bonding method are shown schematically in Figure 1 for the generic case of a DSB wafer having a base Si substrate with a (j'k'l') crystal orientation and a DSB layer with a (jkl) crystal orientation. DSB wafers with a (100) base substrate and a (110) DSB layer are denoted as DSB-A, while DSB wafers with a (110) base substrate and a (100) DSB layer are denoted as DSB-B. The donor wafer of Figure 1A may be either bulk or SOI; SOI is preferable when the donor wafer removal step of Figure 1C is performed by a process such as grinding because the buried oxide may be used as an etch stop.

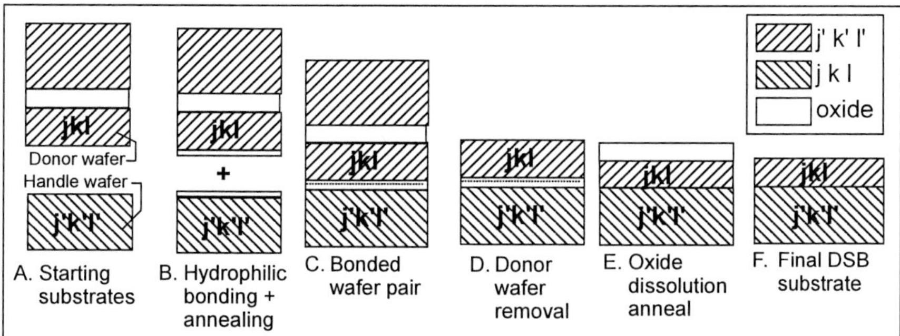

Figure 1. Process steps for DSB wafer fabrication.

Experimental Procedure

Oxide dissolution anneals were performed on both DSB-B wafers bonded in-house and DSB-A wafers provided by a vendor. All handle wafers and DSB layers were derived from Cz-grown Si. For DSB-B, the bulk handle wafer and donor SOI wafer were first cleaned with a SC1/SC2 sequence (11) to leave a thin (~1 nm) chemical/native oxide layer on all Si surfaces, then bonded at room temperature and annealed at 1050 °C for 2 hr to produce the structure of Figure 1C. After removing the bulk portion of the donor SOI wafer (by grinding followed by etching in KOH), the oxide etch-stop layer was removed (by etching in HF) to produce the structure of Figure 1D. The bonded wafer pair is then subjected to an oxide dissolution anneal (typically 1320 to 1325 °C for 3 to 5 hr) in a slightly oxidizing ambient (typically atmospheric pressure Ar with 0.1 to 1% O_2) to produce the structure of Figure 1E. Slow ramp rates were used to minimize thermal stresses; heating rates were typically 2 °C/min between 1000 and 1150 °C, 1 °C/min between 1150 and 1310 °C, and 0.5 °C/min above 1310 °C.

Surface oxides produced during the anneal are then removed by HF to produce the final wafer structure of Figure 1F. The use of protective capping layers can reduce the amount of the DSB layer consumed by surface oxidation during the oxide dissolution anneal. Our standard capping layer, used as noted, comprised 200 nm of a low temperature oxide (LTO) deposited by low pressure chemical vapor deposition (SiH_4/O_2, 400 °C). Vendor-supplied DSB-A wafers with 2-3 nm of interfacial oxide as-received received essentially the same annealing treatments and capping layers as the DSB-B wafers fabricated in-house.

Interface Characterization

DSB-B wafers formed by the above method were characterized by optical reflectance, cross-sectional transmission electron microscopy (XTEM), and secondary ion mass spectroscopy (SIMS). Traces A and B of Figure 2 show the reflectance of two

DSB-B wafer pieces prior to an oxide dissolution anneal. Both trace A (for the initial, pre-anneal DSB layer thickness of 155 - 160 nm) and trace B (for a DSB layer thinned to the post-anneal thickness of 120 nm by reactive ion etching, RIE) show distinct oscillations in the wavelength range 400 to 900 nm, the result of interference between reflections from the wafer surface and reflections from the oxide layer at the DSB interface. After the oxide dissolution anneal, these oscillations are strongly attenuated (trace C) and the reflectance is very close to that seen for a bulk (100) Si wafer (trace D).

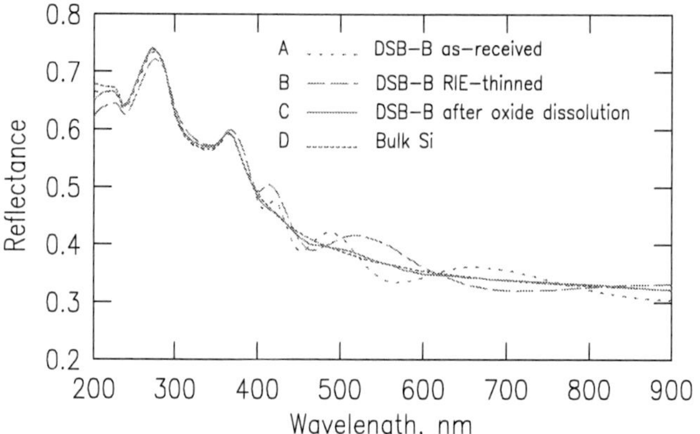

Figure 2. Reflectance vs. wavelength data for DSB-B and bulk Si substrates: DSB-B before oxide dissolution, (A) and (B); DSB-B after oxide dissolution annealing (C), and bulk Si (D).

Figure 3 shows XTEM images for the DSB-B wafer before (3A) and after (3B) a 1325 °C/5 hr oxide dissolution anneal. The 2-3 nm oxide layer is no longer detectable after annealing, though SIMS indicates a residual interfacial oxygen content of 3.5×10^{13} O cm^{-2}.

<u>Surface Oxidation</u>

At the high >1300 °C temperatures of our oxide dissolution anneals, trace amounts of O_2 in the gas ambient can lead to surface pitting, the result of uncontrolled SiO formation and desorption at the Si surface. To avoid this, small amounts of O_2 are deliberately added to the inert ambient to ensure the presence of a less-volatile and thermally stable SiO_2 surface layer. However, this introduces concerns about Si consumption.

Si consumption can be greatly reduced with the use of protective capping layers. After our standard 1325 °C/5 hr anneal, a DSB-B layer with an initial thickness of 155 - 160 nm was reduced to a thickness of 100-120 nm when no cap was used; with our standard LTO cap the Si thickness loss could be reduced to as little as 10 - 20 nm.

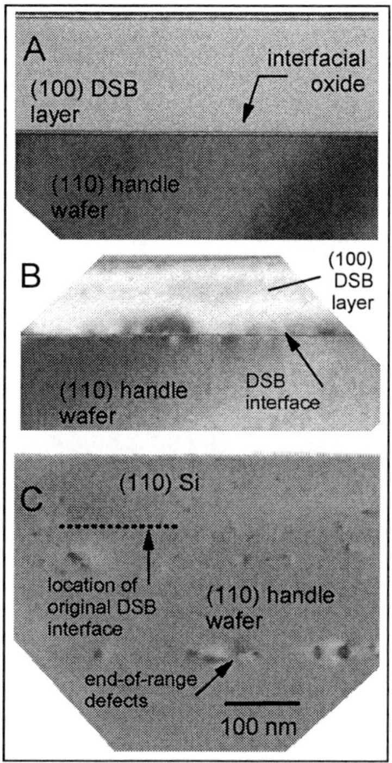

Figure 3. XTEM images of DSB-B samples (A) prior to oxide dissolution anneal, (B) after oxide dissolution anneal and surface oxide strip, and (C) after DSB layer orientation conversion by ATR.

Oxide Dissolution at Poly-Si/Si Interfaces

Oxide dissolution was also studied in poly-Si/Si interfaces, since samples of the structure Si(substrate)/SiO$_2$/poly-Si could readily be prepared with different interfacial oxide thicknesses. We were particularly interested in the evolution of interfacial oxide morphology with annealing, and the dependence of this evolution on the initial interfacial oxide thickness.

<u>Experimental Procedure</u>

Si wafers with a (100) orientation were cleaned with SC1/SC2 and then oxidized in a furnace (O$_2$/HCl ambient, 750 - 900 °C) to form gate-oxide-quality thermal oxide layers 3.5 to 15 nm in thickness. A thickness of 150-180 nm of poly-Si was deposited on the oxide layer by low pressure chemical vapor deposition (SiH$_4$, ~625 °C), followed by the standard 200 nm LTO cap layer. Oxide dissolution anneals were performed on samples with the LTO cap, samples without the LTO cap, and samples from which the LTO cap had been removed (by a wet etch in dilute HF).

Samples were examined by cross section scanning electron microscopy (XSEM) after removing any residual surface oxides with dilute HF. Samples were typically coated with Cr, and then etched after cleaving with either dilute HF (to highlight buried oxide regions) or with a Secco etch (to highlight grain boundaries and defects). Bragg-Brentano x-ray diffraction (XRD) was performed on selected samples to confirm the presence or absence of poly-Si grains.

Results and Discussion

Figure 4 shows optical reflectance traces vs. annealing for samples with a 3.5 nm interfacial oxide and no LTO cap. For the most part, the amplitude of the oscillations in traces A through E decreases with increasing annealing temperature, indicating a decreasing interfacial oxide thickness. Trace B, for the sample given the least aggressive anneal (1200 °C/15 min), is the one exception, showing oscillations that are even stronger than those in the as-deposited sample of trace A. While this might suggest an increase in the average oxide thickness (a possibility that cannot entirely be ruled out), it is more likely the result of changes in poly-Si layer absorption, since annealing reduces the poly-Si extinction coefficient k (a grain growth effect) as well as the poly-Si thickness (via surface oxidation). Oxide dissolution is not inhibited by the presence of the LTO cap layer, since the same interfacial-oxide-free reflectance was seen for 1300 °C ramp-annealed samples with and without the LTO cap.

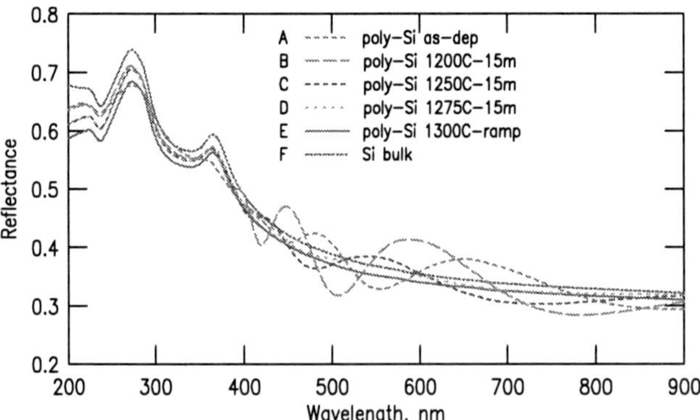

Figure 4. Reflectance vs. wavelength data for uncapped Si/SiO$_2$(3.5 nm)/poly-Si(160 nm) samples after the indicated oxide dissolution anneals, compared to data for bulk Si.

Figure 5 shows XSEM images of oxide layer morphology as a function of annealing temperature and initial oxide thickness. Initially, the interfacial oxide is continuous (Figure 5A). After annealing at 1200 °C for 15 min, the oxide has broken up into large patches separated by oxide-free regions (Figure 5B); after annealing at 1250 °C for 15 min, the patches have shrunk laterally and thickened up vertically (Figure 5C).

After annealing at 1275 °C for 15 min, the oxide patches are typically 0.6 to 1.0 µm apart and have the shape of slightly elongated spheroids with a width of 40 - 90 nm and a thickness of ~30 nm (Figure 5D). After the last anneal shown, a 1300 °C ramp, the oxide precipitates appear to be completely gone.

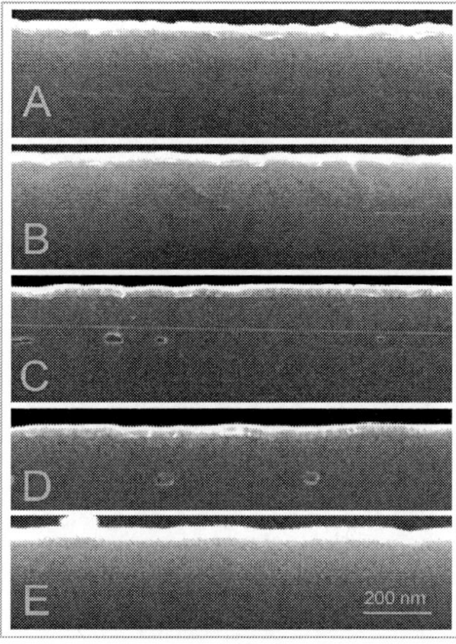

Figure 5. XSEM images (HF decoration) for the uncapped poly-Si samples of Figure 4: before annealing (A), and after anneals of 1200 °C/15 min (B), 1250 °C/15 min (C), 1275 °C/15 min (D), and 1300 °C ramp (E).

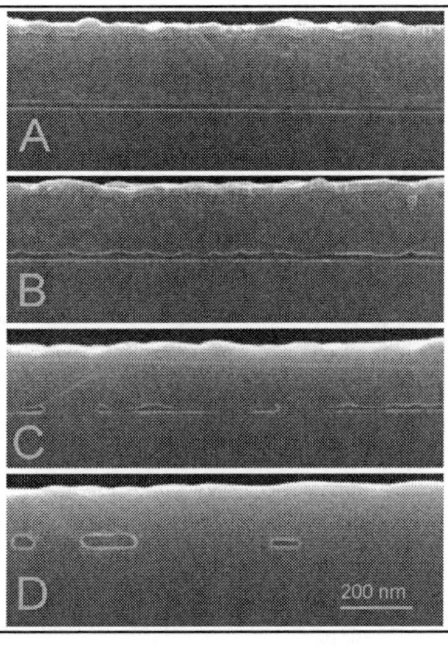

Figure 6. XSEM images (HF decoration) for capped poly-Si samples with different initial interfacial oxide thicknesses: a 15 nm interfacial oxide sample before (A) and after (B) a 1300 °C ramp anneal; 9.7 nm interfacial oxide (C) and 5.3 nm interfacial oxide sample after a 1300 °C ramp anneal.

Figure 6 shows XSEM images of interfacial oxide morphology after a 1300 °C ramp anneal for LTO-capped Si/poly-Si samples having initial oxide thicknesses of 5.3, 9.7, and 15 nm. The image of Figure 6A shows that the 15 nm oxide is smooth prior to any annealing. The oxide is still continuous after annealing, but that it has clearly roughened on the poly-Si side of the interface (Figure 6B). Comparing the images of Figures 6C (for the 9.7 nm oxide) and 6D (for the 5.3 nm oxide) to the images of Figure 5 for the 3.5 nm interfacial oxide, we see that after 1300 °C ramp annealing the 5.3 nm sample has a morphology similar to that seen in Figure 5B after 1200 °C/15 min annealing, while the 9.7 nm sample has a morphology similar to that seen in Figure 5C after 1250 °C/15 min annealing. This suggests that all thin oxides go through the same morphological stages during dissolution, but that the progression through the stages is slows with increasing oxide thickness.

Interfacial oxide at Si/poly-Si interfaces seems to disappear more quickly than comparably thick oxide at Si/Si interfaces, as can be seen from the XSEM images of Figure 7 for a no-cap DSB-B sample before and after the same 1300 °C ramp anneal that was sufficient to remove a 3.5 nm interfacial oxide in the Si/poly-Si samples. Before annealing (Figure 7A), the oxide at the Si/Si interface is present as a continuous layer; after annealing (Figure 7B), the oxide is still there, though it has started breaking up into discontinuous patches with balled-up edges.

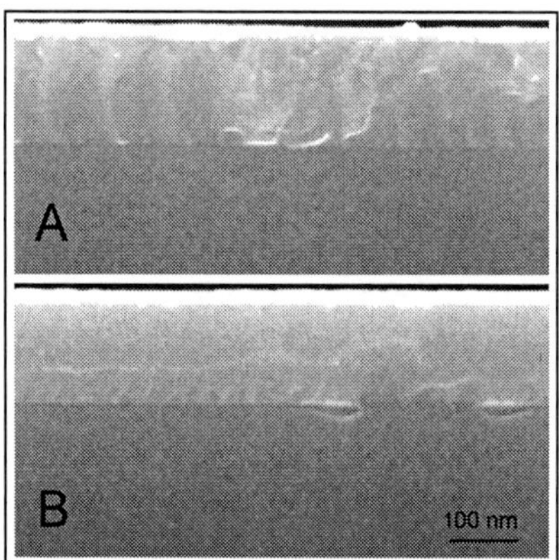

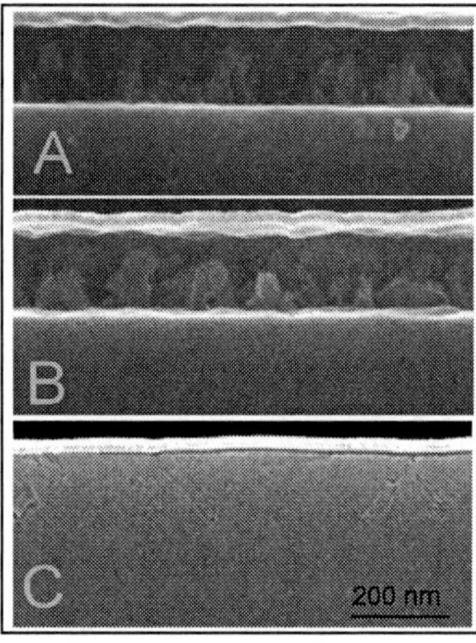

Figure 7. XSEM images (HF decoration) for uncapped DSB-B samples before (A) and after (B) a 1300 °C ramp anneal.

Figure 8. XSEM images (Secco decoration) of capped poly-Si samples: 15 nm interfacial oxide samples before (A) and after (B) a 1300 °C ramp anneal; a 3.5 nm interfacial oxide sample after the same 1300 °C ramp anneal (C).

The one-sidedness of the oxide roughening seen in the 15-nm-oxide Si/poly-Si sample of Figure 6B (i.e., roughening at the oxide/poly-Si interface but not the Si/oxide interface) suggests a reason for the faster dissolution in the poly-Si samples: oxide layer breakup (an essential component of the dissolution process) occurs more easily at interfaces of poly-Si/Si than Si/Si. This easier breakup probably results from the fact that the required atomic diffusion near the oxide interface can occur more readily in poly-Si than in single-crystal Si. In particular, Si diffusion related to poly-Si grain growth might make it easier for the Si in the poly-Si to conform to the changing shape of the interfacial oxide. In addition, grain boundaries in the poly-Si might be expected to provide a fast diffusion path for O.

An even more interesting observation is apparent from the three Secco-highlighted XSEM images of Figure 8. Poly-Si layers having oxide-free interfaces with Si can undergo an orientation-changing grain growth to produce a single crystal material indistinguishable from the base substrate. Figures 8A and 8B compare capped poly-Si layers deposited on a 15 nm thick interfacial oxide before and after a 1300 °C ramp anneal. As expected, annealing increases grain size, but leaves the interfacial oxide layer intact. The grain size increase is also suggested by a doubling of the Si (111) XRD peak intensity. This contrasts with the image of Figure 8C for a similarly capped poly-Si layer on a 3.5 nm interfacial oxide after the same 1300 °C ramp anneal. As expected from the HF-highlighted XSEM image of Figure 5E, the interfacial oxide is gone. However the grain boundaries have disappeared, stacking faults are visible, and the XRD spectra no longer show the presence of any poly-Si peaks (despite the consumption of only 10 nm of Si, as determined from the optically-measured increase in the oxide cap layer thickness). Since the only difference between the samples of Figures 8B and 8C is the interfacial oxide thickness, all evidence points to a conversion of the poly-Si layer into single-crystal Si. However, such a conversion is not entirely unexpected, given reports that DSB-layer islands surrounded laterally by ATR'd material having the orientation of the substrate can undergo a spontaneous conversion from their original orientation to the orientation of the substrate. The exact reasons for these instabilities are not entirely clear, but a likely explanation is that conversion to the substrate orientation produces an energetically favorable reduction in Si grain boundary interface area.

Mechanisms

Literature data on the diffusivity and solubility of O interstitials (O_i) in Si allows one to estimate the maximum "dissolvable thickness" (X^{diss}_{ox}) of oxide that might be expected for low-O_i content (FZ) Si wafers annealed at a given temperature and time. With the diffusivity of O_i in Si given by

$$D_i(T) = 0.07 \exp(-2.44 \text{ eV}/ kT) \text{ cm}^2 \text{ s}^{-1}, \tag{1}$$

where k is the Boltzmann constant and T is the absolute temperature, and the solubility of O_i in Si given by

$$C^{eq}_i (T) = 1.53 \times 10^{21} \exp(-1.03 \text{ eV}/ kT) \text{ cm}^{-3}, \tag{2}$$

one obtains (8)

$$X^{diss}_{ox} (t, T) = (D_i(T) \, t /\pi)^{0.5} [4 \, C^{eq}_i(T)/n_{ox}], \tag{3}$$

where t is the time and $n_{ox} = 4.4 \times 10^{22}$ cm^{-3} is the concentration of oxygen in SiO_2. Equation [3] predicts that 2-3 nm of interfacial oxide should completely dissolve in 4 - 9 hr at 1325 °C, in reasonable qualitative agreement with our observations. However this simple analysis is not strictly applicable to our situation, since the O_i concentration in our Cz wafers is already fairly close to the O_i solubility limit (as C^{eq}_i at 1325 °C is about 9 × 10^{17} cm^{-3}) and would therefore be expected to significantly reduce the oxide dissolution rate. In addition, the situation is complicated by the presence of the SiO_2 capping layer (present initially, or formed by reaction with the O_2 in the ambient) and the presence of O_2 in the annealing ambient.

For the purposes of the present discussion, a qualitative understanding of buried oxide dissolution is sufficient. Only two factors need to be considered for structures with the geometry of Figure 9, comprising a planar buried oxide layer sandwiched between an underlying Si substrate and an SOI/DSB overlayer capped by a superficial thermal oxide. First, all planar Si/SiO_2 interfaces reflect the balance in the competition between oxide growth and oxide dissolution. Oxide growth results when new SiO_2 (produced by interfacial reactions of Si with oxygen interstitials and/or with O that has diffused through the oxide layer) accumulates at the Si/SiO_2 surface faster than bonded O in the oxide layer can be dissolved back into the adjacent Si. The balance between the forward and backward reactions depends on temperature and O_2 concentration in the ambient. The oxide growth regime for buried oxide layers has been well studied in connection with SIMOX wafer fabrication (7) and is known as ITOX (internal oxidation); the oxide dissolution (or "inverse ITOX") regime has also been observed for annealing at similar temperatures but with a lower concentration of O_2 in the ambient (12).

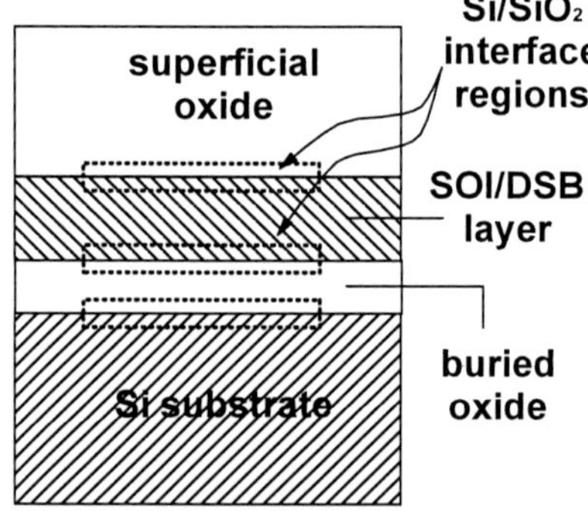

Figure 9. A schematic of buried oxide dissolution in a model substrate. The dotted boxes highlight the Si/SiO_2 interfaces that can move as the SiO_2 layer grows or dissolves.

The second factor to be considered is the phenomenon of oxide spheroidization (9) or "balling up." The high surface energy of the Si/SiO_2 interface makes thin interfacial oxides unstable with respect to breaking up and reforming at the Si/cap layer interface. In principle, this redistribution should occur without any net gain or loss of oxide volume. However, sufficient O diffusivity requires high temperatures, typically (but not always) in ambients that include oxygen in trace amounts. As a result, the disappearance of buried oxide layers would typically be accompanied by superficial oxide growth in excess of the amount contributed by the buried oxide layer. Interestingly, Si/SiO_2 interface energies would be expected to be crystal-orientation dependent, an effect that has been used to explain the experimentally observed dependence of oxide dissolution rates on Si bonding twist angle (8).

Applications for DSB Wafers

In this section we will review how orientation-changing ATR may be used with DSB wafers to fabricate hybrid orientation substrates for device applications.

<u>ATR on DSB wafers formed by quasi-hydrophobic bonding</u>

The DSB interfaces produced by this quasi-hydrophobic bonding method are clean enough to allow orientation conversion by ATR. The DSB-B sample of Figure 3B was subjected to an implantation of Ge^+ ions at a dose of 2×10^{15} cm^{-2} and an energy of 220 keV to produce an amorphous Si (a-Si) layer whose 270 nm thickness extended past the DSB interface located approximately 120 nm below the top surface of the DSB layer. After recrystallization by annealing at 900 °C for 1 min, it can be seen from Figure 3C that the DSB layer has the orientation of the handle wafer; end-of-range defects can be seen near the expected position of the a-Si/Si boundary, but there is no sign of the DSB interface.

It should be noted that ATR-DSB processing is preferably performed with DSB-A wafers (with a (110)-to-(111) conversion via (100) solid phase epitaxy) rather than with DSB-B wafers (with a (100)-to-(110) conversion via (110) solid phase epitaxy) because (100) solid phase epitaxy is faster and less defective. In addition (as discussed below), device applications require the ATR to be in selected areas only, rather than the maskless (or blanket) ATR of illustrated in Figure 3A.

<u>Patterned ATR on DSB for hybrid orientation device applications</u>

Figure 10 shows a preferred arrangement of n-channel field effect transistors (nFETs) and p-channel FETs (pFETs) on a DSB-A wafer; the nFETs are on (100) Si and the pFETs are on (110) Si. For device applications, ATR must be in selected areas only (rather than blanket or maskless).

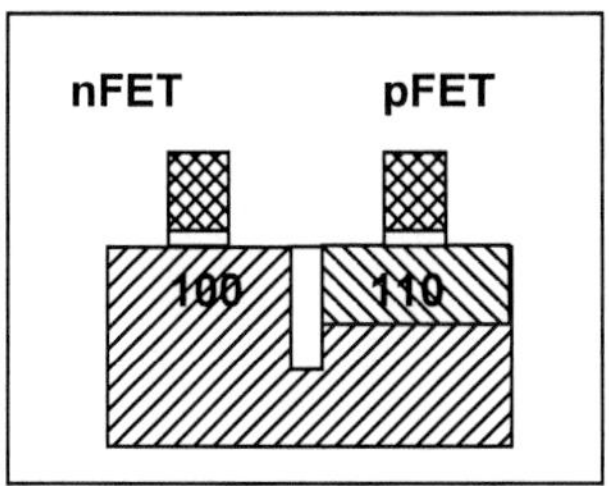

Figure 10. The preferred arrangement of nFETs and pFETs on a hybrid orientation substrate.

Figure 11 shows two integration approaches to using ATR with DSB substrates. In the "ATR first" approach (upper half of Figure 11), the ATR is performed before shallow trench isolation (STI). In the "STI first" approach (lower half of Figure 11), the STI trenches are formed prior to ATR. A priori, the STI-before-ATR approach might appear preferable because the trenches provide alignment marks (a necessity, since the differently-oriented Si regions are optically indistinguishable) and eliminate the possibility of lateral templating. However the ATR-before-STI approach eliminates concerns about trench-edge defects (13).

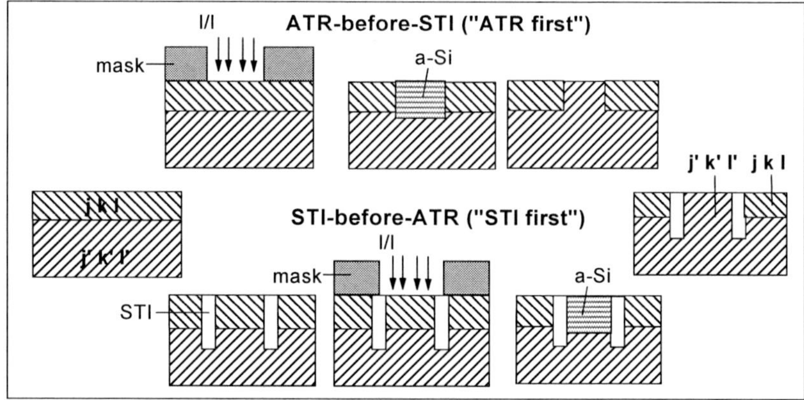

Figure 11. Two integration approaches for patterned ATR.

Results for complementary metal oxide semiconductor (CMOS) circuits with the geometry of Figure 10, the ATR-before-STI process flow, and Lpoly = 45 nm showed good performance (14). In particular, pFET off current (Ioff) vs. saturation current (Ion) curves were found to be the same in both the original (110) DSB layers and (110) bulk controls, implying that the DSB layer was not degraded by processing, and that the bonded interface did not interfere with the device. nFET performance characteristics such as Ion vs. Ioff and reverse source/drain junction leakage current were found to be the same in changed-orientation (100) ATR'd layers, (100) bulk controls, and (100) bulk wafers undergoing ATR with no orientation change (i.e., amorphization followed by recrystallization back into the substrate's original (100) orientation), implying no degradation from defects generated by the ATR process. Ring oscillators fabricated on these same wafers showed improvements of more than 20% compared with the same devices on the (100) control wafers.

Conclusions

We have demonstrated that interfacial oxides of 2-3 nm in thickness at bonded interfaces between Si (100) and Si (110) may be dissolved away by 1325 $^{\circ}$C annealing to produce direct-silicon-bonded (DSB) wafers suitable for the use with the amorphization/templated recrystallization (ATR) technique fabricating hybrid orientation bulk silicon substrates. Our experiments with oxide dissolution at Si/poly-Si interfaces showed that oxide dissolution is faster than with Si(100)/Si(110) interfaces, and that poly-Si on Si (100) is thermally unstable with respect to conversion to Si (100).

Acknowledgments

Paul Ronsheim is thanked for the SIMS; L. Shi and the Microelectronics Research Laboratory staff are thanked for their contributions to sample preparation.

References

1. K.L. Saenger, J.P. de Souza, K.E. Fogel, J.A. Ott, A. Reznicek, C.Y. Sung, D.K. Sadana, and H. Yin, Appl. Phys. Lett., **87**, 221911 (2005).
2. K.L. Saenger, J.P. de Souza, K.E. Fogel, J.A. Ott, A. Reznicek, C.Y. Sung, H. Yin, and D.K. Sadana, in *Transistor Scaling–Methods, Materials and Modeling, edited by S. Thompson, F. Nouri, W. Tsai, W-C. Lee (Mater. Res. Soc. Symp. Proc.* 913, Warrendale, PA, 2006), C1.1.
3. Q.-Y. Tong and U. Gösele, *Semiconductor Wafer Bonding: Science and Technology*, John Wiley (New York, 1999).
4. Q.-Y. Tong, Q. Gan, G. Hudson, G. Fountain, and P. Enquist, R. Scholz and U. Gösele, Appl. Phys. Lett., **83**, 4767 (2003).
5. Y. Li, J.A. Kilner, R.J. Chater, A. Nejim, P.L.F. Hemment, C.D. Marsh, and G.R. Booker, Appl. Phys. Lett., **63**, 2812 (1993).
6. S. Nakashima, T. Katayama, Y. Miyamura, A. Mattsuzaki, M. Kataoka, D. Ebi, M. Imai, K. Izumi, and N. Ohwada, J. Electrochem. Soc., **143**, 244 (1996).
7. P. McCann, S. Byrne, and W.A. Nevin, *Semiconductor Wafer Bonding Science: Technology and Applications*, H. Baumgart, C. E. Hunt, eds., Proceedings Volume 2001-27, 106 (The Electrochemical. Society, Pennington, 2001).
8. K.-Y. Ahn, R. Stengl, T.Y. Tan, U. Gosele, and P. Smith, J. Appl. Phys., **65**, 561 (1989).
9. K.-Y. Ahn, R. Stengl, T.Y. Tan, U. Gosele, and P. Smith, Appl. Phys. A, **50**, 85 (1990).
10. L. Ling and F. Shimura, J. Electrochem. Soc., **140**, 252 (1993).
11. W. Kern, J. Electrochem. Soc., **137**, 1887 (1990).
12. J.P. de Souza, unpublished.
13. K.L. Saenger, J.P. de Souza, K.E. Fogel, J.A. Ott, C.Y. Sung, and D.K. Sadana, J. Appl. Phys., **101**, 024908 (2007).
14. C.-Y. Sung et al, IEDM Tech. Dig. 224 (2005); H. Yin et al, IEDM Tech. Dig. Paper 3.5 (2006), in press.

ECS Transactions, 6 (4) 309-313 (2007)
10.1149/1.2728875, ©The Electrochemical Society

Fabrication of SOI MOSFET by "Separation by Bonding Silicon Islands (SBSI)" method

K. Kanemoto, H. Oka, H. Hisamatsu, Y. Matsuzawa, Y. Kitano, T. Hara, M. Hoshina, S. Ohmi* and J. Kato

Seiko Epson Corporation, Fujimi Plant, 281 Fujimi, Fujimi-machi, Suwa-gun, Nagano-ken, 399-0293, Japan
*Interdisciplinary Graduate School of Science and Engineering, Tokyo Institute of Technology, J2-72, 4259 Nagatsuta, Midori-ku, Yokohama-shi, Kanagawa-ken, 226-8502, Japan

Local SOI MOSFETs were fabricated in the designed area on a Si bulk substrate using the separation by bonding silicon islands (SBSI) method. We describe the key fabrication process technology and show the electrical characteristics. The SBSI method consists of three key technologies: epitaxial growth of SiGe having high germanium concentration, lateral selective etching of SiGe, and bonding of buried oxides (BOX) grown in the gap. Epitaxial growth of 30 nm-thick $Si_{0.63}Ge_{0.37}$ without crystalline defects was achieved by reducing the growth temperature to as low as 450 °C. The $Si/Si_{0.63}Ge_{0.37}$ structure allowed lateral etching of $Si_{0.63}Ge_{0.37}$ with a length of 1 μm in 2 minutes and with negligible etching of Si in a solution of hydrofluoric acid and nitric acid. Fabricated MOSFETs showed excellent characteristics, which were comparable to those of SOI wafers, although a small gap still remained in the BOX.

Introduction

MOSFETs fabricated on silicon on insulator (SOI) substrates have attracted much attention for their superior characteristics, i.e. higher speed operation and lower power consumption, than those on bulk substrate. In spite of their excellent characteristics, however, SOI substrates are not widely utilized but limited to regions such as microprocessor. Major reasons for this are that SOI wafer costs much higher than bulk silicon wafer, and that useful intellectual properties (IP) for conventional bulk Si such as electrostatic discharge (ESD) circuits cannot be easily applied to SOI structures. It is important to inexpensively utilize both advantages of SOI structure IP and bulk Si IP. Local SOI structure on bulk Si substrate solves this problem.

Recently, a novel method to fabricate an SOI structure has been proposed [1]. This method is called "Separation by Bonding Silicon Islands (SBSI)" and provides local SOI structures in the designed area on a silicon substrate by epitaxial growth of Si/SiGe stacked layers and selective etching of SiGe layer, and oxidation to bond the upper silicon island to the substrate by filling the gap with silicon oxide layers. The advantage of this method is that SOI structures and isolation regions are able to be fabricated simultaneously on bulk silicon wafers with inherent high controllability of the thickness of SOI and buried oxide (BOX).

In this paper, we report for the first time the characteristics of an SOI MOSFET fabricated by the SBSI method. The property of the epitaxially grown SiGe layer with a relatively high concentration of germanium, and selective etching of SiGe are also shown.

Fabrication Process

Figure 1 shows the fabrication process of the SBSI-MOSFET. Firstly, the SiGe and Si layers are epitaxially grown on a Si (100) substrate in an ultra-high vacuum chamber using Si_2H_6 and GeH_4. The thicknesses are 30 nm and 100 nm, respectively. Germanium concentration in the SiGe is 37 %. Using mask pattern A, the Si/SiGe layers are dry-etched (Figs. 1 (a) and (b)). SiO_2 with a thickness of 400 nm is deposited by plasma-enhanced chemical vapor deposition (PECVD) using tetraethoxysilane (TEOS) and O_2 as source gases. SiO_2/Si/SiGe is then dry-etched using pattern B (Figure 1 (c)). SiGe is selectively etched in the directions indicated by the arrows in Figure 1 (d) by dipping the wafer in a solution of hydrofluoric acid and nitric acid. The Si/gap/substrate structure is maintained with support from the pattern-B SiO_2 (Figure 1 (e)). The gap is filled with thermal oxide layers, which are grown on the silicon substrate and on the underside of the epitaxially grown Si. The isolation region is formed by PECVD-TEOS and chemical mechanical polishing (CMP) (Figure 1 (f)). After that, MOSFETs are fabricated by a simplified conventional CMOS process (Figure 1 (g)). Channel-dopes are performed by ion implantation of BF_2^+ (25 keV, 2.1e+12 cm^{-2}) for nMOS and P^+ (25 keV, 2.1e+12 cm^{-2}) for pMOS. Gate oxide thickness is 10 nm. Source and drain are formed by P^+ (10 keV, 2e+15 cm^{-2}) for nMOS and B^+ (4 keV, 2e+12 cm^{-2}) for pMOS, followed by furnace anneal at 850 °C. Silicides are not formed. After deposition of a 200 nm-thick PECVD-TEOS and formation of the contact opening, aluminum interconnects and pads are formed. For comparison of electrical characteristics, MOSFETs are also fabricated on the conventional SOI wafers. The initial thicknesses of SOI and BOX were 100 nm and 200 nm, respectively. After thinning the SOI layer to 60 nm by sacrificial oxidation and etching, mesa isolation was formed. The rest of the fabrication process is the same as the SBSI-MOSFET formation process.

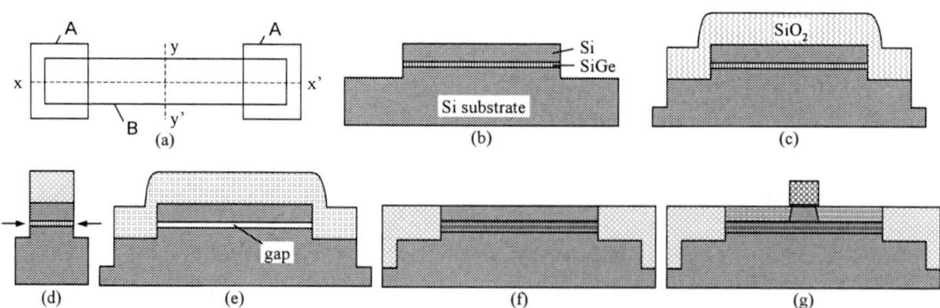

Figure 1. Schematics of SBSI-MOSFET fabrication process; (a) Mask patterns A and B used in the process, (b) epitaxial growth of Si/SiGe layers and trench formation using pattern A (x-x' cross section), (c) SiO_2 deposition and dry etching of SiO_2/Si/SiGe layers using pattern B, (d) y-y' cross section of (c), (e) selective etching of SiGe layer, (f) BOX forming oxidation, SiO_2 deposition and CMP, (g) gate oxide, gate electrode, source and drain formation.

Results and Discussion

<u>Epitaxial Growth of Silicon Germanium</u>

In order to obtain a SiGe with a higher etching selectivity than that of Si, enabling a larger SOI area in the SBSI method, SiGe of a higher germanium concentration is required (shown below). On the other hand, in order to obtain small parasitic capacitance, which is a major feature of the SOI structure, a thick BOX is preferable. In the SBSI method, BOX thickness is determined by the SiGe thickness. Therefore, epitaxial growth of SiGe with the necessary thickness and germanium concentration is required in accordance with the design of the SOI MOSFET.

We have succeeded in epitaxial growth of a 30 nm SiGe that has a germanium concentration of up to 37 %, a smooth surface and no crystalline defects, which was confirmed with a scanning electron microscope (SEM) and a dark-field optical microscope. This is achieved by decreasing the growth temperature to as low as 450 °C while sacrificing the growth rate, which resulted in 1.5 nm/min. Crystalline integrity is analyzed also by x-ray diffraction (XRD). Table I shows lattice constants of Si and SiGe in a 30 nm-$Si_{0.63}Ge_{0.37}$/Si (100) substrate structure. Lattice constant of SiGe in horizontal axis has the same value as a Si substrate of 0.543 nm, and that in the vertical axis has a larger value (2.6 %) than that of Si substrate. This means that the ideal hetero-epitaxial growth is performed. We have achieved not only epitaxial growth of a 30 nm $Si_{0.63}Ge_{0.37}$ but also of Si/$Si_{0.63}Ge_{0.37}$ stacked layers. The growth rate of Si is smaller than that of SiGe, which becomes as small as 0.04 nm/min at 450 °C (obtained from the extrapolation of the arrhenius plot of growth rate). Practically, it is necessary that the growth temperature of Si is increased from that of SiGe. However, crystalline defects such as dislocations tend to occur, probably due to releasing the strain in SiGe. It was confirmed that, no crystalline defects occurred and the growth rate became up to 9.2nm/min at growth temperature of 600 °C for Si.

Table I. Lattice constants of Si and SiGe in a 30 nm-$Si_{0.63}Ge_{0.37}$/Si (100) substrate structure measured by XRD

	Lattice constant [nm]	
	Horizontal axis	Vertical axis
Si (Substrate)	0.543	0.543
$Si_{0.63}Ge_{0.37}$	0.543	0.557

<u>Selective Etching of Silicon Germanium</u>

SiGe can be selectively etched by a solution of hydrofluoric acid and nitric acid. Figure 2 (a) shows the etching rate of SiGe as a function of germanium concentration. The values indicated by the blank symbols are obtained by measuring the change of the SiGe thickness by dipping in a solution. The filled symbols are obtained by measuring the lateral etching length from the SEM image in the Si/SiGe structure. As can be seen, the etching rate is enhanced as the germanium concentration increases. At a germanium concentration of 54 %, indicated by the filled triangle, an etching length of 2.0 μm was obtained in 2 minutes. However, many dislocations were observed in the epitaxial layers. Figure 2 (b) is an SEM image after 2 minutes of etching of the Si/$Si_{0.63}Ge_{0.37}$ structure. The $Si_{0.63}Ge_{0.37}$ is etched and a gap of about 1.1 μm in length is formed with negligible Si

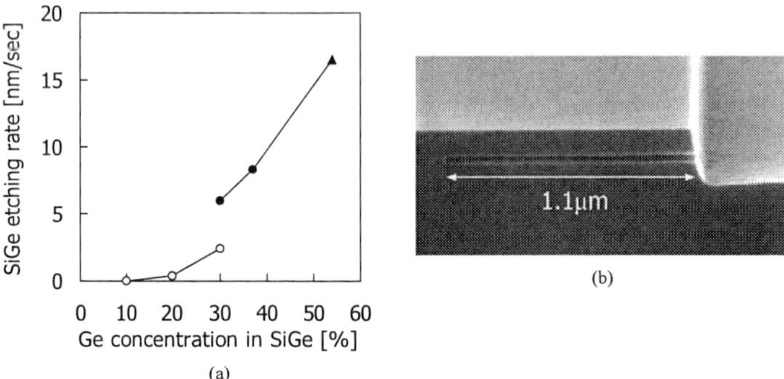

Figure 2. (a) Etching rate of SiGe as a function of germanium concentration. (b) SEM image of Si/Si$_{0.63}$Ge$_{0.37}$ structure after 2-minutes of etching.

etching. Since lateral SiGe etching proceeds from both sides in the SBSI method as shown in Figure 1 (d), an SOI structure having a width double that of the etching length can generally be formed.

<u>MOSFET Characteristics</u>

Figure 3 shows a cross-sectional transmission electron microscope (TEM) image of the SBSI-MOSFET. MOSFET is fabricated on SOI (~60 nm)/BOX (~100 nm). It is found that a small gap remains in the BOX. Though it is speculated, after review of several reports on SON structure MOSFET [2, 3], that the remaining gap does not cause severe influence to the MOS characteristics, perfect BOX bonding is preferable. From the TEM image, it is also confirmed that no crystalline defects occurred in the SOI region.

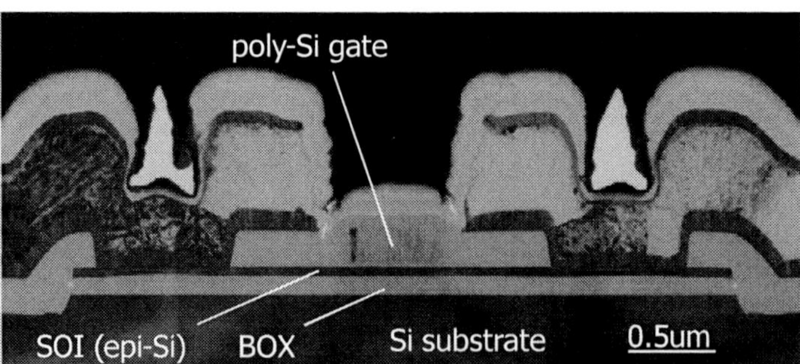

Figure 3. Cross-sectional TEM image of SBSI-MOSFET.

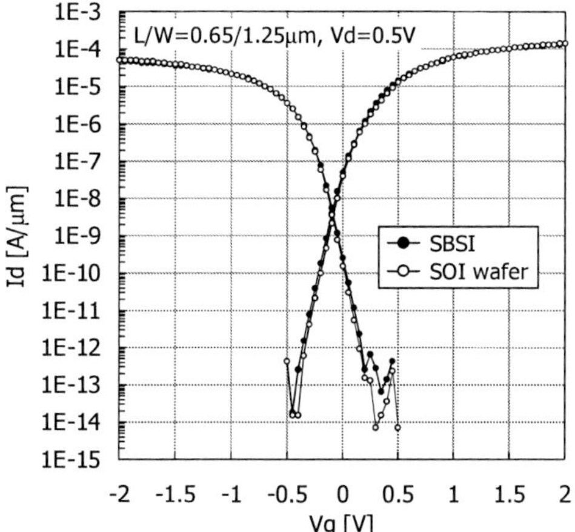

Figure 4. V_g-I_d characteristics of SBSI-MOSFET and an SOI wafer MOSFET.

Figure 4 shows V_g-I_d characteristics of the SBSI-MOSFET. Gate length and width are 0.65 μm and 1.25 μm, respectively. The channel doping conditions mentioned above used for a 60 nm-SOI achieve fully-depletion (FD) in the channel region. As can be seen, excellent characteristics comparable to those of MOSFET on conventional SOI wafers are obtained.

The property of BOX is investigated by measuring the breakdown voltage (V_{bd}) between the drain and substrate. A V_{bd} exceeding around 100 V, which corresponds to 10 MV/cm (V_{bd} divided by BOX thickness of 100 nm), is obtained.

Summary

An SBSI-MOSFET is fabricated on the basis of SiGe epitaxial growth technology and SiGe selective etching technology. Excellent MOSFET characteristics which are comparable to the device fabricated on conventional SOI wafer are achieved for the first time.

References

1 T. Sakai et al., 2nd Intl. SiGe Technology and Device Meeting, Meeting Abstract, p. 230 (2004).
2 M. Jurczak et al., Symp. VLSI Tech. Dig., p. 29 (1999).
3 T. Sato et al., IEDM Tech. Dig., p. 809 (2001).

ECS Transactions, 6 (4) 315-320 (2007)
10.1149/1.2728876, ©The Electrochemical Society

Epitaxial Regrowth of Ge on SGOI and GeOI Substrates
Obtained by Ge Condensation

J-F. Damlencourt[1], Y. Campidelli[2], M-C. Roure[1], B. Vincent[1], E. Martinez[1], F. Fillot[1], Y. Morand[2], B. Arrazat[1], T. Nguyen[3], S. Cristoloveanu[3] and L. Clavelier[1]

[1] CEA-LETI, Minatec / D2NT & DPTS, 17 rue des Martyrs, 38054 Grenoble, France
[2] STMicroelectronics, 350 avenue J. Monnet, 38926 Crolles Cedex, France
[3] IMEP-INP Grenoble-Minatec, BP 257, 38016 Grenoble Cedex 1, France

In this paper, we have investigated the influence of process parameters (bake temperature, bake duration and growth temperature) on the morphology of GeOI substrate thickened by Ge epitaxial regrowth. We have seen that the optimum bake temperature in terms of layer roughness and interfacial contamination is 600°C. For higher bake temperatures, the layer morphology is 3D due to a layer dewetting. The efficiency of the surface preparation prior to the epitaxial regrowth has been assessed by Angle resolved X-Ray Photoelectrons Spectroscopy. No residual oxygen contamination has been detected. Furthermore, we have investigated the influence of the growth temperature on the layer morphology and resistivity. For temperatures higher than 450°C, some thermal grooving assisted by the crystal defects (cross hatch) occurred and 3D Ge layers were obtained. Finally, GeOI substrates thickened with the best growth condition were evaluated by Pseudo MOSFET technique. An exceptional high hole mobility, beyond 400 $cm^2/V.s$ was observed.

Introduction

The challenges imposed by the scaling of Si devices make mandatory the study of new materials to overcome the physical limitations of Si. Ge was actually used in the very first transistors but was then abandoned in favour of Si due to the instability of Ge oxide. However, the recent introduction of high-K gate dielectrics makes the use of Ge possible in an advanced technology (1, 2). Its benefits for MOSFET applications are important: better transport properties than silicon hence higher saturation currents; lower band gap enabling lower supply voltages and lower power dissipation; finally its lattice parameter is compatible with GaAs for mixed circuit.

Germanium On Insulator (GeOI) substrates benefit from both Ge properties and "On Insulator" advantages for microelectronic applications: better electrostatic control i.e. less short channel effects, reduced junction capacitances and lower substrate coupling in RF circuits. Another main advantage of GeOI substrates is the limited quantity of Ge used to obtain such a substrate but also its mechanical behavior, which is close to the one of a Si wafer.

Two fabrication methods for GeOI and SGOI (Silicon Germanium On Insulator) substrates are used: the SMARTCUT[TM] process and the Ge condensation method (3-6). The latter technique consists in a Si selective oxidation done after a SiGe layer epitaxial growth on a SOI wafer. The SiGe layer is initially grown with a low Ge content. Its thickness decreases during the oxidation due to the Si consumption. During this Si selective oxidation performed under dry O_2 at high temperature, Ge pills up at the oxide interface. Competitively due to the high temperature used, Ge diffuses through the top silicon layer of the SOI substrate and then through the remaining SGOI layer (7). Ge diffusion and SiGe thickness decrease involve a Ge content enrichment. Ultra thin SGOI layers with enrichments up to 100% can be achieved after long enough oxidation times.

Today, ultra thin GeOI are not adapted for MOSFET fabrication due to a Ge consumption which may occur during wet processes. Therefore, the capability to regrow Ge layer on SGOI and GeOI substrates is an important issues and motivates our work. We focused especially on the influence of process parameters on the layer qualities.

Experimental details

The SGOI prestructure for condensation has been prepared by growing a $Si_{0.9}Ge_{0.1}$ layer on a SOI wafer with an initial 20nm top Si thickness. Growth has been performed at 650°C, by Reduced Pressure - Chemical Vapor Deposition under 20 Torr. The SiGe thickness target is 75nm. This thickness is lower than the critical thickness for plastic relaxation (8). A silicon cap, 2nm thick, is added to the structure to prevent any GeO_x formation during the early oxidation stages. The cap is oxidized at a low temperature (700°C) to form an initial SiO_2 layer, acting as an upper Ge diffusion barrier. Oxidations have been done under dry oxygen using a two steps temperature process. The first temperature used is 1050°C involving a high oxidation rate. 1050°C being the melting temperature of the $Si_{0.35}Ge_{0.65}$ alloy, a lower temperature step at 900°C is added to reach Ge content higher than 65%. Enrichments up to 100% are obtained with the second step. In order to favor Ge diffusion and get highly homogeneous Ge profiles within the SGOI layers obtained, some annealing steps are added periodically all along the process (9, 10).

The silicon dioxide formed during the oxidation is removed in a concentrated HF solution (10%). The wafers are then loaded in the RPCVD epi tool and a 40 nm thick Ge layer is deposited on a SGOI substrates. The global and local roughness are respectively assessed by laser scattering on a SP2UV (KLA-Tencor™) and by Atomic Force Microscopy (AFM) imaging. The interfacial contamination is analyzed both by Secondary Ions Mass Spectroscopy (SIMS) and Angle Resolved X-Ray Photoelectron Spectroscopy (AR-XPS).

Results and discussions

<u>1-Influence of bake conditions :</u>

The influence of the H_2 bake temperature as well as the bake duration has been investigated. Figure 1 depicts the influence of the bake temperature on the layer's morphology. The growth is 2D for low bake temperatures; a cross-hatch pattern is observed on the layers AFM pictures due to the strain relaxation occurring during the condensation. For temperatures higher than 700°C, the growth shifts in a 3D mode due to the layer dewetting during the H_2 bake. This phenomenon has already been observed on ultra thin SOI (11). Dornel *et al.* explained that dewetting generally occurs on patterned SOI wafers and is initiated at the edge of the pattern (12). In our case, the dewetting is activated on full sheet wafers; we assumed that the cross hatch pattern is the starting point of this phenomenon.

Figure 2 shows the evolution of the RMS roughness extracted from 5x5 μm^2 AFM pictures as a function of the bake temperature. Due to the 2D-3D morphology shift, the local roughness obviously increases with the bake temperature. Focusing on the low bake temperature area (figure inset) the RMS roughness reaches a minimum value for a bake temperature of 600°C. The same behavior is observed on global roughness measurements performed with SP2 tool (not presented here). The optimum temperature is then determined by using this criterion.

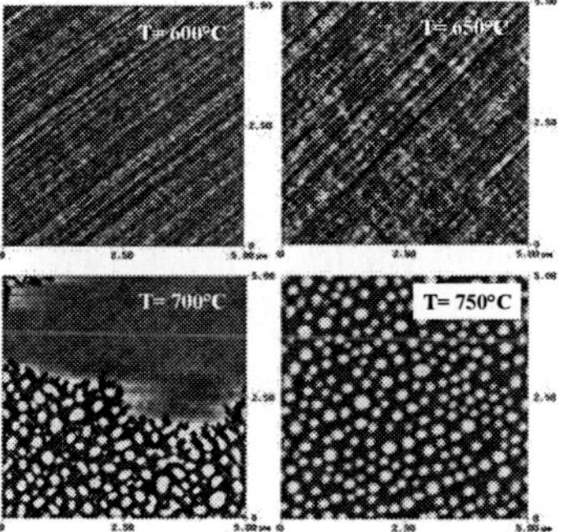

Figure 1: 5x5 μm^2 AFM pictures of the GeOI layers as a function of the bake temperature.

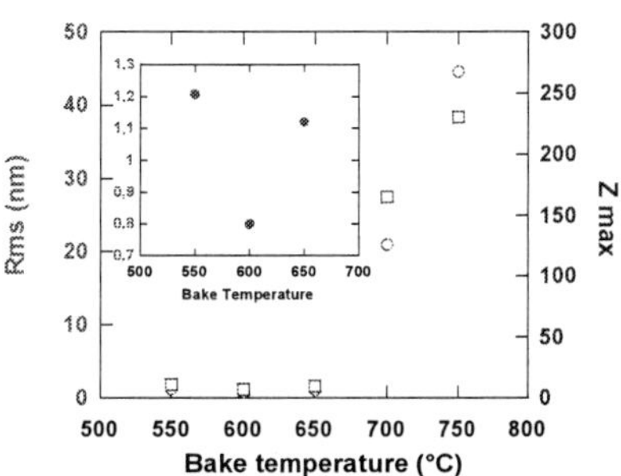

Figure 2: Rms Roughness (red circles) and Zmax in nm (blue squares) extracted from AFM pictures of the GeOI layers as a function of the bake temperature

The efficiency of the cleaning procedure prior the epitaxial regrowth (i.e. HF treatment and bake condition) has been assessed by AR-XPS at miscellaneous take-off angles (the lower the take-off angle is, the more sensitive to the surface the analysis is). Prior to this measurement, a 2nm thick poly-silicon layer is deposited on the wafers after the H_2 bake. This layer protects the surface from non intentional oxygen contamination.

Figure 3 and 4 depict respectively the Si 2p and Ge 3d core level spectra obtained on the GeOI wafer at a take-off angle of 20°. On one hand, two components were used to fit the Si 2p spectrum.

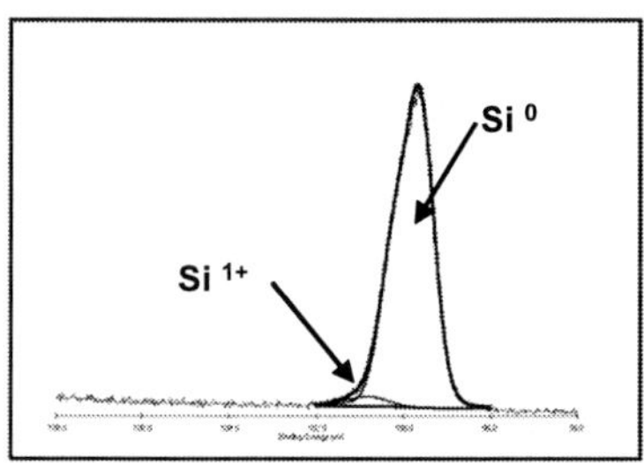

Figure 3: Si 2p core level spectrum obtained by AR-XPS at a take-off angle of 20° on a GeOI wafer baked under H_2 at 600°C and capped with a 2nm thick pol-silicon layer.

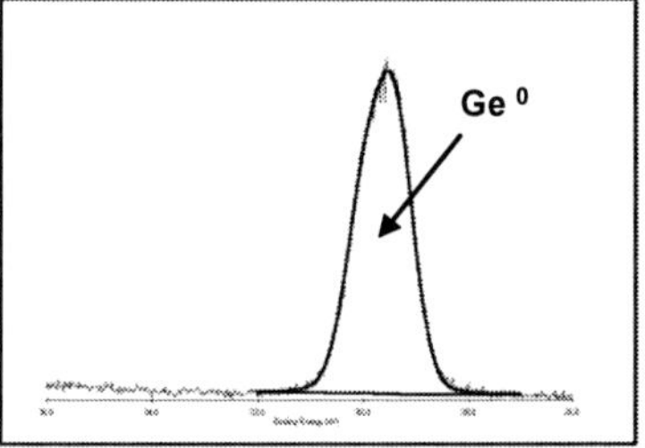

Figure 4: Ge 3d core level spectrum obtained by AR-XPS at a take-off angle of 20° on a GeOI wafer baked under H_2 at 600°C and capped with a 2nm thick poly-silicon layer

The first component at a binding energy of 99.5 eV is related to the Si-Si/Si-Ge bonds while the second one (100.5 eV) is the fingerprint of a sub-stoechiometric SiO_x oxide (Si^{1+}). On the other hand, only one component (Ge^0) was used to fit the Ge 3d core level spectrum which led us to conclude that the poly-Si/Ge interface is free of oxide.

2-Influence of the growth temperature :

The influence of the growth temperature on the layer roughness and on the layer resistivity measured by Four Points Probe (not presented here) has been studied. Figure 5 shows the morphology of the GeOI film obtained by AFM after 40 nm thick epitaxial regrowth at various growth temperature. At low temperature (< 500°C) the growth is 2D with a clear "cross hatch" pattern while at higher temperature (≥ 500°C) the growth switches to a 3D growth mode. Moreover, the Rms roughness has been estimated as a function of the growth temperature (Figure 6). The layer roughness increases strongly from 400°C to 450°C and then monotonically from 450°C to 600°C. Actually, this typical surface morphology is linked to thermal grooving effects occurring during the growth. This phenomenon even exists at low growth temperatures (13). We assume that it is assisted by the crystalline defects observed on the layer surface as a cross-hatch pattern.

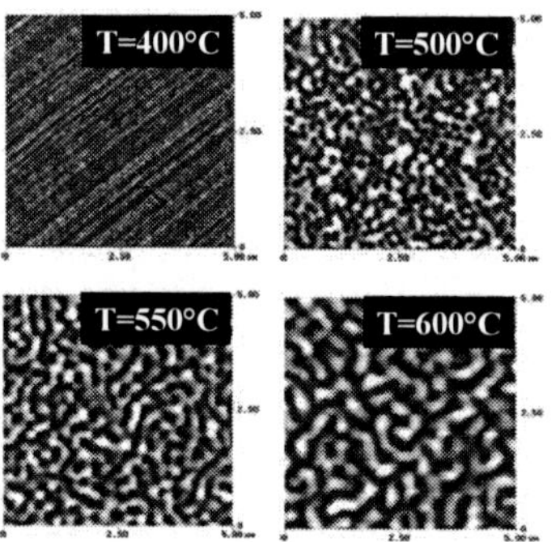

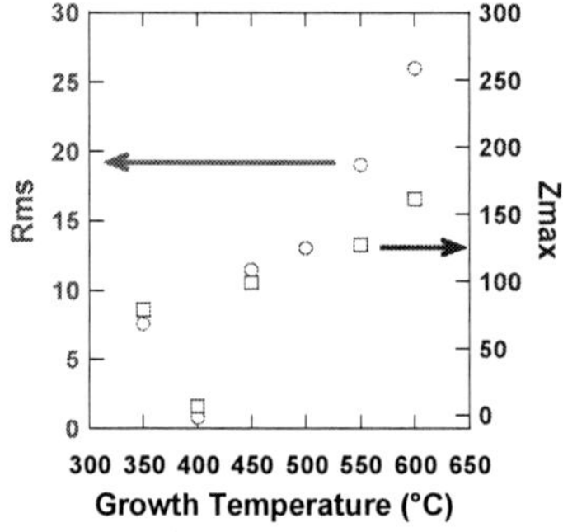

Figure 5: 5x5 µm² AFM pictures of the GeOI layers as a function of the growth temperature.

Figure 6 : Rms Roughness and Zmax (in nm) extracted from AFM pictures of the GeOI layers as a function of the growth temperature

3- Mobility extraction :

Two sets of GeOI wafers with different Ge layer thickness, 10 nm and 100 nm, have been analyzed. The thick Ge films (100nm) were prepared by thickening the GeOI layers using Ge epitaxial regrowth after condensation. Square-shape islands ($8x8$ mm^2) were etched on the conductive film down to the buried oxide.

Electrical characterization was carried out by Pseudo-MOSFET (ψ-MOS), which is a very simple and efficient method to characterize the Ge-buried oxide interface (14, 15). This method uses the natural upside-down MOS configuration of the structure. The GeOI film serves as the transistor body and buried oxide (BOX) acts as gate oxide. The substrate bias V_G induces a conducting channel (inversion or accumulation) at the film-BOX interface. Two pressure-adjustable probes form the source and drain (Figure 7). The low-field mobility μ_0 is extracted by drawing $I_D/g_m^{0.5}$ as a function of V_G . These GeOI wafers were compared to unstrained and strained SOI materials.

Typical Ψ-MOS current and transconductance characteristics are shown in Figs. 8 and 9. They are well behaved, with low leakage current, enabling an accurate extraction of the carrier mobility. Our 100-nm thick GeOI wafers feature an exceptional high hole mobility, beyond 400 $cm^2/V.s$, which exceeds recently reported values (16). The mobility enhancement over unstrained and strained SOI is 350% and 175%, respectively.

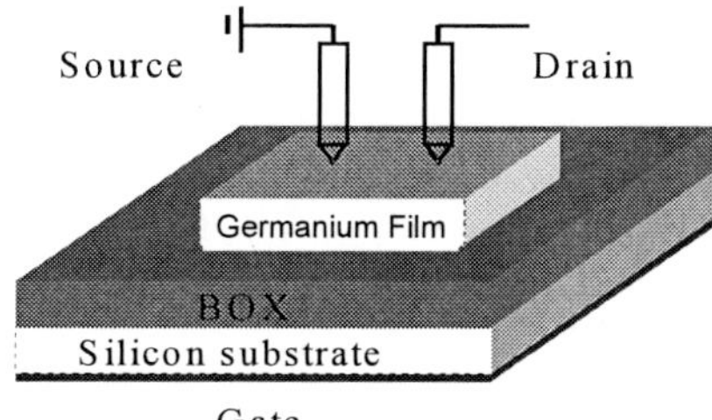

Figure 7: *Schematic configuration of Pseudo-MOS transistor.*

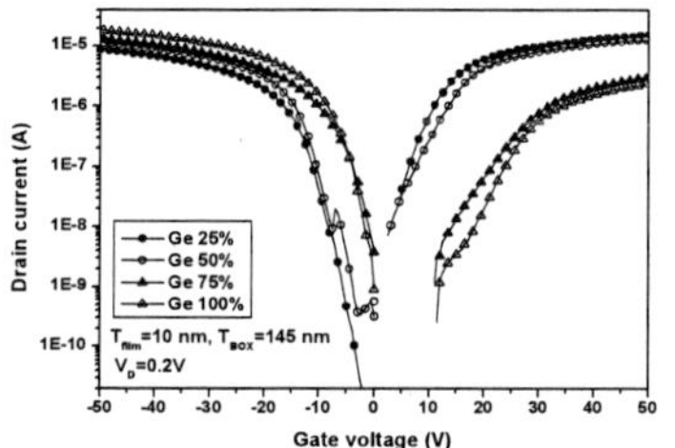

Figure 8: Drain current versus gate voltage in 10-nm SGOI films with different Ge percentage. Note the opposite variations in drain current for electrons and holes as the Ge content increases.

Figure 9: Transconductance versus gate voltage in 10-nm SGOI films. Increasing the Ge percentage improves the transconductance peak for holes and degrades that of electrons.

Conclusion

In this paper, we have investigated the influence of process parameters (bake temperature, bake duration and growth temperature) on the morphology of GeOI substrate thickened by Ge epitaxial regrowth. An optimum bake temperature in terms of layer roughness and interfacial contamination was found. The efficiency of the surface preparation prior to the epitaxial regrowth has been assessed by Angle resolved X-Ray Photoelectrons Spectroscopy. No residual oxygen contamination has been evidenced. The influence of the growth temperature on the layer properties has also been investigated. For temperatures higher than 450°C, thermal grooving assisted by crystal defects (cross hatch) occurred and 3D Ge layers were obtained. Finally, 10 nm and 100 nm thick GeOI substrates, thickened with the best growth condition (bake T= 600 °C and growth T= 400°C) were evaluated by Pseudo MOSFET technique. An exceptional high hole mobility, beyond 400 $cm^2/V.s$, was observed.

Acknowledgments

This work was carried out in the frame of CEA-Léti/ALLIANCE collaboration. The authors would like to thank Mme Sophie Descombes for her technical support and Dr.

François Bertin for his help in performing FFT analysis and subsequent fruitful discussions.

References

1. L.Clavelier, C. Le Royer, C. Tabone, J.M. Harthmann, C. Deguet, V. Loup, C. Ducruet, C. Vizioz, M.Pala, T. Billon, F. Letertre, C. Arvet, Y. Campidelli, V. Cosnier, Y. Morand, *Proceedings of SNW Symposium*, p18 (2005)
2. N.Wu, Q. Zhang, C. Zhu, C. Shen, M.F. Li, D. S. H. Chan, N. Balasubramanian, *Int. Electron Device Meet. Tech. Dig.*, p. 563 (2005),
3. M. Bruel, *IEE Electron. Lett.*, **31**, 1201 (1995).
4. C. Deguet, J. Dechamp, C. Morales, A.M. Charvet, L. Clavelier, V. Loup, J. M. Hartmann, N. Kernevez, Y. Campidelli, F. Allibert, C. Richtarch, T. Akatsu, F. Letertre, *Proceedings of ECS Symposium*, p.78 (2005)
5. T.Tezuka, N. Sugiyama, S.Takagi, *Appl. Phys. Lett.*, **79**, 1798 (2001).
6. T.Tezuka, N. Sugiyama, T. Mizuno, M. Suzuki, S. Takagi, *Jpn. J. Appl. Phys*, **40**, 2866 (2001).
7. N. Sugiyama, T. Tezuka, T. Mizuno, M. Suzuki, *J. Appl. Phys.*, **95**, 8 (2004).
8. J.Huang, Z.Ye, H.Lu and D.Que, *J. Appl. Phys.*, **83**, 171 (1998).
9. J.F. Damlencourt, R. Costa, Patent n° E.N. 0601850.
10. B. Vincent, J-F. Damlencourt, P. Rivallin, E. Nolot, C. Licitra, Y. Morand and L. Clavelier, *Semicond. Sci. Technol.*, **22** (2007).
11. E. Dornel, J-C. Barbé, J. Eymery and F. De Crécy, *Mater. Res. Symp. Proc.*, **910**, A04-05 (2006).
12. E. Dornel, J-C. Barbé, F. De Crécy, G. Lacolle and J. Eymery, *Phys. Rev. B*, **73**, 115427 (2006).
13. N. W. Mullins, *J. Appl. Phys.*, **28**, 333 (1957).
14. S. Cristoloveanu, D. Munteanu, M.S.T. Liu, *IEEE Transaction of Electronic Devices*, **47**(5), p. 1018 (2000).
15. S. Cristoloveanu and S. S. Li, Boston, MA, Kluwer, (1995).
16. T. Akatsu, C. Deguet, L. Sanchez, C. Richtarch, F. Allibert, F. Letertre, C. Mazure, N. Kernevez, L. Clavelier, C. Le Royer, J.M. Hartmann, V. Loup, M. Meuris, B. De Jaeger, G. Raskin, *Proceeding of IEEE International SOI Conferrence*, p. 137 (2005).

ECS Transactions, 6 (4) 321-326 (2007)
10.1149/1.2728877, ©The Electrochemical Society

Ordered Lattices of Quantum Dots on Ultrathin SOI Nanomembranes

Clark Ritz,[a] Frank Flack,[a] Donald Savage,[a] Douglas Detert,[a] Paul G. Evans,[a] Max Lagally,[a] Zhonghou Cai[b]

[a] University of Wisconsin, Madison, WI 53706, USA
[b] Advanced Photon Source, Argonne National Laboratory, Argonne, IL 60439, USA

Undercut or fully released silicon template layers of ultrathin silicon-on-insulator are structurally compliant and allow long-range mechanical interactions that are impossible on supported SOI, thick SOI, or on bulk surfaces. SiGe quantum dots create and respond to strain in these freestanding Si substrates. We show that elastic effects lead to long-range order in the positions of quantum dots grown on both sides of a free-standing thin SOI substrate. A statistical analysis of the distribution of quantum dots shows that the ordered lattice of dots is coherent over distances of at least several hundred of nanometers. X-ray microdiffraction probes the structure of the SOI and finds that curvature in the Si lattice is consistent with the proposed deformation of the substrate by quantum dots.

INTRODUCTION

Recent advances in epitaxial growth, lithographic patterning and processes for thinning and releasing the template layer of silicon-on-insulator (SOI) and similar materials have allowed the creation of ultrathin Si membranes. This development opens up new possibilities in for Si-based nanoelectronics, mechanics, optoelectronics, and nanosensor technologies (1,2). Si nanomembranes can be completely free-standing or partially attached, and either flat or curled into various configurations, depending on the engineering of the strain (3).

Unique features of the membranes arise from their thinness and the purely elastic nature of strain induced in them by thin films and nanostructures. With this in mind, partially freestanding released ultrathin Si membranes (i.e. with the buried oxide layer partially undercut) are unique substrates for epitaxial growth. From these starting materials, there is a range of opportunities to create nanostructures and thin films on both sides of a crystallographically perfect template.

Lattice-mismatched homogeneous heteroepitaxial thin films grown on both sides of released SOI structures result in a uniform distortion of the Si membranes. The strain due to the lattice mismatch is elastically partitioned between the thin silicon substrate and the film (4, 5). Growing 3D "hut" quantum dots (QDs) rather than films, however, results in large and highly localized strains in the region beneath individual QDs. Thin membranes are unique in comparison to bulk surfaces in that QDs on nanomembranes can lead to a large overall bending of their substrate (6).

Depositing QDs on both sides of partially released SiNMs results in a novel and potentially useful self-organization of nanostructures. SiGe hut nanostructures self-assemble in a manner that results in an anticorrelation in their positions on the two sides of the membrane and yields a high degree of long range order. Such a configuration minimizes the elastic energy of the system (7).

Other routes to strain engineered self-assembly in heteroepitaxial growth of thin films have provided attractive methods for fabricating semiconductor QDs, but have been limited by manipulating strain induced elastic interaction between QDs only above the substrate. Producing QDs on thin substrates overcomes these limitations and enables new phenomena.

QUANTUM DOTS ON SILICON NANOMEMBRANES

We have created partially released freestanding 25-nm-thick Si membranes in the form of cantilevers and bridges. These structures are formed by patterning features onto SOI using lithography, etching to remove exposed silicon, and subsequently etching in hydrofluoric acid (HF) to remove the buried oxide. Fully released structures can be formed by allowing the HF etching process to continue to completion. Partially released structures attached by the buried oxide can be formed by monitoring the duration of the exposure to HF and stopping the etching process before it completely consumes the buried oxide. A partially released membrane formed using this procedure is shown in Figure 1. The lithographic patterning can be accomplished using photolithography or electron beam lithography techniques.

The patterned structures were chemically cleaned and loaded into an ultrahigh vacuum chemical vapor deposition (CVD) system. CVD allows material to be deposited on both sides of the undercut regions of the template layers. Using this approach, SiGe hut QDs were grown with Ge compositions ranging from pure Ge to $Si_{0.36}Ge_{0.64}$. At lower Ge concentrations, these pyramidal QDs typically have base widths of approximately 80 nm and heights of 8 nm. For these Ge concentrations, QDs are relatively large in comparison with pure Ge QDs, but still reveal the typical hut shape consisting of four {105} facets. These 3D islands are deposited on faces of the partially released membrane in Figure 1.

Quantum Dot Ordering on Released Membranes

When QDs are formed on a single face of the membranes, as for example when the deposition is performed using a line-of-sight technique such as molecular beam epitaxy, QDs form on the surface in a random arrangement. When QDs are formed on both faces simultaneously, however, the elastic interaction between them favors an arrangement in which the QDs on the top and bottom surfaces are form a superlattice to minimize the elastic energy stored in the deformation of the template layer. Electron beams used for scanning electron microscopy (SEM) penetrate the entire 25 nm thickness of the released SOI and allow QDs on both faces of the membrane to be imaged simultaneously. The QDs on the top surface in Figure 2(a) appear as bright areas near the center of the squares. QDs on the bottom surface are dark in this image and are located at the corners of the squares. In this arrangement, the QDs have nearest neighbors on opposite faces, always along <110> directions.

The QDs deposited on cantilevered membranes organize into square lattices with nearest neighbors along <110>. The direction of this vector separating nearest neighbors is constant regardless of the orientation of any nearby edges. Other strategies for organizing QDs have employed lithographic patterning to define the orientation of the lattice. The ordering of QDs on membranes relies instead on the elastic anisotropy of (001)-oriented silicon sheets (8).

The lack of dependence of the direction of the QD superlattice on the lithographic boundary conditions is illustrated in Figure 2, which shows an inside corner at which the

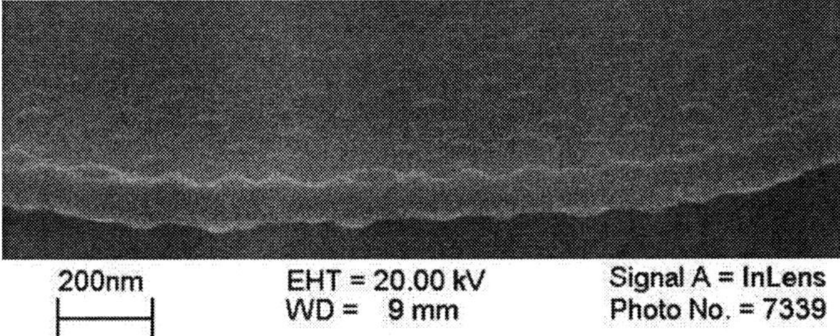

Figure 1. Edge-on view of an undercut cantilever with SiGe quantum dots on both sides.

direction of an undercut cantilever changes by 45°. In Figure 2, one edge of the cantilever is along a <110> direction (horizontal in the image) and another is oriented along a <100> direction (diagonal in the image). The Si template layer is affixed to the buried oxide in the lighter region in the upper and rightmost parts of the image, and released in the regions near the edges. Two sets of lines in Figure 2 point along <100> directions and illustrate the arrangement of second-nearest-neighbor QDs on the top of the silicon membrane. Each of these QDs on the top surface is displaced along <110> from a QD on the bottom surface.

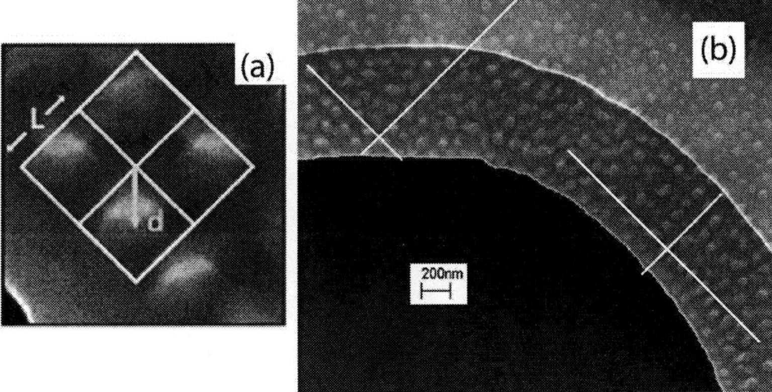

Figure 2. (a) SiGe QDs on a silicon nanomembrane formed from an undercut SOI template layer. A group of four islands on the top surface are each separated by a distance d from a central island on the bottom surface. The square unit cell of this superlattice is L long on each side. (b) SiGe QDs on a undercut Si cantilever edge of a 25-nm-thick Si membrane. The islands arrange themselves such that the nearest neighbors are along <110> directions, leading to a long range order. The QDs exhibit the same directional ordering independent of the direction of the edge.

<u>Statistical Descriptions of Correlation</u>

The degree of order and the dimensions of the hut superlattice can be quantified by computing the two-dimensional pair correlation functions (PCFs) associated with the locations of huts on the top and bottom surfaces. The PCF captures the directions and distances associated with the arrangement of the QDs. We have used a standard definition of the PCF, and computed it using of the positions of QDs measured in a series of SEM images acquired along the length of a cantilever extending along a <100> direction (9). The PCF for freestanding membranes can be compared with one computed for QDs on a supported silicon surface.

Figure 3 uses the PCFs to compare the degree of ordering of QDs on an undercut Si membrane and a region of supported SOI. For both geometries, the PCF involves all of the QDs. For the freestanding membrane, the PCF thus includes quantum dots on both the top and bottom surfaces. In the case of the freestanding Si membrane, shown in Figure 3(a), the arrangement of QDs exhibits a high degree of both lateral and directional order. The bright peaks of the PCF near 125 nm in both horizontal and vertical directions represent the relative nearest-neighbor positions of QDs on the top and bottom surfaces. The PCF for the huts on the supported SOI surface, in Figure 3(b), shows considerably less order. The correlation of QD positions on the supported surface resembles that of hard-sphere packing: there is a minimum center-to-center distance close the value of the diameter of the islands. The correlation shows almost no angular dependence.

Horizontal slices through the 2D PCFs along a line at zero vertical separation are shown in Figure 3(c). For both the freestanding and supported membranes, the value of the correlation function has been normalized to the number of QDs. This allows a quantitative comparison between the relative degree of ordering of QDs on the freestanding membrane and the unreleased SOI. The arrangement of islands on the

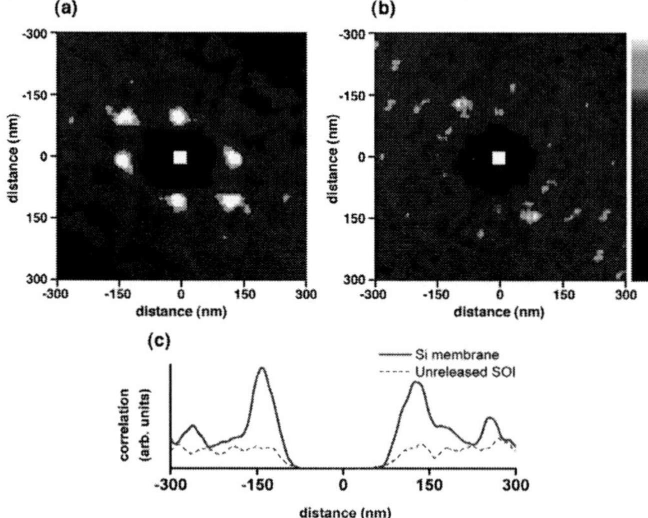

Figure 3. Pair correlation functions for QDs on (a) a released membrane and (b) a supported membrane. The correlation function for the released membrane includes QDs on both surfaces of the membrane. (c) Horizontal slices through the center of (a) and (b) show strong correlation at nearest-neighbor and third-nearest neighbor locations in the released case and a phenomenon analogous to hard-sphere repulsion in the supported case. The central self-correlation peak in both images has been suppressed.

freestanding membrane produces a 1st and 2nd nearest-neighbor peak in the correlation function, while QDs on the supported surface possess a low degree of order, independent of distance. In each case, there is a minimum nearest-neighbor distance of approximately 100 nm.

<u>X-ray Microdiffraction</u>

X-ray microdiffraction studies of SiNMs distorted by correlated arrays of huts provide insight into the elastic ordering. The microdiffraction measurements were performed at station 2ID-D of the Advanced Photon Source using a beam of 10 keV x-ray photons focused to a ~200 nm diameter spot by a Fresnel zone plate. This arrangement is shown schematically in Figure 4 for the case where huts are deposited on only the top surface of a freestanding membrane (10).

The microdiffraction results can be compared with predictions from continuum elasticity theory. We have earlier shown that an isolated pure Ge hut QD locally bends the underlying Si membrane because the 4% lattice mismatch causes significant local strain. We have furthermore made quantitative microdiffraction studies of hut clusters deposited only on the top surface of a freestanding edge of a partially released membrane to quantify the effect of the local strain created by these nanostressors (10). If the density of dots becomes sufficiently high, a uniform global bending results, as if a uniform compressively stressed film had been deposited (11). Microdiffraction studies of the (004) x-ray reflection of the a freestanding membrane with SiGe QDs on both sides show that the membrane has a local curvature introduced by the QDs that is consistent with what is expected from elastic calculations (7).

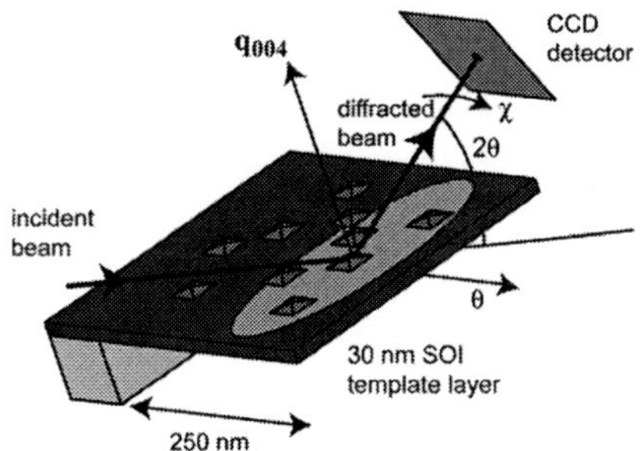

Figure 4. Schematic of x-ray microdiffraction experiments probing the distortion introduced in a silicon nanomembrane by Ge hut QD nanostressors, after ref. 10.

CONCLUSION

The problem of ordering quantum dots for applications in optics and electronics has persisted for many years primarily due to the weakness of interactions mediated by processes on mechanically rigid surfaces. Thin silicon membranes change this situation, allowing long-range interactions that lead to a new degree of precision in ordering QDs. These structures can potentially serve themselves to confine carriers for optical or electronic applications, or can be useful in creating a well-defined spatial distribution of strain in the mediating freestanding silicon. The deformation of silicon sheets, ribbons, and nanowires creates new opportunities to control the electronic structure of large areas of silicon devices.

ACKNOWLEDGMENTS

This work was supported by the NSF through the University of Wisconsin Materials Research Science and Engineering Center (DMR-0520527), and by the Department of Energy. Use of the Advanced Photon Source was supported by the U. S. Department of Energy, Office of Science, Office of Basic Energy Sciences, under Contract No. DE-AC02-06CH11357.

References

1. M. M. Roberts, L. J. Klein, D. E. Savage, *et al.*, Nature Mat. **5**, 388 (2006).
2. P. P. Zhang, E. Tevaarwerk, B. N. Park, *et al.*, Nature **439** (7077), 703 (2006).
3. H. Qin, N. Shaji, N. E. Merrill, H. S. Kim, R. C. Toonen, R. H. Blick, M. M. Roberts, D. E. Savage, M. G. Lagally, and G. Celler, New J. Phys. **7**, 241 (2005).
4. P. M. Mooney, G. M. Cohen, J. O. Chu, and C. E. Murray, Appl. Phys. Lett. **84**, 1093 (2004).
5. M. M. Roberts, D. Tinberg, P. Evans, M. Lagally, C.-H. Lee, A. Lal, Y. Xiao, B. Lai, and Z. Cai, in Silicon-on-Insulator Technology and Devices XII, edited by G. K. Celler, S. Cristoloveanu, J. G. Fossum, F. Gamiz, K. Izumi, and Y-W. Kim, Electrochemical Society Proceedings vol. 2005-03, (2005), pp. 225-230.
6. F. Liu, M. Huang, P. P. Rugheimer, D. E. Savage, and M. G. Lagally, Phys. Rev. Lett. **89**, 136101 (2002).
7. C. S. Ritz, *et al.*, submitted (2007).
8. W. A. Brantley, J. Appl. Phys. **44**, 534 (1972).
9. P. M. Chaikin and T. C. Lubensky, *Principles of Condensed Matter Physics*, Cambrdige University Press, (1995).
10. P. G. Evans, D. S. Tinberg, M. M. Roberts, M. G. Lagally, Y. Xiao, B. Lai, and Z. Cai, Appl. Phys. Lett. **87**, 073112 (2005).
11. M. Huang, M. Cuma, M.G. Lagally, and F. Liu, Mat. Res. Soc. Symp. Proc. **791**, Q6.4.1-6 (2004)

ECS Transactions, 6 (4) 327-332 (2007)
10.1149/1.2728878, ©The Electrochemical Society

LEGO Process for Mixed Power Applications: Fabrication at Low Cost of Localized Thick SOI Layers

I. Bertrand[a,b,c], J-M. Dilhac[b], P.Renaud[a], M.Bafleur[b], C.Ganibal[b]

[a] Freescale Semiconducteurs France SAS, av. du Gal Eisenhower, 31023 Toulouse Cedex, France
[b] Université de Toulouse, LAAS-CNRS, 7 av. du Colonel Roche, 31077 Toulouse Cedex 4, France
[c] now with SOITEC, Grenoble, France

In this paper, we present the low cost fabrication process of a partial and thick silicon-on-insulator (SOI) substrate for mixed power integration (low and high voltage devices on the same chip). It is based on the Lateral Epitaxial Growth over Oxide process (LEGO). It is shown that the proposed process provides a crystalline quality compatible with a CMOS technology and high voltages devices.

Introduction

Silicon-on-Insulator (SOI) represents a key technology for future development in ultra high-level CMOS or Bi-CMOS integration. SOI is also used in MEMS, sensors, smart power and any technology that requires a high level of electrical or, in some cases, thermal isolation. BCD (Bipolar CMOS DMOS) processes are examples of power technologies where SOI leads to dramatic device performance improvements. SOI has also been successfully used in high voltage ICs. In general, all SOI technologies show a very significant improvement in the switching speed of bipolar devices. For example, the increased speed of LIGBT (Lateral Insulated Gate Bipolar Transistors) is linked to a perfect dielectric isolation between the substrate and the drift region. Carriers are only stored in the SOI layer and virtually no currents (other than displacement currents) flow through the substrate. Moreover, the availability of a partial SOI technology would allow integrating high voltage power devices (DMOS or IGBT) with their driving and protection circuitry perfectly isolated into an SOI island. To build such a technology, the main issue is the fabrication cost, particularly when the starting substrate is a full SOI substrate.

Partial SOI Architecture

This paper focuses on the low cost fabrication of a partial and thick silicon-on-insulator (SOI) substrate for mixed power integration (low and high voltage devices on the same chip) (Fig.1). Such applications specifically require silicon substrates having, on the one hand, localized thick SOI patterns associated with lateral isolation for control modules, and on the other, bulk areas for power devices. Most of the existing fabrication methods lead to the fabrication of full and thin SOI layers. Consequently the global fabrication is not cost-effective due to the needed additional patterning, deposition, and related steps. This paper demonstrates that the Lateral Epitaxial Growth over Oxide process (LEGO) is a cost-effective solution to obtain such a substrate with a good SOI layer crystalline quality. This process, firstly developed by G. Celler (1), is based on the fusion and recrystallization of a deposited thick poly-silicon layer (2). It is being reconsidered today because of a new market demand for partial SOI.

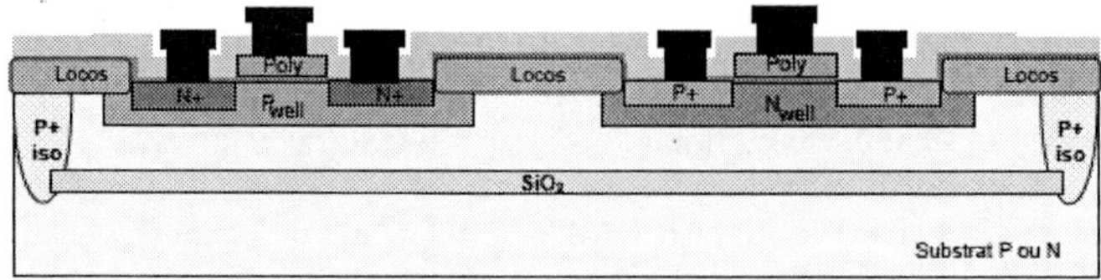

Figure 1. Partial thick SOI schematic cross-section including lateral isolation, here by junction isolation (alternatively using deep trenches). CMOS devices are shown on the partial SOI.

Fabrication Technology and Apparatus

Lateral Epitaxial Growth over Oxide (LEGO) is directly inspired by the Zone Melting Recrystallization (ZMR) method which provides SOI patterns through a rapid melting, followed by a controlled crystallization, of poly-silicon films deposited over SiO_2 patterns grown on a silicon wafer. The thermal gradient that helps control the melting and recrystallization is created by a scanning heater in this case. SOI layers obtained with this process are inexpensive but the crystalline quality is poor because of the thermal strain accumulation in the scanning direction. The main improvement with LEGO is that the heat source is uniform all over the entire wafer surface; hence all poly-silicon patterns are simultaneously recrystallized.

More precisely, the 4" N-type (10^{15} at.cm^{-3}) wafers were pre-processed with a thermal oxidation (1 μm) and a polysilicon seed layer deposition (530 nm). A photolithography, a deep reactive ion etching for polysilicon and a wet etching for oxide defined the SOI patterns (Fig.2). During the following non-selective epitaxy (thickness varying from 10 to 40 μm, depending on sample), poly-silicon was deposited on the oxide patterns and mono-crystalline silicon was grown on the bulk silicon wafer. The grown/deposited silicon layer was doped (8.10^{15}at.cm^{-3}) on purpose with arsenic. A 1 μm thick oxide-capping layer was finally added to prevent the evaporation of the melted silicon during the recrystallization phase.

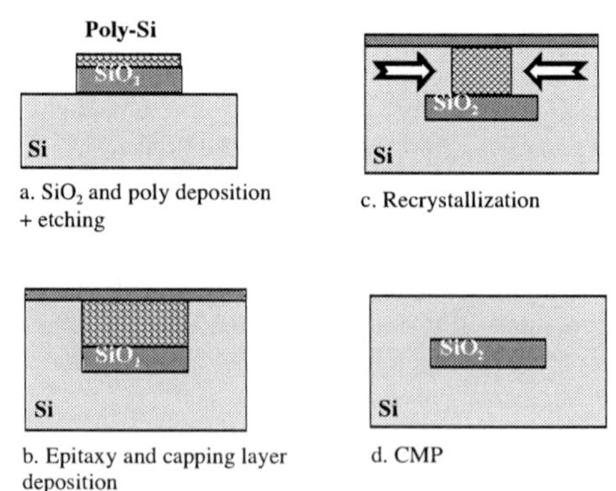

Figure 2. Main steps of a LEGO process

A Rapid Thermal Processing (RTP) furnace with tungsten halogen lamps (3) provides enough energy to reach the melting temperature (1415°C) in poly-silicon areas while maintaining the bulk mono-crystalline area solid. As the temperature ramps down, a thermal gradient induced by the difference in the thermal conductivities of the silicon and oxide layers generates a recrystallization front spreading from the mono-crystal seed over the buried oxide layer, and thus creating mono-crystalline SOI patterns (Fig.2).

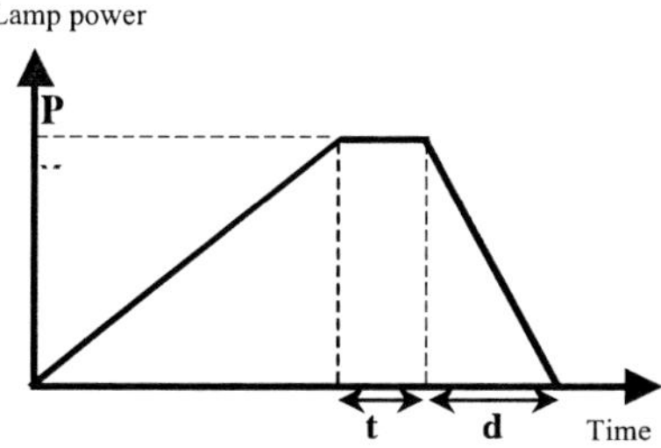

Figure 3. Recrystallization thermal cycle. Key parameters are identified (see text).

After recrystallization, the wafer surface roughness is important because of the liquid movements. Therefore, the planarity has to be restored with a Chemical Mechanical Polishing (CMP).

The critical parameters to be optimized for this process are the SOI patterns geometries, silicon epitaxy and capping layer thicknesses together with the recrystallization cycle parameters, such as heating power maximum level P, plateau and ramp down durations, t and d, respectively (Fig.3).

Experimental Results

<u>Physical Characterization</u>

With respect to physical characterization, SEM cross-sections, chemical revelations, and TEM observations, were carried out and show that most of the residual crystallographic defects are located at the centre of the patterns, where the different recrystallization fronts meet. More precisely, the main residual crystallographic defects after recrystallization are cross-shaped in the centre of the patterns: these defects are due to the recrystallization front mismatch. Indeed, the crystal lattice propagation during solidification is initiated at the four sides of the SOI pattern. Thus, the crystal lattice recovering is perfect at short distances, for example on the smallest recrystallized patterns; in contrast, a longer front propagation over the oxide layer introduces mismatches along the lattice recovering line. The cross in the centre of the pattern exactly reveals where the recrystallization fronts met.

Test vehicles were realized by designing SOI patterns of various sizes ranging from 100×100µm to 1000×2000µm rectangles, and by varying the thickness of the polysilicon deposited films. A design of experiment helped understand that the most important parameters for a good recrystallization quality are the power plateau level P and the ramp down duration d.

Indeed, the power plateau assures a good melting of the poly-silicon layer and must be adapted for the different SOI thicknesses and pattern widths: the polysilicon layer needs to receive enough energy to immediately melt the entire layer. A lower power plateau during a longer time does not provide enough energy because of the change in emissivity between solid and molten silicon. It would provide superficial melting only, leading to a lower emissivity and thus lower energy absorption. A high power during a short time results in a deeper melting of the poly-silicon layer, and the change in emissivity does not alter the whole layer melting anymore.

The slope of the ramp down is also important for the recrystallization front propagation. A slow ramp down could help the germination phenomenon to be dominant during recrystallization, thus creating poly-crystalline patterns because of propagation from different grains. For example with a 45 sec ramp down, the recrystallization front propagation only reaches 80 μm horizontally from the side of the SOI area before germination takes place, compared to 150 μm for a 30 sec ramp down, the other parameters being unchanged. The best results were obtained with a 0 sec ramp down, which helped obtain 1000×2000μm mono-crystalline SOI patterns with an initial 30μm-thick polysilicon film (Fig. 4 and 5).

Figure 4. SEM cross-section after chemical revelation of defects (Wright etch) of a mono-crystalline1000μm×2000μm SOI pattern before CMP. A 30 μm silicon film was deposited prior to LEGO.

Figure 5. 1000μm×2000μm planarized SOI pattern after chemical revelation. Extended defects are located where the four recrystallization fronts met. Deposited Si film was 30 μm thick.

SIMS profiles firstly clearly helped confirm that there is neither carbon nor oxygen contamination of the recrystallized silicon layers during the fusion and recrystallization steps. Secondly, the arsenic doping level appeared to be substantially lower in the SOI area compared to the bulk area. This discrepancy was corroborated by spreading resistance profiles: bulk areas exhibit a 1 Ω.cm resistivity whereas SOI areas a 4 Ω.cm one. This is mainly due to a different arsenic incorporation process in the monocrystalline and polycrystalline areas during the epitaxy taking place before the recrystallization step. It can be consequently concluded that there is no contamination of the SOI patterns due to the recrystallization process.

Finally, X-ray diffraction revealed a good matching of the SOI pattern crystalline lattice and bulk.

Electrical Results

Considering electrical characterization, low power devices (such as diodes and MOS transistors) together with high power devices (such as lateral IGBT and LDMOS) were realized on the partial SOI layers. To evaluate the impact of the recrystallization on their electrical behaviour, MOS transistors and diodes were placed in various regions: over the defective zone where recrystallisation fronts meet and outside the defective zone of the recrystallized SOI areas as well as over the seed area (Fig.6). For reference, test devices were also realized on bulk substrates that did not suffer the high temperature LEGO process. Electrical measurements of the main parameters of these MOS transistors (Table 1) show that the electrical properties are the same wherever the transistors are located.

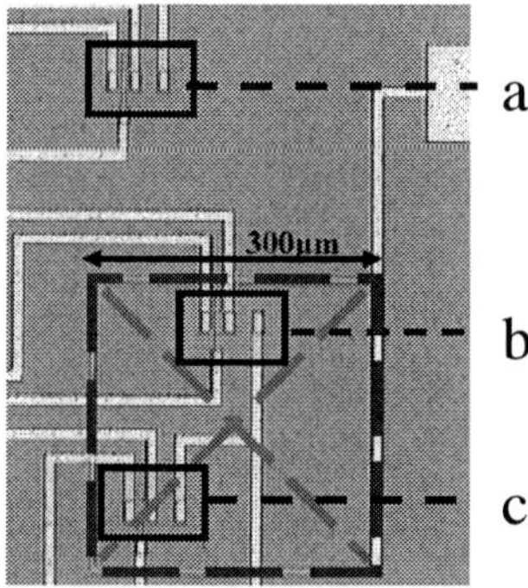

Figure 6. MOS transistor locations – a) bulk – b) SOI, defect free – c) SOI, with defects. The rectangles materialize the MOS transistors and the SOI area. The cross materializes the residual defect zone (see text).

Table I. PMOS and NMOS electrical performances (threshold voltage, mobility, subthreshold slope and breakdown voltage). Subthreshold slope missing value is due to a technical measurement problem.

	PMOS			NMOS		
	bulk	SOI	SOI + defects	bulk	SOI	SOI + defects
V_T(V)	-1.3	-1.3	-1.3	0.6	0.6	0.6
μ(cm²/V.s)	160	159	159	421	433	420
S(mV/dec)	87	94	108	125	105	--
BV(V)	-26	-26	-25.5	14.7	14.7	15

We also compared the behaviour of the reverse biased diodes. A large number of diodes were tested. As shown in Table 2, there is a large variation of the breakdown voltages between the reference substrate and the partial SOI substrate. The lower breakdown voltage value for the bulk substrate is due to a difference of doping level compared to the epitaxial layer. The important result here is that the diodes realized on the seed areas and on defect free SOI areas exhibit the same breakdown voltage and leakage current. Over the defective area, a slight decrease in the breakdown voltage on some of the diodes could be observed but the related leakage current stays within the range of the reference one. Finally, in a previous work (3), it was also demonstrated that the device quality of the SOI layers allowed successfully implementing high voltage LIGBT and LDMOS devices. They exhibited a measured breakdown voltage of 300V as expected from simulations. For these devices, the partial SOI offers the possibility to greatly improve the breakdown voltage and the thermal performance compared to a full SOI process.

Table II. Diodes measurement data obtained on bulk substrate and on different locations of the partial SOI.

	Bulk Substrate	Partial SOI				
Average value	Reference	Seed area	SOI no defect	SOI defects		
Breakdown voltage at 10µA (V)	-31	-39,5V	-39V	-39,5V	-32V	-33,7V
Leakage current at 5V (nA)	-11	-14	-12	-14	-10	-13

Conclusion

To summarize, LEGO process enables the fabrication of monocrystalline SOI patterns up to at least 2 mm^2 with an epi-layer thickness up to 30µm before planarization and very few residual defects. MOS transistors placed onto these SOI patterns show good electrical performances; for the case of LIGBT and LDMOS, when the SOI is laid out appropriately, the performances are even superior to those of a full SOI technology. LEGO represents therefore today the most cost effective process for the fabrication of partial SOI wafers and is particularly adapted for mixed power applications.

Acknowledgments

This project was partially carried out within the framework of the European Community funded program IMProving multidisciplinary ACcess to microsystems Technologies (IMPACT) and of the joint lab LISPA between FREESCALE Semiconductor and LAAS-CNRS funded by the French Ministry of Industry and the Midi-Pyrénées Regional Authorities.

References

1. G.K. Celler, McD. Robinson, L.E. Trimble, D.J. Lischner *J. Electrochem. Soc.*, **132**, 211 (1985).
2. I. Bertrand, J.M. Dilhac, P. Renaud, C. Ganibal, *Microelectronics Journal*, **37(3)**, 257 (2006).
3. I. Bertrand, V. Pathirana, E. Imbernon, F. Udrea, M. Bafleur, R. NG, H. Granier, B. Rousset, J.M. Dilhac, *Bipolar/BiCMOS Circuits and Technology Meeting,* Santa Barbara – USA (2005).

ECS Transactions, 6 (4) 333-338 (2007)
10.1149/1.2728879, ©The Electrochemical Society

A Novel Method to Fabricate Multiple-layer SOI -- Single-Crystal Si Nanomembrane Transfer and Stacking

Weina Peng[a], Michelle M. Roberts[b], Eric P. Nordberg[a], Frank S. Flack[b], Paula E. Colavita[c], Robert J. Hamers[c], Donald E. Savage[b], Max G. Lagally[b], Mark A. Eriksson[a]

[a] Department of Physics
[b] Department of Material Science
[c] Department of Chemistry
University of Wisconsin Madison, Madison, Wisconsin 53706

Multiple-layer SOI, in which there are multiple device layers per unit area, is of great interest because it potentially enables fully 3D integration. It also offers the prospect of more complete integration between optics and microelectronics than is possible with purely 2D structures. Here we present a technique for transfer and stacking of multiple single crystal Si nanomembranes with intervening oxide layers, enabling fabrication of multiple-layer SOI. We explain how to overcome transfer challenges, the most important of which is unintentional bonding of nanomembranes to their host substrate during lift-off. We examine the surface roughness of the membranes using atomic force microscopy (AFM). Intriguingly, the intermediate spin-on-glass layers used in this work are rougher than the silicon layers above and below, indicating that membrane transfer can ameliorate roughness introduced by such layers. We have used this process to create multilayer Si/SiO_2 heterostructures, i.e., multiple-layer SOI. These structures naturally function as Bragg reflectors, and thus their optical reflectivity can be used as a measure of structure quality. The reflectivity of one, two, and three membrane structures has been measured, and reflectivity above 99% has been achieved for three-layer samples.

Introduction

The use of silicon-on-insulator (SOI) substrates has enabled a number of advances in microelectronics technology. These include increases in device speed and decreases in power consumption, reduction or elimination of latch-up effects, and improvement in the short channel effect (SCE) [1]. Extending single layer SOI to multilayer SOI opens the prospect of fully 3D integration. At the simplest level, 3D integration offers more working area per wafer and therefore increased device densities. Moreover, multilayer SOI potentially enables increased integration of optics, mechanics (MEMS), and electronics. Some of these examples are already becoming visible in photonics applications, such as Si based waveguides and Bragg reflectors [2].

Although growth techniques, such as CVD, are capable of low-cost, large-scale deposition of Si, they cannot achieve crystalline Si on SiO_2. Grain boundaries and high surface roughness in poly and amorphous Si degrade electrical properties and lead to problems for optical applications. As a result, complicated Smart Cut[TM] processes, involving hydrogen implantations and successive annealing steps to restore the crystal structure from implantation damage, have to be applied in order to fabricate prime multiple-SOI wafers of high quality. Maleville et al. have used multiple Smart Cut steps to form multilayer SOI with up to 3 layers[3].

We report here a different approach to the fabrication of multilayer SOI. By transferring silicon nanomembranes onto amorphous oxide, multilayers of single-crystal Si and amorphous silicon dioxide can be fabricated. The technique is easy to implement, offering freedom in substrate choice, including flexible substrates such as plastics for applications in flexible electronics [4, 5]. Further, the bonding described here between a thin membrane and the supporting substrate is in some ways simpler to implement than conventional wafer bonding between two rigid, thick wafers, because the thin membrane conforms to the substrate. Nanomembrane transfer also enables bonding of single crystal layers to curved substrates, something not possible with conventional bonding approaches. Finally, nanomembranes need not be large in their lateral dimensions. While in this work we have transferred membranes as wide as 1 cm by 1 cm, and further development may offer much larger dimensions, it is quite possible that the most important application of this technique may be in the creation of small, local heterostructures of single crystal silicon with its amorphous oxide.

Fabrication

The membrane transfer process shown schematically in Fig. 1 results in Si/SiO_2 heterostructures in which each layer is single crystal. We use SOI as the source material

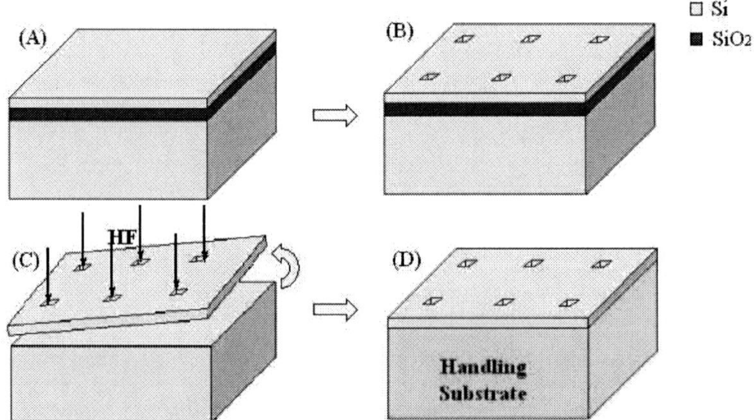

Figure 1. Process for membrane transfer. A) SOI consisting of 108nm silicon on 3μm buried oxide. B) After lithography and plasma etching, holes are opened on the top silicon layer. C) HF passes through the holes and etches away the oxide. D) The released membrane is transferred and bonded to a new substrate. See text for additional details.

for the formation of silicon nanomembranes, because it is a convenient starting point for research. However, membranes can also be made from conventional silicon wafers using etching techniques [6, 7]. Such methods offer dramatically reduced cost and are thus very appealing for potential applications. Here, for rapid under-etching and release of the membrane from the underlying oxide, we pattern an array of square holes into the SOI using photolithography followed by SF_6 reactive ion etching, an etch that is selective to Si over SiO_2. In our experiments, the hole size ranged from 5μm to 20μm while the spacing between neighboring holes varied from 60μm to 200μm. Following patterning, the membranes are released by immersing the patterned SOI chip into a solution of 10% HF. In principle, the holes are not necessary, but they decrease the time required for complete release of large membranes. In practice, we have released membranes as large as 400 μm on a side without holes. We have successfully made and transferred silicon nanomembranes as large as 1 cm × 1 cm, the largest size attempted in the present experiments.

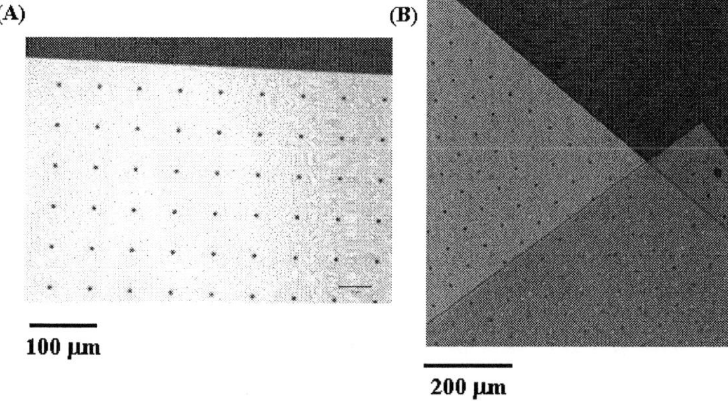

Figure 2. Optical micrograph of the transferred and stacked membranes. A) A single transferred membrane on a silicon dioxide substrate. B) A two-layer SOI stack. The green membrane was first transferred to a substrate pre-deposited with dielectrics (spin on glass, dark green). Next a second SOG layer was spun-on and cured, then the second membrane (appearing orange in the figure) was stacked onto the area where the first one sat. Two sets of holes with different orientation and almost no line-up are visible. All colors are due to interference effects from the multiple layers and interfaces.

Several techniques have been used to increase the rate of under-etching and lift-off of epitaxial films, including the application of wax to the surface [8] and the use of applied forces [9], both of which open a gap between the lift-off layer and the substrate, reducing lift-off time. Recently, strained epitaxial films have been transferred [10]. In this latter case, it appears that strain relief during the release process causes mechanical motion of the film, reducing the likelihood that the membrane will stick to the underlying substrate after release. Further, the tendency for such strained films to curl or roll up into tubes [11-13] enhances the diffusion of etchant underneath the membrane. For the case of the unstrained silicon membranes studied here, membrane adhesion to the substrate after release is much more likely. In order to release unstrained silicon membranes successfully, we leave the photoresist (PR) on the surface after the plasma-etching step that defines holes in the membrane. Immersing the PR covered sample into HF releases the membrane, which curls slightly, presumably due to the stress built into the PR film during pre- and post-baking. In this way the PR both reduces adhesion and enhances diffusion of the etchant in a way similar to the effect present in films with built-in strain. The baking procedure is a key factor in adjusting the stress in the PR layer and must be adjusted for different types of PR and atmospheric moisture conditions. After release, agitation of the solution causes the membrane to rise to the surface, at which point it can be transferred to DI water for rinsing and removed by lifting it out of the water with the desired final substrate. Although the membranes curl while underneath the water surface, after agitation surface tension flattens the PR-covered membranes, enabling membrane transfer. Fig. 2(A) shows an optical image of a single, transferred Si nanomembrane. After the water evaporates and the membrane is dry, immersion in acetone plus IPA followed by O_2 plasma etching is effective in removing PR from the membrane. Finally, high-temperature annealing (100 $^\circ$C for 10 min., followed by 500 $^\circ$C for 30 minutes) enhances the bonding of the membrane to the new substrate.

In this work we use spin-on-glass as a convenient, rapid method for generating amorphous insulating layers. However, the fabrication of insulating layers in stacked heterostructures such as those we demonstrate here can be accomplished by a wide variety of methods. Dry oxidation in a furnace is compatible with the methods we use here, as are other growth techniques, including sputtering, evaporation and plasma enhanced chemical vapor deposition (PECVD). After the deposition of a dielectric layer, membrane transfer can be repeated. The end result is a heterostructure of single crystal and amorphous layers (Fig. 2(B)).

Heterostructure Characterization

X-ray diffraction was conducted on the transferred membranes to verify the crystalline structure after transfer and to extract thickness information. A typical

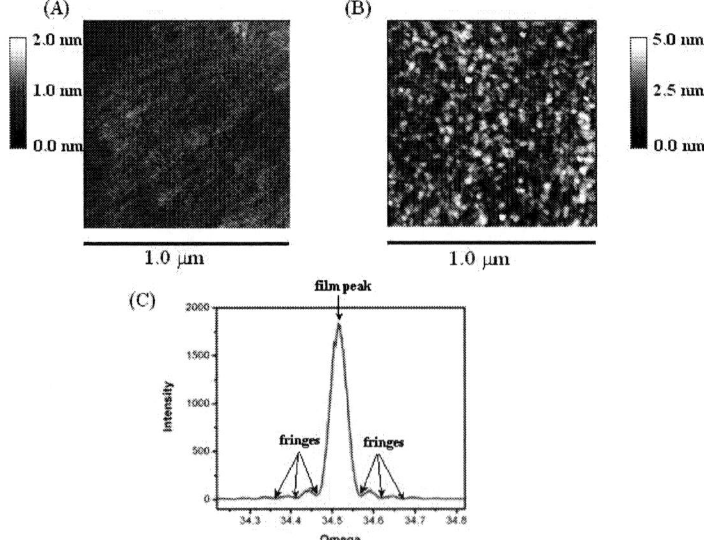

Figure 3. A) AFM image of the surface of a transferred membrane. B) AFM image of an SOG surface C) X-ray diffraction data from a single transferred membrane.

theta/two theta scan is shown in Fig. 3(C), indicating a central peak as well as clear fringes on both sides. The thickness fringes reveal the thickness of the layer, 103 nm, in this case.

Surface roughness is important for many applications. Interface scattering is frequently a limiting factor determining carrier mobility. For optical applications, roughness leads to scattering losses. Here we use AFM to examine the surface morphology of the membranes, both before and after transfer, and of the spin-on-glass films. From AFM images of the transferred membrane, one of which is shown in Fig. 3(A), the RMS roughness of transferred membranes is found to be approximately 0.13nm, similar to the RMS roughness of the prime SOI wafers used as source material in our experiments. Interestingly, the surface of the spin-on-glass layer is almost five times rougher (Fig. 3(B)) than the transferred membranes, even when the membrane is transferred onto the spin-on-glass. Instead of following the behavior of the surface of the supporting substrate, the membrane actually decreases the surface roughness. It is believed that the stiffness of the membrane helps to suppress the short-range roughness, whereas the flexibility of the membrane on longer length scales allows good conformation to the substrate and, as a result, simple yet effective bonding.

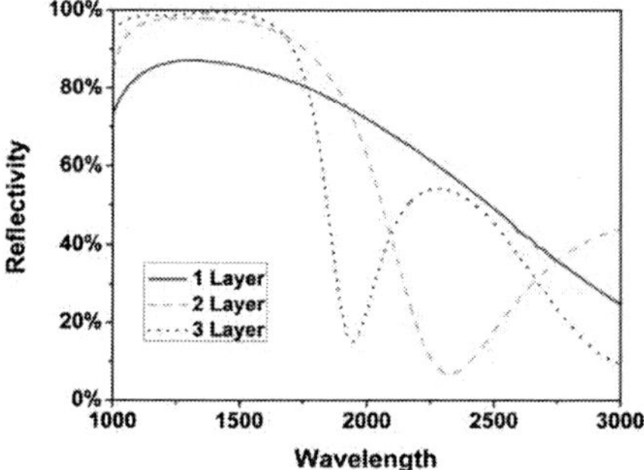

Figure 4. The data shows specular reflectivity of Bragg reflectors with one, two, and three membranes, for s-polarized radiation and an incident angle of 35 degrees from the normal. In our experiments SOG (spin on glass) is used to replace silicon dioxide. Due to its low index of refraction, similar to SiO_2, SOG performs well in optical applications requiring high index contrast with Si. For electronics, thermal oxide with better electrical properties can be fabricated using wet or dry oxidations.

The multilayer Si/SiO_2 structures we have created are essentially Bragg reflectors with one, two and three Si overlayers. Silicon is actually of considerable practical interest for such applications, due to the large difference between the index of refraction of Si (n=3.5) and that of its thermal oxide (n=1.45). Such a large contrast is beneficial because it results both in high peak reflectivity and also in a broad operational bandwidth and omnidirectional properties. To characterize the performance of the structures, FTIR was used to measure the reflectivity, as shown in Fig. 4. A simulation based on the matrix formulation agrees well with the measurements; details can be found in reference[14]. The peak reflectivity increases from 86% for one single membrane (i.e., common SOI) to more than 99% for three stacked membranes (three layered SOI), and, with increasing number of layers, the whole reflectance spectrum acquires a broad, high-reflectivity plateau characteristic of Bragg reflectors.

Conclusions

We have presented a method to fabricate multiple-layer single-crystal silicon and amorphous oxide heterostructures, or multilayer SOI. The technique is easy to implement and has good compatibility with existing growth and fabrication techniques. From AFM analysis, it was found that membranes transferred onto rough substrates decrease the surface roughness from one step to the next. Multilayer heterostructures, with several single crystal device layers and smooth surfaces and interfaces, may find application in electronics, optics, or hybrid technologies. As a simple example application, we have demonstrated the use of multilayer SOI as a Bragg reflector, and in this context we have demonstrated high reflectivity. The membrane-based approach to multilayer systems that we describe can be extended to many materials systems, and the smoothing properties of the transferred membranes opens a wide variety of options for the intervening dielectric layers. For example, optical properties for reflectivity can be optimized in the bottom layers of the device using inexpensive techniques that yield a good optical index but poor electrical quality. The top layer could then be formed with a high-quality electrical oxide, for electronics integration. The membrane transfer technique presented here seems particularly well suited to applications requiring one or several local regions of multiple layer SOI, integrated with other functional regions across a substrate.

References

1. G. K. Celler and S. Cristoloveanu, *J. Appl. Phys.*, **93**, 4955 (2003).
2. S. Akiyama, F. J. Grawert, J. Liu, K. Wada, G. K. Celler, L. C. Kimerling and F. X. Kaertner, *IEEE Photonics Techno. Lett.*, **17**, 1456 (2005).
3. C. Maleville, T. Barge, B. Ghyselen, A. J. Auberton, H. Moriceau and A. M. Cartier, *2000 IEEE Int. SOI Conf.*, 134 (2000).
4. E. Menard, R. G. Nuzzo and J. A. Rogers, *Appl. Phys. Lett.*, **86**, 093507 (2005).
5. H. C. Yuan, Z. Q. Ma, M. M. Roberts, D. E. Savage and M. G. Lagally, *J. Appl. Phys.*, **100**, 013708 (2006).
6. H. C. Ko, A. J. Baca and J. A. Rogers, *Nano Lett.*, **6**, 2318 (2006).
7. S. Mack, M. A. Meitl, A. J. Baca, Z. T. Zhu and J. A. Rogers, *Appl. Phys. Lett.*, **88**,(2006).
8. E. Yablonovitch, T. J. Gmitter, E. Kapon, C. P. Yun and R. Bhat, *J. Electrochem. Soc.*, **135**, C451 (1988).
9. J. J. Schermer, G. J. Bauhuis, P. Mulder, E. J. Haverkamp, J. van Deelen, A. T. J. van Niftrik and P. K. Larsen, *Thin Solid Films*, **511**, 645 (2006).
10. M. M. Roberts, L. J. Klein, D. E. Savage, K. A. Slinker, M. Friesen, G. Celler, M. A. Eriksson and M. G. Lagally, *Nat. Mater.*, **5**, 388 (2006).
11. V. Y. Prinz, V. A. Seleznev, A. K. Gutakovsky, A. V. Chehovskiy, V. V. Preobrazhenskii, M. A. Putyato and T. A. Gavrilova, *Physica E*, **6**, 828 (2000).
12. O. G. Schmidt and K. Eberl, *Nature*, **410**, 168 (2001).
13. H. Qin, N. Shaji, N. E. Merrill, H. S. Kim, R. C. Toonen, R. H. Blick, M. M. Roberts, D. E. Savage, M. G. Lagally and G. Celler, *New J. Phys.*, **7**, 241 (2005).
14. W. Peng, M. M. Roberts, E. P. Nordberg, F. S. Flack, P. E. Colavita, R. J. Hamers, D. E. Savage, M. G. Lagally and M. A. Eriksson, *submitted for publication*.

ECS Transactions, 6 (4) 339-344 (2007)
10.1149/1.2728880, ©The Electrochemical Society

Strained Silicon-On-Insulator – Fabrication and Characterization

M. Reiche[a], C. Himcinschi[a], U. Gösele[a], S. Christiansen[a], S. Mantl[b], D. Buca[b], Q.T. Zhao[b], S. Feste[b], R. Loo[c], D. Nguyen[c], W. Buchholtz[d], A. Wei[d], M. Horstmann[d], D. Feijoo[e], and P. Storck[e]

[a] Max Planck Institute of Microstructure Physics, Halle, Germany
[b] Institute of Bio- and Nanosystems, Research Centre Juelich, Juelich, Germany
[c] IMEC, Leuven, Belgium
[d] AMD Saxony LLC & Co. KG, Dresden, Germany
[e] Siltronic AG, München, Germany

SSOI substrates were successfully fabricated using He^+ ion implantation and annealing to relax thin (< 500nm) SiGe buffer layers, bonding and layer transfer processes to realize strained-Si layers onto oxide layers. The reduced thickness of the SiGe buffer possess numerous advantages such as reduced process costs for epitaxy and for reclaim of the handle wafer if the layer splitting is initiated in the SiGe/Si interface.

The electron mobilities in the fabricated SSOI layers were measured using transistors with different gate lengths. An electron mobility of ~530 cm^2/Vs was extracted, being much higher than in non-strained SOI substrates. Furthermore, an 80% drive current (I_{DSAT}) improvement has been measured for long channel devices.

Introduction

The performance improvements in CMOS circuits during the last decades result primarily from reductions in the dimensions of the individual transistors. The smaller device size permits a higher device density that has, for example, resulted in higher clock speeds of logic devices. As device dimensions approach values below 100nm, scaling becomes increasingly difficult. Strain engineering and material innovations have been identified as the main contributors to the continued performance improvement in CMOS devices. One example implemented recently is the silicon-on-insulator (SOI) material. Further improvements of the performance are obtained by an increased carrier mobility which has been reported for devices fabricated on strained silicon layers (for example [1-3]). Combining the advantages of SOI and strained silicon results in strained silicon on insulator (SSOI) substrates connecting the properties of both materials.

For fabrication of SSOI wafers strained silicon (sSi) layers grow on a relaxed SiGe virtual substrate and were then transferred to oxidized Si handle wafers by direct wafer bonding. The Ge content in the SiGe alloy and the degree of plastic relaxation in the SiGe with respect to unstrained Si determines the degree of strain in the sSi layer. The relaxation of the SiGe is mediated through misfit dislocations near the SiGe/Si interface. The misfit dislocations are connected to the free surface by dislocation segments threading through the layer. These threading dislocations (TDs) are penetrating not only through the SiGe but also through the sSi layer and may thereby deteriorate the device performance. The density of TDs is reduced by slow compositional grading of the SiGe layer. Typical values are 10% Ge content grading per μm [4,5] resulting in thick buffer layers increasing the process costs. An alternative is the relaxation of a thin pseudomorphic SiGe layer (<500 nm) induced by hydrogen or helium implantation and subsequent annealing [6,7].

Experimental

The epitaxial SiGe layers were deposited on 200 mm Si wafers (orientation [100]), using a horizontal cold wall, load locked Reduced Pressure Chemical Vapor Deposition System (RP-CVD, ASM Epsilon-2000 reactor), which has been developed for production applications. Epitaxial SiGe growth was done at 600°C and 40 Torr, using SiH_4 and GeH_4 (1% in H_2) as Si and Ge source gases. H_2 was used as carrier gas.
A $Si_{0.77}Ge_{0.23}$ layer, about 200 nm thick, was grown first on the Si substrate. The layer was covered with a 6 nm thick Si film. An implantation with He ions (energy 40 keV, dose $7{\cdot}10^{15}$ cm^{-2}) and a subsequent annealing in Ar at 850°C for 600 sec were applied to relax the SiGe layer. During a second epitaxial growth a 200 nm thick strain-adjusted SiGe layer having a Ge content of 16 at% was deposited. In this case the overgrowth layer is fully relaxed on top of the 70% relaxed $Si_{0.77}Ge_{0.23}$ layer. Finally a 20 nm thick strained silicon layer was grown.
After deposition of an oxide layer (thickness 190 nm) the virtual substrates were implanted with hydrogen using 115 keV H_2^+ ions with a dose of $3{-}5{\cdot}10^{16}$ cm^{-2}. The energy causes the peak of implantation induced damage to be about 0.4 μm below the sSi/SiO_2 interface according to simulations. Standard chemical cleaning (SC1 + SC2) followed by de-ionized (DI) water rinse was employed for all the wafers prior to bonding. The virtual substrates were then bonded to Si handle wafers in a Süss CL 200. In order to increase the bond strength and to induce the layer splitting, the wafer pairs were annealed at moderate temperatures.
The remaining SiGe layers were finally removed by selective etching in $HF:H_2O_2$: CH_3COOH (Hac) solutions.
After SSOI fabrication a further epitaxial growth process was applied to increase the thickness of the strained Si layer. The SSOI wafers were wet chemically cleaned including an HF dip to remove the native oxide. After loading into the epitaxy reactor, the SSOI wafers received a pre-epi bake at 800°C to remove traces of remaining oxide. Si growth was done at 700°C and 40 Torr using $SiCl_2H_2$ as Si source gas.

SSOI Fabrication

The virtual substrate consists of a 200 nm thick $Si_{0.77}Ge_{0.23}$ layer pseudomorphically grown on a Si(100) wafer. The SiGe layer is initially fully strained and the strain relaxation occurred by performing a He implantation and subsequent annealing at 850°C for 10 min. During this process a narrow defect band underneath the SiGe/Si substrate interface is generated (Fig. 1). It provides a high density of dislocation loops as sources for misfit dislocations resulting in efficient strain relaxation during annealing. The dislocations propagate from the bottom of the relaxed SiGe towards the surface into the Si cap layer. On top of the $Si_{0.77}Ge_{0.23}$ layer a strain adjusted $Si_{0.84}Ge_{0.16}$ layer was grown acting as substrate for the deposition of a strained silicon layer about 20 nm thick (Fig. 2).

The dislocation density was determined by optical microscopy and selective chemical etching (dilute Secco etch). A TD density of about $6 \cdot 10^5$ cm^{-2} and a dislocation pile-up density of 25 cm^{-1} were measured [7]. Furthermore, AFM analysis proved a surface roughness (root mean square) RMS < 1 nm. In contrast, graded buffers cause the generation of a cross-hatch pattern on the surface that gives rise to a RMS value > 1 nm [8]. In this case surface planarization is required prior to wafer bonding process.

Layer transfer processes based on hydrogen implantation and wafer bonding were employed for the SSOI fabrication. Designed PE-CVD layers were deposited first on the virtual substrates acting as protection layer during the hydrogen implantation and as buried oxide layer in the final SSOI wafer. The hydrogen implantation was carried out with H$_2^+$ at 115 keV using doses of 3 to 5 x10^{16} cm^{-2}. The energy causes the peak of implantation induced damage to be about 0.4 μm below the strained silicon surface, i.e. close to the interface between the underlying Si$_{0.77}$Ge$_{0.23}$ layer and the Si substrate. Therefore, the layer splitting is initiated at this interface resulting in the complete transfer of the sSi- and SiGe layers (Fig. 3). This means that all SiGe layers are removed from the Si wafer of the virtual substrate which can be reclaimed without expensive processes.

A perfect layer transfer is obtained during a 2-step annealing process, which consists of a first annealing at 300 °C (in order to increase the bonding strength (interface energy) to about 1.5 J/m²) followed by a second annealing step at 450 °C to initiate the layer splitting. The surface roughness of the transferred layer stack is about 7.5 nm (RMS) measured by AFM.

The SiGe layers were removed after the layer transfer by selective spin etching (Fig. 4).

Figure 1: TEM cross sectional (XTEM) image of the pseudomorphically grown Si$_{0.77}$Ge$_{0.23}$ layer after relaxation by He implantation and subsequent annealing.

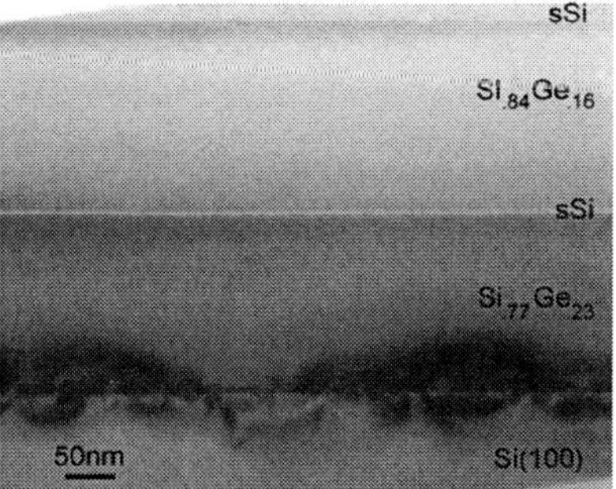

Figure 2: Typical XTEM image of the SiGe virtual substrates.

Figure 3: XTEM image of the stack after layer transfer.

Figure 4: XTEM image of the final SSOI wafer.

Characterization

The application of thin SiGe buffer layers instead of the conventionally applied thick buffers possess numerous advantages. Besides the reduced effort for epitaxy, virtual substrates having a thinner SiGe buffer are characterized by a lower compressive stress of the whole layer system causing the bow of the wafers. A low value of the bow (or the compressive stress) is especially important for the layer transfer process, because large stresses result in debonding instead of layer splitting during annealing. Figure 5 shows the bow of a virtual substrate wafer having a thin SiGe buffer during an annealing cycle up to 500°C. The measured deviation of the bow is only 1.5 μm corresponding to a low compressive stress of about 25 MPa. On the other hand, conventionally applied thick buffer layers are characterized by an alteration of the bow of more than 16 μm during the same annealing cycle.

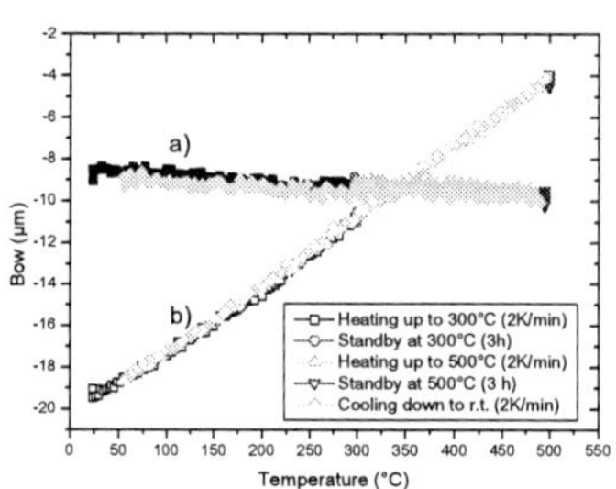

<u>Figure 5:</u> Results of measurements of the bow of different virtual substrates during annealing (wafer diameter 200 mm). a) Virtual substrate used for present investigations (thin SiGe buffer layer, full circles). b) Conventional substrate having at thick (4 μm) SiGe buffer layer (open circles).

Using virtual substrates with thin SiGe buffer layers SSOI wafers with homogeneous strained silicon layers are obtained. The surface roughness of the sSi layer is below 1 nm (RMS) after the final spin etching. UV-Raman measurements proved that the whole strain of the initial sSi layer (grown on the virtual substrate) is transferred to the SSOI wafer.

Measurements over large areas (a quarter of a 200 mm wafer) result in variations of the strain of $\Delta\varepsilon = \pm 0.005\%$ [9]. Variations of the strain are found in smaller areas correlate to the cross hatch pattern. Fig. 6 shows a 2-dimensional map obtained by UV-Raman spectroscopy. Areas of 300 μm x 300 μm were measured after the layer transfer and selective etching and after an additional epitaxial step in order to increase the thickness of the sSi layer to about 60 nm.

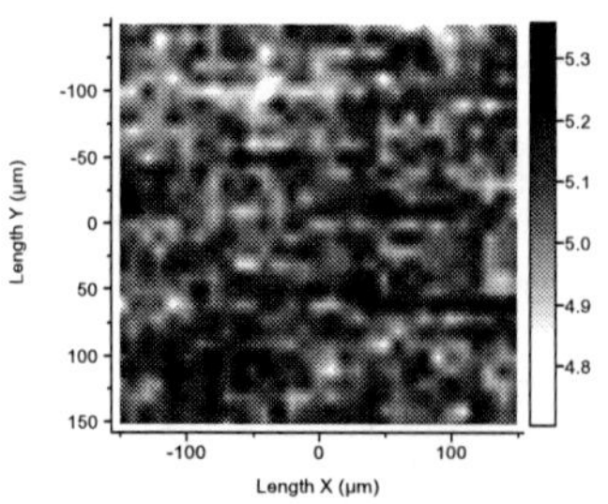

<u>Figure 6:</u> Raman mapping of the surface area of a SSOI wafer after layer transfer and selective etching of SiGe. Measured area 300 μm x 300 μm corresponding to 961 points.

According to these measurements a mean value of the strain of $\varepsilon = 0.62 \pm 0.013$ % result after the layer transfer (statistics of 961 measured points), while a mean value of $\varepsilon = 0.59 \pm 0.014$ % is obtained after the final epitaxial growth.

MOSFETs were fabricated on SSOI substrate with a thickness of the top strained-Si layer of 58 nm. Fig. 7 shows the transfer characteristics of the device with a gate length of 5μm and a gate width of 20 μm. The sub-threshold slope is 75mV/dec. The insert in Fig.7 shows the

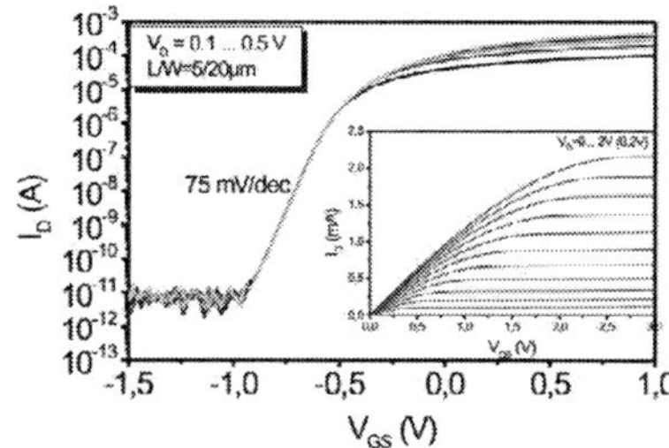

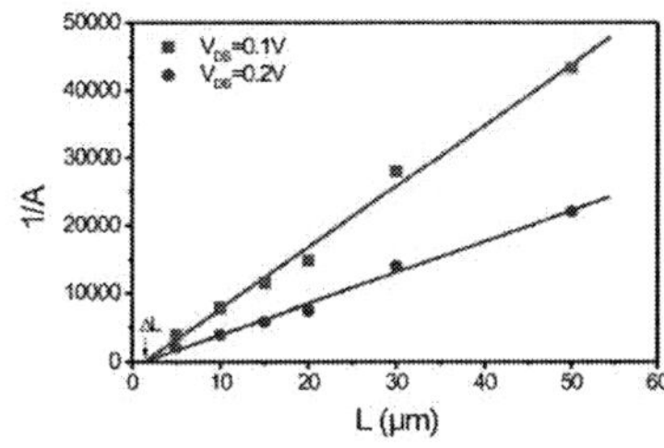

Figure 7: Transfer and output (insert) characteristics of an n-MOSFET .

Figure 8: Mobility extraction from the curve of 1/A as a function of the mask channel length L.

output characteristics of the device. The device exhibits a quite large S/D resistance because no silicide is applied at S/D before the deposition of the Al contacts. The S/D resistance could be easily lowered by using NiSi silicided contacts.

The mobility of the electrons is calculated from the transfer characteristics of the devices. using the following equations

$$\frac{I_D}{\sqrt{g_m}} = \sqrt{A}\left(V_G - V_{TH}\right) \tag{1}$$

$$A = \mu_0 C_{ox} \frac{W}{L_{eff}} V_D, \tag{2}$$

where g_m is the transconductance of the device which is defined by dI_D / dV_{GS}. V_{TH} is the threshold voltage, C_{ox} the gate oxide capacitance, μ_0 the mobility of the carriers, and L_{eff} the effective channel length. A linear dependence should be obtained by drawing 1/A as a function of the gate length, as shown in Fig.8. We extracted an electron mobility of 530 cm^2 /Vs, which is much larger than the value in the non-strained SOI substrates of 330 cm^2 /Vs. Analogous results were also obtained on pseudo-MOSFET structures resulting in the same mobility value [10].

A state of the art CMOS process with 4 uniaxial stressors (NMOS: stress memory and tensile overlayer, PMOS: embedded-SiGe and compressive overlayer) used in volume manufacturing [11] was run on the SSOI wafers. Fig. 9 shows a cross sectional TEM image of the NMOS with stress memory and tensile overlayer. Long channel devices show clearly the benefit of the biaxially-strained SSOI wafers. An 80% drive current (I_{DSAT}) improvement at same source-to-drain leakage (I_{OFF}) has been measured (Fig. 10). Short channel devices with Lg=40nm show an I_{DSAT} improvement of up 10% limited by source-to-drain resistance components [12, 13].

Conclusions

SSOI substrates were successfully fabricated using He$^+$ ion implantation and annealing to relax thin (< 500nm) SiGe buffer layers, bonding and layer transfer processes to realize strained-Si layers onto oxide layers. The low surface roughness allows wafer bonding without any pretreatments for the virtual substrates. A perfect layers transfer is obtained during annealing at 450 °C. The reduced thickness of the SiGe buffer possess numerous

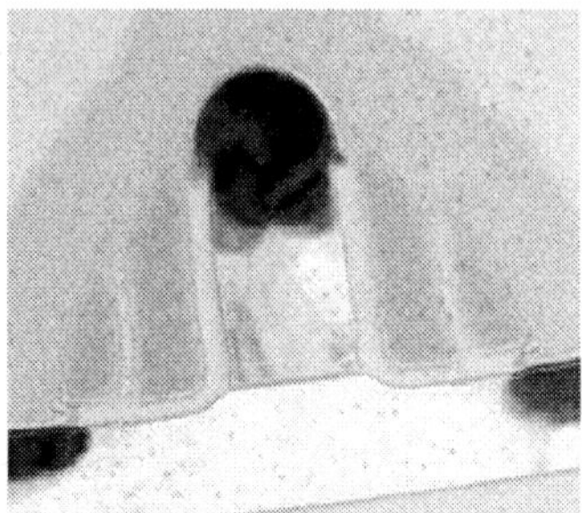

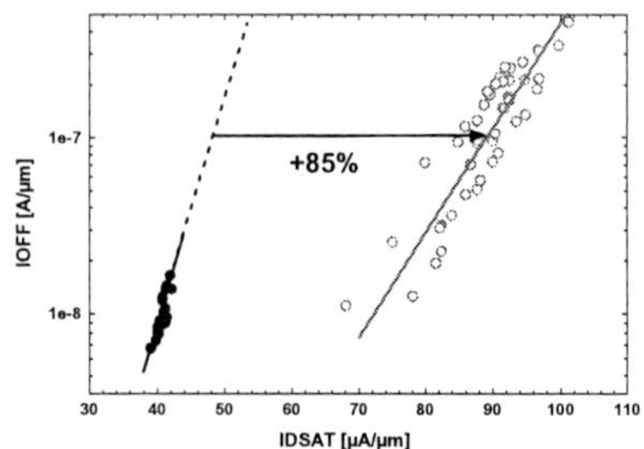

Figure 9: TEM cross sectional image of an NMOS short channel transistor (L_g = 40 nm) fabricated on SSOI.

Figure 10: Off current versus saturated drive current (long channel device, L_g = 1 µm). An improvement of about 80% results to control wafers (SOI, full circles).

advantages such as reduced process costs for epitaxy and for reclaim of the handle wafer if the layer splitting is initiated in the SiGe/Si interface.The electron mobilities in the fabricated SSOI layers were measured using transistors with different gate lengths. An electron mobility of ~530 cm^2 /Vs was extracted, being much higher than in non-strained SOI substrates of 330 cm^2 /Vs. Furthermore, an 80% drive current (I_{DSAT}) improvement has been measured for long channel devices, while an I_{DSAT} improvement of up 10% result for short channel devices.

Acknowledgments

We are thankful to S. Hopfe and R. Scholz for the sample preparation and XTEM investigation. This work was financially supported by the German Federal Ministry of Education and Research in the framework of the TeSiN project (contract no. V03110).

References

1. I. Cayrefoureq et al., *ECS Transactions* **3**(7) p. 399, The Electrochemical Society, Pennington, NJ (2006).
2. G. Taraschi et al., *J. Electrochem. Soc.* **151**, G47, (2004).
3. L. J. Huang et al., Symp. *VLSI Techn. Dig.* **57**, (2001).
4. P.M. Mooney, *Mater. Sci. Eng.*, **R 17**, 105, (1996).
5. E. A. Fitzgerald et al., *Appl. Phys. Lett.*, **59**, 811, (1991).
6. S. Mantl et al., *Nucl. Instr. and Meth. Phys. Res. B*, **147**, 29, (1999).
7. S. Mantl et al., *ECS Transactions* **3**(7) p. 1047, The Electrochemical Society, Pennington, NJ (2006).
8. I. Radu et al., *ECS Transactions* **3**(7) p. 317, The Electrochemical Society, Pennington, NJ (2006).
9. B. Ghyselen, private communication
10. Q.T. Zhao et al., paper submitted to the 8th Intern. Conference on Ultimate Integration on Silicon (ULIS), Leuven March (2007)
11. M. Horstmann et al., *IEDM Tech. Digest*, 243 (2005).
12. A. Wei et al.., *ECS Transactions* **3**(7) p. 719, The Electrochemical Society, Pennington, NJ (2006).
13. A. Wei et al., *SSDM*, 32 (2005).

ECS Transactions, 6 (4) 345-350 (2007)
10.1149/1.2728881, ©The Electrochemical Society

Advanced heterostructure Si-InSb on insulator formed by bonding of hydrogen transferred Si layer and implanted SiO_2 film

V. P. Popov[a], I. E.Tyschenko[a], A. G. Cherkov[a], G. P. Pokhil[b], V. M. Fridman[b], M. Voelskow[c]

[a] Institute of Semiconductor Physics, Novosibirsk 630090, Russia
[b] MSU Research Institute of Nuclear Physics, Moscow, 119992, Russia
[c] Institute of Ion Beam Physics and Material Research, 01314 Dresden, Germany

Using bulk silicon may be limited for 22 nm technological node due to silicon mobility limitation. New type of substrates needs for further scaling in CMOS microelectronics. We speculate that this new type of materials can be semiconductor heterostructure on insulator (HOI) compatible with current silicon planar CMOS technology. In this work an effect of interface mediated endotaxial (IME) growth of thin InSb film at Si/SiO_2 bonded interface was experimentally observed and investigated for the first time. Joint semiconductor material stack obtained by hydrogen transfer of one layer material (silicon) and endotaxially grown second one (indium antimonide) placed initially into amorphous silicon dioxide film is presented. Thermodynamic, kinetic and lattice mismatch parameter influences on IME process are considered.

Introduction

Transfer to tera-scale integration with the CMOS transistors channel length, lower than 10 nm and based on silicon, is difficult due to decreasing a carrier mobility, increasing leakage currents, and respective growth of heat-generation caused by the low dimensions. Last few years, a number of the structures, that promote an increase in the carrier mobility in the channel, were proposed. There are stressed silicon-on-insulator film and pure germanium instead of silicon. Recent publications (1-4) shown, that SiGe quantum well (QW) device structure and its drain current gate voltage characteristics have great advantages relative to the volume silicon. SiGe thin film dual channel (TF DC) MOSFETs may provide even higher parameters that needs for current technological 45 nm node (2). Other publications (5,6) shown that QW FET based on A3B5 structures may have higher characteristics, than even SiGe FETs. Among them there is InSb based FETs, where the electrons have 50 times higher mobility at RT compared to silicon (5).

We have suggested earlier (6) to use hydrogen transfer of Si layer on implanted with Ge, oxidized, and bonded wafer to produce semiconductor heterostructure on insulator (HOI). Using transferred layer as a matrix for monocrystalline heterolayer endotaxy allows obtaining Ge film coherent with Si lattice. Film growth, as suggested, may be mediated by implanted hydrogen and vacancy-rich bonded interface.

Physical approach for InSb is based on high thermal mobilities of In and Sb atoms in Si/SiO_2 heterostructures and their segregation at the grain boundary. Particularly as it was shown in the works (7,8) In atoms implanted in Si or SiO_2 layers segregate to the interface between them at the annealing temperatures >800°C. The same behavior was observed for Sb atoms, but only in the case of implantation in silicon, because in SiO_2 these atoms are relatively immobile (9,10). Subsequent implantation of Sb and In atoms in silicon at high concentration allows to nucleate crystalline InSb embryo's at the defects in implanted layer (11). Using of SOI structure can promise nucleate InSb layer at Si/SiO_2 grain boundary. The same approach was used by us (6) for monocrystalline Ge layer formation at the silicon – silicon dioxide grain bounadary, where the role of silicon crystalline template matrix was provided by Si layer in SOI structure.

Using IME method for InSb monocrystalline layer formation is connected with the few problems that should be solved during the investigation. Even in the case of germanium where the segregation observed (6,12-15), there is not an answer about the roles of: strain near the interface; weaken and broken covalent bonds here; other impurities like hydrogen or carbon; internal electric field in the heterostructure; applied external electric field during impurity diffusion on the segregation of impurities and the nucleation of new phase at the Si/SiO_2 grain boundary. We did not find also in the literature the data about InSb segregation to this interface during their implantation into silicon. The main goal of the work was a study of nucleation and growth of monocristalline indium antimonide thin film at the Si/SiO_2 grain boundary. Using our DeleCut method (16) we can separate diffusion of Sb atoms from the silicon film, whereas In atoms can segregate to the interface from silicon dioxide, grown on another Si wafer before bonding.

Experiment and results

The scheme used for Si-InSb heterostructure on insulator (HOI) formation was the next. In and Sb ions were implanted with flunce $(0.5\text{-}1.5)\times10^{16}$ cm^{-2} and energy 100 keV in thermally grown silicon dioxide with 300 nm thickness on handle wafer before bonding. Transferring of Si film on Ge implanted dioxide on handle wafer was made by direct bonding and slicing. Different thermal treatments at $500 \div 1100^{\circ}C$ in inert or oxidation atmosphere during $0.5\div5$ h were used after Si film transfer to syntheses InSb semiconductors at the bonding interface.

Random and aligned RBS spectra were measured for InSb position and crystalline properties determination. According to the spectra on Figure 1 the InSb crystal part, coherent with upper Si lattice, has not monotonic dependence on the temperature contrary to Ge atoms (6). This part is small and growths slightly at $700^{\circ}C$, has a highest value at $900^{\circ}C$ and practically coincides after annealing at 800 and $1100^{\circ}C$ in quantities and in forms with the most of InSb atoms in the middle of BOX, while there is clear accumulation of InSb atoms at both interfaces at $900^{\circ}C$. RBS data coincide with TEM/HREM cross-section images as it is shown on Figure 5. A lot of InSb nanocrystalls looked as bright spots an dark field images (Figure 2 b,c) with size lower than 20 nm are placed inside the BOX with other orientations than upper Si lattice. They are partially coherent with Si lattice only at Si/SiO_2 interface after annealing at higher temperature $900^{\circ}C$ (Figure 2 c).

There is not any evidence of InSb atoms at Si/SiO_2 interface after higher temperatures (1000 or $1100^{\circ}C$). All InSb nanocrystals with larger sizes (up to 30 nm) were placed at the middle of BOX similar to the case at $800^{\circ}C$.

Strong difference in behavior of InSb in comparison with Ge atoms (6) may be attributed to extremely high lattice mismatch of Si and InSb atoms (>18%), while for Si and Ge lattices this value is near 4%. To increase upper Si lattice parameter we use Sb implantation and solid state epitaxy of upper Si layer done before bonding. We suggested that this increase in Si lattice constant due to Sb atom incorporation would serve for stress accommodation at InSb film growth at Si/SiO_2 interface. All other treatments were the same as in the case of In and Sb ion implantation into BOX.

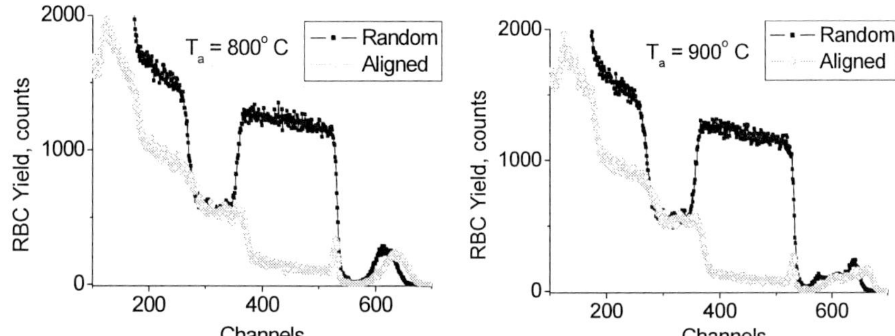

Figure 1. RBS/channeling spectra of SOI with InSb in buried oxide SiO_2 (BOX) after annealing at 800°C (a) and 900°C (b).

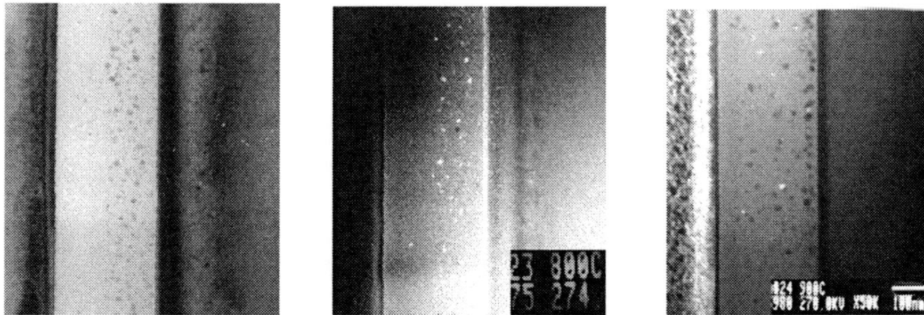

Figure 2. XTEM bright (a) and dark (b,c) field images of SOI with InSb atoms in buried oxide (BOX) after annealing at 800°C (a,b) and 900°C (c). Bright strip on image (b) is due to InSb film incoherent with upper Si lattice.

The RBS/channeling results are presented on Figure 3 for this approach. An evidence of strong InSb atom accumulation at upper Si/SiO_2 interface is seen as peak at channel number 750 on Figure 3 b. In this case we observed higher coherent part (~30%) of InSb atoms with upper Si lattice. They are equal to 5-7 monolayers of InSb film at Si/SiO_2 interface. High resolution image on Figure 4a confirms this estimation. InSb film looks like dark strip of coherent atoms at the interface. Dark spots in BOX are due to incoherent InSb nanoclusters near the interface. HREM image on Figure 4b shows a transition from flat to rough InSb film and a growth of strain in Si crystal after higher temperature treatment at 900°C. Break of pseudomorphic growth is also observed in thicker parts of InSb film with highest contrast. Higher temperature 1000 and 1100°C completely destroy pseudomorphic InSb films and lead to semicoherent InSb islands placed exactly at top Si/SiO_2 interface (Figure 5).

BOX and bottom Si/SiO_2 interface are completely free from InSb nanoparticles. Semicoherent InSb islands partially placed at the both sides of top Si/SiO_2 interface and has (111) planes declined from (111) planes of upper Si layer with interplane space equal to 0.365 nm, which corresponds to this value for pure InSb.

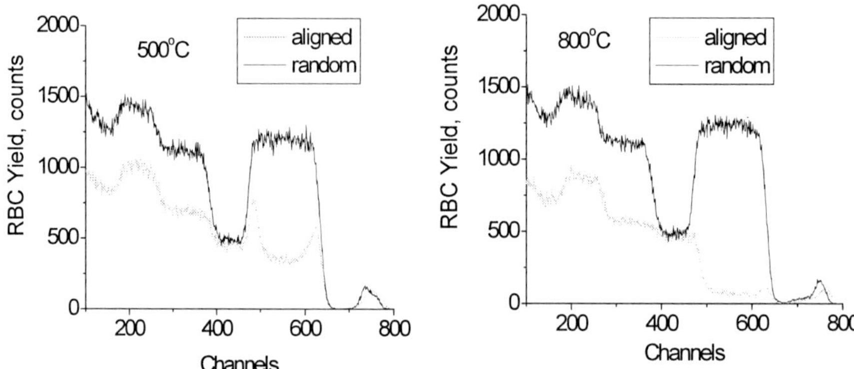

Figure 3. RBS/channeling spectra of SOI with InSb in BOX and at upper Si/SiO$_2$ interface after annealing at 500°C (a) and 800°C (b).

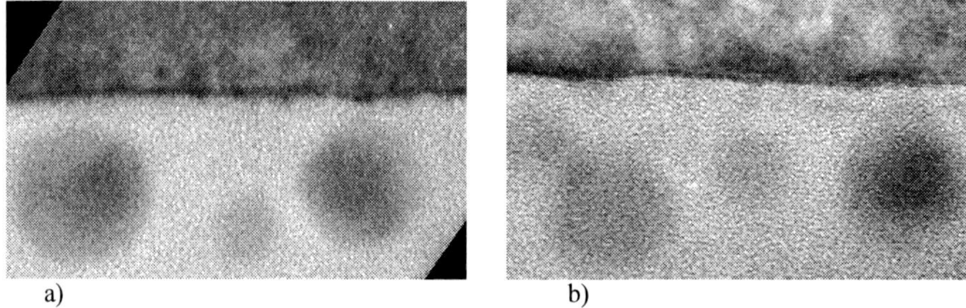

a)   b)

Figure 4. HREM image of SOI with InSb atoms at upper Si/SiO$_2$ interface and in buried oxide SiO$_2$ (BOX) after annealing at 800°C (a) and 900°C (b) for tensile strained upper Si layer.

These InSb islands have an elliptic form with dimensions 20-30 nm along Si/SiO$_2$ interface and 10-20 nm across its with moiré fringes on Si side (Figure 5 insert). Using tensile strain in top Si layer allows obtaining quite different InSb atom distribution including thin pseudomorphic film growth in SOI structure (see Figures 2, 4 and 5) and semicoherent InSb island formation at top Si/SiO$_2$ interface. This approach may be useful in the case of huge lattice mismatch for HOI formation by IME process.

Discussion

Monotonic increase of Ge film thickness (6) with increase in time or temperature is in contrast with disappearance of InSb film during increase in time. It may be connected with more thermodynamically stable configuration of Ge film in SOI at Si/SiO$_2$ interface than in the case of InSb. Let consider thermodynamic quantities connected with new layer formation. As it is known from epitaxial growth of new film in vapor or in liquid in order to be stable the next relation between interfacial energies should take place (Frank – Van der Merwe growth):

$$\gamma_{BC} > \gamma_{AB} + \gamma_{AC} \, , \qquad\qquad (I)$$

Here γ_{BC} is an initial interface energy between B and C substances, and A is a new substance with interfacial energies γ_{AB} and γ_{AC} between them. We suggested that the same relation is true for the case of endotaxy or oriented growth inside solid state without elastic stresses.

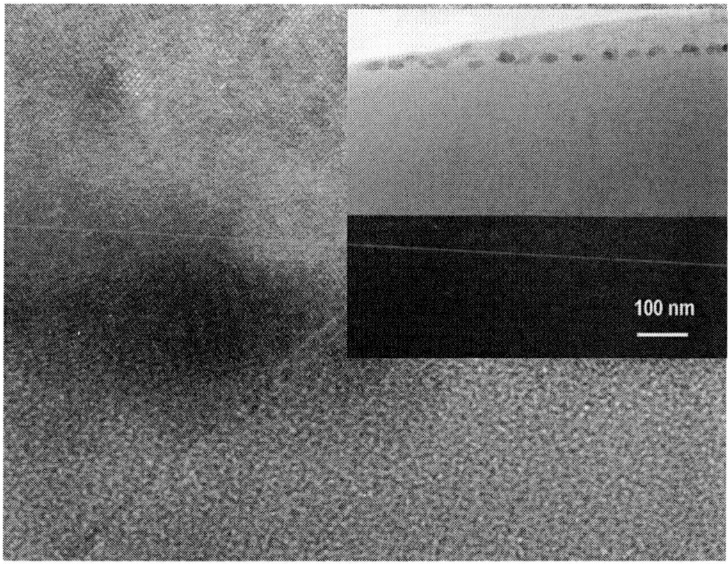

Figure 5. HREM and Bright XTEM (insert) images of SOI with InSb nanocrystals at top Si/SiO₂ interface after annealing at 1100°C during 1 h. Dark strips on HREM image is due to InSb lattice moiré fringes with Si.

According to the published data (see work (17) and references therein) γ_{BC} (or interfacial Si/SiO₂ energy) inequation I does not satisfy and driving forces for Ge film formation is rather kinetics than thermodynamics (6). Unfortunately, interfacial energy data for InSb, Si and SiO₂ system do not complete in literature to give a conclusion about mechanism, but it is clear that BOX has preference for InSb nanocrystal growth than Si/SiO₂ interface may be due to large lattice mismatch for pseudomorphic growth.

Pseudomorphic growth of strained heterofilm with n monolayer thickness is thermodynamically favorable when elastic energy contribution σ(n) to the total energy is still lower than previous interfacial one:

$$\gamma_{BC} > \gamma_{AB} + \gamma_{AC} + \sigma(n) \, , \qquad\qquad (II)$$

If inequation II does not satisfy, metamorphic growth of InSb film with different than (100) silicon orientation is still possible. It can be obtained using patterned implantation or tensile stress in top Si film. Even small tensile stress (few percents) obtained by Sb implantation and solid state crystallization allowed us to grow extra thin (5-7 ML) InSb film at moderate temperature 800°C. Such tensile deformation can be introduced using silicon nitride capping layer or silicide growth on thin top Si layer. Metamorphic InSb nanoislands with other orientation than Si can form during high temperature treatment by their melting and recrystallization at Si/SiO_2 interface.

Conclusion

We proposed in this work a simple CMOS compatible approach named Interface Mediated Endotaxy (IME) for silicon - indium antimonide heterostructure-on-insulator (HOI) formation. Upper Si film with built-in few nanometer thin A3B5 layer allows obtaining of new material for CMOS production. Presented here results allow us to conclude that suggested IME method can be attractive alternative for integration of compaund semiconductors into new generations of CMOS technological nodes.

References

1. C. W. Leitz, M. T. Currie, M. L. Lee, Z.-Y. Cheng, D. A. Antoniadis, E. A. Fitzgerald, *J. Appl. Phys.* **92**, 3745 (2002).
2. S. Takagi, T. Mizuno, T. Tezuka et al., N. Sugiyama, S. Nakaharai, T. Numata, J. Koga, K. Uchida, *Solid-State Electronics,* **49**, 6844 (2005).
3. Z. Cheng, J. Jung, M. L. Lee, A. J. Pitera, J. L. Hoyt, D. A. Antoniadis and E. A. Fitzgerald, *Semicond. Sci. Technol.,* **19,** L48 (2004).
4. M. L. Lee, E. A. Fitzgerald, M. T. Bulsara, M. T. Currie, and A. Lochtefeld, J. Appl. Phys. **97**, 011101 (2005).
5. S. Datta, T. Ashley, J. Brask, L. Buckle, M. Doczy, M. T. Emeny, D. G. Hayes, K. P. Hilton, R. Jefferies, T. Martin, T. J. Phillips, D. Wallis, P. J. Wilding, and R. Chau, IEDM Technical Digest, 783 (2005).
6. V. P. Popov, I. E. Tyschenko A. G. Cherkov, M. Voelskow. ECS Trans. **3**, (7) 303 (2006).
7. Peng Chen, Zhenghua An, Ming Zhu et al. Materials Science and Engineering B, **114–115,** 251 (2004).
8. Peng Chen, Ming Zhu, Ricky K. Y. Fu et al. J. Appl. Phys., **96**, 3217 (2004).
9. J. S. Williams, M. Petravic, Y. H. Li, J. A. Davies and G. R. Palmer. NIM B, **64**, 156 (1992).
10. [V. A. Ignatova, O. I. Lebedev, U. Wätjen, L. Van Vaeck, J. Van Landuyt, R. Gijbels F. Adams. J. Appl. Phys., **92**, 4336 (2002).
11. C. W. White, J. D. Budai, S. P. Withrow, J. G. Zhu, S. J. Pennycook, *et al.* NIM B, **127-128**, 545 (1997).
12. L. Rebohle, I. E. Tyschenko, J. von Borany, B. Schmidt, R. Grötzschel, A. Markwitz, R. A. Yankov, *et al..* Mater. Res. Soc. Symp. Proc. **486**, 175 (1998).
13. T. Gebel, L. Rebohle, W. Skorupa, A. N. Nazarov, I. N. Osiyuk, V. S. Lysenko. Appl. Phys. Lett. **81**, 2575 (2002).
14. A. N. Nazarov, T. Gebel, L. Rebohle, W. Skorupa, I. N. Osiyuk, V. S. Lysenko. J. Appl. Phys. **94**, 4440 (2003).
15. M. Klimenkov, J. von Borany, W. Matz, R. Gro"tzschel, and F. Herrmann. J. Appl. Phys. 91, 10062 (2002).
16. V. P. Popov, I. E. Tyschenko. Method of silicon-on-insulator structure creation. Invent of Russian Federation, №2003100747 from 14.01.2003.
17. Q. Xu, I. D. Sharp, C. W. Yuan, D. O. Yi, C. Y. Liao, A. M. Glaeser, A. M. Minor, J. W. Beeman, M. C. Ridgway, P. Kluth, J. W. Ager III, D. C. Chrzan, E. E. Haller. Phys. Rev. Lett. **97**, 155701 (2006).

SESSION 7

COMPUTER SIMULATIONS

ECS Transactions, 6 (4) 353-362 (2007)
10.1149/1.2728882, ©The Electrochemical Society

Atomic-scale simulations of electron mobilities in ultrathin SOI MOSFETs

Sokrates T. Pantelides,[a,b]
G. Hadjisavvas,[a] M. H. Evans,[a*] L. Tsetseris,[a]
M. Caussanel,[c†] and R. D. Schrimpf[c]

[a]Department of Physics and Astronomy, Vanderbilt University, Nashville, TN 37235,
USA
[b]Oak Ridge National Laboratory, Oak Ridge, TN 37831, USA
[c]Department of Electrical Engineering and Computer Science, Vanderbilt University,
Nashville, TN 37235, USA

Carrier mobilities in metal-oxide-semiconductor field-effect
transistors (MOSFETs) are conventionally modeled in a semiclassical approximation, i.e., particles are treated as classical particles with a suitable effective mass or with a kinetic energy given
by the crystalline energy bands. Here we review a recent method
based on atomic-scale models of the Si-SiO$_2$ interface. Carriers are
described by wave functions and scattering potentials are constructed by parameter-free quantum-mechanical calculations. The
effect of primitive atomic-scale roughness (Si-Si bonds on the oxide side and Si-O-Si bonds on the Si side of the interface) in ultra-
thin SOI MOSFETs has been investigated. It is shown that this
mechanism naturally accounts for strain-induced enhancement in a
quantitative way. In addition, the method has been used to determine the impact of stray Hf atoms in the SiO$_2$ interlayer in alternate-dielectric Si-SiO$_2$-HfO$_2$ structures. Finally, some features of
the method are compared with the "density gradient" method.

INTRODUCTION

Conventional methods for modeling mobilities in Metal-Oxide-Semiconductor
Field-Effect Transistors (MOSFETs) are based on a semiclassical approximation. The
underlying atomic structure in the channel and the Si-SiO$_2$ interface is suppressed.
Electrons are treated as a classical Maxwell-Boltzmann gas with the kinetic energy
expressed in terms of a suitable effective mass derived from the energy bands or directly
in terms of the bulk-crystal energy bands [1,2]. In this context, the Si-SiO$_2$ interface is
treated as an infinite potential barrier, confining electrons in the Si channel (no
penetration into the gate oxide). Empirical models for various scattering mechanisms are
adopted and the Boltzmann transport equation is solved at various levels of sophistication
(the drift-diffusion approximation, the hydrodynamic approximation, or by solving the
full Boltzmann equation using Monte-Carlo methods [3]).

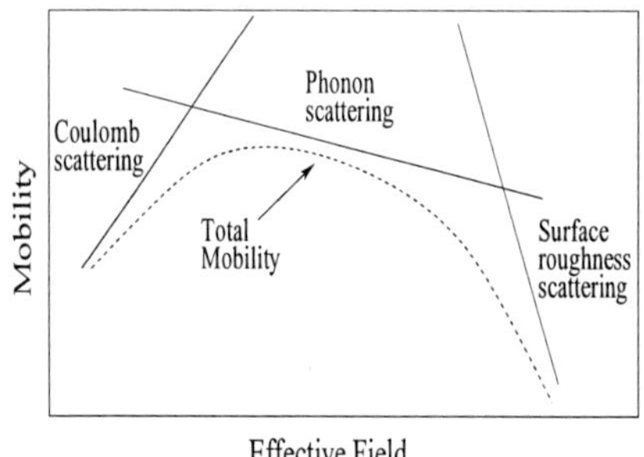

Fig. 1 Schematic diagram showing the different mechanisms that dominate scattering for different values of the effective field in a MOSFET (see Ref. 4)

Over several decades, experiments established that mobilities in Si-based MOSFETs fall on a "universal mobility curve" (UMC) as functions of the effective gate field or, equivalently, the electron density in the channel (the two are related by a simple relation) [4]. Conventional modeling has successfully reproduced the UMC. It is generally accepted that the dominant mechanisms at different values of the effective field are as shown in the schematic diagram of Fig. 1 (see Ref. 4).

In recent years, however, with device features in the nanoscale and with the advent of new materials and novel device structures, it has been widely recognized that the conventional mobility models are reaching their limits of applicability. It has been recognized that, for nanoscale devices, carrier penetration into the oxide can no longer be neglected. Phenomenological models to incorporate the effect have already been proposed [5]. Inclusion of the effect becomes more urgent as alternative high-k dielectrics are being explored because the barrier resulting from the Si-dielectric band alignments is generally smaller. It has also been found that strained Si leads to mobility enhancement. Current technologies take advantage of the effect by using strained-Si channels. Conventional modeling, however, cannot account for the effect. In a careful analysis, Fischetti et al. [6] demonstrated that the experimental data for strain-enhanced mobilities can be reproduced only by assuming that scattering from interface roughness is reduced significantly in a strained-Si channel, but it was not possible to obtain a physical justification for such an assumption. Finally, mobilities in ultrathin double-gate MOSFETs, fabricated using Silicon-on-Insulator technology (UTSOI MOSFETs) have also been found to be off the UMC [7] and accurate modeling of the effect has remained elusive.

In 2005, the present principal author and his collaborators developed a new approach to parameter-free quantum-mechanical calculations of carrier mobilities [8]. The approach is based on *atomic-scale models of the Si-SiO$_2$ interface*. Electrons are described by wave functions so that penetration into the gate dielectric is automatic. Scattering potentials are calculated in a parameter-free way using Density Functional Theory (DFT), the state-of-the-art method for quantum mechanical calculations in solids. The corresponding scattering rates (carrier lifetimes) are then calculated in the Born approximation and mobilities are calculated using the linearized Boltzmann equation. In the initial implementation reported in Ref. 8, a UTSOI MOSFET structure (Fig. 2) was adopted. Additional results were reported at the IEEE International Electron Devices Meeting (IEDM) in 2005 [9]. The work was summarized at the 11th International Workshop on Computational Electronics in Vienna in 2006 [10]. Since then new results have been obtained [11]. Here, we will give an overview of the theoretical method and a summary of results available to date.

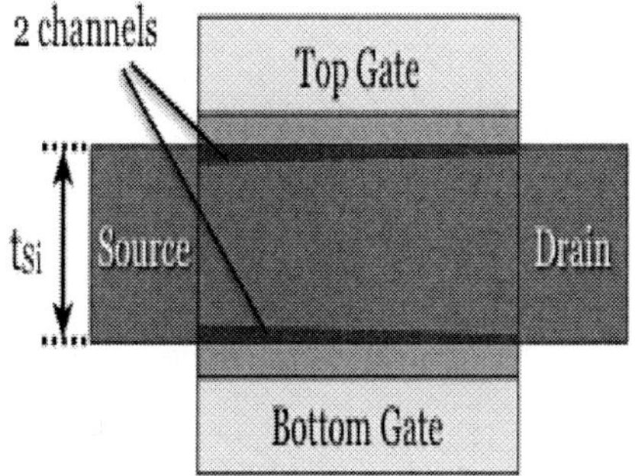

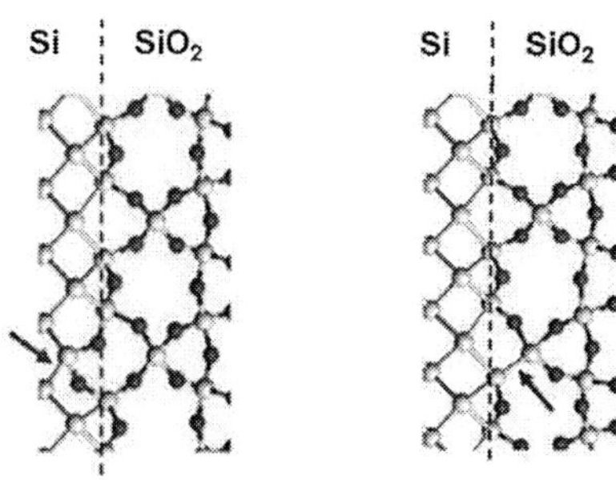

Fig. 2. Schematic diagram of a double gate SOI MOSFET. In ultrathin devices (channel thickness t_{Si} only a few nanometers), the two channels merge and the majority of carriers flow in the middle of the film (see Figs. 3 and 8).

Fig. 3. Schematics of Si-SiO$_2$ interfaces showing an oxygen protrusion (left) and a suboxide bond (right). Grey balls are Si atoms and red balls are O atoms.

THE METHOD

The foundation of the method is an atomic-scale model of the Si-SiO$_2$ interface as shown in Fig. 3. Such structures with abrupt interfaces were first constructed in Ref. 12. The Si and SiO$_2$ layers are repeated periodically and form a superlattice to facilitate the calculation. The SiO$_2$ regions are made up of either biaxially-strained quartz or biaxially-strained tridymite, which are crystalline forms. The Si layer represents the channel of a UTSOI MOSFET. Thicknesses of 1, 1.5 and 2 nm have been used, but for thicker channels the computational time becomes prohibitive. We expect that the crystalline nature of SiO$_2$ does not affect our results because electrons in the channel penetrate only ~0.5 nm into the oxide (see below).

One may wonder how much of the bulk Si properties are retained in the channel of UTSOI MOSFETs with channel thicknesses as small as 1 or 2 nm. In Fig. 4 we show the energy gap of our SiO$_2$-Si-SiO$_2$ structure when the Si layer is 1 or 2 nm thick (the gap is calculated by projecting conduction band wave functions on planes of Si atoms). It is clear that the interface transition regions are thin enough so that a well-defined Si energy gap is maintained in the middle of the channel. The interface transition regions in this figure correspond to the true barriers experienced by electrons in the channel (these are the barriers that are modeled as infinite abrupt potentials in conventional mobility models).

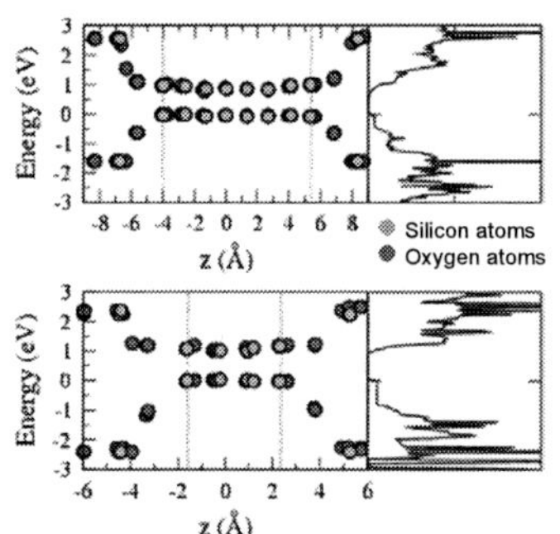

Fig. 4. Plot of the effective energy gap as projected on planes of atoms in a SiO$_2$-Si-SiO$_2$ structure along the z-axis perpendicular to the structure. The Si layer is 1 nm (top) and 0.5 nm (bottom).

For calculations in a UTSOI MOSFET with a particular channel thickness, we start with ideally abrupt interfaces as the *reference state*. In the absence of any defects, conductivity in the channels of our "superlattice" is infinite. The key objective is to introduce defects that act as scattering centers

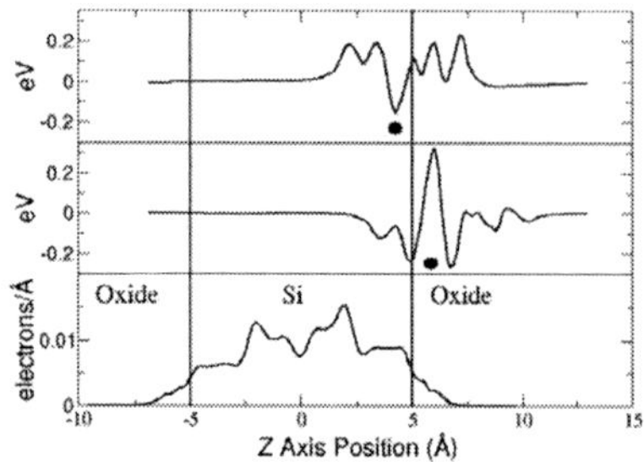

Fig. 5. Scattering potentials and carrier electron density (bottom) for oxygen protrusion (top) and suboxide bond (middle) defects. Black dots indicate defect centers. Penetration of the cannel electron density is evident (From Ref. 8).

and result in finite mobilities. We can then compare the effect of different defects on mobilities by performing parameter-free calculations as described below.

All calculations are performed using DFT [13]. Pseudopotentials are used to describe the atomic cores so that the calculation entails only valence electrons. The local-density approximation with generalized gradient corrections is used for the exchange-correlation potential. The electron wave functions and the corresponding electronic density are expanded in terms of plane waves. The pertinent Schrödinger equation is solved self-consistently. Convergence of the basis set, the number of $\mathbf{k}$ points in the Brillouin zone for summations, and all other relevant quantities is checked.

First, calculations are performed for the ideally abrupt interfaces. Forces on atoms are computed and all atoms are allowed to relax until the forces are smaller than 0.1 eV/Å. In the end, the self-consistent potential of the reference structure, $V_{ref}(\mathbf{x})$, is obtained. After a defect is introduced, the calculations are repeated, all atoms are relaxed, and the new self-consistent potential, $V_{def}(\mathbf{x})$ is obtained. The difference, $V_{scat}(\mathbf{x}) = V_{ref}(\mathbf{x}) - V_{def}(\mathbf{x})$, is the parameter-free scattering potential. Once the scattering potential is known, the scattering rate is calculated using the Born approximation and then the mobility is calculated using the linearized Boltzmann transport equation. The pertinent equations are given in Ref. 8.

EFFECT OF ATOMIC-SCALE ROUGHNESS ON MOBILITIES

The primitive deviations from interface abruptness are a) a missing O atom on the oxide side, which we shall refer to as a suboxide bond, and b) an O atom in an Si-O-Si configuration on the Si side of the interface, which we shall refer to as an oxygen protrusion. They represent atomic-scale roughness, a source of scattering that so far has not been included in mobility modeling (macroscopic roughness, treated usually by a phenomenological model, is produced by steps at the interface).

The scattering potentials of isolated suboxide bonds and oxygen protrusions, as first reported in Ref. 8, are shown in Fig. 5a and 5b. Also shown is the self-consistent electron density for a particular gate field. Carrier penetration into the oxide is evident. The density peaks at the center of the channel, which is a well-known feature of UTSOI devices (the would-be triangular wells at the two interfaces merge and the net potential resembles a "box" so that the electron density takes the form of a "particle in a box". The "box" is not flat-bottomed but is characterized by atomic pseudopotentials that cause the oscillations superposed seen in Fig. 5c.

Mobilities produced by a particular density of suboxide bonds and, separately, of oxygen protrusions as first reported in Refs. 8 and 9 are shown in Fig. 6. as a function of

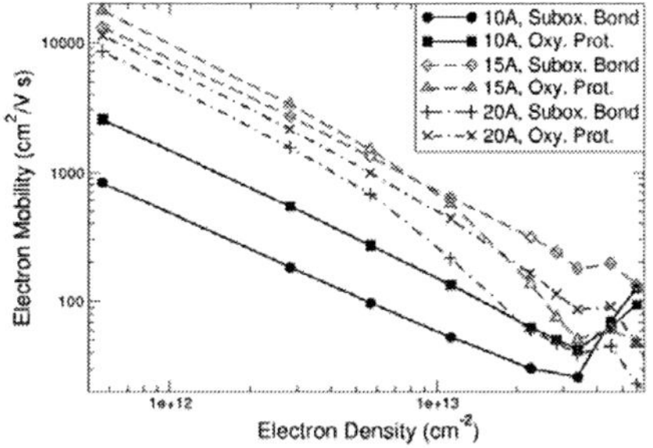

Fig. 6. Calculated mobilities for oxygen protrusions and suboxide bonds in 10Å-20Å channels. Defect density is 5.6×10^{11} cm^{-2}. No other scattering mechanisms are included.

the two-dimensional electron density in the channel. Three different channel thicknesses are shown. The trend is clear: when mobilities are controlled by the primitive atomic-scale roughness, they degrade as the thickness is decreased from 2 to 1 nm. It is noteworthy that, even though only scattering from atomic-scale roughness is included, the calculated mobilities are in the range of measured values, which suggests that the role of atomic-scale roughness may be important. We will return to this point when we examine mobilities in strained-Si channels.

EFFECT OF STRAY Hf ATOMS IN Si-SiO$_2$-HfO$_2$ STACKS

One of the most promising "high-k dielectrics" that may be adopted to replace SiO$_2$ as the gate dielectric is a composite structure of a thin SiO$_2$ interlayer followed by HfO$_2$. Typically, HfO$_2$ is first deposited in Si by atomic-layer epitaxy and the sample is then flash annealed. A thin (0.5 – 0.8 nm) SiO$_2$ interlayer grows during the annealing step, producing a "good" Si-SiO$_2$ interface.

Atomically-resolved images of Si-SiO$_2$-HfO$_2$ alternate-dielectric structures recently obtained by van Benthem and Pennycook (Fig. 7) revealed the presence of "stray" individual Hf atoms in the thin SiO$_2$ interlayer. DFT calculations were used to obtain relaxed structures of such Hf atoms near the interface. Mobility calculations find a significant reduction caused by scattering from neutral Hf atoms (Fig. 8) at a density equal to the observed density. The resulting finite mobility is significantly smaller than the mobilities produced by typical densities of primitive atomic-scale-roughness defects. The result is in agreement with the fact that measured mobilities in Si-SiO$_2$-HfO$_2$ structures are lower than in conventional Si-SiO$_2$ structures. The possible presence of such stray atoms in

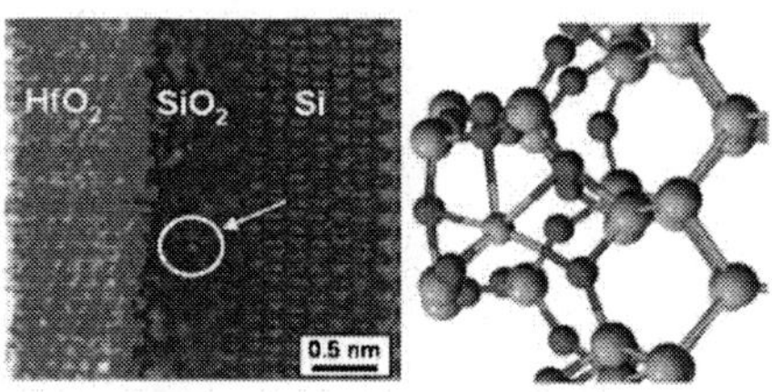

Fig. 7. Left: Z-contrast STEM image of Si-SiO$_2$/HfO$_2$ interface. The arrow highlights a stray Hf atom in the SiO$_2$ interlayer (K. Van Benthem and S. J. Pennycook, unpublished); right: schematic of the stray Hf atom and its neighbors.

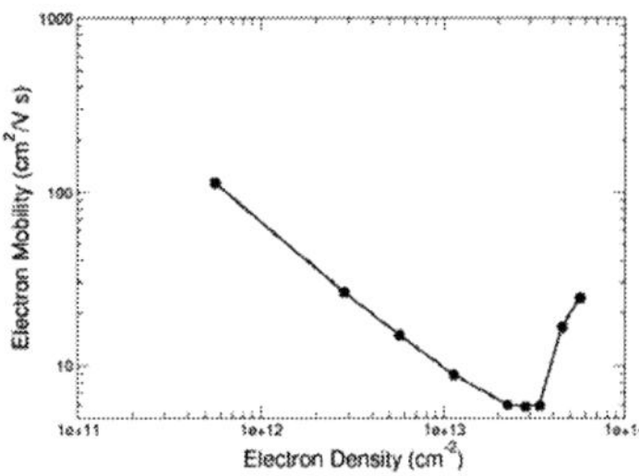

Fig. 8. Calculated mobility due to a neutral Hf defect as shown in Fig. 6. Hf sheet density is 10^{12} cm^{-2}. No other scattering mechanisms are included.

composite high-k stacks makes it clear that conventional mobilities need to be augmented to include wave function penetration.

MOBILITY ENHANCEMENT IN STRAINED-Si UTSOI MOSFETs

The fact that strained-Si channels result in enhanced mobilities has been known for more than 20 years [14]. The effect has been observed in both conventional single-gate MOSFETs and in double-gate SOI MOSFETs [15-21]. Initially, strain-induced electron mobility enhancement was attributed to the fact that biaxial strain raises the energy of four of the six equivalent conduction-band minima, thus reducing intervalley scattering [14]. In a detailed analysis of the effect of strain on mobilities in 2002, Fischetti et al. [6] pointed out the splitting of the valleys is already present for sufficiently large gate fields so that it cannot be responsible for the observed mobility enhancement. The authors then reported calculations of electron mobilities that incorporated the effect of strain in conventional models. They found that the only mechanism that is sufficiently sensitive to strain is scattering from macroscopic roughness. However, the values of roughness reduction that are necessary to account for the data are quite large and could not be justified. The authors concluded that our current understanding of transport properties in this system is not sufficient to explain the observed phenomena.

It is worth also pointing out that mobility enhancement is observed relatively uniformly across all values of the electric field or, equivalently, the electron density [6]. As shown in Fig. 1, however, conventional modeling has identified different dominant scattering mechanisms in three different ranges of effective field values. The question why enhancement should be present relatively uniformly in all three regions has remained open.

A detailed study of the effect of strain on electron mobility using the present first-principles method is reported in Ref. [11]. Here we give a summary of the main results for the effect of atomic-scale roughness in unstrained and strained Si channels.

The scattering potentials for a suboxide bond and an oxygen protrusion in unstrained and strained Si in a 1-nm-thick UTSOI MOSFET are shown in Fig. 9. For the

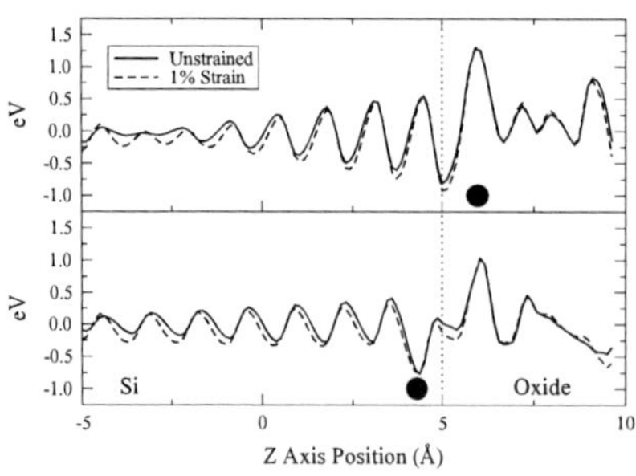

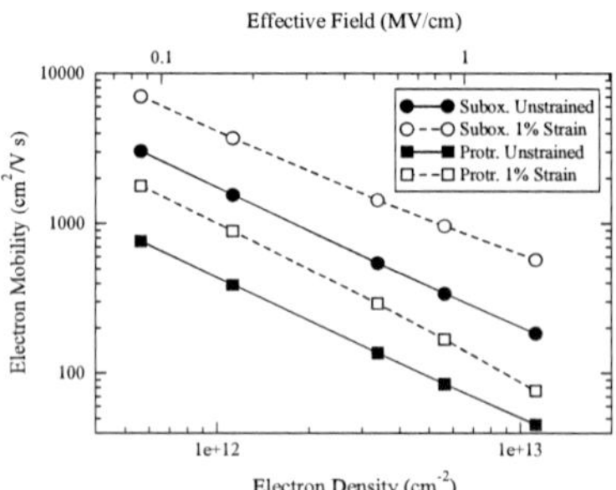

Fig. 9. The scattering potentials for a suboxide bond (top) and an oxygen protrusion (bottom) in an SOI MOSFET with unstrained and strained Si.

Fig. 10. Electron mobilities limited by suboxide bonds or oxygen protrusions in an SOI MOSFET with unstrained and strained Si.

corresponding electron mobilities, it was found that, under strain, both suboxide bonds and oxygen protrusions yield enhancement that is uniform across all values of the effective field (Fig. 10), as indeed has been observed. The result further allows us to define a unique enhancement factor for each value of strain. Note that the density of defects drops out when we calculate the enhancement factor, which is a ratio of two mobility values.

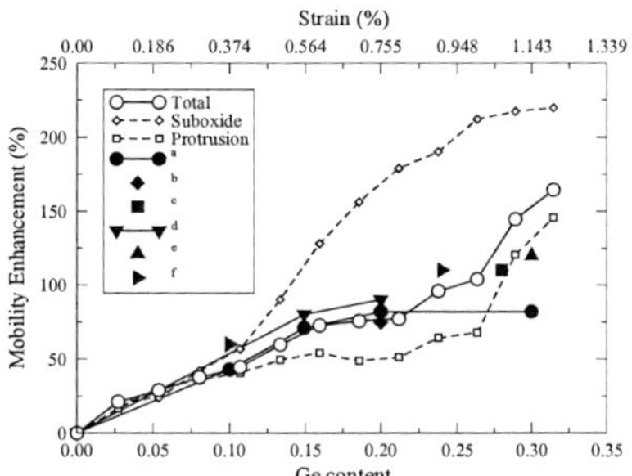

Fig. 11. Calculated mobility enhancement factor for electrons in an SOI MOSFET with strained Si channel as a function of strain (see text).

In Fig. 11 we show the calculated enhancement factors for the two primitive defects as a function of strain (equivalently, the value of x in the $Si_{1-x}Ge_x$ substrate that is used to produce strained Si). An intriguing result is that for small strains (up to ~0.4%) the enhancement is linear and essentially the same for the two primitive defects. Thus, the relative densities of the two defects drop put and we have a parameter-free prediction for the mobility enhancement factor. It is evident that the prediction is in excellent quantitative agreement with the available data. For larger strains, O protrusions dominate. In order to combine the effects of the two primitive defects we estimated the ratio of their densities from available data on bonding at the Si-SiO_2 interface [22]. We obtained a ratio of 1:3 suboxide bonds to oxygen protrusions. The combined enhancement factor is also shown in Fig. 11, again in quantitative agreement with the data. In general, the ratio of defect densities may depend on processing conditions and can be treated as a fitting parameter (first-principles calculations [12,23] suggest that suboxide bonds are not favored energetically at the Si-SiO_2 interface and appear only due to entropic reasons).

The explanation for strain-induced mobility enhancement when mobilities are limited by atomic-scale roughness can be traced to the nature of the primitive defects and the respective scattering potentials. In the case of an O protrusion, it is clear that the overall scattering potential is weaker in strained Si because the Si-Si bonds are already stretched somewhat so that insertion of the O atom results in smaller atomic relaxations. Similarly, for a stretched suboxide bond, the Si atoms are closer to the positions they would have if O were present.

It is noteworthy that the calculated mobility enhancement factors are in agreement with data in both conventional MOSFETs and fairly thick SOI MOSFETs (data on UT-SOI MOSFETs are not available). The calculations are of course for a 1-nm-thick UTSOI MOSFET. The scattering potentials shown in Fig. 9 suggest an explanation for the seemingly universal enhancement factor. The reduction in the scattering potential is uniform over the range of 1 nm. At most fields, the electron density in single-gate and thick double-gate SOI MOSFETs is confined by the triangular well in small distances from the interface, typically 2-5 nm [24]. If the uniformity seen in the reduction of the scattering potential in Fig. 9 extends over such regions, one would expect the same enhancement factor in both UTSOI and conventional MOSFETs. Experimental data on UTSOI MOSFETs would of course be useful to check the conclusion.

The net conclusion is that atomic-scale roughness can naturally account for the observed strain-induced electron mobility enhancement across all values of the effective field, i.e., in regions where conventional modeling assigned different dominant scattering mechanisms, namely Coulomb scattering, phonons, and macroscopic roughness. The new result suggests that atomic-scale roughness – which is definitely present – must be a co-dominant mechanism across all values of the effective field. If so, in conventional modeling the effect of atomic-scale roughness is masked by the choice of fitting parameters for the other mechanisms.

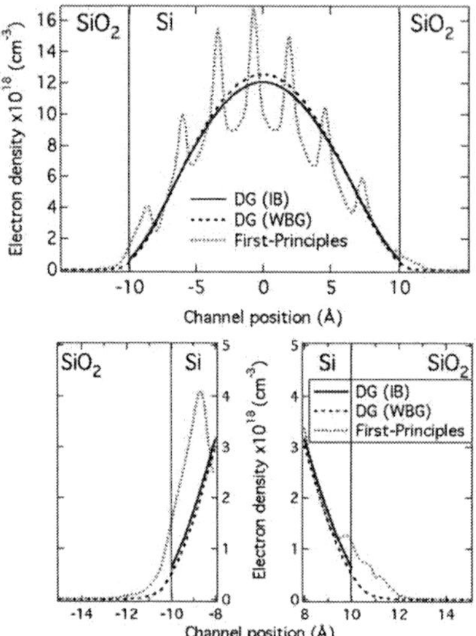

Fig. 12. Conduction electron densities in a 20Å ultrathin SOI double-gate MOSFET calculated from first principles (dotted red line) and using the Density Gradient (DG) method (black and blue lines). The solid black line is a DG simulation treating SiO_2 as an impenetrable barrier (IB). The dashed line treats SiO_2 as a "wide band gap" material (WBG), allowing penetration into the oxide. The lower frame highlights the near-interface region to emphasize oxide penetration.

COMPARISON WITH "DENSITY-GRADIENT" METHOD

The "density-gradient" (DG) method [25] is a way to go beyond the usual drift-diffusion approximation to Boltzmann's transport equation. Instead of modeling the electron system as a classical Maxwelllian gas, one adopts an equation of state that depends not only on the electron density but on the density gradient as well. In this theory, Newton's second law takes the form of a macroscopic "Schrödinger equation". Planck's constant $\hbar$ enters the picture, even though the method is not fully quantum mechanical. It has been shown [26] that this method can also be derived microscopically by doing a gradient expansion of the quantum mechanical kinetic energy operator and keeping just the lowest order term. The method is available in commercial device simulators, which provides a new level of accuracy in simulating nanoscaled devices.

We have used the DG correction combined with drift-diffusion transport to calculate electron densities in double-gate UTSOI MOSFETs. The results are compared with our fully quantum-mechanical parameter-free results in Fig. 12. The DG method reproduces well the overall shape of the electron-density spatial distribution, but it smooths out all the atomic details of the fully quantum mechanical treatment. From the comparison of the two methods, an interesting feature shows up: the DG correction underestimates carrier penetration into the oxide, which reveals a limit of the DG method applied to ultra small potential wells. The results of first-principles calculations can be used to optimize the range of applicability and accuracy of the DG method for device modeling.

SUMMARY

We have given a brief review of recent developments in calculations of carrier mobilities in MOSFETs using a first-principles, parameter-free quantum-mechanical method. The results so far show that atomic-scale roughness, which has been neglected in conventional mobility modeling, is in fact a co-dominant scattering mechanism across all values of the effective field. Our calculations show that atomic-scale roughness can naturally account for the observed strain-enhanced mobilities and that the limited mobilities in Si-SiO_2-HfO_2 structures may in part be caused by stray Hf atoms in the SiO_2 interlayer.

Acknowledgements

We would like to thank S. Cristoloveanu for valuable discussions and K. Van Benthem and S. J. Pennycook for providing the image in Fig. 7. The research was supported in part by the National Science Foundation grant ECS-0524655, by the Air Force Office of Scientific Research Grant FA9550-05-1-0306, and by the McMinn Endowment at Vanderbilt University.

References

* Present address: Lockheed-Martin, Sunnyvale, CA 94089.

† Present address: Laboratoire de Physique Appliquee et d'Automatique, Universite de Perpignian Via Domitia, 66860 Perpignian Cedex, France

1. T. Ando, A. B. Fowler, and F. Stern, Rev. Mod. Phys. **54**, 437 (1982).

2. M. V. Fischetti, and S. E. Laux, Phys. Rev. B **48**, 2244 (1993).

3. C. Jiacoboni and L. Reggiani, Rev. Mod. Phys. 44, 645 (1983); M. V. Fischetti and S. E. Laux, Phys. Rev. B **38**, 9721 (1988); IEEE Electron Devices **38**, 650 (1991); W. Wagner, J. Stat. Phys. **66**, 1011 (1992).

4. S. Takagi, A. Toriumi, M. Iwase, and H. Tango, IEEE Trans. Electron Devices **41**, 2357 (1994).

5. I. Polishchuk, and C. Hu, VLSI Technology Digest of Technical Papers, 51 (2001); D. Mounteanu and J. L. Autran, Solid State Electron. **47**, 1219 (2003); D. Esseni, IEEE Trans. Electron Dev. 51, 394 (2004).

6. M. V. Fischetti, F. Gamiz, and W. Hansch, J. Appl. Phys. **92**, 7320 (2002).

7. K. Uchida, J. Koga, and S. Takagi, IEDM Technical Digest, 805 (2003); D. Esseni et al., IEEE Electron Dev. **50**, 802 (2003); D. Esseni and A. Sangiorgi, Solid State Electron. 48, 927 (2004).

8. M. H. Evans, X.-G. Zhang, J. D. Joannopoulos, and S. T. Pantelides, Phys. Rev. Lett. **95**, 106802 (2005).

9. M. H. Evans, M. Caussanel, R. D. Schrimpf, and S. T. Pantelides, IEDM Tech. Digest, 597 (2005).

10. M. H. Evans, M. Caussanel, R. D. Schrimpf, and S. T. Pantelides, Proceedings of 11th International Workshop on Computational Electronics (IWCE 11), Vienna, Austria, May 2006 [J. Comp. Electron., in press].

11. G. Hadjisavvas, L. Tsetseris, and S. T. Pantelides, to be published.

12. R. Buczko, S. J. Pennycook, and S. T. Pantelides, Phys. Rev. Lett. **84**, 943 (2000).

13. M. C. Payne, M. P. Teter, D. C. Allan, T. A. Arias, and J. D. Joannopoulos, Rev. Mod. Phys. **64**, 1045 (1992).

14. G. Abstreiter, H. Brugger, T. Wolf, H. Jorke, and H. J. Herzog, Phys. Rev. Lett. **54**, 2441 (1985).

15. M. T. Currie, C. W. Leitz, T. A. Langdo, G. Taraschi, E. A. Fitzgerald, and D. A. Antoniadis, J. Vac. Sci. Technol. B **19**, 2268 (2001).

16. K. Rim, J. L. Hoyt, and J. F. Gibbons, IEEE Trans. Electron Devices **47**, 1406 (2000).

17. K. Rim et al., VLSI Technology Digest of Technical Papers, 98 (2002).

18. K. Rim et al., Solid-State Electron **47**, 1133 (2003).

19. N. Sugii, D. Hisamoto, K. Washio, N. Yokoyama, and S. Kimura, IEEE Trans. Electron Devices **49**, 2237 (2002).

20. S. H. Olsen et al., IEEE Trans. Electron Devices **50**, 1961 (2002).

21. T. Mizuno, N. Sugiyama, T. Tezuka, T. Numata, and S. Takagi, IEEE Trans. Electron Devices **50**, 988 (2003).

22. F. J. Himpsel, F. R. McFeely, A. Taleb-Ibrahimi, J. A. Yarmoff, and G. Hollinger, Phys. Rev. B **38**, 6084 (1988).

23. L. Tsetseris, and S. T. Pantelides, Phys. Rev. Lett. **97**, 116101 (2006).

24. F. Stern and W. E. Howard, Phys. Rev. **163**, 816 (1967); F. Stern, Phys. Rev. B **5**, 4891 (1972).

25. M. G. Ancona, and H. F. Tiersten, Phys. Rev. B **35**, 7959 (1987).

26. M. G. Ancona and G. J. Iafrate, Phys. Rev. B **39**, 9536 (1989)

ECS Transactions, 6 (4) 363-368 (2007)
10.1149/1.2728883, ©The Electrochemical Society

A comprehensive study of the Corner Effects in Pi-Gate SOI MOSFETs

F. G. Ruiz, A. Godoy, F. Gámiz, C. Sampedro, L. Donetti
Dpto. de Electrónica y Tecnología de los Computadores, Universidad de Granada
Facultad de Ciencias, 18071, Granada (Spain). Email: franruiz@ugr.es

In this work, a simulation-based study of the Pi-Gate SOI
MOSFETs was carried out, using a 2D self-consistent
Schrödinger-Poisson solver. Corner effects have been studied as a
function of different parameters such as the doping density, silicon
fin dimensions, corner rounding and gate oxide thickness.

Introduction

Traditional bulk MOSFETs face various problems when the channel length is reduced
to below 100nm. In order to keep the short channel effects (SCEs) under control,
Multiple Gate Silicon-on-Insulator Metal-Oxide-Semiconductor Field-Effect-Transistors
(MuG-SOI-MOSFETs) such as FinFETs, Trigates, Pi and Ω-gates or Gate All Around
(GAAs) have been proposed (1-3). These devices have demonstrated a better electrostatic
control of the channel with a reduction of the short channel effects (2). In spite of their
benefits, multiple gates can give rise to the formation of independent channels with
different threshold voltages (4). This phenomenon is known as 'corner effects'. In this
work, we have carried out a thorough study of the corner effects on a Pi-Gate SOI
MOSFET. The main difference between this device and a Trigate is that in this device,
the gate gets into the buried oxide, allowing a more effective control of the electrostatics
in the silicon body.

Due to the small size of the devices under study, quantum effects are important and
therefore we need the accurate simulations provided by the self-consistent solution of
Poisson and Schrödinger equations in two dimensions. To get a fast convergence, both
equations were self-consistently solved using the predictor-corrector scheme proposed by
Trellakis et al (5).

Corner effects on a reference Pi-Gate device

Corner effects appear when threshold voltages vary acroos the device. This means
that independent channels located in different regions of the transistor are activated at
different gate voltages. There are several ways to define the threshold voltage (V_{Th}) of a
device. Here, the quantum electron density $n(x,y)$ in the device is obtained from
simulations, and V_{Th} is calculated as the maximum of the second derivative of the total
charge density $N_T \equiv \int n(x,y)\,dxdy$ (cm^{-1}) as a function of the gate voltage $\left(d^2 N_T / dV_G^2\right)$.

First, we defined a reference device, the parameters of which were selected to
manifest a double threshold voltage. Fig. 1 shows the geometry of a Pi-Gate device where
the most important parameters are defined. W_{Si} is the silicon width, H_{Si} the silicon height,
T_{ox} the gate oxide thickness, N_A the substrate doping density, d_{BG} the extension of the
buried gate, and r_{curv} the parameter that determines the corner rounding. The values

selected for this reference device are the following: $W_{Si} = H_{Si} = 30nm$; $T_{ox} = 2nm$; $N_A = 5\times10^{18}cm^{-3}$; no corner curvature ($r_{curv}=0$), buried gate length of 10 nm and a gate work function $\Phi_m = 4.05$ eV.

Fig. 2 represents $\left(d^2 N_T / dV_G^2\right)$ in this reference device. As can be observed in the figure, the double peak indicates that two different channels are activated at different gate voltages, the first for $V_G = 0.39V$ (V_{Th1}) and the second for $V_G = 0.56V$ (V_{Th2}). As is shown in the figure insets, V_{Th1} corresponds to the formation of a channel in the corners of the device, while V_{Th2} is the gate voltage at which channels are created in the lateral sides of the device.

To check this behavior, we have represented in Fig. 3 the quantum electron density in the top gate along a line where $n(x,y)$ reaches its maximum. The figure inset shows this line separated from the Si-SiO$_2$ due to the quantum effects. The solid line was calculated when the gate voltage was equal to the first threshold voltage of the device (V_{Th1}) and the dashed-line shows the second threshold voltage (V_{Th2}). At $V_G=V_{Th1}$ the corners of the device reach the inversion, while the lateral sides will not become inverted until they reach $V_G=V_{th2}$.

Dependence of corner effects on device parameters

Previous works in the literature have studied the dependence of corner effects on some device parameters. Specifically, a reduction of the doping density has been proven as an efficient way to eliminate the corner effects (6). In Fig. 4, we have chosen several devices similar to the one used as a reference but with one parameter having been modified each time. First the doping density was reduced to 2×10^{18} cm^{-3} (o), second the device dimensions were shrunk to $W_{Si} = H_{Si} = 20nm$ ($\square$), third, rounded corners were considered ($\Diamond$), and fourth the gate oxide thickness was modified to $T_{ox} =1nm$ (*). In all these modified devices only one threshold voltage was observed. The following subsections will examine how all of these parameters influence the corner effects.

<u>Substrate doping dependence</u>

Fossum et al. (6) studied the corner effects in Trigate MOSFETs, demonstrating that devices with low substrate doping concentrations do not show corner effects. According to their results, devices with high N_A have an effective doping concentration $N_{A(eff)} < N_A$ near their corners. This effective doping concentration is obtained from the 2D Poisson equation. With regard to depleted devices, $n = p = 0$, Poisson equation can be written as

$$\frac{dE_X}{dx} = -\frac{dE_Y}{dy} - \frac{q}{\varepsilon_{Si}} N_A = -\frac{q}{\varepsilon_{Si}} N_{A(eff)} \quad . \tag{1}$$

In the calculation of V_{Th} (6), as long as the value of N_A is not negligible, variations of $N_{A(eff)}$ in the device volume give rise to different values of V_{Th}, and therefore to corner effects. However, when very low doping concentrations are used (virtually zero), their effect on the threshold voltage calculation can be neglected and therefore small variations of $N_{A(eff)}$ can be neglected too. As a consequence, threshold voltage can be considered almost constant in the whole device, and no corner effects are observed.

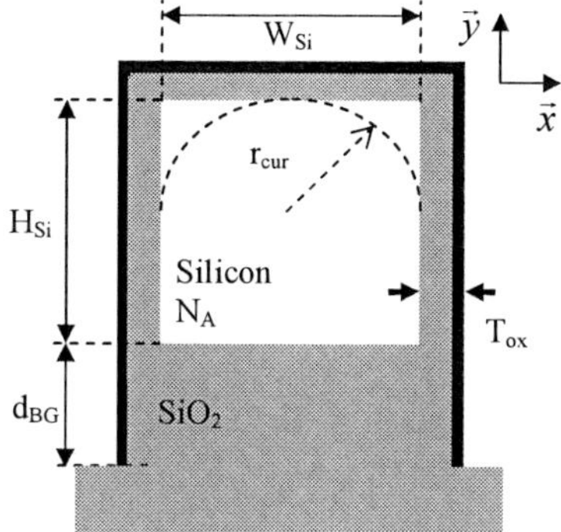

Figure 1. Geometry of a Pi-Gate SOI MOSFET and definition of its geometry parameters.

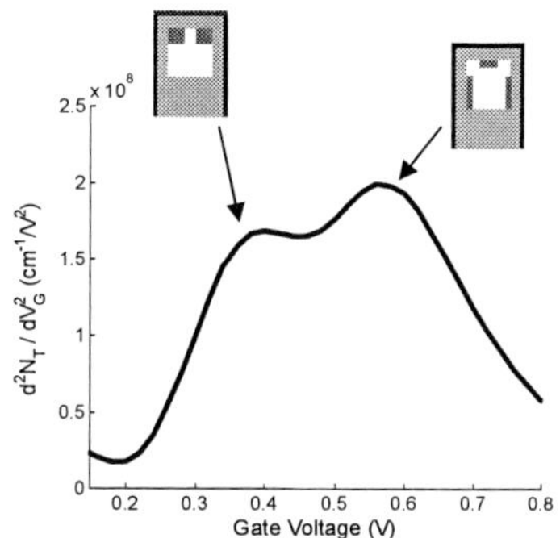

Figure 2. Duplication of the threshold voltages in a Pi-Gate device.

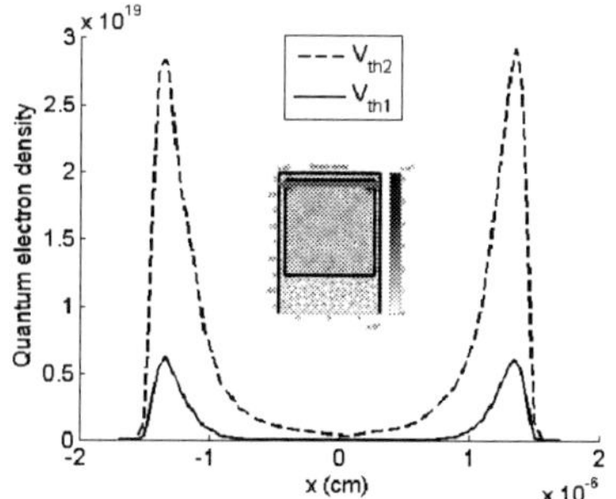

Figure 3. Quantum electron density in the reference device along the top interface.

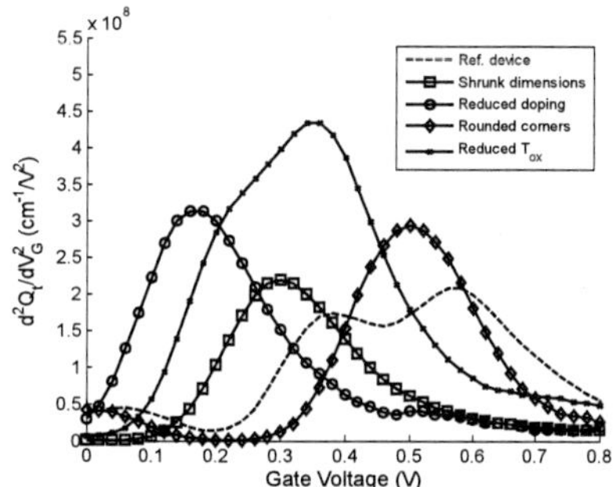

Figure 4. Reduction of corner effects is demonstrated when modifying some of the parameters of the Pi-Gate SOI MOSFET.

Nevertheless, we consider that using such a small value of N_A to get rid of the corner effects is too restrictive a condition. To check this assumption, we studied the behavior of Pi-Gate devices with different doping concentrations and demonstrated that corner effects are cancelled when reducing N_A. We observed (Fig. 4) that such corner effects are cancelled even if $N_A = 2\times10^{18}$ cm^{-3}, which cannot be considered close to zero.

In Fig. 5 we have represented, for the reference device, the derivatives of the electric field components dE_x/dx and dE_y/dy on a horizontal slice close to the top gate (similar to that in Fig. 4 inset). Two different values of gate voltage V_G have been selected: $V_G < V_{Th1}$ and $V_G = V_{Th1}$. Fig. 6 shows the same derivatives when N_A is reduced to 2×10^{18}cm^{-3}. Again, two values of gate voltage V_G have been used: $V_G < V_{Th}$ and $V_G = V_{Th}$. The threshold voltage of the second device ($V_{Th} = 0.3$V) has been extracted from Fig. 4.

When a device is depleted, from Eq. [1], the average value of dE_x/dx and dE_y/dy is equal to $qN_A/2\varepsilon_{Si}$. This value has been used as a reference in the previous figures (solid line). In figures 5 and 6, when $V_G < V_{Th1}$ both derivatives are balanced (symmetrical with respect to $qN_A/2\varepsilon_{Si}$) and therefore quantum electron density is negligible. When gate voltage reaches the threshold, the balance is lost next to the corners of the reference device (Fig. 5), while it remains in the center of the slice. On the other hand, with reduced doping concentration (Fig. 6) the whole channel is inverted at the same gate voltage.

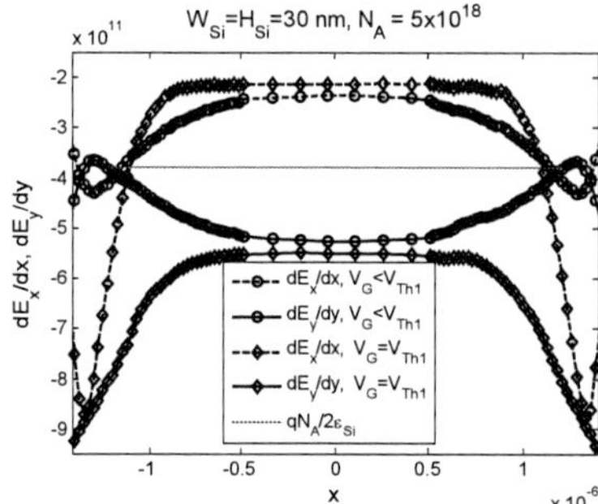

Figure 5. Electric field derivatives dE_x/dx and dE_y/dy for the reference device. Gate voltage: $V_G < V_{Th1}$ and $V_G = V_{Th1}$.

Figure 6. Electric field derivatives dE_x/dx and dE_y/dy for a device with reduced doping concentration ($N_A = 2\times10^{18}$ cm^{-3}). Gate voltage: $V_G < V_{Th}$ and $V_G = V_{Th}$.

It is interesting to take a closer look to each of the derivatives of the components of the electric field. The behavior of dE_x/dx is quite similar in the two previous devices: once the device has reached the threshold voltage, $|dE_x/dx|$ is at its maximum next to the corners and then it quickly decreases to its value in the center of the top gate. Some differences can be observed for dE_y/dy in the two devices. For the reference device, at the threshold voltage V_{Th1}, $|dE_y/dy|$ has a maximum value near the corner and then decays when $|x|$ is reduced, reaching its equilibrium value at a certain distance from the center of the top gate. However, when the doping concentration is reduced and $V_G = V_{Th}$, the plateau region observed in the center of the channel disappears for dE_y/dy. Moreover, its value is higher than qN_A/ε_{Si} . This means that the contribution of the electron density is noticeable.

We can define a parameter termed the corner effect width, W_{CE}, as the width of the region in which dE_y/dy is modified due to the presence of lateral gates (Fig. 7). This width is related to N_A^{-k}, $k > 0$. Therefore, when $W_{CE} > W_{Si}/2$ both corner regions interact along the whole channel in the top gate and the corner effects disappear.

Finally, we would like to point out that although the case has been made using a horizontal slice next to the top gate, similar results would have been obtained had we considered a vertical slice.

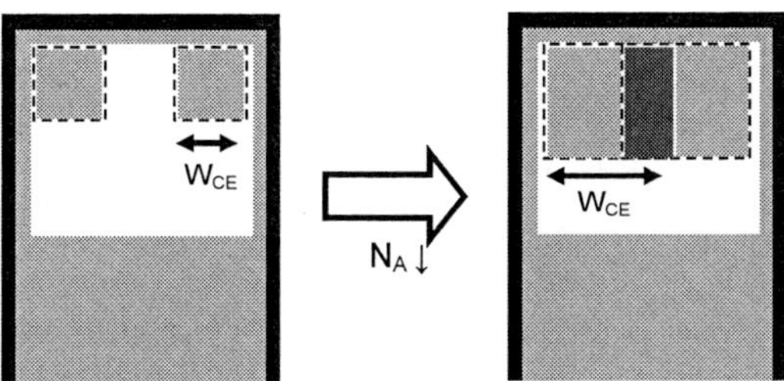

Figure 7. Schematic representation of corner area. Left: devices with corner effects, $W_{CE} < W_{Si}/2$. Right: device with reduced doping density and no corner effects, $W_{CE} > W_{Si}/2$.

Silicon body dimensions dependence

According to the previous results, corner effects can be avoided when the influence of different corner regions covers the whole interface. Therefore, an alternative way of canceling corner effects would be by merely reducing the device dimensions. From Fig. 7 (left), and considering a constant doping density, the corner region width W_{CE} does not depend on the device size. Thus, if device dimensions are shrunk, different corner regions overlap again. To show this, a device similar to the reference one has been simulated with a reduced lateral size, $W_{Si} = H_{Si} = 20$nm, and as can be seen in Fig. 4, the double threshold voltage disappears.

Fig. 8 shows the electric field derivatives dE_x/dx and dE_y/dy in this device where two different values of the gate voltage V_G have been used: $V_G < V_{Th}$, and $V_G = V_{Th}$. V_{Th} was calculated and a value of $V_{Th} = 0.3$V was found. As can be seen, the behavior of the electric field derivatives with the normalized distance is similar to that shown in Fig. 6, which is consistent with our previous hypothesis.

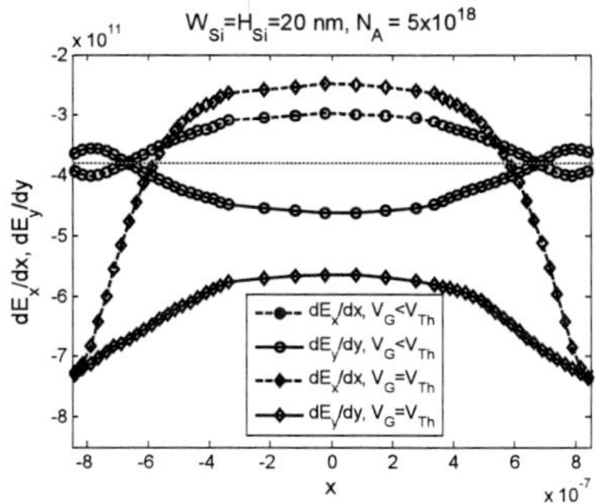

Figure 8. Electric field derivatives dE_x/dx and dE_y/dy for the device with reduced dimensions ($W_{Si} = H_{Si} = 20$nm) for the two different gate voltages applied: $V_G < V_{Th}$ and $V_G = V_{Th}$.

<u>Corner rounding and oxide thickness dependence</u>

These two techniques have been shown (Fig. 4) to reduce corner effects. However, while previous studies have demonstrated the dependence of corner effects with doping and the dimensions of the device to be related to the size of the corner regions relative to the device width, corner rounding and oxide thickness reduction cancel the corner effect by the elimination of electrostatically favorable areas in the device.

In the case of corner rounding, reduction of the corners produces a much more uniform potential along the Si-SiO$_2$ interface, and therefore inversion is reached at the same gate voltage. On the other hand, at the limit $T_{ox} \rightarrow 0$ there would be no potential drop in the oxide and the applied gate voltage would be directly transferred to the silicon substrate. Again the interface potential becomes more uniform, eliminating the potential wells in the corners, which are responsible for the accumulation of electrons there.

Conclusions

A thorough study of the corner effects in Pi-Gate SOI MOSFETs was carried out and the influence of different parameters was analyzed. It was observed that the extension of corner regions has an inverse dependence on doping concentration. Thus, a reduction of doping density and silicon width would help to prevent the presence of undesirable double threshold voltages. Moreover, the influence of corner rounding and a reduction of the gate oxide thickness would help mitigate corner effects.

Acknowledgments

This work has been supported by the Spanish Government (project FIS-2005-6832) and Junta de Andalucía (project TIC-00831).

References

1. D. Hisamoto et al., *IEEE Trans. Electron Dev.*, 47(12), Nov. 2000, pp. 2320-2325.
2. J.T. Park et al., *IEEE Electron Device Letters,* vol. 22(8), Aug. 2001, pp. 405-406.
3. J. P. Colinge et al., *IEEE Electron Device Letters,* 24(8), Aug. 2003, pp. 515-517.
4. R. Ritzenthaler et al., in *7th European Workshop on Ultimate Integration of Silicon, ULIS,* 2006, pp. 25-28.
5. A. Trellakis et al., *J. Applied Physics*, 81(12), June 1997, pp. 7880-7884.
6. J. G. Fossum et al., *IEEE Electron Device Letters,* 24(12), Dec. 2003, pp. 745-747.

ECS Transactions, 6 (4) 369-374 (2007)
10.1149/1.2728884, ©The Electrochemical Society

IMPACT OF PHONON-LIMITED MOBILITY SUPERIORITY IN DOUBLE-GATE OR FINFET WITH A (111) SILICON AND (001) GERMANIUM SURFACE CHANNEL ON DEVICE SCALING

T. Yamamura, S. Sato and Y. Omura

High-Technology Research Center, Kansai University, 3-3-35, Yamate-cho, Suita, Osaka, 564-8680 Japan

Abstract. 1D self-consistent calculations and relaxation time approximations are used to study the phonon-limited electron mobility of the inversion layer at room temperature for ultra-thin Si (111), Ge (001) and Ge (111) layers in single-gate (SG) and double-gate (DG) MOSFET's. Assuming a 5-nm-thick SOI layer, it is shown that intra-valley phonon scattering in the DG SOI MOSFET inversion layer is strongly suppressed within a range of medium E_{eff} value; DG devices have higher phonon-limited electron mobility than SG devices. The suppression of intra-valley phonon scattering in a 5-nm T_{SOI} DG device primarily stems from the reduction of the form factor (F_{00}) value within medium E_{eff} values. Similar aspects of electron mobility expected for SG and DG GOI MOSFET's are also discussed.

INTRODUCTION

The phonon-limited electron mobility on the (001) Si surface of ultra-thin body (UTB) single-gate (SG) and double-gate (DG) silicon-on-insulator (SOI) metal-oxide-semiconductor field-effect transistors (MOSFET's) has been widely analyzed [1-5], and its merits are well known. Recently, the electron mobility on the (110) Si surface of SG and DG SOI MOSFET's has been analyzed, and various strain techniques have been proposed and experimentally verified [6, 7]. However, the results yielded by these techniques are not always reproduced and fabrication costs are high because strain device fabrication requires new fabrication processes and/or materials. We still think that another approach is needed to improve the performance of UTB DG SOI MOSFET's or FinFET's [8].

In this paper, we examine the features of the phonon-limited electron mobility of DG and SG SOI MOSFET's on (111) Si surface with SOI layer thickness (T_{SOI}) ranging from 1 to 30 nm. Self-consistent simulations are performed assuming intra-valley scattering (acoustic phonon scattering) and inter-valley scattering (optical phonon scattering). The phonon-limited mobilities of SG SOI MOSFET and DG SOI MOSFET are compared. Similar aspects of electron mobility expected for SG and DG GOI MOSFET's are also discussed for Ge (001) and Ge (111) surfaces.

SIMULATION METHOD

The simulations assume SG and DG n-channel SOI MOSFET's. It is further assumed that the gate oxide thickness (T_{ox}) of the SG and DG SOI MOSFET's is 3 nm and the buried oxide layer thickness (T_{BOX}) of the SG SOI MOSFET is 200 nm. Impurity concentrations (N_A) in the SOI-layer and Si substrate are taken to be 5×10^{15} cm^{-3}. This paper simulates the phonon-limited electron mobility in the inversion layers of SG and DG SOI MOSFET's on (111) Si surfaces at 300 K using a relaxation time approximation based on a one-dimensional self-consistent solution of the Schrödinger and Poisson equations; the physical parameters assumed in the simulations refer to past papers [4, 9-12]. Electron mobility is derived by the following formulae.

$$\mu_i = \frac{q \int (E - E_i) \tau_i(E) \frac{\partial f}{\partial E} dE}{m_{c,i} \int (E - E_i) \frac{\partial f}{\partial E} dE}, \tag{1}$$

where $m_{c,i}$ is the conductivity effective mass of electrons at the specific i-th subband, $\tau_i(E)$ is the relaxation time of phonon scattering, and E_i is the i-th subband energy level.

E_{eff} for the SG SOI MOSFET is defined as

$$E_{eff} = \frac{\int_{Z_0}^{T_{soi}} n(z) E(z) dz}{\int_{Z_0}^{T_{soi}} n(z) dz}, \tag{2}$$

where $n(z)$ is the local electron density, $E(z)$ is the local transverse electric field, and z_0 stands for the position of the front Si/SiO$_2$ interface. On the other hand, E_{eff} for the DG SOI MOSFET is defined as

$$E_{eff} = \frac{\int_{Z_0}^{\frac{T_{soi}}{2}} n(z) E(z) dz}{\int_{Z_0}^{\frac{T_{soi}}{2}} n(z) dz}. \tag{3}$$

Integration is stopped at $T_{SOI}/2$ because we assume a symmetric DG SOI MOSFET.

RESULTS AND DISCUSSION

Simulated phonon-limited electron mobility on the (111) Si surface is shown in Fig. 1 as a function of T_{SOI} for various E_{eff} values; simulated mobility values of the DG and SG SOI MOSFET's are compared. It is seen that the DG SOI MOSFET offers superior mobility for T_{SOI} values ranging from 4 to 8 nm; it is very important that the mobility superiority of DG SOI MOSFET appears in a range of medium and high E_{eff} values. The maximal mobility enhancement appears around T_{SOI}= 5 nm for E_{eff}> 0.5 MVcm^{-1}. It is known in simulations that the mobility superiority of DG SOI MOSFET's with (001) Si surfaces appears in a low-E_{eff} range [1-3]. In past articles on the simulated electron mobility of (001) Si surface inversion layers, SG SOI MOSFET's have higher electron mobility than DG SOI MOSFET's in a range of high E_{eff} values (~1 MVcm^{-1}) for T_{SOI} values ranging from 2 to 5 nm [1-3, 5]; at other values of T_{SOI}, it is known that DG SOI MOSFET's have slightly higher electron mobility than SG SOI MOSFET's at low E_{eff} values (~0.1 MVcm^{-1}) [1, 2]. The expected phonon-limited electron mobility of a 5-nm T_{SOI} DG SOI MOSFET with (111) Si surfaces is 448 cm^{2}V^{-1}s^{-1} at E_{eff}= 1 MVcm^{-1}; this is 83 % of that of the equivalent DG SOI MOSFET with (001) Si surface. This may be the merit of using the (111) Si surface configuration in device applications such as FinFET's. In real devices, it is anticipated that surface-roughness scattering should strongly influence the electron mobility [13]; however, progress of modern etching technique will yield an almost flat surface in the future [14].

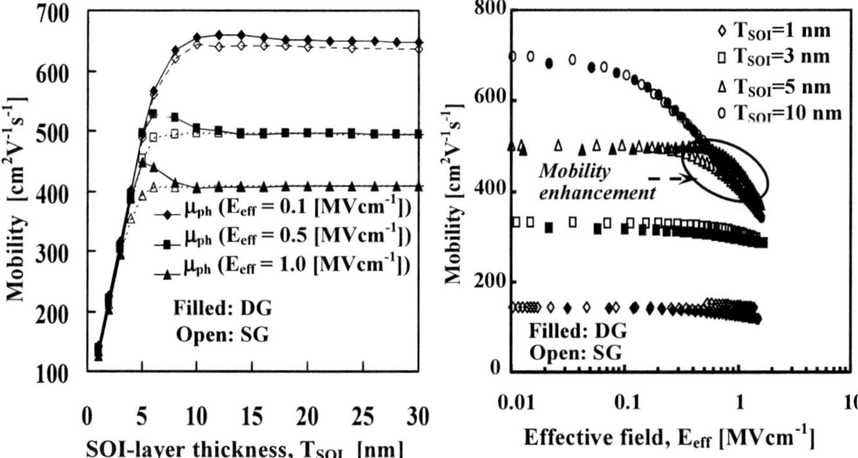

Figure 1. T_{SOI} dependencies of simulated phonon-limited electron mobility with parameter of E_{eff} for (111) Si surface.

Figure 2. E_{eff} dependencies of simulated phonon-limited electron mobility with parameter of T_{SOI} for (111) Si surface.

Phonon-limited electron mobility on the (111) Si surface is shown as a function of E_{eff} for DG and SG SOI MOSFET's in Fig. 2; the parameter is T_{SOI}. The electron mobility of the DG SOI MOSFET with 5-nm T_{SOI} exceeds that with 10-nm T_{SOI} for E_{eff}> 0.5 MV/cm; in contrast, the SG SOI MOSFET does not exhibit such behavior. As shown in Fig. 2, the DG SOI MOSFET with 5-nm T_{SOI} has higher electron mobility than the SG SOI MOSFET for E_{eff} > 0.2 MVcm^{-1}. These results are very interesting because they are not seen in SOI MOSFET's with (001) Si surfaces for medium and high E_{eff} values [2, 3, 5].

Figure 3 shows the mobility of electrons sharing the lowest subband as a function of E_{eff} for a 5-nm-T_{SOI} SG and DG SOI MOSFET with (111) Si surface channel; acoustic-phonon-limited mobility (i. e., primarily the intra-band transition) ($\mu_{0,intra}$) and optical-phonon-limited mobility (i. e., primarily the inter-band transition) ($\mu_{0,inter}$) are also shown. Figure 4 plots E_{eff} dependencies of occupation fractions of electrons sharing the lowest subband (f_0) and the 2nd subband (f_1) for 5-nm T_{SOI} SG and DG SOI MOSFET's; form factors (F_{ij}) of electrons are also shown for comparison. Form factor (F_{ij}) influences the probability of a transition from subband i to j.

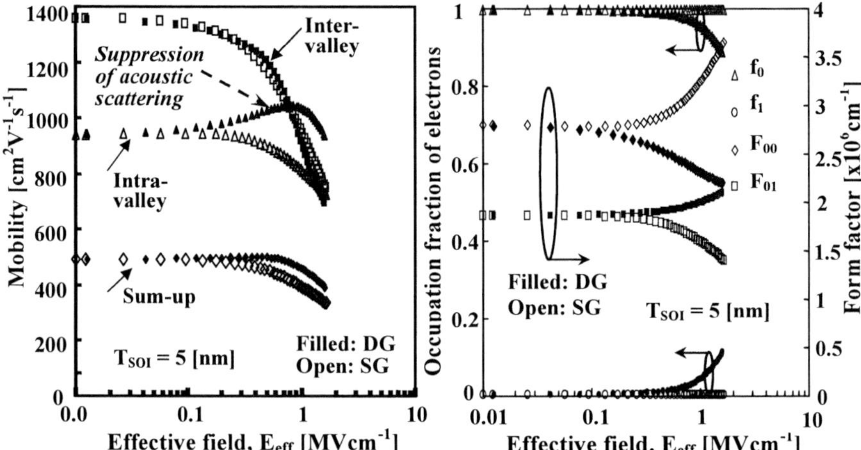

Figure 3. E_{eff} dependencies of lowest-subband phonon-limited electron mobility for (111) Si surface.

Figure 4. E_{eff} dependencies of occupation fractions of electrons sharing the lowest subband. Respective form factors are also shown for comparison.

Figure 3 suggests that the inter-band scattering (i. e., primarily optical-phonon scattering) of electrons sharing the lowest subband is not suppressed at high E_{eff} values in SG and DG SOI MOSFET's, while intra-band scattering (i. e., primarily acoustic-phonon scattering) of electrons sharing the lowest subband is well suppressed in a medium E_{eff} value for DG SOI MOSFET's. Intra-band scattering of electrons sharing the lowest subband is not suppressed at high E_{eff} values for SG SOI MOSFET's. In addition, as shown in Fig. 4, most electrons share the lowest subband for SG and DG SOI MOSFET's with 5-nm T_{SOI} over a wide range of E_{eff} values.

Intra-band scattering of the lowest-subband electrons is well suppressed at medium E_{eff} values for the 5-nm T_{SOI} DG SOI MOSFET, while it isn't suppressed at high E_{eff} values as seen in Fig. 3. The F_{00} value of the 5-nm T_{SOI} DG SOI MOSFET decreases as E_{eff} increases (see Fig. 4), which results in the suppression of intra-band scattering of the lowest-subband electrons; however, the suppression of intra-band scattering becomes modest at high E_{eff} values because the decrease in subband-to-subband energy difference promotes the transition between the lowest subband and the 2nd subband as is anticipated from the increase in F_{01} at high E_{eff} values [1, 2]. Therefore, it can be concluded that the suppression of intra-band scattering in 5-nm T_{SOI} DG SOI MOSFET's primarily stems from a reduction of F_{00} value at medium E_{eff} values.

Finally, simulated phonon-limited electron mobility on the (001) Ge surface is shown in Fig. 5 as a function of T_{GOI} for various E_{eff} values. It is seen that the DG GOI MOSFET on the (001) surface offers superior mobility for T_{GOI} values ranging from 5 to

8 nm with medium E_{eff} values; the maximal mobility enhancement appears around $T_{SOI}=7$ nm for $E_{eff}> 0.1$ MVcm^{-1}.   However, the expected phonon-limited electron mobility of 7.2-nm T_{GOI} DG GOI MOSFET with (001) Ge surface is 950 cm^{2}V^{-1}s^{-1} at $E_{eff}=1$ MVcm^{-1}; this is about 65 % of that of the equivalent DG GOI MOSFET with (111) Ge surface. This inidicates the using the (001) Ge surface configuration in device applications is not appealing because of the very low mobility.

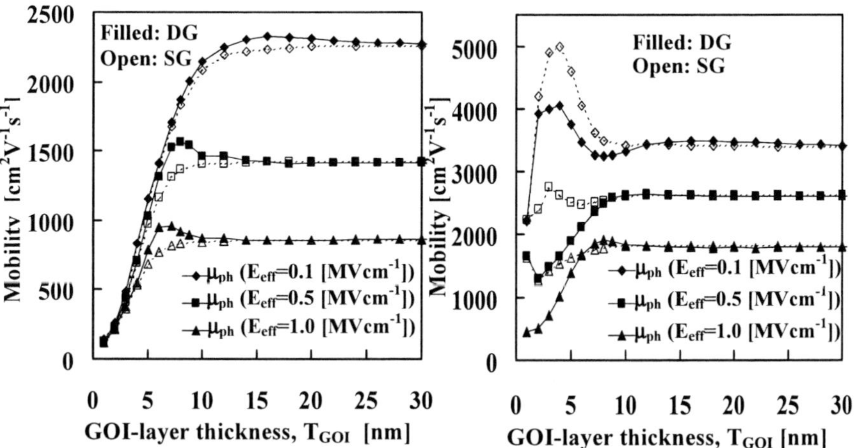

Figure 5. T_{GOI} dependencies of phonon-limited electron mobility on the (001) Ge surface as a parameter of E_{eff}.

Figure 6. T_{GOI} dependencies of phonon-limited electron mobility on the (111) Ge surface as a parameter of E_{eff}.

Simulated phonon-limited electron mobility on the (111) surface is shown in Fig. 6 as a function of T_{GOI} for various E_{eff} values; simulated mobility values of the DG and SG GOI MOSFET's are compared. In contrast to the (001) Ge surface, it is seen that the SG GOI MOSFET offers superior mobility for T_{GOI} values ranging from 2 to 7 nm in a low to high E_{eff} range. Maximal mobility enhancement appears around $T_{GOI}=3$ nm for $E_{eff}> 0.1$ MVcm^{-1}. The expected phonon-limited mobility of a 3-nm T_{GOI} SG GOI MOSFET with (111) Ge surface is 2700 cm^{2}V^{-1}s^{-1} at $E_{eff}=0.5$ MVcm^{-1} [15, 16]; this is about 700 % of that of the equivalent SG GOI MOSFET with (111) Ge surface. This represents a significant merit in using the (111) Ge surface configuration, rather than the Ge (001) surface in device applications.

CONCLUSION

1D self-consistent calculations and relaxation time approximations were used to study the phonon-limited electron mobility of the inversion layer at room temperature for ultra-thin body Si (111), Ge (001), and Ge (111) layers in SG and DG SOI and GOI MOSFET's. Assuming a 5-nm-thick SOI layer, it has been demonstrated that intra-valley phonon scattering (acoustic phonon scattering) in the DG SOI MOSFET inversion layer is strongly suppressed within a range of medium and high effective field (E_{eff}) values; DG SOI MOSFET's have higher phonon-limited electron mobility than SG SOI MOSFET's. It is suggested that the suppression of acoustic-phonon scattering in a 5-nm-thick DG SI

MOSFET primarily stems from the reduction of the form factor (F00) at medium Eeff values. Though similar penomena were discovered in about 7-nm-thick GOI layers with Ge (001) surface, it has been shown that the use of Ge (001) surfaces in DG GOI MOSFET's offers little merit. On the other hand, we have demonstrated that the superior electron mobility of SG GOI MOSFET's with Ge (111) surface indicates the great merit of this structure with regard to applications.

ACKNOWLEDGMENT

A part of the computations made for this research were performed on the Large-Scale Computer System at the Osaka University Cyber-media Center.

REFERENCES

1. M. Shoji and S. Horiguchi, *J. Appl. Phys.*, **82**, 6096 (1997).
2. M. Shoji and S. Horiguchi, *J. Appl. Phys.*, **85**, 2722 (1999).
3. D. Esseni, A. Abramo, L. Selmi, and E. Sangiorgi, *IEEE Trans. Electron Devices*, **50**, 2445 (2003).
4. S. Takagi, A. Toriumi, M. Iwase, and H. Tango, *IEEE Trans. Electron Devices*, **41**, 2363 (1994).
5. S. Takagi, J. Koga, A. Toriumi, in *Tech. Dig. IEEE IEDM*,(Washington. D. C., 1997) p.219.
6. J. Cai, K. Rim, A. Bryant, K. Jenkins, C. Ouyang, D. Singh, Z. Ren, K. Lee, H. Yin, J. Hergenrother, T. Kanarsky, A. Kumar, X. Wang, S. Bedell, A. Reznicek, H. Hovel, D. Sadana, D. Uriarte, R. Mitchell, J. Ott, D. Mocuta, P. O'Neil, A. Mocuta, E. Leobandung, R. Miller, W. Haensch and M.Ieong, in *Tech. Dig. IEEE IEDM*, (San Francisco, 2004) p.165.
7. G. Tsutsui, M. Saitoh, T. Saraya, T. Nagumo and T. Hiramoto, in *Tech. Dig. IEEE IEDM*, (Washington D. C., 2005) p.729.
8. Y. Liu, E. Sugimata, K. Ishii, M. Masahara, K. Endo, T. Matsukawa, H. Yamauchi, S. O'uchi and E. Suzuki, *Jpn. J. Appl. Phys.*, **45**, 3084 (2006).
9. F. Stern and W. E. Howard, *Phys. Rev.*, **163**, 816 (1967).
10. F. Stern, *Phys. Rev. B* **5**, 4891 (1972).
11. I-H. Tan, G. L. Snider, L. D. Chang, and E. L. Hu, *J. Appl. Phys.*, **68**, 4071 (1990).
12. C. Jacoboni and L. Reggiani, "The Monte Carlo method for the solution of charge transport in semiconductors with applications to covalent materials," *Rev. Mod. Phys.*, **55**, 645 (1983).
13. D. Esseni, M. Mastrapasqua, George K. Celler, C. Fiegna, L. Selmi and E. Sangiorgi, *IEEE Trans. Electron Devices*, **50**, 802 (2003).
14. K. Endo, S. Noda, M. Masahara, T. Ozaki, T. Kubota, S. Samukawa, Y. Liu, K. Ishii, Y. Ishikawa, E. Sugimata, T. Matsukawa, H. Takashima, H. Yamauchi, and E. Suzuki, in *Tech. Dig. IEEE IEDM* (Washington, D. C., 2005) p. 859.
15. G. Du, X. Liu, Z. Xia, J. Kang, Y. Wang, R. Han, H.-Y. Yu, and D.-L. Kwong, *IEEE Trans. Electron Devices*, **52**, 2258 (2005).
16. T. Low, Y. T. Hou, M. F. Li, C. Zhu, A. Chin, G. Samudra, L. Chan, and D.-L. Kwong, in *Tech. Dig. IEEE IEDM* (Washington, D. C., 2003) p. 691.

ECS Transactions, 6 (4) 375-380 (2007)
10.1149/1.2728885, ©The Electrochemical Society

Significance of Gate Underlap Architecture in FinFETs for Low–Voltage Analog/RF Applications

Abhinav Kranti and G. Alastair Armstrong

School of Electrical and Electronic Engineering (NISRC), Queen's University Belfast, Ashby Building, Stranmillis Road, Belfast BT9 5AH, Northern Ireland, UK.
Email: a.kranti@ee.qub.ac.uk

In the present work, we analyze the enormous potential of source/drain extension (SDE) region engineering in FinFETs for Ultra Low–Voltage analog/rf applications. We show that SDE region design can simultaneously improve two key analog figures of merit (FOM) – intrinsic dc gain (A_{VO}) and cut-off frequency (f_T) for 60 nm FinFETs operated at low drive current (I_{ds} = 5 μA/μm). The results are analyzed in terms of spacer–to–gradient ratio, a new design parameter for devices with non–abrupt source/drain regions. The present work provides new opportunities for realizing future ultra low-voltage/low-power analog/rf design with nanoscale SDE engineered FinFETs.

Introduction

Over the past few years, low-power low-voltage silicon-on-insulator (SOI) MOS technology has emerged as a leading candidate for highly integrated low–voltage mixed-mode circuits for wireless applications. However, in ultra short gate length regime, upcoming CMOS technologies face many technological challenges [1], the most crucial being the short-channel effects (SCEs) that tend to degrade analog figures of merit (FOM) such as Early voltage (V_{EA}), transconductance-to-current ratio (g_m/I_{ds}), intrinsic dc gain ($A_{VO} = g_m/g_{ds} = g_m/I_{ds} \times V_{EA}$) and cut-off frequency ($f_T = g_m/2\pi C_{gg}$ where g_m is the transconductance and C_{gg} is the total gate capacitance) [2]. To overcome the degradation in analog FOM, certain techniques such as HALO implants and laterally asymmetric channel (LAC) or graded-channel (GC) design [2-4] have been proposed. However, in ultra short gate length devices, the control of dopant profile at the source end of the channel to enable a feasible LAC/GC concept is a technological challenge. Therefore, alternative and innovative techniques are required for improving analog FOM of sub-100 nm MOS devices. We use the concept of source/drain extension (SDE) region engineering [5-9] to significantly improve the analog FOM of 60 nm gate length FinFETs. In this work, we focus specifically on Ultra Low–Voltage (gate and drain voltages are lower than the threshold voltage) analog applications focusing on device design and optimization, analyzing important device and technological parameters, targeting applications for the ultra low–voltage/low–frequency applications.

Simulations

FinFETs (Fig. 1(a)) analyzed here have been simulated using 3D simulator, ATLAS [10]. The doping (N_a) of p-type SOI layer of 10^{15} cm^{-3}, gate length (L_g) of 60 nm, gate oxide thickness (T_{ox}) = 2.2 nm, and fin height (H_{fin}) = 60 nm were chosen for the devices. Fin width (T_{fin}) was varied from 22 nm to 42 nm. Drain voltage (V_{ds}) was fixed at 0.2 V whereas gate bias (V_{gs}) was always maintained below threshold voltage (V_{th}) in order to analyze potential of FinFETs for Ultra Low–Voltage (ULV) analog applications. The simulations have been performed with CVT mobility model and analog FOM are extracted at $I_{ds} = 5$ µA/µm. As our region of interest is below threshold i.e. V_{gs}, $V_{ds} < V_{th}$, quantum effects will not be significant in our devices [11]. Source/drain profile was modeled using the expression

$$N_{SD}(x) = (N_{SD})_{peak}\, e^{-x^2/\sigma^2}$$

[1]

where $(N_{SD})_{peak}$ is the peak source/drain doping, σ (lateral straggle) defines the roll–off [8] of the source/drain profile as

$$\sigma = \sqrt{2sd/\ln(10)}$$

[2]

where s is the spacer width and d the source/drain doping gradient [6-9] at the gate edge, was varied from 3 to 9 nm/decade. The lateral straggle parameter (σ) was varied from 7.5 to 15 nm and the spacer widths (s) corresponding to these values of σ lie in the range of 3 to 90 nm.

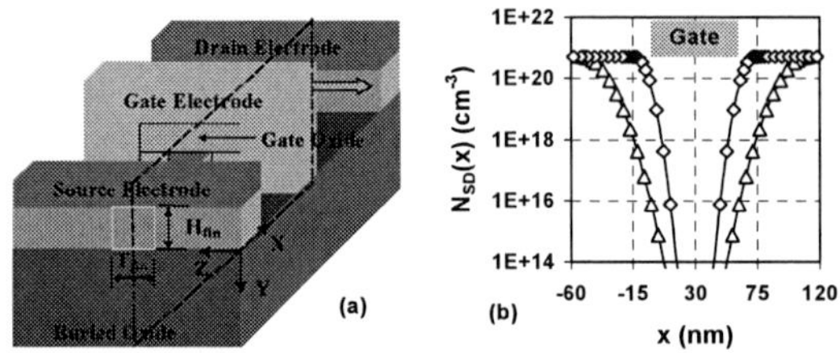

Figure 1. (a) Schematic diagram of a FinFET analyzed in the present work and (b) Variation of source doping profile for various σ values along the cut-plane indicated by dashed lines. Notations: ◇—◇ $\sigma = 7.5$ nm, and △—△ $\sigma = 15$ nm with $d = 5$ nm/dec.

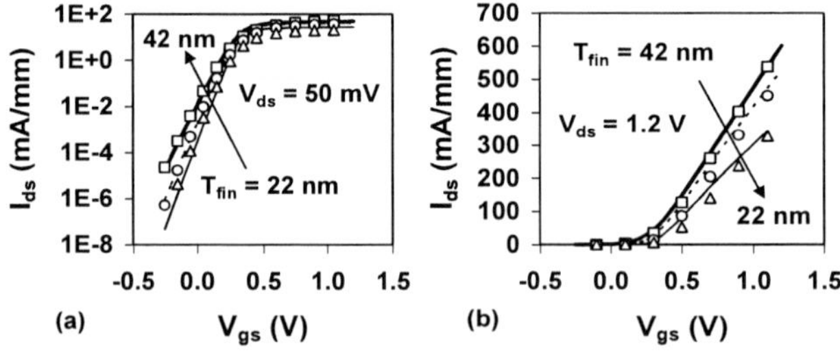

Figure 2. I_{ds}–V_{gs} curves of FinFET at (a) $V_{ds} = 50$ mV and (b) $V_{ds} = 1.2$ V. T_{fin} is varied from 22 nm to 42 nm in 10 nm steps. Symbols indicate experimental data [12] whereas lines refer to 3D simulation results. Device parameters: $H_{fin} = 60$ nm, $L_g = 60$ nm and $T_{ox} = 2.2$ nm.

Results and Discussion

Fig. 2(a) and (b) show the simulated and experimental I_{ds}–V_{gs} characteristics for 60 nm FinFETs at low (50 mV) and high (1.2V) drain voltages. The agreement of our simulations with experimental results [12] provides a sound basis for our analysis. FinFETs with thinner fins achieve excellent immunity from SCEs and show lower S–

slope value. As our interest is to analyze FinFETs for ULV analog applications, we focus our attention to designing with low drain bias at lower drive currents with T_{fin} = 32 nm.

Fig. 3(a) and (b) show the variation of f_T and A_{VO} with s for various values of lateral straggle (σ) at I_{ds} = 5 μA/μm and a typical circuit operating voltage (V_{ds}) of 0.2V. f_T was extracted as the frequency at which short circuit current gain (h_{21}) is 0 dB. Results show that SDE region optimization in 60 nm FinFETs can result in high values of f_T and A_{VO} of 40 GHz and 35 dB, respectively, an improvement of ~ 1.5 – 2 compared to FinFETs with designed with abrupt SDE regions. Devices designed with gradual source/drain doping gradients (d = 7 – 9 nm/dec) at lower σ values perform worse than FinFETs with abrupt SDE regions due to SCEs as $L_{eff} < L_g$. An increase in σ from 7.5 nm to 15 nm leads to an improvement in both A_{VO} and f_T. This is somewhat surprising as it indicates an increase in f_T with increase in L_{eff}. Please note that L_{eff} increases with increase in spacer width for a constant σ [8]. We will show in subsequent discussion that the key issue to improve f_T is the effectiveness of gate underlap in reducing the fringing capacitance, without compromising g_m. However, devices designed with large values of σ along with much larger spacer widths (σ = 15 with s = 86 nm) show a reduction in cut–off frequency due to the additional parasitic series resistance associated with very wide spacers.

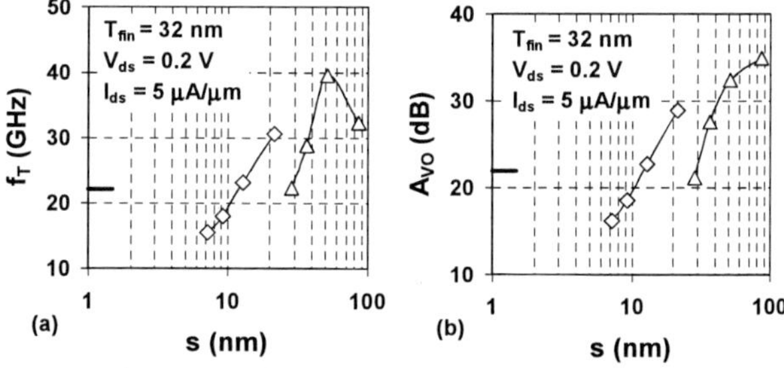

Figure 3. Dependence of (a) cut–off frequency (f_T) and (b) voltage gain (A_{VO}) on spacer widths (s) for various values of lateral straggle (σ) extracted at I_{ds} = 5 μA/μm and V_{ds} = 0.2 V. Notations: ◊—◊—◊ σ = 7.5 nm, and △—△—△ σ = 15 nm. The horizontal line on the y axis represent f_T and A_{VO} values of FinFET designed with abrupt SDE regions. For each σ curve, the lowest A_{VO} or f_T value corresponds to d = 9 nm/dec whereas the highest value represents d = 3 nm/dec. d is varied from 3 to 9 nm/dec in steps of 2 nm/dec. Device parameters: H_{fin} = 60 nm, L_g = 60 nm and T_{ox} = 2.2 nm.

It is important to note that in a conventional design with abrupt SDE regions, it is not possible to attain a simultaneous improvement in A_{VO} and f_T, as an increase in A_{VO} requires a longer L_g (assuming a linear dependence of V_{EA} on L_g), which would compromise f_T because of a reduction in g_m (g_m is approximately inversely proportional to L_g). As shown in Fig. 3(a)–(b), f_T and A_{VO} values appear to saturate at about 40 GHz and 35 dB, respectively. The limitation of too wide a spacer or too large a straggle is parasitic series resistance effect. These values of spacer widths suggest that for achieving considerable improvement in both A_{VO} and f_T, the maximum value of spacer width is limited to about 60–75 nm i.e. (1.00–1.25)L_g. However, devices designed with optimal σ values and wider spacers at low currents (~ 5 μA/μm) result in higher values of both f_T and A_{VO}, provided the additional series resistance associated with wider spacers is not significant at low drain current. The results of f_T and A_{VO} can also be understood in terms of spacer–to–gradient ratio ($\chi = s/d$), an important technological parameter introduced in [8], which indicates the contribution of SDE regions to the channel. Spacer–to–gradient ratio serves as an important FOM for gate–source/drain underlap devices as the effective channel length is a function of gate bias and thus different in weak and strong inversion regions. The values of χ in figure 3 vary from 0.8 ($= \chi_{min}$) to 28 ($= \chi_{max}$). In order to achieve substantial improvement in f_T and A_{VO}, FinFETs should be designed with χ values lying between 7 ($\sigma = 7.5$ nm with $d = 3$ nm/dec) and 10 ($\sigma = 15$ nm with d lying between 5 and 7 nm/dec). FinFETs designed with $\chi \leq 4$ perform worse than devices with abrupt SDE regions because a large d at lower σ causes dopant spill into the channel, leading to lower A_{VO} and f_T. As the value of d depends on thermal budget and diffusivity, very small values (< 3 nm/dec) may be difficult to achieve (very small values of d require non-standard process [13-14] such as solid phase epitaxy or laser thermal annealing), as it may ultimately require control of individual atoms. Therefore, in order to design devices with higher χ values, it is more appropriate to increase the spacer width in an optimal design.

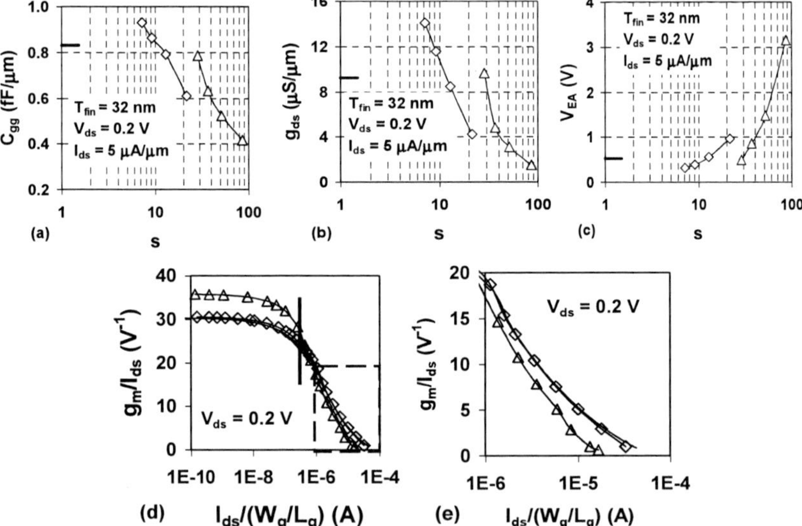

Figure 4. Variation of (a) total gate capacitance (C_{gg}), (b) output conductance (g_{ds}) and (c) Early voltage (V_{EA}) extracted at $I_{ds} = 5$ µA/µm and $V_{ds} = 0.2$ V with spacer width. (d) Variation of transconductance-to-current ratio (g_m/I_{ds}) with normalized drain current at $V_{ds} = 0.2$ V, $d = 5$ nm/dec and $T_{fin} = 32$ nm. (e) Enlargement of the region denoted by dashed rectangle shown in (e) showing g_m/I_{ds} values in the strong inversion region. The vertical line in (d) at $I_{ds}/(W_g/L_g) = 3 \times 10^{-7}$ A represents $I_{ds} = 5$ µA/µm. Notations and device parameters are same as in figure 3.

Fig. 4(a)–(d) shows the variation of total gate capacitance ($C_{gg} = C_{gs} + C_{gd} + C_{gb}$, where C_{gs}, C_{gd} and C_{gb} represent gate-to-source, gate-to-drain and gate-to-substrate capacitances, respectively), output conductance (g_{ds}) and Early voltage (V_{EA}) with spacer widths for various σ values. An increase in the spacer width shifts the source/drain doping away from the gate edge, resulting in a significant reduction in parasitic fringing capacitance that leads to the decrease in total gate capacitance as shown in Fig. 4(a). This reduction is nearly 40 % ($\sigma \sim$ 15 with s $\geq$ 40 nm) when compared with abrupt SDE FinFETs. It is important to note that SDE engineered devices designed with shorter spacer widths and gradual doping gradients achieve nearly same value of C_{gg} as compared to devices designed with abrupt SDE regions. The data corresponding to $C_{gg} >$ 0.83 fF/μm i.e. $C_{gg\ SDE}/C_{gg\ ABRUPT} > 1$, represents gate–overlap architecture rather than the desirable gate–underlap design which is beneficial for the reduction of total gate capacitance. Thus the optimization of SDE regions is extremely important to minimize the capacitances associated with the SDE region. It is important to note that such a drastic reduction in parasitic gate capacitance is not possible in other approaches (graded channel and laterally asymmetric channel design) that have been suggested for improving analog FOM.

As shown in Fig. 4(b)–(c), the most significant aspect of SDE region engineering is the substantial reduction in g_{ds} or the improvement in V_{EA} with an increase in s. Due to an increase in the spacer width (s), reduction in the net dopant concentration at the gate edge causes a significant reduction in the peak electric field at the gate edge, yielding higher values of V_{EA} [9]. The reduction in g_{ds} (when compared with abrupt SDE regions) by a factor of $\sim$ 3 for $\sigma \sim$ 15 nm with $s > 40$ nm due to SDE region engineering explains the increase in Early voltage. Fig. 4(d)–(e) shows the variation of g_m/I_{ds} with normalized drain current ($I_{ds}/(W_g/L_g)$) for various σ values. g_m/I_{ds} ratio is a measure of the efficiency to translate current (hence power) into transconductance; i.e., the greater the g_m/I_{ds} value, the greater the transconductance at a constant current value. Therefore, g_m/I_{ds} ratio is sometimes interpreted as a measure of the "transconductance generation efficiency" [2] of the device. An increase in σ shifts the source/drain doping away from the gate edge, thus minimizing the influence of drain on the channel region under the gate, leading to higher values (in weak inversion region) of g_m/I_{ds} ($\sim$ 35 V^{-1}) as compared to $\sim$ 30 V^{-1} for abrupt SDE regions. This reduction in SCEs is reflected in an increase in current ratio ((I_{ds})$_{SDE}$/(I_{ds})$_{Abrupt}$) to 2 at g_m/I_{ds} = 25 V^{-1} for σ = 15 nm. This factor of $\sim$ 2 times improvement in drain current translates into a higher A_{VO} at lower I_{ds} (@ V_{ds} = 0.2 V). SDE region optimization is particularly advantageous at low gate voltage, as the current flow is mainly due to diffusion of carriers where a wider spacer region ($\sim$ 45 nm; σ = 15 nm) does not significantly degrade the device performance. However, as shown in Fig. 4(e), in strong inversion region ($\sim g_m/I_{ds}$ = 5 V^{-1}), large σ values associated with wider spacer yielding introduces additional parasitic resistance, which degrades g_m at higher current level yielding a current ratio ((I_{ds})$_{SDE}$/(I_{ds})$_{Abrupt}$) $<$ 1. These results show that SDE engineering will be **even more useful** to increase bandwidth for low–voltage analog/rf applications for sub 50 nm gate length FinFETs.

Conclusions

The enormous potential of source/drain extension region engineering in FinFETs for analog applications has been extensively analyzed in 60 nm FinFETs. SDE region engineered devices operated at low current levels are particularly useful for Ultra Low–Voltage analog applications as **both gain and speed** of devices can be significantly improved. For a typical device designed with d = 5 nm/dec, the spacer widths in the range of 20–50 nm and lateral straggle values (independent of gate length) in the range of 10–15 nm represents the optimal range to achieve significant improvement in analog FOM.

The enormous potential of SDE region engineered FinFETs for ultra low–voltage analog applications is attributed to

(i)   Improvement in transconductance–to–current ratio (g_m/I_{ds}) and transconductance (g_m) due to reduction in short channel effects.

(ii)   Reduction in drain conductance (g_{ds}) and consequent improvement in Early voltage (V_{EA}) due to reduction in peak electric field at the gate edge towards the drain.

(iii)   The drastic reduction in parasitic fringing capacitance due to an optimal gate–source/drain under lap design.

Acknowledgments

A. Kranti is grateful to Engineering and Physical Sciences Research Council (EPSRC), UK for the financial assistance for the research work.

References

1. International Technology Roadmap for Semiconductor 2004 edition (http://public.itrs.net).
2. F. Silveira, D. Flandre and P.G.A. Jespers, *IEEE J. Solid-State Circuits*, **31**, 1314 (1996).
3. V. Kilchytska, A. Neve, L. Vancaillie, D. Levacq, S. Adriaensen, H. van Meer, K. De Meyer, C. Raynaud, M. Dehan, J.-P. Raskin and D. Flandre, *IEEE Trans. Electron Devices*, **50**, 577 (2003).
4. A. Kranti, T.M. Chung, D. Flandre and J.-P. Raskin, *Solid-State Electronics*, **48**, 947 (2004).
5. V.P. Trivedi and J.G. Fossum, p. 192, *IEEE SOI Conference*, USA (2004).
6. A. Kranti and G.A. Armstrong, p. 96, *IEEE SOI Conference*, USA (2005).
7. A. Kranti and G.A. Armstrong, *Semiconductor Science and Technology*, **21**, 409 (2006).
8. A. Kranti and G.A. Armstrong, *Solid-State Electronics*, **50**, 437 (2006).
9. A. Kranti, T.C. Lim and G.A. Armstrong, p. 141, *IEEE SOI Conference*, USA (2006).
10. ATLAS Users Manual, SILVACO, 2006.
11. J.-P. Colinge, J.C. Alderman, W. Xiong and C.R. Cleavelin, *IEEE Trans. Electron Devices*, **53**, 1131 (2006).
12. D. Lederer, B. Parvais, A. Mercha, N. Collaert, M. Jurczak, J.-P. Raskin and S. Decoutere, p. 8, *Silicon Monolithic Integrated Circuits in RF Systems*, (2006).
13. S. Gannavaram, N. Pesovic and M. C. Öztürk, p. 437 *IEDM Tech. Dig.* 2000.
14. B. Yu, Y. Wang, H. Wang, Q. Xiang, C. Riccobene, S. Talwar and M.-R. Lin, p. 508, *IEDM Tech. Dig.* (1999).

ECS Transactions, 6 (4) 381-386 (2007)
10.1149/1.2728886, ©The Electrochemical Society

Non-Vertical Sidewall Angle Influence on Triple-Gate FinFETs Corner Effects

R.Giacomini[1,2], J. A. Martino[2]

[1] Centro Universitário da FEI, 09850-901 – São Bernardo do Campo – Brazil
e-mail: renato@fei.edu.br
[2] LSI/PSI/USP, University of Sao Paulo, Brazil

Some fabricated FinFET devices present width variations along the vertical direction due to fabrication process limitations. These variations lead to non-rectangular cross-section shapes. One of the most frequent shapes is the trapezoidal (plane and inclined sidewalls). Another identified phenomenon in multiple-gate devices such as FinFETs is the corner effect, which occurs due to the overlapping of the influences of two gate planes near the device corners. This paper addresses the variation of the corner effect as a function of the sidewall inclination angle, through 3-D numeric simulation. A set of devices of several inclination angles and body doping levels were simulated. The corner effect depends on the inclination angle, specially for higher doping levels.

Introduction

The FinFET is one of the most promising multiple-gate alternative structures for scaling down MOS technology, because of its better current drive and smaller short-channel effects (1). Due to limitations of process uniformity, some fabricated FinFETs have width variation along the vertical direction, resulting in several fin cross-section shapes, other than the designed rectangular shape. In (2), the author presents some fin shapes from different references. The most frequent resulting cross sections are trapezoidal, concave, and convex. Figure 1 shows this shapes for Triple-gate FinFETs.

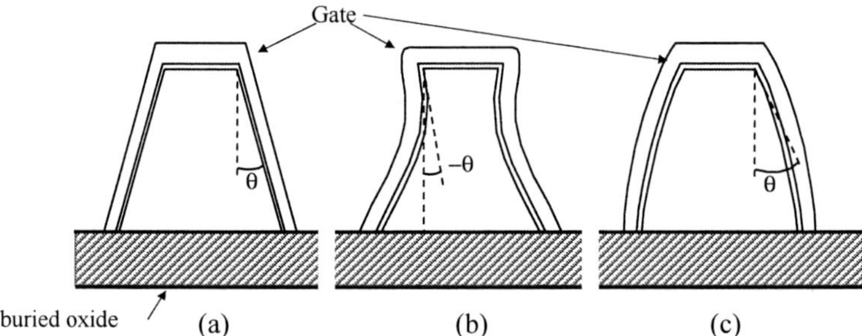

Figure 1 – FinFET cross-section shapes: (a) trapezoidal, (b) concave, (c) convex, and reference angle θ.

An identified phenomenon in multiple-gate devices such as Triple-Gate FinFETs is the corner effect (3),(4), which occurs due to the overlapping of the influences of two gate planes near the device corners. The corner effect is characterized by the increase of inversion charge in the proximity of the corners, with immediate consequences to current density distribution along the fin cross-section. Figure 2 shows an electron concentration isosurface curve of a simulated 3-Gate FinFET in which the top-corner effect can be clearly identified. This curve was obtained from the numeric simulator, using BQP approximation for quantum effects, drain to source voltage V_{DS}=0.05 V, gate to source voltage V_{GF}=0.42V (threshold voltage), buried oxide thickness t_{box}=200nm, gate width W_{Fin}=50nm, and gate length L=60 nm.

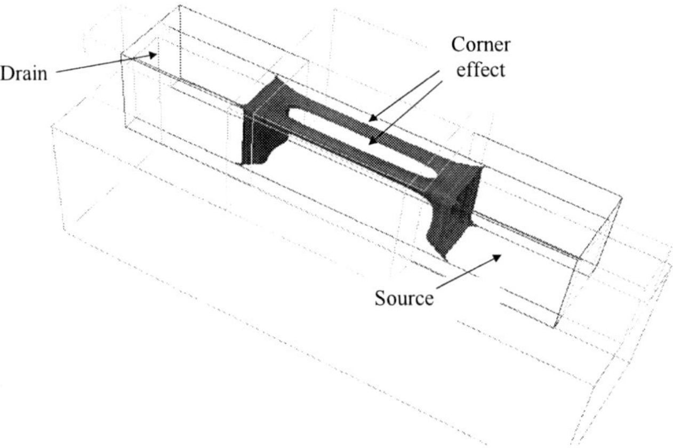

Figure 2 – Electron concentration isosurface, from numeric 3-D simulation showing corner effect. V_{DS}=0.05 V , V_{GF}=0.42V, t_{box}=200nm, W_{Fin}=50nm, and L=60 nm.

A similar effect occurs at the bottom corners, depending on the substrate bias, due to the influence of substrate potential and the gate bottom edge. Figure 3 shows the electron concentration curves for a 3-Gate device similar to that of figure 2, but with t_{box}=100nm and trapezoidal cross-section shape. It can be clearly seen a greater carrier concentration near the bottom corners, as well as at the top corners.

This work investigates the influence of the sidewall angle on the corner effect, through three-dimensional numeric simulation.

Simulated Devices

The basic device for simulation is a trapezoidal cross-section 3-Gate FinFET, as shown in figure 4. The θ angle is defined for trapezoidal devices as the sidewall inclination in degrees, referenced to a vertical plane. The other geometric parameters and main physical dimensions are defined as in the figure. For convex and concave shapes, the sidewall geometry near the corners is locally approximated by a tangent plane, with θ angle defined as in Figure 1.

The simulated devices present gate oxide thickness t_{ox} = 2 nm, buried oxide thickness t_{box} = 100 nm, silicon doping concentration (N_a) ranging from 10^{15} to 10^{17} cm^{-3}, and interface charge densities of 3.0 x 10^{10} cm^{-2}. The gate material used was Tungsten, a near midgap material. The channel lengths (L) are 60 and 200 nm. The 3-D simulator used was ATLAS (5).

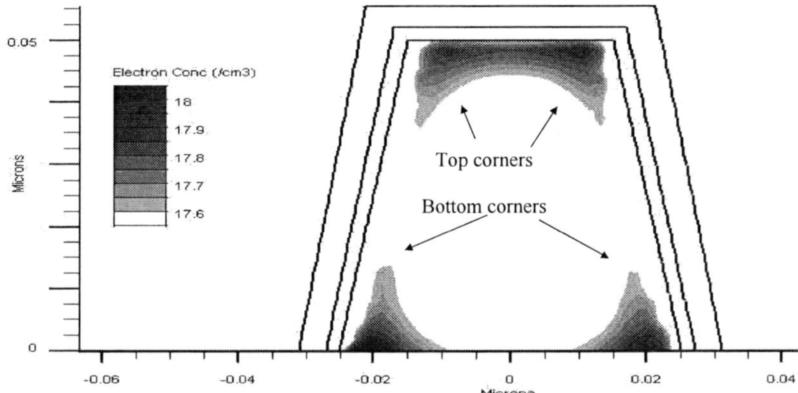

Figure 3 – Electron concentration isosurface, from numeric 3-D simulation showing corner effect. V_{DS}=0.05 V , V_{GF}=0.42V, t_{box}=100nm, and L=200 nm.

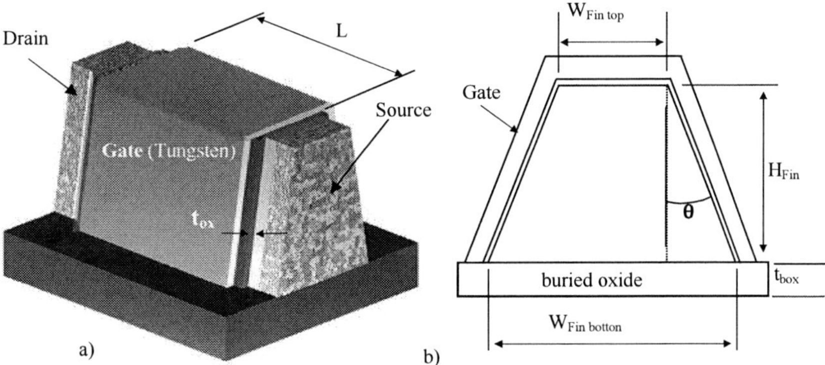

Figure 4 – Trapezoidal 3-Gate FinFET perspective view (a) and cross-section (b).

The angle (θ) variation was obtained through the variation of the top silicon width ($W_{Fin\,top}$) from = 30 to 70 nm, while the botton silicon width ($W_{Fin\,botton}$) and the silicon height (H_{Fin}) were maintained constant at 50nm. The inclination angle, θ, ranges from 0 to 11 degrees. The threshold voltage was extracted directly from the drain current (I_{DS}) versus gate voltage (V_{GF}) curve, using the maximum transconductance change (MTC) method. MTC method defines threshold voltage as the gate voltage at which d^2I_{DS}/dV_{GF}^2 reaches a maximum value (d^3I_{DS}/dV_{GF}^3=0).

A set of simulated devices with L=200 nm was biased with V_{DS}=0.05 V and V_{GF}=0.44V. The maximum current density that crosses the plane determined by L/2 was taken as a parameter for the analysis, as well as the inversion charge density at the same plane. The current density was normalized by dividing the values by the average current density at the device cross-section. The obtained values for the top corner are plotted in

Figures 5 and 6, as a function of inclination angle, for silicon doping levels of 10^{15}, 10^{16}, 10^{17} and 5×10^{17} cm^{-3}. The curves show that the corner effect at the top of the devices decreases as the sidewall inclination angle increases. This fact can be explained by considering that the resulting vector obtained from the sum of the electric field components at the corner, originated by the gate charge in the two planes, depends on this angle. At the limit situation (θ=90^0) the relative maximum current density would be the same as for a single-gate SOI MOS device.

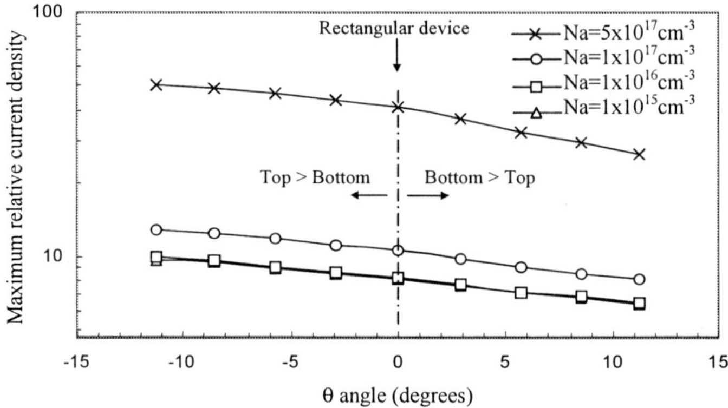

Figure 5 – Maximum relative current density as a function of inclination angle, at top corners, for body doping levels of 10^{15}, 10^{16}, 10^{17} and 5×10^{17} cm^{-3} (V_{DS}=0.05V, V_{GF}=0.44V, L=200 nm)

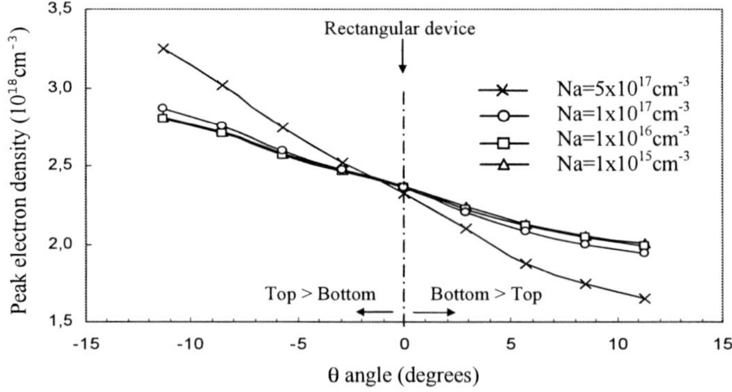

Figure 6 – Maximum electron concentration as a function of inclination angle, at top corners for body doping levels of 10^{15}, 10^{16}, 10^{17} and 5×10^{17} cm^{-3} (V_{DS}=0.05V, V_{GF}=0.44V, L=200 nm).

Comparing the four simulated doping levels it can be seen that the corner effect is less meaningful at lower doping levels. This is a reasonable result, once the current distribution along the cross-section of such devices is more uniform and a large part of the current flows far from the silicon/gate-oxide interface, and consequently far from the corners.

The bottom corner effect has a weaker dependence on the inclination angle than the top, as can be seen comparing figures 7 and 8 to figures 5 and 6 respectively.

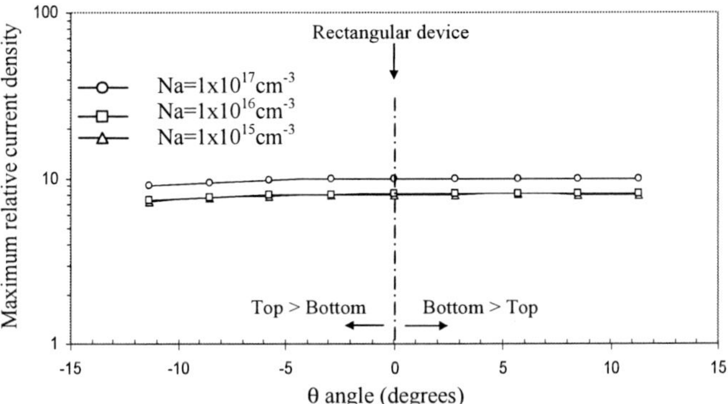

Figure 7 – Maximum relative current density as a function of inclination angle, at bottom corners, for body doping levels of 10^{15}, 10^{16}, and 10^{17} cm^{-3} (V_{DS}=0.05V, V_{GF}=0.44V, L=200 nm)

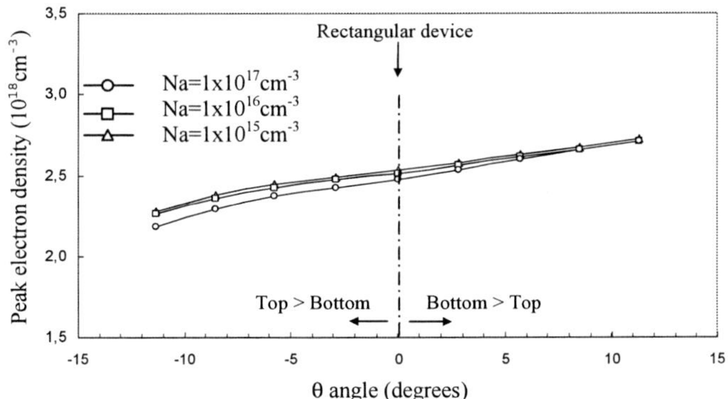

Figure 8 – Maximum electron concentration as a function of inclination angle, at bottom corners for body doping levels of 10^{15}, 10^{16}, and 10^{17} cm^{-3} (V_{DS}=0.05V, V_{GF}=0.44V, L=200 nm).

Conclusions

A set of trapezoidal FinFETs with different sidewall inclination angles were simulated, using a three-dimensional numeric device simulator. The current density and the charge distribution at the device cross section were evaluated. The maximum relative current density as well as the maximum electron density were adopted as comparison factors for the corner effects. The results show that the corner effect at the top of the devices decreases as the sidewall inclination angle increases. The corner effect at the bottom corner has a weaker dependence on the inclination angle than at the top.

Comparing the simulated doping levels it can be seen that the corner effect is less meaningful at lower doping levels. This fact occurs because the current distribution along the cross-section of such devices is more uniform and a large part of the current flows far from the silicon/gate-oxide interface, and consequently far from the corners.

References

1. J. P. Colinge, *SOI Technology: Materials to VLSI*, 3rd Edition, p.2, Kluwer, Norwell, Massachusetts (2004).
2. X. Wu, P.C.H. Chan, M. Chan, *IEEE Trans on Electron Devices*, v.52, no. 1, p.63 (2005).
3. J. G. Fossum *et al.*, *IEEE Elec Dev Let*, v.24 p.745 (2003).
4. W. Xiong, J.W. Park, J.P. Colinge, Proc of the IEEE International SOI Conference, pp. 111-113, (2003)
5. Atlas Device Simulator User's Manual, v. 5.10.0.R, Silvaco Int. Santa Clara, CA (2005).

ECS Transactions, 6 (4) 387-392 (2007)
10.1149/1.2728887, ©The Electrochemical Society

Improved Model to Determine the Generation Lifetime in Short Channel SOI nMOSFETs

M. Galeti [a,*], J. A. Martino [a], E. Simoen [b], C. Claeys [b, c]

[a] LSI/PSI/USP, University of Sao Paulo - Brazil
Av. Prof. Luciano Gualberto, trav. 3, n° 158 – São Paulo
[*] mgaleti@lsi.usp.br
[b] IMEC, Kapeldreef 75, B-3001 Leuven, Belgium
[c] KU Leuven, Electr. Eng. Dept., B-3001 Leuven, Belgium

An improvement of the one-dimensional analytical model for the extraction of the generation lifetime in partially depleted SOI nMOSFETs is proposed. This refined technique, based on the drain current switch-off transient, considers the influence of the halo implanted region and the channel length reduction. The results obtained are compared with measurements and two-dimensional numerical simulations, showing a good agreement. The improved model can be applied as a simple tool for the electrical characterization of short channel devices.

Introduction

One of the main factors limiting the performance of partially depleted SOI nMOSFET devices is the strong influence of the floating body effects. However, there are situations in which one can take advantage of the floating body effect, as in case of electrical characterization for extracting the carrier lifetime.

The generation-recombination properties are important for the performance of the SOI MOS devices, due to their influence on the leakage current in *pn* junctions, the dynamic random access memories (DRAMs) refresh time, the floating body effect and the gain of the parasitic bipolar transistors (1). Several carrier generation lifetime measurement methods consist in applying a step voltage at one of the gates and observing the transient drain current due to the floating body nature of the partially depleted devices (2-5).

This paper reports an improvement of the one-dimensional analytical model for hole generation lifetime extraction, published in reference (4), which can quantitatively predict the impact of the floating body effect on the drain current switch-off transients. In this refined technique, the influence of the halo implanted region and of the channel length reduction are considered.

Methods and Devices

The method presented in ref. (4) consists in the application of a negative voltage step (from above to below the threshold voltage) in the front gate of a partially depleted SOI MOSFET device. A transient drain current is observed due to the decrease of the floating body potential during the transient, which causes an increase in the threshold voltage. While the body potential relaxes back to zero and the threshold voltage to its steady state value, the drain current gradually reaches the steady state value due to thermally activated carrier generation (7). The method employed here assumes that the dominant generation mechanism is the generation in the space charge region.

Two-dimensional numerical simulations (MEDICI) (8) have been performed, using the Shockley-Read-Hall recombination mechanism with a concentration dependent lifetime model (CONSRH). The simulations were carried out for PD SOI nMOSFET devices, with a 100 nm thick silicon film (t_{si}), 5.5x 10^{17}cm^{-3} channel region doping (N_a), a gate oxide thickness (t_{ox}) of 2.5 nm and the total channel lengths (L) varying from 0.2 to 10 μm. For the simulations a value of $1.17x10^{-7}$s is used for the generation lifetime.

The PD n-channel MOSFETs used in this study were fabricated in a 0.13 μm-CMOS technology using a PELOX isolation scheme, a 2.5 nm Nitrided gate Oxide (NO), a 150 nm n+ polysilicon gate and 80nm nitride spacers. Processing was performed on UNIBOND wafers with a 400nm thick buried oxide, resulting in a final p-type silicon film thickness and channel doping concentration of 100 nm and 5.5x10^{17}cm^{-3} respectively.

Improved Model Proposed

The impact of a laterally non uniform channel doping, which is length-dependent due to the influence of the halo region on the floating body effect, may be represented by an approximated average channel doping (N_{eff}) as shown in equation [1] (9).

$$N_{eff} = \frac{N_a(L-2L_h)}{L} + \frac{2N_h(L_h)}{L}$$

[1]

where L_h, N_h and N_a are the halo implanted region length, halo region and body region concentration, respectively, as shown in the figure 1.

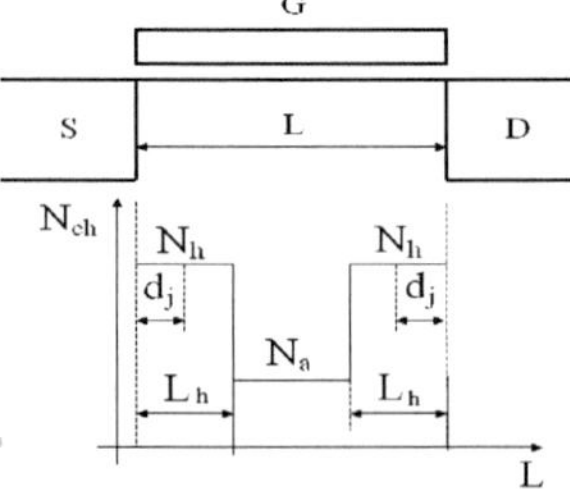

Figure 1 - Cross section of a device with a doping profile along the channel.

Considering only the depletion charge controlled by the front gate, without the contribution of the charge controlled by the source/drain junction, or the reverse, allows to write:

$$Q_D(t) = -qx_d(t)N_{QD}$$

[2]

where,
$$N_{QD} = \frac{N_a(L-2L_h)}{L} + \frac{2N_h(L_h - d_j)}{L}$$

[3]

$x_d(t)$ and d_j is the channel depletion region controlled by the front gate and the source/drain junctions, respectively.

Immediately after applying the negative gate step bias, the body potential decreases due to the coupling gate capacitance since the body is floating. The carriers (holes in the case of an nMOSFET) are generated in the space charge region, filling a depletion region between the maximum depletion width (X_{dmax}) and the steady state depletion width ($X_{d\infty}$), and the body potential comes back to steady state. The depletion charge $Q_D(t)$ variation due to the bulk generation inside the space charge region is given by [4].

$$-qN_{QD}\frac{dx_d}{dt} = q\frac{n_i}{\tau_g}[x_d(t) - x_{d\infty}]$$ [4]

The solution to Eq. [4] is given by

$$x_d(t) = [x_{d\,max} - x_{d\infty}]exp\left(-\frac{t}{\dfrac{\tau_g N_{QD}}{n_i}}\right) + x_{d\infty}$$ [5]

With the new considerations, described in Eqs. [1]-[3], $x_{d\infty}$ can be calculated from the one-dimensional Poisson equation:

$$x_{d\infty} = \frac{-3N_{QD}t_{ox} + \sqrt{9(N_{QD}t_{ox})^2 + \dfrac{2\varepsilon_{si}N_{eff}}{q}(V_{Glow} - V_{FB})}}{N_{eff}}$$ [6]

Figure 2 describes for different channel lengths the channel depletion region behavior normalized to the maximum depletion width as a function of time. It is observed that for devices with smaller channel length there is a reduction in the variation of the depletion region due to the proportional reduction in the charges controlled by the front gate.

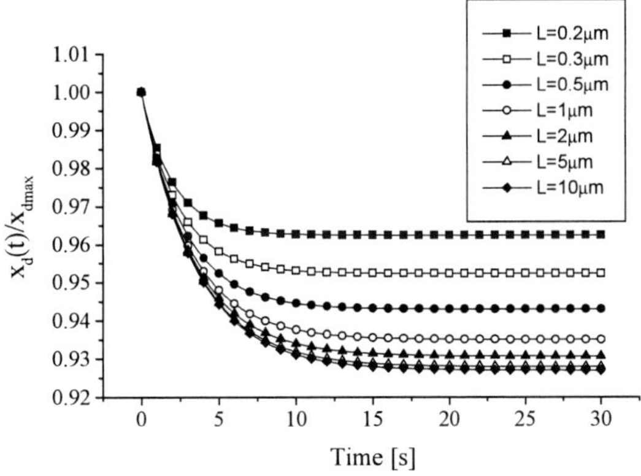

Figure 2 - Channel depletion region behavior normalized to the maximum depletion width as a function of time.

The transient drain current ($I_D(t)$) normalized to the steady state current ($I_{D\infty}$) is represented by the semi-empirical model [7].

$$\frac{I_D(t)}{I_{D\infty}} = exp\left\{\frac{q\gamma}{\xi kT}\left(\sqrt{2\phi_F} - \sqrt{2\phi_F - V_{BS}(t)}\right) + \frac{\xi V_{BS}(t)}{2}\right\} \qquad [7]$$

where $\qquad V_{BS}(t) = -\dfrac{qN_{QD}}{2\varepsilon_{si}}\left\{[x_d(t)]^2 - [x_{d\infty}]^2\right\} \qquad$ and $\qquad \xi = 1 + [C_D(t)/C_{ox}]$

Figure 3 shows the drain current $I_d(t)$ transients (experimental and improved model curves) normalized to the steady-state current level $I_{d\infty}$ for SOI nMOSFET devices after switching off the gate from $V_{Ghigh}=0.7V$ to $V_{Glow}=0.1V$ for different channel lengths ranging from 0.2 to 5 μm. The curves show that just after the application of the negative gate voltage step the drain current is suppressed and then gradually increases with time to the steady state value due to thermally activated carrier generation (10).

The lower $x_d(t)$ excursion during the transient, for devices with smaller channel length, causes a reduction in the variation of the floating body potential and at the same time in the drain current transient.

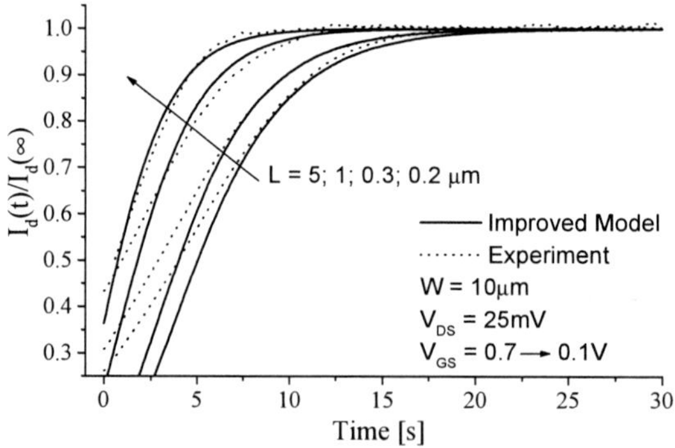

Figure 3 - Transient drain current for a PD SOI nMOSFET with different channel lengths.

The carrier generation lifetime (τ_g) can be calculated using Eqs. [5] and [7] resulting in equation [8], whereby the transient time (T_0) is defined as the time required for the drain current to reach 90% of the steady state current value.

$$\tau_g = \frac{T_0 n_i}{N_{QD} \ln\left(\dfrac{x_{d\,max} - x_{d\infty}}{x_d(90\%) - x_{d\infty}}\right)} \qquad [8]$$

Results and Discussion

Figure 4 compares the τ_g obtained by the improved model (Eqs. [7] and [8]) with the experimental and numerical simulation data and shows a good agreement.

As expected the results point out that τ_g decreases when the channel length decreases due to the increment of the effective channel doping caused by the proximity of the source and drain halos regions.

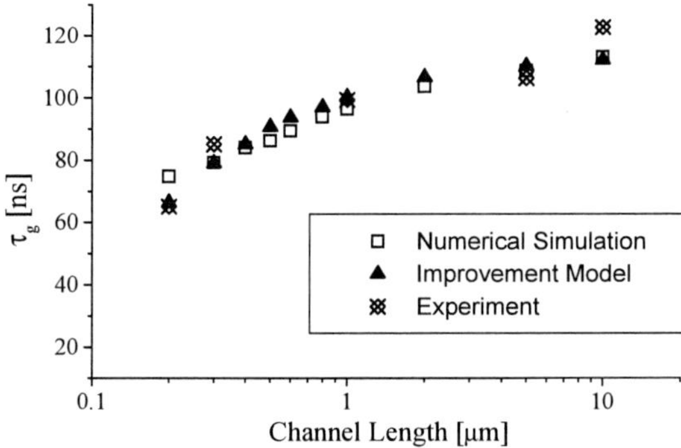

Figure 4 - Generation Lifetime (τ_g) as function of the channel length.

As shown in the table 1 for three different values of generation lifetime substituted in equation [5], the obtained transient time (T_0) values perfectly fit with the numerical simulation. These T_0 values were used as input in the equation [8] in order to extract the generation lifetime for several channel lengths. The results demonstrate that the improved model shows good ability to track decades of lifetime with error of 17% in the worst case.

TABLE I. Extracted generation lifetime for different input generation lifetime values and channel lengths.

Channel Length [μm]	$\tau_{g\,(input)}=1\times10^{-8}$ [s]	$\tau_{g\,(input)}=1\times10^{-7}$ [s]	$\tau_{g\,(input)}=1\times10^{-6}$ [s]
0.2	0.848×10^{-8}	0.826×10^{-7}	0.825×10^{-6}
0.3	0.887×10^{-8}	0.863×10^{-7}	0.862×10^{-6}
1	0.983×10^{-8}	0.963×10^{-7}	0.956×10^{-6}
5	0.976×10^{-8}	0.980×10^{-7}	0.978×10^{-6}

Conclusion

This work reports an improvement of the one-dimensional analytical model reported in the literature, which can quantitatively predict the impact of the floating body effect on the drain current switch-off transients.

The laterally non-uniform channel profile due to the presence of a HALO implanted region and the amount of charge controlled by the drain and source junctions were inserted in the existing model, whereby it became possible to study the drain current transient variation with channel length reduction.

The obtained results were compared with both experimental and two-dimensional numerical simulation data and show a good adjustment with the channel length variation. The improved model was also evaluated for three decade difference in generation lifetime input values and it was observed that the newly extracted generation lifetime values perfectly fit with the numerical simulation, yielding an error of 17%, in the worst case.

Acknowledgments

Milene Galeti and João Antonio Martino would like to thank CNPq for the financial support.

References

1. S. Venkatesan, R. Pierret, G. Neudeck, in *SOI Conference/1992*. IEEE International, p. 120   (1992).
2. H. K. Lim, J. G. Fossum, *IEEE Trans. on Electron Devices*, **31**, 1251 (1984).
3. D. Ioannou, S. Cristoloveanu, M. Mukherjee, B. Mazhari, *IEEE Electron Device Letters*, **11**, 409 (1990).
4. H. Shin, M. Racanelli, W. M. Huang, J. Foerstner, T. Hwang, D. K. Schroder, *Solid-State Electronics*, **43**, 349 (1999).
5. A. M. Ionescu, D. Munteanu N. Hefyene, C. Anghel, *J. Electrochem. Soc.*, **151**, 396 (2004).
6. J.A. Martino, M. Galeti, J.M. Rafí, A. Mercha, E. Simoen, C. Claeys , *J. Electrochem. Soc.*, 153, G502 (2006).
7. M. Galeti, J.A. Martino, E. Simoen, C. Claeys, *High Purity Silicon 9, ECS Trans.*, **31,** 351 (2006).
8. TMA MEDICI, version 4.0 (1997).
9. P. Su, W. Lee, *IEEE Trans. on Electron Devices*, 52, 1662 (2005).
10. J.M. Rafí, A Mercha, E. Simoen, C. Claeys, *Solid-State Electronics*, **48**, 1211 (2004).

ECS Transactions, 6 (4) 393-395 (2007)
©The Electrochemical Society

Author Index